Y0-CUN-185

STOCK ASSESSMENT

Quantitative Methods and Applications for Small-Scale Fisheries

Edited by

Vincent F. Gallucci Saul B. Saila
Daniel J. Gustafson Brian J. Rothschild

Boca Raton New York London Tokyo

The material in this volume is an outcome of the Fisheries Stock Assessment Collaborative Research Support Program, sponsored in part by Grant No. DAN-4146-G-SS-5071-00 of the United States Agency for International Development, intended to support collaborative research between the United States and developing countries' universities and research institutions on fisheries stock assessment and management strategies. The chapters in this book were written, submitted, and revised over the period of time from 1992 to 1995.

LIMITED WARRANTY

CRC Press warrants the physical diskette(s) enclosed herein to be free of defects in materials and workmanship for a period of thirty days from the date of purchase. If within the warranty period CRC Press receives written notification of defects in materials or workmanship, and such notification is determined by CRC Press to be correct, CRC Press will replace the defective diskette(s).

The entire and exclusive liability and remedy for breach of this Limited Warranty shall be limited to replacement of defective diskette(s) and shall not include or extend to any claim for or right to cover any other damages, including but not limited to, loss of profit, data, or use of the software, or special, incidental, or consequential damages or other similar claims, even if CRC Press has been specifically advised of the possibility of such damages. In no event will the liability of CRC Press for any damages to you or any other person ever exceed the lower suggested list price or actual price paid for the software, regardless of any form of the claim.

CRC Press specifically disclaims all other warranties, express or implied, including, but not limited to any implied warranty of merchantability or fitness for a particular purpose. Specifically, CRC Press makes no representation or warranty that the software is fit for any particular purpose and any implied warranty of merchantability is limited to the thirty-day duration of the Limited Warranty covering the physical diskette(s) only (and not the software) and is otherwise expressly and specifically disclaimed.

Since some states do not allow the exclusion of incidental or consequential damages, or the limitation on how long an implied warranty lasts, some of the above may not apply to you.

Library of Congress Cataloging-in-Publication Data

Stock assessment: quantitative methods and applications for small-scale fisheries/ edited by Vincent F. Gallucci ... [et al].

p. cm.

Includes bibliographical references and index.

ISBN 1-56670-151-1

Fish stock assessment—Developing countries. 2. Fisheries—Developing countries.

I. Gallucci, Vincent F.

SH329.F56S59 1995

333.95´611—dc20 95-34212

CIP

Lewis Publishers is an imprint of CRC Press

International Standard Book Number 0-56670-151-1
Library of Congress Card Number 95-34212
Printed in the United States of America 1 2 3 4 5 6 7 8 9 0
Printed on acid-free paper

CONTENTS

PREFACE

This book was written as part of a research project funded by the U.S. Agency for International Development. Therefore, a research flavor permeates the chapters. This flavor is reflected in many ways, from the organizational choice of chapters to the material included within the chapters. In terms of the presentation, however, the chapters are written in a tutorial style. Thus, in greater sense than is usually the case in a multiauthored volume, this book is as suitable for the scientist in a research or management institute working more or less alone as it is for use as a textbook in a class on fisheries stock assessment, population dynamics, or even applied aquatic ecology.

This book fills a particular existing niche. To practitioners in the field of fisheries management and stock assessment, a new book which is said to be on the subject of *stock assessment* would immediately call to mind any number of other books or scientific papers which deal with mathematical models of the dynamics of population growth, of capture rates, etc. What is new or different in each such book would be the types of models presented, their mathematical detail and perhaps their interpretation of the mathematical results. However, if one were to approach the same practitioner and ask her or him: What subjects do you need to deal with if you are to do a successful stock assessment?", it is quite possible that the response would be: "Well, a careful sampling design is most important; and then maybe, if it's possible, we must age the fish (since most traditional models require age data); and then we would need to apply various mathematical representations, or models, as a basis for making decisions about the effects of exploitation on the stocks, their environment and the fishermen."

This book is about **all** of these things. It appears to be the only book available which brings together most of the elements of what constitutes stock assessment, as it is literally carried out. After an introductory chapter, the next chapter is, in fact, on mathematical models, for when age data do *not* exist; followed by a chapter on age determination and growth, when it is possible; followed by three chapters on sampling fish populations, one mostly statistical, one mostly methodological and the last mostly on acoustical sampling. These are followed by chapters on species diversity when multiple species are captured, on time series analysis and finally, a holistic, systems perspective of stock assessment.

Most of the applications presented are to the stock assessment of small-scale or artisanal fisheries in tropical developing countries. However, this is barely a limitation. In fact, most of the limitations on our ability to assess developing country fisheries are exaggerations of problems that have always existed in temperate fisheries but which can be ignored or considered to be second order effects. To be sure, there are more than quantitative or scale differences between tropical ocean biological processes and temperate ocean processes, but not enough is known about these differences to drastically affect the stock assessment methods applied. The differences include, e.g., difficulties determining the ages of fish, the absence of consistent sampling data, concerns over varied recruitment patterns, and so on. There is little doubt that the perspectives described in this text will be as applicable to temperate fisheries as they are to tropical fisheries and to large scale fisheries as to small scale fisheries.

FOREWORD

A realistic picture of the status of any natural resource is a critical pre-condition for improving its management. Obtaining an accurate reading of living aquatic resources is an especially daunting task. The multidimensional character of a fishery and the dynamic nature of its populations constant renewal and recruitment process make fisheries stock assessment a unique challenge in natural resource management. Over the years, good stock assessment statistical models and methods have been developed and have provided sound scientific advice that feeds into the management and politics of large fisheries. These methods, however, have been developed almost exclusively for large-scale, temperate fisheries, generally with one or two major, high-value commercial species. Many of the world's fishers, however, operate in small-scale, sometimes low-value, multispecies fisheries. In terms of improving management of the planet's living aquatic resources and increasing the incomes and welfare of the world's fishers, these artisanal fisheries are clearly where the future challenges lie.

The present volume represents the culmination of a 9-year program financed by the United States Agency for International Development (USAID) to provide new or improved methods of stock assessment for artisanal fisheries. The contributions to improved stock assessment described in this volume are not limited in applicability to tropical waters however, and should be of considerable value to students of temperate fishery management problems. The USAID approach to involve United States universities has been through Title XII of the Foreign Assistance Act, which established the Collaborative Research Support Program (CRSP). This was a unique concept which involves U.S and host country scientists in a collaborative mode to address issues of regional and global concern through innovative research.

The Stock Assessment CRSP addressed stock assessment utilizing two innovative approaches. The conventional wisdom when the program started was that simple methods were preferable to sophisticated mathematical models for developing countries' fishery managers. The CRSP stock assessment scientists took the alternative approach of simultaneously developing appropriately sophisticated methods and the computer software to make them usable, without the need for a complete scientific understanding of all that goes into them. The rapid spread of microcomputers throughout the world over the past decade shows this to have been the appropriate course.

The other unique approach was to combine the efforts of universities in the United States and in developing countries to develop, test and adapt the new methodologies. Although this combination contains inevitable difficulties, the effective collaboration that developed overcame idiosyncrasies and took advantage of the strengths of all partners, to create a research whole that was truly greater than the sum of its parts.

A great deal remains to be done to improve stock assessment methods, to systematically undertake assessments that make use of them, and most importantly, to incorporate the resulting science-based advice into concrete fishery management decisions and actions. The hope of the editors and authors of this volume is to have contributed to this enormous, but absolutely vital challenge.

Daniel Gustafson, University of Maryland — Lamarr B. Trott, USAID

ACKNOWLEDGMENTS

Many people contributed to this book but do not appear as authors. They made invaluable contributions to the book as a whole over its long gestation period. Contributors to specific chapters are acknowledged at the end of that chapter. The authors wish to express appreciation for their wise council to the members of the Board of Institutional Advisors for the United States Agency for International Development (USAID) research project: Drs. Robert Stickney, Kenneth Tenore, Gerald Donovan, Spiros Constantinides, Robert Miller and Peter Larkin; as well as Drs. Edgardo Gomez and Efram Flores of the University of the Philippines and Manual Murillo of the University of Costa Rica. USAID's External Evaluation Panel included at various times: Drs. Michael Sissenwine, Albert Tyler, Serge Garcia and John Caddy, all of whom provided valuable insights and ideas. Drs. Murillo, then director of CIMAR, and later Jose Vargas, at the University of Costa Rica and Donald McCright, director of ICMRD at the University of Rhode Island, provided resources and opportunities for research. The students, faculty and staffs of the Bolinao Laboratory in the Philippines; in particular, its director Dr. Helen Yapp, and the Punto Morales Laboratory, in Costa Rica, all helped make field work fun and research results a reality. Dr. John Rowntree managed the finances for this cooperative research and teaching grant freeing principal investigators as much as possible from budgetary and organizational problems and he guided both the scientists and the granting agency through the formative years of the project with skill and dedication. He now has more than enough material for his own book, if he chooses to write one, on managing science and scientists on international research projects. His role was later assumed by Dr. Gustafson, one of the book's editors. Dr. Garcia has consistently been of assistance to the research group and he and Dr. Harlan Davis were central to the success of the project during one of its reviews. Dr. Lamar F. Trott was the granting agency's (USAID' s) project manager for the 9 years of the project. His active support within the agency, his patience over issues ranging from government-wide Gramm-Rudman budget cuts to the exercise of diplomatic skills *par excellence* (especially after meetings of the troika) are greatly appreciated. He sagely guided, encouraged, cajoled and kept the faith as the research and the volume moved through their multiple lives from research working papers, to journal papers, to handbook chapters and ultimately, to textbook form.

As usual, there is another group who did the literal midwifery work to bring this book into the world. These are the linchpins who often go unrecognized for their patience and diligence. At the University of Washington (UW), work-study student typists Katherine Peterson, Julie Watanabe and Erin Moline labored over the drafts of the working papers and chapters from the UW authors. Steven Anderson piloted the UW chapters through their stages of existence from lives as handbook chapters in a binder to chapters in a book, a significant rebirth. He also coordinated the synthesis of all of the book's chapters putting them into a coherent whole. He worked unerringly with good humor, when there was good reason to lose it, to make the book a reality. This, despite surgery that would have caused a less dedicated person to at least fumble if not drop the book! Steve was assisted with the coordination among authors from different locations by Annie Baumann, at the University of Maryland, College Park. Annie worked until and past the last minute.

Some final notes are in order. First, Dr. Stickney's role as chair of the Board of Directors does not do justice to the importance of his contribution both to the research and to the book. Bob brought an appreciation for fisheries in its fullest, from ecology to management to aquaculture and guided the projects with innate sensitivity and capability over sometimes trying terrain. He was also instrumental in bringing the publisher and the editors together. Second, Dr. Tyler voluntarily drew upon his extensive stock assessment experience to provide wise and sympathetic advice and assistance. His concern for professionalism and people was always evident. And third, Dr. Caddy could always be counted upon to bring fresh and creative perspectives to any problem we presented to him. Finally, recognition and appreciation goes to USAID. The agency's funding of research and writing in fisheries, especially in the tropical and artisanal areas, illustrates recognition of the escalating importance of fish protein to the people in the developing world and an appreciation for the current world-wide declines in fish availability and catches, as described in Chapter 1.

CONTRIBUTORS

Porfirio M. Aliño
Marine Science Institute
University of the Philippines
Diliman, Quezon City
Philippines

Benyounes Amjoun
Center for Quantitative Science
University of Washington
Seattle, Washington 98195

Nygiel P. Armada
College of Fisheries
University of the Philippines
in the Visayas
Miag-ao, Iloilo
Philippines

Jerald S. Ault
Rosenstiel School of Marine
and Atmospheric Science
University of Miami
Miami, Florida 33149

Robert Burr
Math Sciences Computing Center
University of Washington
Seattle, Washington 98195

Jorge Campos
CIMAR, Faculty of Biology
University of Costa Rica
San Jose, Costa Rica

Juan B. Chavarría
CIMAR, Department of Statistics
University of Costa Rica
San Jose, Costa Rica

Loveday Conquest
School of Fisheries
Center for Quantitative Science
University of Washington
Seattle, Washington 98195

Annabelle G.C. del Norte
Marine Science Institute
University of the Philippines
Diliman, Quezon City
Philippines

Robert Donnelly
School of Fisheries
University of Washington
Seattle, Washington 98195

Sonia Formacion
College of Arts and Sciences
University of the Philippines
in the Visayas
Miag-ao, Iloilo
Philippines

William W. Fox Jr.
National Marine Fisheries Service
Washington, D.C.

Vincent F. Gallucci
School of Fisheries
Center for Quantitative Science
Management Assistance for
Artisanal Fisheries
University of Washington
Seattle, Washington 98195

Edgardo D. Gomez
Marine Science Institute
University of the Philippines
Diliman, Quezon City
Philippines

Donald R. Gunderson
School of Fisheries
University of Washington
Seattle, Washington 98195

John B. Hedgepeth
Biosonics Inc.
Seattle, Washington 98103

Han Lin Lai
National Marine Fisheries Service
Woods Hole, Massachusetts

Helen Li
Chesapeake Biological Laboratory
P.O. Box 38
Solomons, MD 20688

Richard McGarvey
South Australian Research and
Development Institute
Henley Beach, Australia

James.E. McKenna, Jr.
Florida Department of
Environmental Protection
Florida Marine Research Institute
St. Petersburg, Florida, 33701

John W. McManus
ICLARM
Makati, Metro Manila
Philippines

Cleto L. Nañola
Marine Science Institute
University of the Philippines
Diliman, Quezon City
Philippines

Joseph N.P. Pasamonte
Marine Science Institute
University of the Philippines
Diliman, Quezon City
Philippines

Daniel Pauly
University of British Columbia
Vancouver, British Columbia

Rodolfo B. Reyes, Jr.
Marine Science Institute
University of the Philippines
Diliman, Quezon City
Philippines

Brian J. Rothschild
Chesapeake Biological Laboratory
P.O. Box 38
Solomons, MD 20688

Saul B. Saila
Graduate School of Oceanography
University of Rhode Island
Narragansett, RI 02882

Geronimo T. Silvestre
College of Fisheries
University of the Philippines
in the Visayas
Iloilo City, 5901
Philippines

Steven G. Smith
Chesapeake Biological Laboratory
P.O. Box 38
Solomons, MD 20688

Richard E. Thorne
Biosonics, Inc.
Seattle, Washington 98103

CHAPTER 1

OVERVIEW AND BACKGROUND

Saul B. Saila[1] and Vincent F. Gallucci[2]

[1]Graduate School of Oceanography, University of Rhode Island,
[2]School of Fisheries, University of Washington

1. OVERVIEW

1.1 THE IMPORTANCE OF STOCK ASSESSMENT FOR FISHERIES MANAGEMENT AND CONSERVATION

The production and consumption of fish and other aquatic organisms has historically been vital to the economic and social well-being of many developing countries. Fish, including finfish, crustaceans, mollusks, and seaweeds, are the fifth most important agricultural commodity based on global agricultural statistics. Developing countries catch and produce about 52 million tons annually, which is more than one half of the total global production. It is estimated that at least 50 million persons are involved in small-scale fisheries, which are typically found in the developing world.

Fish harvests increased very rapidly in the 1950s and 1960s as a result of improved gear, technology and expanded fishing grounds. However, these harvests have increased at much slower rates (less than 1% per year) since about 1970. It is evident that new fishing grounds which can be developed economically are no longer found easily. However, the number of persons seeking a livelihood in fisheries is rapidly increasing, especially in the developing world where fishing is frequently a livelihood of last resort for the poorest of the poor.

1-56670-151-1/96/$0.00+$.50

Fisheries in the developing nations are frequently overexploited, and catches of many species have either leveled off or are declining. Even in the western Atlantic, New England groundfish fishery, which is primarily on cod, haddock and flounder, catches dropped 39% from 1990 to 1993. In overexploited capture fisheries, increases in inputs (fishing effort) will not result in further increases in output (yield) and may even cause a decrease in output. The problem of overfishing forms a basis for one of the important concepts of fishery resource management, namely, optimum sustainable yields. Many of the methods described in this book are directed toward the goal of optimizing yields by means of a better understanding of the vital statistics and the dynamic behavior of exploited populations. The value of this effort is of considerable significance. For example, FAO estimated that 20–30 million tons could be added to the world catch, and that at least one half of that would be achieved by better management of the fisheries. Although the potential for increases in the harvest from artisanal fisheries is not well known, due to limited resource surveys and unreliable landing records, under better management policies it is reasonable to anticipate increased yields of several million metric tons from tropical small-scale fisheries.

It must be noted, however, that the benefits of fisheries management are not always measured in terms of increases in yields. The most important quality of all is not the increase in yield but the sustainability of yield in the face of environmental variability. Good management based on sound scientific stock assessments will usually embrace sustainability of the fishery, at whatever yield. This is one of the many reasons why it is said that good management always increases the overall benefits from a fishery, but these benefits may be difficult to quantify. Examples of additional benefits which may result from imposition of rational fisheries management include:

(a) A resource is underfished and the fishery is managed according to a development plan which is realistic relative to long-term resource potential;

(b) A resource is severely overexploited and management action is taken to reduce short-term catch rates to allow resource recovery and long-term catch increases;

(c) A fishery could produce about the same yield with considerably less fishing effort, and management actions are taken to increase the profitability of the fishery; or

(d) A fishery is being rationally exploited and management decisions are taken to prevent overexploitation which would be inevitable otherwise.

The benefits can be substantial if effective fisheries management is based on sound stock assessment. For example, it is not unusual for a stock assessment to demonstrate that the same amount of catch can be taken at one half the current cost (item c above). This benefit does not result from building more efficient fishing vessels. Instead, it results from effectively utilizing the correct number of vessels as established by stock assessment procedures.

1.2 OVERVIEW OF THIS BOOK AND STOCK ASSESSMENT

Among the many books and manuals that deal with stock assessment and fisheries management, those particularly relevant to the current text are: *Quantitative Fisheries Stock Assessment* by Hilborn and Walters (1992), *Bioeconomic Modeling and Fisheries Management* by Clark (1985), Dahlem conference proceedings on: *The Exploitation of Marine Communities* (May, 1984) and the two classics, *Fish Stock Assessment, a Manual of Basic Methods* by Gulland (1983) and *Computation and Interpretation of Vital Statistics of Fish Populations* by Ricker (1975). These books focus on the use of mathematical models for describing and predicting the dynamic behavior of fish populations. They represent the content of what is usually meant by "quantitative stock assessment". Not only do they describe the background for the models, most of them also provide insight into the estimation of the parameters. It would not be possible to write the present book without the groundwork and framework presented by these volumes. Although the current book contains related material, it also contains most of the other elements of stock assessment that either are not available elsewhere, or are not available in a stock assessment format. Thus, there is relatively little overlap in content between this book and the above volumes.

For example, age is a kind of independent variable that underlies much of the mathematical structure of contemporary stock assessment and all of the above volumes' contents are formulated primarily in terms of age. However, in many stock assessment experiments age data are not available and the proxy measurement of size must be used. For this reason, Chapter 2 introduces the critical concepts and quantitative models of stock assessment, expressed in terms of size. Although the mathematical models in Chapter 2 are primarily formulated in terms of size, many of the derivations begin with formulations in terms of age, thus fostering a continuity with the contents of other works. The substitution of size for age is ripe with complexity and remains one of intensive active research.

Since, in some cases, size data are an intermediate step in the estimation of age and growth, the current book has a chapter on the estimation of fish age and growth presented with the constraints and objectives of a stock assessment in mind. Successive chapters deal with the most basic issue of all, the collection of data for stock assessment, via sampling designs. Data collection is an especially important part of stock assessment because the method of collection and the sampling design dictate the type of assessment techniques that may be applied and underlie the estimation of a model's parameters. This point is made repeatedly in the three sampling chapters which divide, in a broad sense, into chapters on designs for survey and/or catch sampling and for hydroacoustic sampling. In fact, sampling issues take many forms in stock assessment. While survey, catch and acoustic sampling designs address the larger issues, it is frequently necessary to subsample the data set, e.g., for age, growth, fecundity or related population dynamic data. The chapter on age and growth addresses subsampling questions in detail, thus supplementing the three sampling chapters.

Another chapter deals with data collected over time, providing the appropriate time series models for sequential data that might include covariates and/or periodicities. The penultimate chapter deals with the multispecies issue which represents one of the principal limitations of the stock assessment models in use today and which is fundamental to major problems such as "by-catch". It addresses questions of species diversity in multispecies environment when it would be useful to predict species composition and yield under alternative management strategies. The last chapter to this book presents a systems science approach to stock assessment. A conceptual framework is developed demonstrating how the multidisciplinary activities of the preceding chapters are integral to stock assessments in complex biological environments such as bays, estuaries and seas. A prototype decision support system designed to reduce the inherent complexities of the activities and the uncertainties of the biological environment is described.

Of the above volumes most relevant to this text, none focus on age estimation, on theoretical or practical sampling issues, nor on acoustics sampling. Size-based analyses have a very minor role, if noted at all in these volumes, and only the Dahlem conference volume (May, 1984) has significant content in the multiple species area.

A common characteristic of this book and of the other five volumes is their dependence upon mathematics. Some of the chapters in the current text make use of a high level of mathematics with the advantage that it opens the door to the consideration of an array of sophisticated models in common use today. These models are also the stepping stones to the models which will be in use in the future. In practice, the higher level of mathematics is not a serious problem because many of the models discussed are available in the form of computer programs that come either as part of the text or can be obtained by writing to the appropriate chapter authors. These programs are like "black boxes" in the sense that they are easily transferable and do not require mathematical sophistication to apply. It is our belief, however, that users should try to have as firm an understanding of the analytical aspects of the mathematics as they can and, for this reason, the mathematics is presented and not "skimmed-over". Nevertheless, in a few cases, when the mathematical details are very extensive, the reader is referred to the appropriate CRSP Working Paper. A complete list of the CRSP Working Papers is in Annex A (see below).

In further contrast to the above volumes, the current text has a significant focus upon fisheries in tropical environments and in developing countries. This suggests that many of the applications are drawn from small scale or artisanal fisheries. Other volumes which also have a focus on tropical fisheries include the manuals by (Pauly, 1984) and Sparre et al. (1989). The focus on developing country, tropical environments stems from the geographical areas of application of the authors, viz., the Philippines and Costa Rica, which were part of the project that led to this book. Since most of the issues and assumptions that are central to modeling fish dynamics in tropical environments are also important in numerous temperate environments, the book has a much wider breadth of application than it would otherwise. Expanded treatments of the tropical, developing country applications can be found in the CRSP Working Papers in Annex A.

We recognize that computer software related to stock assessment is developing very rapidly so it is difficult to make specific software recommendations. The BASICA programs of Saila et al. (1988) may have some utility, but they are not included herein. The current text does include a diskette with a collection of programs and their documentation, which are part of the standard repertoire of fishery managers in industrialized temperate fisheries where these techniques were first developed. These are described in Annex B. These programs have been compiled for IBM-PC or compatible computers with the Microsoft FORTRAN compiler (Copyright Microsoft Corporation), an ANSI subset of FORTRAN 77. The programs will not run on personal computers that do not have a math coprocessor but they do accept data stored in ASCII format. Some of these programs have a history, appearing as FORTRAN IV programs in either Abramson (1971), Pienaar and Thomson (1973) or Sims (1985).

Some of the programs used in Chapter 8 are also included on a diskette. These are written in Microsoft Quick BASIC. The source code, executable code and examples are found in the diskette. Other programs in Chapter 8 are available either from the CRSP Working Papers or by contacting the authors.

In addition to these programs, there are additional programs for the stock assessment models in Chapter 2, programs for the analyses in the sampling and acoustics chapters and programs used in other chapters, which are not included with this book. These may be found in the CRSP Working Papers or they may be obtained from sources listed in the appropriate chapter.

It should be noted that, despite the large number of programs on the diskette accompanying the text and the others referred to in the chapters, they are not an all-inclusive collection. For example, the book by Hilborn and Walters (1992) contains programs in Quick BASIC that address some of the same stock assessment topics; Freon et al. (1993) developed the interactive package CLIMPROD for fitting surplus production models with parameters to account for variability assumed due to climate; Seijo et al. (1994) developed a collection of programs which include economic parameters associated with the harvest of spatially dispersed demersal and benthic species; and Prager (1991) developed ASPIC, which models stock production and accounts for covariates.

In addition, CRSP software programs CHTL LBAR, MCARLO and LCANAL, LCAN all deal with length-based analyses. The program LCAN has a focus on sensitivity analyses that follow from length cohort analysis, as does the ANALEN package (Chevaillier and Laurec 1990). A new software package, the FAO-ICLARM Stock Assessment Tools (FiSAT) for stock assessment contains stock assessment programs that also include size-at-age analysis, some of which are traditional assessment programs. The program CASA (catch-at-size analysis) developed under the CRSP research project is a sophisticated size-based stock assessment program based on stochastic optimization techniques. CASA can be “tuned” with other data that may be available to enhance the solution.

This book’s chapters were all written with both research and teaching objectives in mind. It would be a suitable textbook for university classes where stock assessment is viewed as a multidisciplinary activity. The book will also be useful to all researchers concerned with the concepts and the methods of carrying out quantitative stock assessments and management.

2. BACKGROUND AND ORIGINS OF THE VOLUME

Recognizing the benefits to fisheries management from sound stock assessment, the U.S. Agency for International Development (USAID) funded the Fisheries Stock Assessment Collaborative Research Support Program (CRSP). The Fisheries Stock Assessment CRSP was one of eight international programs focused on research and training at sites where research could offer the largest global impact.

The present volume is an outcome of the Fisheries Stock Assessment CRSP which was designed to improve stock assessment methodologies and to provide stock assessment advice to fishery managers who operate under diverse conditions, and to train U.S. and host country scientists. It was particularly concerned with improving stock assessment methodology in small-scale single and multiple species tropical fisheries. There was also great interest in contributing to the theoretical and practical aspects of fisheries stock assessment and management, in general.

In very general terms, the procedure of fish stock assessment consists of the following elements: *inputs* (fisheries data and various assumptions concerning the data and methodology); a process (analysis or analyses of the data); and *outputs* (estimates of population or system parameters). These outputs then provide *inputs* to another *process* which consists of predictions under various alternatives, and there is a final *output* consisting of some form of management strategy, which includes optimization of yield or some other objective function(s). Although there are a number of common elements between temperate and tropical fisheries, our experience suggests that the nature of the input data to the stock assessment process is generally different in tropical fisheries. We are beginning to recognize more clearly the complexities and some of the unique features of tropical fisheries. Our efforts to develop stock assessment methodologies have attempted to take into account some of the above features. However, as is often the case in attempts to push the frontiers of a science in one direction, there are ramifications for the whole field which lead to a better understanding of the science as a whole. This has happened here and, as a consequence, our search for new methods in stock assessment has resulted in an array of methodologies which have applications across a broad geographic range and which have contributed to the progress of fishery science, in general.

Management of fisheries in tropical environments is greatly complicated by the paucity of biological knowledge of tropical marine ecosystems. Such factors as the complex patterns of recruitment and of the presence of a large number of interacting species make prediction and management difficult and sometimes even counter-intuitive. Models of such systems are one way for scientists and managers to try to understand the complex processes involved. However, models require inputs and quantitative expressions of real value, because modeling must be based on firm biological field and capture data. In spite of these realities, current fishery management is largely based on a mixture of theoretical concepts and practical considerations which often do not result in rational bases for policy nor in self-evident criteria for management decisions. Examples include the common use of steady-state solutions to deterministic equations for single species dynamics in which the effects of other species, of stochastic environmental change and of the economic and social motivations of fisherman are regarded as exogenous variables, if they are considered at all. This is unfortunate because it is apparent that human behavior in fishing-related activities can amplify moderate environmental fluctuations, leading to large changes in catch and profitability. Stock assessment of the future must begin to account for the realities of the whole system being assessed. Although this CRSP has not addressed all of these problems, it has incorporated field data, modeling and heuristic approaches in an attempt to balance the abstractions of models with the empirical lens of real data from real scenarios.

Specific objectives for this CRSP program included the following:

- Production of stock assessment guidelines for fishery managers in tropical countries that provide new approaches toward optimal fishery stock assessment and management.

- Testing of existing methodology for stock assessment, especially as it relates to tropical fisheries; our rationale was that there may be opportunities for application and/or modification of existing methods which should not be overlooked.

- Development of new methodologies that utilize alternatives to age-based analyses for stock assessment in tropical developing countries.

- Development and testing of multispecies fishery management models.

- Development of software, that is easily transferable to computers and miscellaneous institutional environments around the world, which can implement the methodologies and models with minimal user training.

The research was carried out primarily by personnel of three U.S. institutions: The University of Maryland, the University of Rhode Island and the University of Washington and scientists from two host country institutions: the University of Costa Rica and the University of the Philippines. Other participating institutions with less central roles included the University of Miami, the University of Delaware and the International Center for Living Aquatic Resource Management.

The research results summarized in this volume, as well as related research undertaken by project scientists, are discussed in greater detail in the program's Working Paper series. A list of the 105 CRSP Working Papers is provided in Annex A, all of which are available from the authors. It is clear from this list of Working Papers that a variety of subjects and approaches related to stock assessment were explored. It will also be evident that numerous working papers evolved into publications in primary journals.

ACKNOWLEDGMENT

The authors thank Dr. Jerald Ault for a critical review of a draft of this chapter.

REFERENCES

Abramson, N.J. 1971. Computer Programs for Fish Stock Assessment. FAO Fisheries Technical Paper No. 101.

Chevaillier, P. and Laurec, A. 1990. ANALEN: Logiciels d'analyse des données de capture par classes de taille et de simulation des pêcheries multi-engins avec analyse de sensibilité. FAO Document Technique sur les Pêches. 101 Supplement 4.

Clark, C.W. 1985. *Bioeconomic Modeling and Fisheries Management.* John Wiley and Sons, New York.

Freon, P., Mullon, L. and Pichou, C. 1993. Experimental interactive software for choosing and fitting surplus production models including environmental variables. Computerized Information Series. Fisheries. Food and Agriculture Organization of the United Nations.

Gulland, J. 1983. *Fish Stock Assessment, a Manual of Basic Methods.* John Wiley and Sons, New York.

Hilborn, R. and Walters, C.J. 1992. *Quantitative Fisheries Stock Assessment.* Chapman and Hall, New York.

May, R. M. (Ed.) 1984. *The Exploitation of Marine Communities*, Life Sciences Research Report 32. Springer-Verlag, Berlin.

Pauly, D. and Sparre, P. 1991. A note on the development of a new software package, the FAO-ICLARM Stock Assessment Tools (FiSAT). Fishbyte, Aug. 1991, pp. 47-48.

Pienaar, L.V. and Thomson J.A. 1973. Three programs used in population dynamics, WVONB-ALOMA-BHYLD. Fish. Res. Bd. Can. Tech. Rept. No. 367.

Prager, M. 1991. *User's Manual for ASPIC: a Stock-Production Model Incorporating Covariates. Program Version* 2.8. U.S. National Marine Fisheries Service, Miami Florida. Miami Laboratory Document MIA-91/91-20.

Ricker, W.E. 1975. *Computation and Interpretation of Biological Statistics of Fish Populations.* Bulletin of Fisheries Research Board of Canada, Ottawa.

Saila, S.B., Recksiek, C.W. and Prager, M.H. 1988. *Basic Fishery Science Programs.* Elsevier, New York.

Sims, S.E., (ed). 1985. Selected Computer Programs in FORTRAN for Fish Stock Assessment. FAO Fisheries Technical Paper No. 259.

Seijo, J.C., Caddy, J.F. and Euan, J. 1994. Space-time dynamics in marine fisheries: a bio-economic software package for sedentary species. Computerized Information Series. Fisheries. Food and Agriculture Organization of the United Nations.

Sparre, P., Ursin, E. and Venema, S.C. 1989. Introduction to tropical fish stock assessment Part 1-Manual. FAO Fisheries Technical Paper No. 306.

CHAPTER 2

SIZE-BASED METHODS OF STOCK ASSESSMENT OF SMALL-SCALE FISHERIES

Vincent F. Gallucci[1], Benyounes Amjoun[1], John B. Hedgepeth[2] and Han-Lin Lai[3]

[1]School of Fisheries, University of Washington, [2]Biosonics Inc., [3]National Marine Fisheries Service

1. INTRODUCTION

In many applications of stock assessment, it is necessary to find a substitute for age-dependent data. Collecting age-specific data can be a problem because of the inherent technical difficulty in determining the age of tropical fish (see Chapter 3) and because, even when possible, it is very expensive. Thus, the proxy measurement of length or size has become popular.

The field of size-based analysis has expanded in recent years from relatively straightforward analytical and *ad hoc* methods to sophisticated models based on stochastic analyses and nonlinear optimization. In this chapter we attempt to get to the core of some of the stock assessment methods to describe their foundations and some aspects of their application. We sometimes linked methods together showing commonalties and other times we tried to resolve statistical and conceptual ambiguities. Ideally, the step after gaining an insight into how the models interrelate conceptually is to see how the models' outputs interrelate. This could be done by generating an artificial data set from a simulation model for which the primary parameters are known. The methods in this chapter would then be applied to these data and the resulting parameter estimates

1-56670-151-1/96/$0.00+$.50

compared to the known input parameters. The alternative of using data from real fisheries presents the obvious problem that the true parameter values are unknown so, even if four out of five methods give similar estimates, four may be quite wrong and the fifth one correct.

To our knowledge no other source is available that draws together and synthesizes the variety of advanced models now available for the estimation of stock assessment parameters based on length data. On the other hand, a volume by Sparre et al. (1989) entitled *Introduction to Tropical Fish Stock Assessment* has drawn together many of the applied methods that require less advanced mathematical concepts. The first part of our work overlaps the contents of Sparre et al. (1989) but, after that, the volumes go their separate ways. Even in the areas of overlap, the examples, focus and philosophy are quite different so there is in fact little redundancy.

Two independent reviewers of the chapter had opposing views about the relatively high level of the mathematics we use. One, quite simply, objected to the high level. Quoting more or less from the other, "...if you are going to do brain surgery or build a bridge you should be trained in surgery or engineering". We tried to steer a path between these polar positions. As a generalization, the users most likely to gain from this volume are those inclined to, or at least patient with, mathematics. Many such individuals who will work on developing country fisheries graduate annually from universities in Europe, the U.S. and in some developing countries. Therefore, some of our work will appeal primarily to recent and future generations of stock assessment/management scientists. We feel, however, that the perspective and philosophy contained herein will benefit any reader, even one who liberally skips over the equations.

This chapter has served as the basis for a one quarter advanced intermediate course in stock assessment. The course material was supplemented by some topics not in the chapter, e.g., gear selection and stock production models. Otherwise, considering the focus on size based models, no comparable course text is available today.

An important consideration in the adoption and application of any method is the response or the sensitivity of its output to error in the various input parameters. The fact that different inputs may have big variances and that some inputs are more critical than others to the accuracy of the output is often overlooked. When possible, we will include information on the sensitivity of the methods. At the very least, simulation or Monte Carlo studies should be carried out to obtain a sense of model response. Alternatively and preferably, analytical equations for sensitivity studies may be developed so that computer simulation is unnecessary, or less necessary. In such cases, direct substitution of the inputs provides estimates of relative error.

There is an important principle of conservation one should note in the beginning: measurements of size, e.g., length, are easier to take than age **and** are much more variable. That is, there exist a much greater number of lengths corresponding to one age than *vice versa*. Therefore, it should be expected that the use of length-based methods will, at the very least, lead to a loss of information in the sense of providing more variable results than age-based methods. It is extremely rare in science, as in life, to get something more for less. Therefore, it is unreasonable to expect to obtain the same quality output when size data are input, as when higher "quality" age data are used in age-dependent methodologies. The art is knowing that there is a trade-off and the gains and losses with each approach. Thus, length-based methods can be satisfactory and extremely useful for stock assessment, especially when age information is scarce but, the quality of the output must be kept in perspective. To emphasize this, we frequently illustrate that the absence of measures of variability around point estimates of parameters should be of major concern and, that simulations to determine the impact of variability in input parameters on projected allowable catches (sensitivity studies) are a crucial part of a stock assessment — in tropical or non-tropical fisheries. The recent manual of Chevaillier and Laurec (1990) has moved significantly in this direction by providing error estimates and simulation programs (in FORTRAN) as a central part of their stock assessments.

The first part of this chapter refers to the parameter Z/K which is the ratio of total mortality and the rate term in the von Bertalanffy growth model. This parameter is considered from the perspective of its interpretation and its estimation. One of the major interpretations is found in its occurrence in the subsequent section on yield-per-recruit analysis which can be carried out in terms of age or size. We replicate the size-based theory showing the emergence of the Z/K and M/K ratios and then discuss the consequences for the choice of an exploitation strategy. Successive sections consider the estimation of other stock parameters using length cohort analysis, where a sensitivity analysis for errors in input parameters is illustrated. This is followed by methods based on algorithms that follow the modes in catch-at-length data collected over successive time periods. Most of these analyses are based on statistical models that allow the estimation of variability around a point estimate. Quite often these models require that assumptions be made about the processes that generate the data, which provide another type of entree into evaluating the validity of a methodology. The last group of methods is based on models which make assumptions about the biological processes that may have generated the data, thus requiring the estimation of parameters for these processes. The projected estimates of yield, mortality, etc. are based on the estimates of the parameters for the relevant biological processes.

In many cases, computer programs and users guides are available to carry out the numerical analyses in a particular model. In some cases, the software is sold commercially and in others, it is currently available free of charge. When software availability is known, it is noted in the section.

Finally, it has become fashionable to decry the use of steady-state models. It is undeniable that few ecological processes are in long term steady states. However, experience from stochastic modeling and the analysis of transients in physics, engineering and biology clearly demonstrates the importance of steady-state solutions (see, e.g., Hertzberg and Gallucci, 1980). Non-steady-state and stochastic models are indeed the next step in sophistication. In many cases the point of demarcation will be the same dynamic model used to find the steady-state solution where the simplifying assumption of the existence of a steady state is not made, thus allowing solutions as functions of time.

2. SIMPLE METHODS OF ESTIMATION OF GROWTH AND MORTALITY PARAMETERS

2.1 CATCH CURVES

In most situations, data are generated from sampling the catch. In the following materials we examine the various analyses that can be carried out using catch data to estimate properties of the population that provided the catch. One of the most basic of these analyses is termed "catch curve" analysis.

The catch curve analysis is based on several assumptions:

1. There is a steady state in the population over time.
2. Recruitment to the fishery is knife-edged, that is, fish are fully recruited to the fishery at age t_c.
3. Population survival is negative exponential given by

$$N_t = Re^{-Z(t-t_c)}$$

where N_t is the population size at time t and R is recruitment. Note that $N_{t_c} = R$.

In the following we consider catch curves from two different perspectives.

2.1.1 Catch Curve in Terms of Age

The early work on catch curve analysis assumed the ability to estimate the ages of the sampled animals, which would usually be ascertained with the help of an age-length key. The analytical work of Chapman and Robson (1960) and Robson and Chapman (1961) set the stage. Seber (1973) has a useful summary and extensions. The analysis is based on the assumption that year class strength and annual survival rates per year class are constant over a limited set of age groups. In other words, we assume that it is equivalent to say "over age groups" or "over time," i.e., we are assuming a steady state between the numbers entering the fishery via recruitment and losses via death from all causes at least over the number of years as far as the oldest age group in the catch.

If the abundance by age-class in the sample is plotted versus age in the logarithmic scale, the result is a linear function in t (Figure 1), called the "catch curve." The slope of the right hand side of the catch curve from $t = t_c$ is the instantaneous rate of total mortality Z, which is related to the survival fraction S by

$$Z = -\ln S$$

The parameter Z is often written as $Z = M + F$, where M and F are constant instantaneous rates of natural and fishing mortality, respectively. Figure 1 shows an age x which relates the age at which there is a turning point in the curve caused by fish of ages less than x not being fully recruited to the fishery. This introduces the concept of

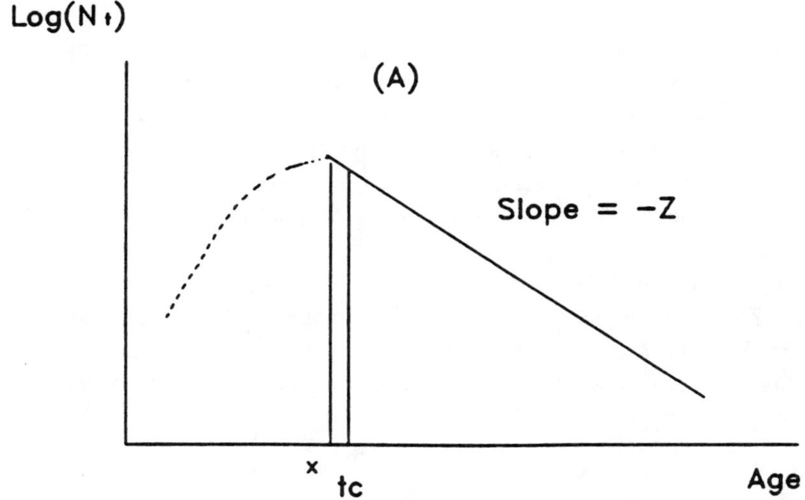

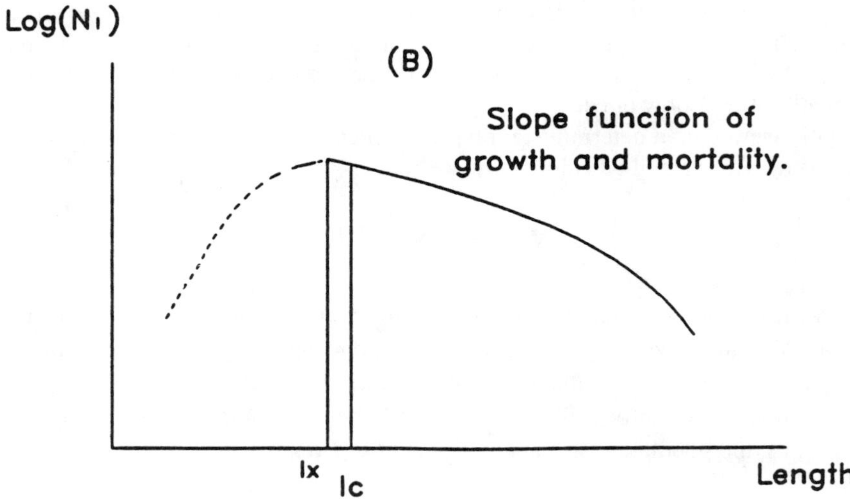

Figure 1. Catch curves in terms of age (A) and length (B). (from Jones. 1984. FAO Fish. Tech. Pap. No. 256. With permission.)

"gear selectivity" where we associate "being fully recruited to the fishery" to mean that "selectivity" is equal to one; for our purposes, selectivity equals one means that the probability of capture is one if the fishing gear makes contact with the fish at age t_c and older. That is, it is assumed that fish aged x and older are equally likely to be captured. However, for statistical reasons, one moves to a slightly older age, t_c, and calls it the "age of full recruitment." Later sections will deal with the more realistic cases when selectivity has a particular functional form and is associated with the effort that is exerted in a fishery.

The theory surrounding the estimation of survival S after age t_c and thus, of the mortality rate Z, is: the number of fish growing into age x is balanced by the number leaving it and S is the fraction that survives over x to $x+1$. That is, let N_x be an estimate of the number of fish entering age x and N_{x-1} be an estimate of the number leaving age $x-1$. Then

$$N_x = SN_{x-1} = S^x N_0 \tag{1}$$

where N_0 is the initial (unknown) number in age group $x = 0$.

The **probability** that an animal is of age x or older is found from above as

$$\frac{N_x}{N_0} = S^x$$

This empirical expression also fits the definition of a geometric probability density function (pdf) where S is a parameter of the pdf and x is a random variable representing age. The advantage of a probability model is that a likelihood function can be written and estimates of S follow directly.

If (1) is viewed as a deterministic relationship and age x as a continuous quantity, the result is an exponential equation

$$N_x = N_0 e^{-Zx}.$$

The variable for age is represented by "x" in the above, but in later sections it also appears as "t" and as "a", depending upon common usage for the model in question. Thus, N_x is N_t or $N(t)$. The abundance at an arbitrary **initial** time t_c will be $N_c = N_{t_c}$ or $N(t_c)$ which frequently will correspond to the abundance at recruitment, i.e., $N_c = R$. Thus, the population size at a given age t is given by,

$$N_t = N(t) = Re^{-Z(t-t_c)} = N_c e^{-Z(t-t_c)} \tag{2}$$

A principal premise of stock assessment is that the catch $C(t)$ is related to abundance $N(t)$. In Figure 1(A) the probability of capture for fish aged $t < t_c$ is less than one but otherwise is unspecified. This means that these younger fish have not been "fully recruited to the fishery." However, we do assume that selectivity, i.e., the probability of capture, increases with age for ages $t < t_c$ so the curve $C(t)$ also increases with t. Since this is considered hypothetical for $t < t_c$ the curve is shown as dotted. After age t_c, capture is guaranteed upon contact with the gear but now the above exponential model acts to make the number available $N(t)$ and the number captured $C(t)$ functions of the mortality rate Z.

2.1.2 Catch Curve in Terms of Size

A corresponding situation can be described for **length** distributions if age data are not available. A generic catch curve as a function of size, e.g., length, is seen in Figure 1(B). In this curve, the horizontal axis is length and length l_c corresponds to the size at which fish become fully recruited to the fishery. Naturally, l_c must correspond to age t_c in a conceptual way. If the correspondence is made specific by choosing a model such as the von Bertalanffy (LVB) growth model to functionally relate age and size, then l_c is the exact length corresponding to age t_c. One of the principal differences between the length and age catch curves is that the right hand side corresponding to $l > l_c$ is curved so that the slope is not a constant but is a function of the growth and mortality parameters and the changing value of fish length. Furthermore, in the size-based formulation, length (l) has an upper limit, e.g., L_∞ from the von Bertalanffy model, whereas age goes on indefinitely.

The development of the equation for catch curve abundance as a function of length uses the negative exponential model, just as in the age case, and the von Bertalanffy (LVB) growth model given by

$$l_t = L_\infty\left(1 - e^{-K(t-t_0)}\right) \qquad (3)$$

where l_t is fish length at time t, L_∞ is the asymptotic length, K is the growth coefficient, and t_0 corresponds to the theoretical time at which fish length is zero.

Solving the LVB model for age t in terms of length and substitution into the exponential model (2) yields

$$N(l) = N_c \exp\left\{\frac{Z}{K}\left[\log\left(\frac{L_\infty - l}{L_\infty}\right) - \log\left(\frac{L_\infty - l_c}{L_\infty}\right)\right]\right\} \qquad (4)$$

The term N_c refers to abundance at age t_c or at length l_c. Simplification yields

$$N(l) = N_c \exp\left(\frac{Z}{K}\log\left(\frac{L_\infty - l}{L_\infty - l_c}\right)\right) \tag{5}$$

or

$$N(l) = N_c\left(\frac{L_\infty - l}{L_\infty - l_c}\right)^{Z/K} \tag{6}$$

Transformation to natural logarithm of both sides of (6) yields

$$\log N(l) = \alpha + (Z/K)\log\left(L_\infty - l\right) \tag{7}$$

which is a straight line with positive slope Z/K when $\log N(l)$ is plotted versus $\log(L_\infty - l)$ with intercept α (Jones, 1984). To make a comparison to the age-based catch curve in Figure 1(A), the graph of log $N(t)$ versus t must be compared to the graph of log $N(l)$ versus l. This is plotted in Figure 1(B), where the part of the curve to the right of l_c is **curved** with a slope that is a function of l, Z/K and L_∞. The catch curve to the left of l_c is hypothetical, as before, and is dotted. It is interesting to note for future reference where and how the parameter Z/K emerges in this length-based analysis.

2.2 GEOMETRIC INTERPRETATION

The preceding section describes the mathematical basis for estimating the total mortality parameter Z from catch curves plotted in terms of either age or length. Beverton and Holt (1959) and others have suggested a relationship between the parameter for natural mortality M and the parameter K of the von Bertalanffy growth model. In other words, the suggestion is that there is a pattern to be seen with respect to high or low K-values (rapid or slower approaches to the asymptotic length) and high or low mortality rates.

The parameter Z/K and its subset M/K evolve naturally as parameters in several different circumstances. For example, we have seen how it is a natural parameter in length-based catch curves and we will see that it arises naturally in Beverton-Holt yield-per-recruit analyses where age-dependent input is minimal, as well as in length-based cohort analysis. We will discuss the estimation of the parameter Z/K, and will see that the efficiency of different estimators of Z/K depends upon the range of values the ratio takes. In fact, different ranges of Z/K have been associated with different geometric forms of the length frequency distribution. That is, different ranges of values of Z/K are associated with patterns of survivorship through, e.g., how the number of survivors decreases as size (and thus age) increases. This section will elaborate on the association between Z/K and the length frequency distribution and the next section will focus on estimation of Z/K.

A rough estimate of the ratio Z/K (Powell, 1976, 1979) can be inferred from the shape of the length distribution of the sampled catch. Figure 2 shows the catch curves of the theoretical length frequency data as a function of different ranges of Z/K. It is explicitly assumed in these plots that length changes in accord with the von Bertalanffy growth model, that the population survival is described by the negative exponential model, and that the asymptotic length is normally distributed random variable with mean L_∞ and variance σ_L^2. These curves are computed by numerically integrating equation (2) in Powell (1979). That is,

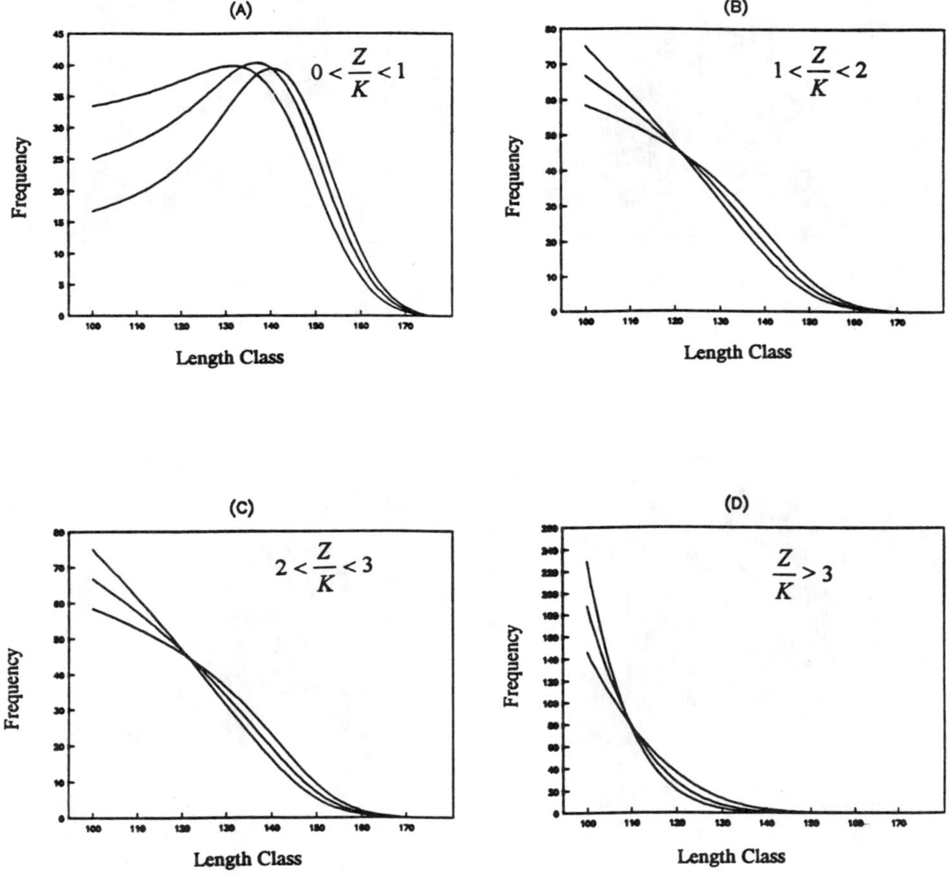

Figure 2. Theoretical length frequencies. (from Powell. 1979. Cons. Int. Explor. Mer. 175. With permission.)

$$f(l) = \frac{Z}{K}\int_{l}^{\infty} (L-l)^{\frac{Z}{K}-1}\left(L-l_c\right)^{-\frac{Z}{K}} f_L(L)\, dL \tag{8}$$

where $f_L(L)$ is the probability density function of the asymptotic length with mean L_∞ and variance $\sigma_L^{\,2}$, and $f(l)$ is the probability density function of length l.

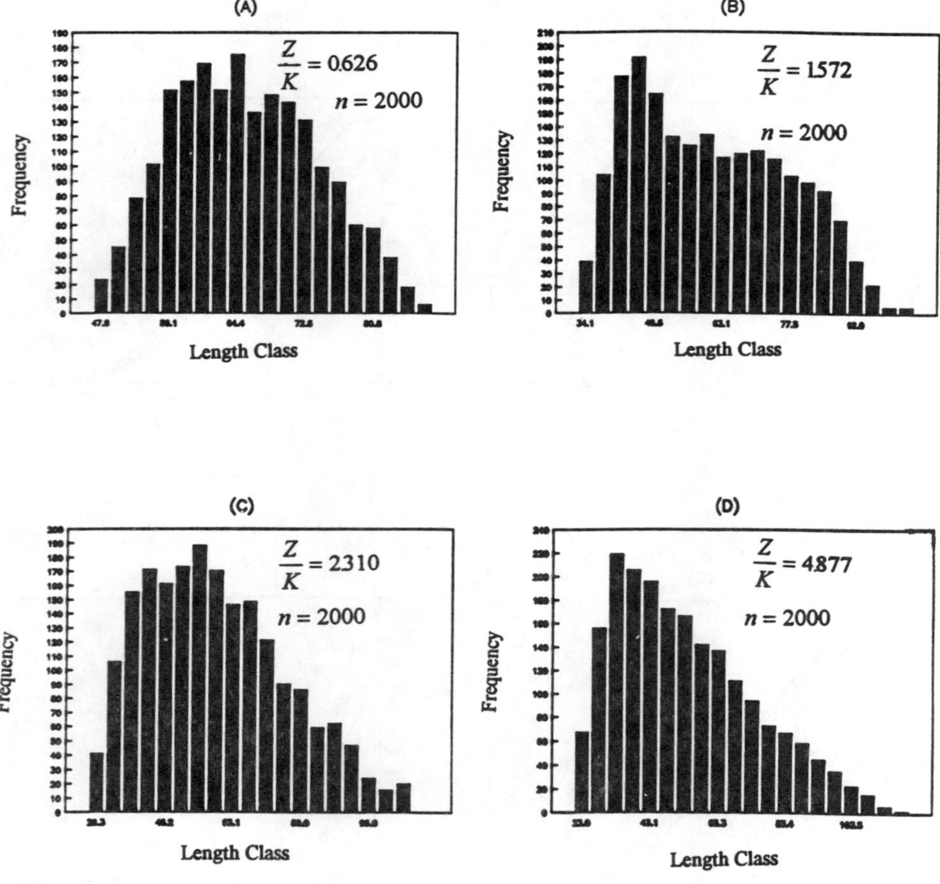

Figure 3. Simulated length frequencies for different values of Z/K ($Z = 0.5$, $t_0 = -0.95$).

We developed a simulation program using the algorithm suggested by Breen and Fournier (1984). This algorithm has two types of variability: (1) deviation of lengths around the mean length at age and (2) deviation of the means around the von Bertalanffy growth curve. The total instantaneous mortality rate Z was kept fixed in all simulations and the parameters K and L_∞ were varied systematically according to the following relationship (Ricker, 1975)

$$K = \log\left(\frac{L_\infty - l_t}{L_\infty - l_{t+1}}\right) \tag{9}$$

where l_t is fish length at time t.

Figure 3 shows catch curves of simulated size distributions for the following values of Z/K: (a) 0.626, (b) 1.572, (c) 2.31, (d) 4.877. Examination of the theoretical length frequencies in Figure 2 suggests that empirical (our simulated data) relative frequency plots can be used to infer estimates of Z/K by noting how well the empirical plots mimic the theoretical curves. Note that as Z/K increases, the length at first capture decreases as can be expected, since high Z/K with Z fixed implies slow growing animals which in turn implies a lower overall average length in the catch. Thus, the length at first capture must be smaller for low values of K than for high values of K, assuming the total mortality Z is constant.

2.3 ESTIMATION OF Z/K

The estimating equations of the ratio Z/K require fish length frequency data. In this section, two types of estimators are considered: (1) estimators based on an assumed probability distribution of lengths such as those suggested by Ssentongo and Larkin (1973), Hoenig (1983), Hoenig et al. (1983), and Powell (1979); and (2) estimators based on regression methods such as those suggested by Jones and van Zalinge (1981), Jones (1984), Wetherall (1986), Wetherall et al. (1987) and Pauly (1983a, 1983b, 1986).

All of these estimation techniques rest upon a set of assumptions common to most simple length-based analyses. These assumptions are listed as part of the catch curve analysis but they bear repeating:

1. It is assumed that the population is in a steady state, not changing with a decided trend over successive time periods.

2. Recruitment is knife-edged into the fishery without selection with respect to size.

3. Survival of the recruited fish is described by a negative exponential function.

Since the processes of recruitment, fishing and growth of fish may be taking place almost throughout the whole year, especially in tropical environments, it may be important to try to "average out" some of the resulting variability. It has been recommended by some authors that this may be done by pooling or averaging catch curves from several areas or over several time periods. This should be done with caution, however, taking into account statistical criteria that must be satisfied before pooling data.

2.3.1 Distributional Methods

2.3.1.a *Beverton and Holt (1956)*:

Beverton and Holt (1956) derived deterministically an estimator of Z based on mean length as follows:

$$\bar{l} \;=\; L_\infty\left(1-\frac{Z}{Z+K}\left(1-\frac{l_c}{L_\infty}\right)\right) \tag{10}$$

$$\hat{Z} \;=\; K\left(\frac{L_\infty-\bar{l}}{\bar{l}-l_c}\right) \tag{11}$$

where $\bar{l}$ is mean length in the sample.

Rearranging (11) yields an estimator of $\theta = Z/K$:

$$\hat{\theta}_{BH} \;=\; \frac{L_\infty-\bar{l}}{\bar{l}-l_c} \tag{12}$$

Powell (1979) and Wetherall et al. (1987) derived the same estimator in a probabilistic framework. Given the assumptions (1) to (3), the probability density function of length l is given by

$$g(l) \;=\; \theta\left(L_\infty-l\right)^{\theta-1}\left(L_\infty-l_c\right)^{-\theta} \tag{13}$$

for $l_c \leq l < L_\infty$.

Estimator (12) is just the moment estimator of θ based on (13):

$$L_\infty-\bar{l} \;=\; \frac{\theta\left(L_\infty-l_c\right)}{\theta\,+\,1} \tag{14}$$

Solving (14) for θ gives (12). Note that (10) and (14) are equivalent.

Wetherall et al. (1987) use Taylor series approximation to show that the bias and variance of (12) are:

$$\widehat{\text{Bias}}\left(\hat{\theta}_{BH}\right) \;=\; \frac{\hat{\theta}\left(\hat{\theta}+1\right)}{n\left(\hat{\theta}+2\right)}$$

$$\hat{\text{Var}}\left(\hat{\theta}_{BH}\right) = \frac{\hat{\theta}\left(\hat{\theta}+1\right)^2}{n\left(\hat{\theta}+2\right)}$$

where n is the number of observations in the sample.

2.3.1.b *Ssentongo and Larkin (1973)*:

The derivation of the catch curve in terms of length was based upon the von Bertalanffy function solved for age t in terms of L_∞, l and t_0 as

$$t = \frac{1}{K}\left(-\log\left(1-\frac{l}{L_\infty}\right)\right)+t_0 \tag{15}$$

and substituted into the negative exponential model.

After replacing t, Ssentongo and Larkin (1973) derived the maximum likelihood estimate (MLE) of θ based on the negative exponential probability density

$$f(y) = \theta e^{-\theta\left(y-Y_c\right)} \tag{16}$$

where $Y=-\log(1-l/L_\infty)$ and $Y_c=-\log(1-l_c/L_\infty)$. Given n observations in the sample, it can be shown that the MLE of θ is

$$\hat{\theta}_{SL} = \frac{1}{\overline{Y}-Y_c}$$

where

$$\overline{Y} = \frac{1}{n}\sum_{i=1}^{n} Y_i$$

The MLE θ_{SL} is slightly biased (Hoenig, 1983). Unbiased estimates of θ and its variance are readily available:

$$\hat{\theta}_{USL} = \left(\frac{n-1}{n}\right)\frac{1}{\overline{Y}-Y_c}$$

$$\text{Var}\left(\hat{\theta}_{SL}\right) = \frac{\left(\hat{\theta}\right)^2}{(n-1)^2(n-2)}$$

since $\sum_{i=1}^{n} Y_i$ is gamma distributed with parameters n and θ. The paper by Ssentongo and Larkin (1973) contains an unbiased estimator of θ that differs from both the $\hat{\theta}_{USL}$ and $\hat{\theta}_{SL}$ derived here. The error introduced by the use of the wrong equation for the estimator is very low; e.g., even if n were very small, $n = 4$, the error is only five percent. For realistic larger *n*, the error is nil.

2.3.1.c *Powell (1979)*:

Powell (1979) suggested the use of the coefficient of variation around l_c as a basis for the estimation of θ and considered the asymptotic length L as a random variable with mean L_∞ and variance $\sigma_L{}^2$. Using the following relationship

$$E\left[\left(l-l_c\right)^n\right] = \frac{\Gamma(n+1)\Gamma(\theta+1)}{\Gamma(n+\theta+1)} E\left[\left(L-l_c\right)^n\right]$$

Powell derives an estimate of θ :

$$\hat{\theta}_P = \left(\frac{2c^2}{1-c^2}\right) - \left(\frac{2\sigma_L{}^2(\theta+1)}{\left(L_\infty - l_c\right)^2\left(1-c^2\right)}\right) \tag{17}$$

where $c^2 = Var(l)/(\bar{l} - l_c)^2$. If $\sigma_L{}^2$ is small compared to $\left(L_\infty - l_c\right)^2$, then (17) reduces to

$$\hat{\theta}_P = \frac{2c^2}{1-c^2}$$

Estimates of the bias and variance of $\hat{\theta}_P$ are not available.

2.3.1.d *Hoenig (1983)*:

Hoenig (1983) suggests the use of the median length as an estimator of Z. As in Ssentongo and Larkin (1973), they use the negative exponential function given by (16) to compute the median of *y*, denoted by y_m. Thus, an estimate of Z based on median length is

$$\hat{Z} = \frac{K \log 2}{y_m - y_c} \tag{18}$$

Simple manipulation of equation (18) yields an estimator of θ :

$$\hat{\theta}_H = \frac{\log 2}{y_m - y_c}$$

Closed forms for the bias and variance of $\hat{\theta}_H$ are unavailable.

2.3.2 Regression Methods

2.3.2.a *Jones and van Zalinge (1981)*:

This linear regression method uses the cumulative number of fish at any age $t \geq t_c$

$$R\int_t^{\infty} e^{-Z\tau} d\tau = \frac{R}{Z} e^{-Zt}$$

Following the analysis for the catch curve in length, age t is replaced by the LVB in terms of length. The resulting linear regression model is

$$\log N_{cum} = \log C + \theta \log\left(L_{\infty} - l\right)$$

where N_{cum} is the number of fish with length greater than or equal to l. This equation is similar to (7) where $\log C$ has an interpretation similar to that of α in the catch curve in length equation. The use of $\log N_{cum}$ instead of $\log N_t$ probably adds stability to the estimation process but then a functional dependence develops amongst the observations in the dependent variable. A weighted linear regression may resolve some of the difficulty.

2.3.2.b *Jones (1984)*:

Using similar procedures as above, Jones proposed the following linear regression model to get an estimate of θ :

$$\log N_t = \log C' + (\theta - 1)\log\left(L_{\infty} - l\right)$$

where N_t is the population size at length l and C' is a constant.

2.3.2.c *Wetherall, Polovina and Ralston Method*:

Wetherall (1986) and Wetherall et al. (1987) suggest the estimation of both L_{∞} and Z/K by regressing an increasing sequence of fish lengths on average lengths. Let $\{l_1, l_2, \cdots, l_m\}$ denote an increasing sequence of fish lengths from the interval $\left(l_c, L_{\infty}\right)$ and let $\bar{l}_i$ be the average length of fish in the interval $\left(l_i, L_{\infty}\right)$, then

$$\bar{l}_i = L_\infty\left(\frac{1}{1+\theta}\right) + \left(\frac{\theta}{1+\theta}\right)l_i + \varepsilon_i$$

This is a simple linear regression of the form

$$\bar{l}_i = \alpha + \beta l_i + \varepsilon_i$$

where α and β are parameters to be estimated by regression and ε_i are normally distributed errors. Then,

$$\hat{L}_\infty = \frac{\hat{\alpha}}{1-\hat{\beta}}$$

$$\hat{\theta} = \frac{\hat{\beta}}{1-\hat{\beta}}$$

Comments about functional dependence among observations similar to those made after the Jones and van Zalinge method need to be made. See, e.g., the discussion in Wetherall (1986) and Wetherall et al. (1987). There also exist related methods such as a Modified Wetherall et al. method (Pauly, 1986) and the length-converted catch curves method (Pauly, 1983b), which are not shown here.

2.3.3 Simulation Results

The above estimators of Z/K are compared in Tables 1 and 2. The simulation program used in Section 2.2 on *Geometric Interpretation* was used to generate length-frequency distributions for different values of the ratio Z/K while keeping the instantaneous total mortality rate fixed at 0.50. Since a negative correlation between L_∞ and K exists (Gallucci and Quinn, 1979), these two parameters were varied in each simulation according to the relation (2). Tables 1 and 2 include the different estimates and their bias and variance, when these have closed forms. For regression methods, the residual sum of squares is also displayed.

Estimators based on distributional assumptions, viz., Beverton-Holt's and Ssentongo-Larkin's, estimate Z/K with greater precision than the two other techniques but can still be as much as 100% wrong. They achieve a relatively low bias for a wider range of Z/K values (1.5 – 4.5). Estimates based on median length achieved their lowest levels of relative bias for true values of Z/K between 1.5 and 3.5. Powell's estimate performed poorly over all Z/K ranges. This may be because the effects of variability in L_∞ were not taken into consideration. Most of the estimates arrived at via Powell's method are positively biased. A quick look at Equation (10) suggests that the positive bias can be reduced if the asymptotic length is assumed to follow a probability

Table 1. Estimators of Z / K based on simulation where $Z = 0.50$ and K varies.

Ratio Z / K ===>	0.6258	0.8645	1.0841	1.5718	1.9084	2.1097	2.7071	3.3025	4.4843	5.6625
Number of Observations	1053	1506	1173	1625	1666	1456	1534	1761	1654	1603
Total Mortality rate	0.5	0.5	0.5	0.5	0.5	0.5	0.5	0.5	0.5	0.5
Asymptotic Length	75	80	85	96.7	105	110	125	140	170	200
Growth Coefficient	0.799	0.5784	0.4612	0.3181	0.262	0.237	0.1847	0.1514	0.1115	0.0883
t-Zero	-0.95	-0.95	-0.95	-0.95	-0.95	-0.95	-0.95	-0.95	-0.95	-0.95
True Value of Z / K	0.6258	0.8645	1.0841	1.5718	1.9084	2.1097	2.7071	3.3025	4.4843	5.6625
BEVERTON and HOLT (1956):										
Estimate	.3417	1.2722	1.7125	1.5776	2.0173	1.9375	3.0404	3.8823	5.8085	7.8384
Variance (Asymptotic)	0.0021	0.0013	0.0029	0.0018	0.0027	0.0029	0.0064	0.0089	0.0208	0.0388
Bias	0.7159	0.4077	0.6284	0.0057	0.1090	-0.1722	0.3333	0.5798	1.3242	2.1759
% Bias	114%	47%	58%	0%	6%	8%	12%	18%	30%	38%
SSENTONGO and LARKIN (1973):										
Estimate	1.3133	1.0441	1.7563	1.5410	1.9788	1.9490	2.9819	3.8807	5.7141	7.7084
Variance	1.56E-06	4.81E-07	2.24E-06	9.00E-07	1.41E-06	1.79E-06	3.78E-06	4.86E-06	1.19E-05	2.31E-05
Bias	0.6875	0.1797	0.6722	-0.0309	0.0704	-0.1607	0.2748	0.5782	1.2298	2.0459
% Bias	110%	21%	62%	-2%	4%	-8%	10%	18%	27%	36%
HOENIG (1983):										
Estimate	1.6842	1.4664	1.9255	1.7239	2.0216	1.9675	3.5361	4.2208	7.8243	10.7110
Variance	-	-	-	-	-	-	-	-	-	-
Bias	1.0584	0.6019	0.8414	0.1521	0.1132	-0.1422	0.8290	0.9183	3.3400	5.0485
% Bias	169%	70%	78%	10%	6%	-7%	31%	28%	74%	89%
POWELL (1979):										
Estimate	1.9109	1.8634	2.0183	2.1329	2.3701	2.0489	4.7981	4.2831	24.9474	-53.2870
Variance	-	-	-	-	-	-	-	-	-	-
Bias	1.2851	0.9989	0.9342	0.5610	0.4617	-0.0608	2.0910	0.9806	20.4631	-58.9495
% Bias	205%	116%	86%	36%	24%	-3%	77%	30%	456%	-1041%

Table 2. Estimators of Z / K based on linear regression models where $K = 0.50$.

Ratio Z / K ===>	0.6258	0.8645	1.0841	1.5718	1.9084	2.1097	2.7071	3.3025	4.4843	5.6625
Number of Observations	1053	1506	1173	1625	1666	1456	153	1764	1654	1603
Total Mortality Rate	0.5	0.5	0.5	0.5	0.5	0.5	0.5	0.5	0.5	0.5
Asymptotic Length	75	80	85	96.7	105	110	125	140	170	200
Growth Coefficient	0.799	0.5784	0.4612	0.3181	0.262	0.237	0.1847	0.1514	0.1115	0.0883
t-Zero	-0.95	-0.95	-0.95	-0.95	-0.95	-0.95	-0.95	-0.95	-0.95	-0.95
True Value of Z / K	0.6258	0.8645	1.0841	1.5718	1.9084	2.1097	2.7071	3.3025	4.4843	5.6625
JONES and VAN ZALINGE (1981):										
Estimate	0.9107	0.5151	1.4400	1.4014	1.4185	1.6089	2.5905	3.5576	5.1724	7.9537
Standard Error	0.0225	0.0621	0.0351	0.0048	0.0950	0.0549	0.0536	0.1288	0.3728	0.5173
Bias	0.2849	-0.3494	0.3559	-0.1705	-0.4899	-0.5008	-0.1166	0.2551	0.6881	2.2912
% Bias	31%	68%	25%	12%	35%	31%	5%	7%	13%	29%
R-Squared	0.9969	0.8844	0.9959	0.9998	0.9409	0.9840	0.9936	0.9794	0.9233	0.9403
JONES (1984):										
Estimate	1.0983	1.1364	1.2396	1.4420	1.8326	1.9951	2.1712	3.5055	4.8693	6.8580
Standard Error	0.0402	0.0164	0.0593	0.0363	0.0780	0.0576	0.0560	0.1054	0.3505	0.4217
Bias	0.4725	0.2720	0.1554	-0.1299	-0.0758	-0.1146	-0.5359	0.2030	0.3850	1.1955
% Bias	43%	24%	13%	-9%	-4%	-6%	-25%	6%	8%	17%
R-Squared	0.5449	0.8850	0.6995	0.9250	0.8907	0.9550	0.9840	0.9725	0.8840	0.9279

distribution with known mean L_∞ and known variance σ_L^2. In practical applications, however, characterization of the distribution of the asymptotic length from length-frequency data alone is difficult.

Even though regression methods have lower precision (Table 2) in the sense of wider confidence intervals, they are robust to departures from model assumptions. Both Jones and van Zalinge's (1981) and Jones' (1984) methods performed well for almost all values of Z/K. Higher relative bias, however, can be observed for very low and very high values of Z/K.

From the above discussion, the following steps can be followed as a guide to estimating the mortality-growth ratio Z/K:

1. Plot the length-frequency data and locate the length of full recruitment l_c.
2. Compare this plot with those of Figure 3 to get an approximate range where the true Z/K may lie.
3. Select the appropriate estimators for this range of Z/K values from Tables 1 and 2. Since the true value of the ratio is unknown, one criterion that can be used to choose among the subset of estimators is to balance lower variance and bias, as suggested by the simulation results in Tables 1 and 2.

The catch curve analysis makes some strong assumptions as to the population equilibrium and constancy of the biological parameters. Furthermore selectivity does not enter the estimation process but rather is assumed nonexistent beyond some arbitrary age or length at first capture. Although these simple methods can be used as an approximation when the available catch data span a few time periods only and/or biological information is limited, it is advisable to use methods that do not restrict the population to be at equilibrium.

3. THE YIELD-PER-RECRUIT

3.1 DEVELOPMENT OF BEVERTON-HOLT MODELS

Until recently, models that were used for predicting yield or production from a fishery could be classified into two types, based on whether age-structure is a part of the model or not. Work in the last decade demonstrates that the additional classification of **size** structure is also needed. These three classifications are related since one would expect a size-dependent model to reduce to an age model if an age-size conversion were applied, as would an age-dependent model reduce to a pure biomass model, if biomass per age group were summed over all age groups.

The most well known yield, age-structure-dependent model is probably that of Beverton-Holt (B-H) (Beverton-Holt, 1957), in which age enters via the von Bertalanffy (LVB) model parameters, K, L_∞, t_0. That is, one needs to have estimated these parameters from size-at-age data. As a first step in demonstrating that size-dependent models are an important classification, we shall begin with the age-dependent B-H yield model and convert it to a form wherein age data are essentially unnecessary. This form of the B-H model is parameterized by ratio parameters that have been developed in preceding sections.

The underlying premise of the B-H yield model is that there are two parameters that a manager controls that can influence yield, the mesh or hook size that determines t_c, age of entry to the fishery, and the rate of fishing mortality F. The mortality rate F is directly proportional to fishing effort f, perhaps expressed in hours or days fishing. These relationships are usually summarized by writing yield Y as

$$Y = Y\left(t_c, F\right) \tag{19}$$

and

$$F = qf \tag{20}$$

where q is the proportionality constant called the catchability coefficient and f and t_c are fishing effort and age of entry to the fishing, respectively. The B-H formulation of yield per recruit models is a relatively simple special case in that the variables t_c and F are both expressed as variables with single values. A more general model would recognize that there is a distribution of sizes around each age. Thus, no single age of entry can actually be defined since larger fish of one age are equally likely to be caught as smaller fish of older ages, which leads to the concept of an age-specific "selection" (by the capture gear) function. The B-H model's use of t_c is said to assume a "knife-edged" selection function. A look back to catch curves reminds us that we used knife-edged selection when we choose t_c or l_c as the age or size of first capture, respectively, choosing it on the basis of the age or size where the fish were "fully recruited" to the fishery (see Figure 1).

Later in this section, we convert the B-H yield per recruit function from an expression in terms of age where t_c is used, to one in terms of size, where l_c replaces t_c. In contrast to (20) we note the alternative formulation where the fishing mortality term F is considered to be separable into components, e.g.,

$$F\left(l, s, f\right) = qfs(l) \tag{21}$$

where $s(l)$ is a selection function in terms of size. A further generalization is found in the section on process models where f becomes a function of time. There are also models where catchability q is a function, e.g., of abundance (Quinn, 1987).

We now develop the Beverton-Holt yield per recruit model in its age and length based formulations. The B-H yield per recruit model assumes that a steady-state condition, in which the yield from multiple cohorts over one year, can be described by following one cohort over its life time, from t_c (or l_c) to death at t_λ (age t_λ is commonly represented by ∞, with little error). In other words, the yield *over one year* from multiple cohorts is

$$Y = \int_{t_c}^{\infty} F\, w(t) N(t)\, dt \tag{22}$$

where it is clear that the "t" here represents age and F is a constant that represents the selection and effort on an average fish of age t, from the time it enters the fishery at t_c or l_c until natural death. For purposes of comparison, note that equation (22) can be written in terms of length as

$$Y = \int_{l_c}^{L_\infty} F\, w(l) N(l)\, dl \tag{23}$$

The standard age-based formulation with knife-edge selection rests upon following one cohort over its lifetime. The population size is represented by

$$N(t) = Re^{-M(t_c - t_r)} e^{-(M+F)(t-t_c)} \tag{24}$$

where R is recruitment, M is the rate of natural mortality, F is the rate of fishing mortality represented as a constant rather than as a function of selection as in (21), and t_c is the age of entry to the fishery. Parameter t_r is the age of availability to the gear, a concept which allows some flexibility to represent "partial recruitment" and t_λ is the effective age of exit from the fishery, so $(t_\lambda - t_c)$ is the number of years in the fishery.

The LVB model in weight is represented by the general expression

$$w(t) = W_\infty \left(1 - e^{-K(t-t_0)}\right)^b$$

Three parameters of interest, W_∞, K, t_0, can be estimated by fitting the LVB model with weight data $w(t)$ if b is set to 3 or, indirectly, using size data $l(t)$, in which case the LVB model is

$$l(t) = L_\infty \left(1 - e^{-K(t-t_0)}\right).$$

The general weight model where all four parameters are estimated should not be fit with weight-at-age data, directly. Unrealistic estimates may result. An allometric length-weight function will provide an estimate of the power parameter b by fitting

$$w = a \times l^b \tag{25}$$

The B-H model can take several forms, two of which differ by whether the exponent "b" in the weight-LVB model is statistically equal to three or not.

In this treatment we maintain the $b = 3$ assumption. The B-H model for $b \neq 3$ involves the incomplete Beta function, which is developed elsewhere (Jones, 1957; Ricker, 1975; Gulland, 1983). If the estimate of b in (25) has a wide confidence interval, the consequences of assuming 3 in contrast to using b-values near the upper and lower ends of the confidence interval must be evaluated. The maximum yield for fixed F and t_c can differ by about 20% for a b-value that is about 15% different than 3.0. The assumption of $b = 3$ is closely associated with the assumption of isometric growth, which is frequently violated. In other words, the reader should note that statistically estimating b is an acceptable procedure whether working with an age or a size-based yield model.

The B-H yield-per-recruit model in (Beverton-Holt, 1957) given by (22) can be expressed using (24) and (25) as,

$$\frac{Y}{R} = Fe^{-M(t_c - t_r)} W_\infty \sum_{j=0}^{3} \Omega_j \frac{e^{-jK(t_c - t_0)}}{(F + M + jK)} \left(1 - e^{-[F+M+jK](t_\lambda - t_c)}\right) \tag{26}$$

where Y/R is used to reflect our lack of knowledge about recruitment, Y/R is measured in units of weight per recruited individual; later we'll write $Y' = Y/RW_\infty$ which is a dimensionless quantity; M, F, K, t_c, t_λ, t_r, t_0 are all presumed known to us and Ω_j = +1, −3, +3, −1 for j = 0, 1, 2, 3, respectively. If t_λ is assumed "large" then (26) is simpler because the last exponential term disappears. That is, if the upper limit in the integral (22) is t_λ, (26) is correct. If the limit is ∞, (26) does not include the last bracket.

We begin the move to express the B-H model in terms of size by noting that, in general, the yield function is defined by (23). Our development, however, is based on transformation of (26). We first define the rate of exploitation fraction E, the fraction of a year class that will be caught, and the ratio F/M,

$$E = \frac{F}{F + M} \quad \text{and} \quad \frac{F}{M} = \frac{E}{1 - E} .$$

Algebraic manipulation of the summation in (26) yields

$$\sum\left(\frac{F}{F+M+jK}\right) = \sum\frac{E}{1+\frac{jK}{M}(1-E)}.$$

The exponential term in (26) is manipulated by solving the length version of the LVB equation for the exponential term to get

$$e^{-K(t-t_0)} = 1-\frac{l(t)}{L_\infty}$$

It follows that if l_c is the length corresponding to t_c, raising both sides to power j yields

$$e^{-jK(t_c-t_0)} = \left(1-\frac{l_c}{L_\infty}\right)^j = (1-c)^j$$

where l_c is the "mean selection length" and c is the proportion of the asymptotic length at which the length of full recruitment occurs. This specification of an l_c without measures of variability around it, or the absence of a selection function, is the "knife edge selection" assumption. For the moment, we continue to assume knife edge selection.

It is thus valid to express t_c in terms of the corresponding biomass, w_c, and then to define c in terms of biomass,

$$c = \left(\frac{w_c}{W_\infty}\right)^{\frac{1}{b}}$$

The exponential term in (26) is rewritten as

$$e^{-M(t_c-t_r)} = e^{M(t_r-t_0)-M(t_c-t_0)}.$$

where the first exponential is a constant R based on the pre-recruitment phase over which management has no control.

From above,

$$e^{-jK(t_c-t_0)} = (1-c)^{+j}$$

so one of the terms on the RHS is

$$e^{-M(t_c-t_0)} = (1-c)^{+\frac{M}{K}} = (1-c)^m .$$

where $m = M/K$. Similar transformations of the exponential terms and algebraic manipulation result in a size-based B-H model derived from the age-based version in (26) with large t_λ

$$Y = R(1-c)^m W_\infty \sum_0^3 \frac{\Omega_j E}{1+j(1-E)/m}(1-c)^j \tag{27}$$

Which can be rewritten as

$$Y' = \frac{Y}{RW_\infty} = E(1-c)^m \sum_{j=0}^3 \frac{\Omega_j \cdot (1-c)^j}{1+j\frac{K}{M}(1-E)} \tag{28}$$

The characteristics of (27) and (28) are:

1. The allometric exponent is assumed to be $b = 3$.
2. Only three parameters need to be estimated, all of which are ratios:

$$E = \frac{F}{F+M}$$

$$c = \left(\frac{w_c}{W_\infty}\right)^{\frac{1}{3}} = \left(\frac{l_c}{L_\infty}\right)$$

$$m = \frac{M}{K}$$

where E and c must be between 0 and 1. The presence of parameter c is the knife-edge selection assumption. The ratio M / K expresses fundamental life history relationships independent of a fishery. Good estimates of M are not easy to determine and without age data, the same applies to K. Fortunately, estimation of the ratio m avoids both issues. However, since K is a measure of how rapidly a fish approaches its larger sizes (perhaps, how rapidly it finds refuge in being big) and M is a mean measure of natural mortality which is most intensive at smaller sizes, m is an important parameter of the model. Beverton and Holt (1959) explore this relationship making species-specific graphs of M versus K and of maximum age (t_λ) *vs* L_∞. The values of M and

K tend to increase or decrease together forming a ratio which is less variable than individual values of M or K.

3. Y' is a dimensionless quantity allowing ease of relative comparisons as either effort or selectivity change.

Generally speaking, the estimation of the three ratios E, c, and m requires less detailed information about the nature of the fish or the fishery than estimation of the parameters for the basic age-dependent B-H, Y/R curve.

The parameters of the model may be estimated in several different ways. It has been demonstrated that the slope of the catch curve versus a function of length, $\log(L_\infty - l)$, is Z/K, suggesting that m can be estimated by subtraction if F is known and thus, E follows directly. The elements of c would be estimated directly by the weights of the lightest fish in the catch or by converting l_c to w_c.

This introduction to the B-H length model is built around the fact that t_c and F are the two parameters a manager can manipulate. Parameter F continues to be visible in this size-dependent version but t_c is submerged into w_c, but can be estimated as noted above. Thus, the new model is now more directly formulated as

$$Y = Y(w_c, F)$$

a surface where one has control of w_c and F.

The preceding derivation is similar to that found in sources such as Beverton and Holt (1966) and Gulland (1983). It raises opportunities for the construction of yield isopleths and the investigation of strategies of harvest based on the estimated ratios and the two control variables w_c and F. Further developments in this area will start with length-frequency distributions of the catch and include consideration of selection functions.

Selection functions can be broadly classified in accord with the type of capture gear. For gears that are towed, e.g., seines, we usually expect sigmoidal type functions, although more recent work has broaden that expectation. Passive and quasi-passive gears such as gillnets typically result in $s(l)$ functions that look quasi-Gaussian. Recent work indicates that it is not acceptable to automatically eliminate selection functions and to replace them by the simpler l_c, thus opting for knife-edge selection. This is especially true for short-lived fish which do not grow to large sizes. They may become fully recruited at, say, 25% – 75% of L_∞, making l_c a big fraction of L_∞.

Methods to use selection functions for towed gear in a yield per recruit formulation are widely known when the application is to an age-based analysis, although the length-based formulation derived above is the easiest version in which to introduce selection. Beverton and Holt (1966) introduce selection by dividing the entire length spectrum into sub-lengths over which fishing mortality is constant. Gulland (1961, 1963) considered changes in yield under changes in selectivity functions and Silvestre et al. (1991)

considered the use of a selection function in those cases where L_∞ is relatively small, as in some tropical circumstances. A variation on the problem of using selection functions in single species yield per recruit models is that of choosing "optimal" mesh sizes in multi-species fisheries. Sainsbury (1984), Murawski (1984) and Pikitch (1987) define "optimal" mesh sizes on biological or economic yield criteria and go on to specify optimal harvest strategies.

The use of selection functions in yield per recruit analyses of gillnet fisheries is less common. The computation of selection functions themselves for gillnets is well explored (e.g., Regier and Robson, 1966; Hamley, 1975) but the fact that (gillnet) selectivity is quasi-Gaussian and drops after a maximum contradicts the usual conceptual model for use in yield analysis. Some approaches worth noting are: the Catch-At-Size Analysis (CASA) model (Sullivan et al., 1990) given in Section 6.4 allows the specification of a general functional form of a selection function, and the program estimates the parameters. Lai et al. (1993) used catch-at-length data from an artisanal tropical corvina fishery in which three different mesh sizes are commonly employed and combines these mesh sizes with the history of effort per mesh size. Ehrhardt and Die (1988) used gillnet selectivity in a size-structured yield per recruit model for Spanish mackerel to determine the gains in yield resulting from an increase in mesh size and fishing mortality rate.

Gallucci et al. (1992) describe the next level of complexity by noting that there are three other corvina species present in the fishery, all of which have different growth characteristics. This formulation makes use of the methodology described for the single species problem to specify an optimal (mesh size, effort) pair that leads to simultaneous optimal yields for all species.

3.2 IMPLICATIONS FOR STRATEGIES OF EXPLOITATION

The natural approach to a B-H yield per recruit analysis is to examine the yield surface to see how yield varies with w_c as F is held constant and how yield varies with F for fixed w_c. This is analogous to the analysis where yield isopleths are used to define harvest strategies in terms of t_c and F.

First, however, the exponent $m = M / K$ should be examined carefully. Gulland (1983, p. 151) considers the M / K ratio, which we need to interpret in the current context. Consider two cases: M / K high and low.

1. If M / K is low, a stock will contain more large fish. To get the "best" yield, exert minimal effort (low F) but increase the size of first captive l_c (end of paraphrase). This does **not** mean that species with small M / K are larger, since small M / K suggests a large K and thus a small L_∞ (Gallucci and Quinn, 1979). Instead, recall the length distribution curves that were generated for different ranges of Z / K ratios and how each range could be related to a particular class of distributions. In a similar way, small M / K means that there will be a relatively large number of larger animals present compared to the number of small ones. It is then clear that it is more efficient to target large

ones by increasing (e.g., by increasing mesh size) l_c with a lower constant effort F and not by chasing smaller ones using a higher F.

2. If M / K is large, K is low relative to M, fish with more effort and a small size of first captive (end of paraphrase). Clearly, M / K large suggests K is low and L_∞ is high. A high M relative to K suggests that mortality from natural causes will kill many fish before they get large, so there will be relatively few big fish. The curve for Z / K high gave an approximately negative exponential decrease so that most of the fish will be small and there will be very few big fish. Thus, there would be little reason to increase l_c (e.g., with a bigger mesh) but an increase in effort would provide increased yield.

The M / K ratio also has a prominent role in the specification of one particular exploitation strategy. There a number of different strategies in common use all of which have in common an optimization of some function involving yield. The yield $Y(l_c, F)$ in the Beverton-Holt Y / R model does not have a global maximum for arbitrary l_c and F so the strategy to find the F that maximizes yield, F_{max}, must be defined for a fixed t_c or l_c.

One of the most well known strategies is to fish at the rate F that generates a maximal sustainable yield (MSY), called F_{MSY}. This rate is defined with reference to stock-production models where Y_{MSY} is the maximum yield. We would be reluctant to move to a stock production-based strategy, however, because such models epitomize situations where control is lost of many of the parameters that permit a manager to take intermediate positions balancing fish size, entry size and effort. But, such a strategy could be utilized if the problem were recast in a stock-production format. When this has been done, F_{max} has been observed to sometimes exceed F_{MSY}, which is a cause for concern. It is on this basis that F_{max} is often considered to be less than a conservative strategy. The related reference point, $F = M$ has also been used to choose an optimal strategy based on the observation that F_{MSY} is frequently close to $F = M$. Although easy to implement in our yield per recruit model, $F = M$ is usually omitted as a strategy because M itself is a poorly defined concept which has yet to be satisfactorily estimated for most stocks. We could never accept the variability in F that we accept in M.

The ratio M / K uniquely specifies "the $F_{0.1}$-criterion", which defines one of the more frequently discussed harvest strategies. The $F_{0.1}$-criterion is a relationship defined in terms of the B-H yield-per-recruit model (Gulland and Boerema, 1973; Deriso, 1987), resulting in an explicit instantaneous rate of fishing mortality $F_{0.1}$ for a fixed age or size of entry to the fishery, t_c or l_c. The definition of the $F_{0.1}$-criterion is somewhat mathematical but it is instructive to examine a small part of it. Let $Y' = Y / R_0 W_\infty$, as in (28), which is yield-per-recruit standardized to a unitless quantity Y'. Then the criterion

$$\left.\frac{\partial Y'}{\partial F}\right|_{F=F_{0.1}} = (0.1)\left.\frac{\partial Y'}{\partial F}\right|_{F=0}$$

says Y' is a function of the two variables which the manager controls, t_c and F, and the partial derivative defines a set of curves $y = y(F)$, each of which is a derivative of Y' and a function of F, each for a different fixed t_c. The slope of each of these y-curves in the neighborhood of $F = 0$ has a numerical value, say x_0, which can be multiplied by 0.1 or 10%; thus, the "0.1 criterion." The solution of the above equation leads to a solution for F which is denoted "$F_{0.1}$" which contains a fixed term t_c, the age of entry (assuming knife-edge recruitment), and defines a new term

$$t_w = \frac{\int_{t_c}^{x} xB(x)\,dx}{\int_{t_c}^{x} B(x)\,dx}$$

where t_w is the mean age of a unit of biomass for a cohort over its lifetime after entering the fishery, $B(x)$ is the biomass per recruit of the cohort at age x and a new term Φ_F is the total biomass generated by a recruit entering at t_c over its lifetime while exploited at a constant level F. Then

$$F_{0.1} = \frac{\left[1 - 0.1\left(\frac{\Phi_{F=0}}{\Phi_{F=F_{0.1}}}\right)\right]}{\left(t_w - t_c\right)}$$

Thus, a $F_{0.1}$ level of exploitation is indeed dependent upon the plane $(Y' - F)$ at t_c. In other words, the $F_{0.1}$ criterion requires the manager to utilize both of the parameters available for control, t_c and F. The method can be generalized to allow specification of a criterion designed to produce a yield at a level of exploitation F_p, suggesting that

$$\left.\frac{\partial Y}{\partial F}\right|_{F=F_p} = p\left.\frac{\partial Y}{\partial F}\right|_{F=0}$$

needs to be satisfied. A corresponding relationship for $F_{0.1}$ and F_p can be derived. This, and associated theory, is currently under development.

Our final point about this strategy of fishing is that the criterion can be rephrased in terms of the exploitation fraction $E = F/(F+M)$ and the ratio $m = M/K$, which we have used consistently in this chapter. It can be shown that when the $F_{0.1}$ criterion is applied, the yield it produces is a decreasing function of *m*.

In summary, the ratio M/K uniquely defines the $F_{0.1}$ rate. The $F_{0.1}$-criterion is now seen to be even more of a **size**-dependent than an **age**-dependent criterion. And we see again, that the ratios M/K and Z/K have great importance in stock assessment models whether they be size cohort analysis, size frequency analysis, or sized-based yield-per-recruit analysis.

A word of caution should be inserted here about the limitations of steady-state models. There is little doubt that models which do not require the steady-state assumption are, in principle at least, preferable. However, non-steady-state models are far more complex and usually less specific. On one hand, this may be an advantage in systems as variable as tropical fisheries. On the other hand, a steady-state analysis should always be carried out as part of an investigation to better understand the dynamic relationships among parameters and to define at least a probable approximation of the assessments.

4. LENGTH-COHORT ANALYSIS

4.1 OVERVIEW

The preceding applications of ratios like Z/K as parameters in the stock dynamics equation of Beverton-Holt were central to the estimation of potential yields and in the formulation of a suitable fishing strategy. We now consider an alternative to this approach to stock assessment where such a detailed prior knowledge of the stock dynamics is not a prerequisite. Cohort analysis, in particular, length-cohort analysis (LCA), is more empirical and less demanding in terms of parameter estimates.

Cohort parameters such as age-specific fishing mortality and cohort strength are often estimated by age-dependent methods such as "virtual population analysis (VPA)" (Gulland, 1965) and "cohort analysis" (Pope, 1972). These methods depend on accurate age determination, which is not always feasible (Lai, 1985). Jones (1979, 1984) proposed a length-cohort analysis (LCA) based on an adaptation of Pope's age-cohort analysis to estimate the cohort parameters in a steady state when the ages of the animals are not available. Sparre (1987) also develops a LCA and provides software to carry it out. Lai and Gallucci (1988) provide an alternative mathematical derivation of LCA and investigate the sensitivity of the LCA to errors in the input parameters.

4.2 THE MODEL

The derivation of the equation for LCA is based on three models: (i) the negative exponential model, which uses the parameter Z; (ii) a length-based catch equation which uses mortalities $F(l)$, M and $Z(l)$, and relates catch to abundance; and (iii) the von Bertalanffy (LVB) model, which contributes parameters K and L_∞, and allows the transformation of age to length. If $N(l)$ is the abundance of fish at length l, then the number of fish growing from length l to length $(l+\Delta l)$ is

$$N(l+\Delta l) \;=\; N(l)\left[\frac{\left(L_\infty-(l+\Delta l)\right)}{\left(L_\infty-l\right)}\right]^{-\frac{Z}{K}} \tag{29}$$

where L_∞ is the asymptotic growth parameter. Equation (29) is rewritten as

$$N(l+\Delta\, l) \;=\; N(l)A(l)^{-\frac{Z}{K}} \tag{30}$$

where $A(l)$ is the indicated function. Substitution into the Baranov catch equation in terms of length yields the canonical equation for LCA,

$$N(l) \;=\; N(l+\Delta\, l)A(l)^{\frac{M}{K}} + C(l)A(l)^{\frac{1}{2}\left(\frac{M}{K}\right)} \tag{31}$$

This equation links abundance to catch via the parameter M/K which can be estimated from Z/K where M is an input parameter and the vector $F = F(l)$ is estimated. These steps are given in Lai and Gallucci (1987).

The input parameters and data are:

i. A vector of catch-at-length data $C(l)$ categorized into units of size Δl (e.g., 2 cm wide).

ii. The parameters of the growth model, L_∞ and K. The LCA method requires that the ages and lengths of animals be interchangeable in a consistent and predictable way which suggests the use of a model, in this case the von Bertalanffy growth equation.

iii. The rate of natural mortality M and the fishing mortality of a single length group $F(l)$. Usually the value of F for the last Δl, $F_\lambda = F(\Delta l_\lambda)$, is a guess.

The output from the analysis are the typical estimates: a vector of estimates of abundance $N(l)$ per length interval Δl and a vector of estimates of the fishing mortalities, $F(l)$, also per Δl.

Lai and Gallucci (1988) investigate the sensitivity of the model by assuming variability in the input parameters. In their model, the input parameters are: (i) the sizes of the length interval in the histogram; (ii) the values of the von Bertalanffy growth parameters, L_∞ and K; (iii) the estimates of the natural mortality rate M over the length groups; (iv) estimates of F_λ, the fishing mortality rate for the last length group Δl_λ; and (v) the size of Δl. Since the natural mortality M is constant over all

Δl it is added to the estimated $F(l)$ vector to obtain the new vector of total mortality rates, $Z(l)$.

The methodology of age-cohort analysis and its parent, virtual population analysis (VPA), are quite commonly used sometimes in forms that allow the use of input information in addition to that noted above, such as trawl or hydroacoustic survey data. The introduction of these "auxiliary data" is called "tuning". Little has been done in the area of tuning a length-cohort analysis to date, however.

One of the major limitations of an LCA is the restrictive assumption of a steady state over time since only one year's catch $C(l)$ is input [e.g., see discussion in Fournier and Doonan (1987)]. The assumption is that the length distribution catch vector $C(l)$ in year t is statistically the same as found in $t+\tau$, viz, $dC\left(l(t)\right)/dt = 0$. This assumption, common to so much of stock assessment, should not be ignored. Sometimes it can be approximated if not directly satisfied by moving averages over years or by other methods.

Lai and Gallucci (1987) present the program LCAN for executing a LCA, and for carrying out the above noted sensitivity analysis. The program and user's guide (CRSP Working Paper No. 12) can be obtained from the authors.

4.3 SENSITIVITY ANALYSIS

Lai and Gallucci (1988) analyzed the LCA methodology to determine the sensitivity of the output to variation or error in the input data. The procedure is similar to that of Sims (1984) for age-based cohort analysis. In particular, analytical expressions for error were derived from the methodology with an error term built in. The canonical Equation (31) shows that for any input vector $C(l)$ different choices of Δl will certainly yield different estimates of the vectors $N(l)$ and $F(l)$. Similar statements apply to the use of different choices of input parameters M, L_∞, and K. The sensitivity of the LCA model to such situations was investigated by deriving closed-form equations for relative error in the output vector $N(l)$, viz.,

$$\rho_N = \rho[N(l)] = \frac{N_{\text{est.}} - N_{\text{true}}}{N_{\text{true}}}$$

and similarly, for relative error in the other output vector $F(l)$, viz.,

$$\rho_F = \rho[F(l)]$$

Different choices of input parameters might result from poor estimates caused by variability in the data suggesting estimates with variances, or to a poor choice of an interval size Δl. Depending upon the gear used in the fishery there will be more or less bias in the size distribution of the catch $C(l)$, relative to the size distribution of the population $N(l)$. Despite the inevitable fact of gear selectivity, these $C(l)$ data might be the only possible source of an estimate for L_∞. There are rules of thumb for "guessing" L_∞ from the largest fish(es) in $C(l)$ but the estimation is complicated because the "best" estimate will depend upon the probability distribution for an unknowable L_∞. A similar but more difficult problem exists for K. But, if L_∞ is not to be inferred from $C(l)$, then it must be estimated from an age determination study, which may or may not have been carried out. In any event, it is clear that the sensitivity of the estimation method to variability in the estimate of L_∞ is a relevant concern (see Isaac, 1990 and Mug et al., 1994 and references therein).

The vectors $\hat{N}(l)$ and $\hat{F}(l)$ can either under- or overestimate the true $N(l)$ and $F(l)$. Sample results that suggest ranges of possible mis-estimation are given in Lai and Gallucci (1986) for a bivalve population. The reconstructed stock size as a function of length $N(l)$ is overestimated as Δl increases. The relative error $\rho[N(l)]$ varies by an order of magnitude from 0.05 to 0.55 as Δl changes from 3 mm to 9 mm. The estimates of fishing mortality F are underestimated as Δl increases from 3 mm to 5 mm. The relative error, $\rho(F)$, varies by an order of magnitude from –0.025 to –0.25. Even when the variability in estimates of L_∞ and K is low, disappointingly high errors occur in $N(l)$ and $F(l)$. For example, if $\hat{V}(L_\infty) = 7.0$ and $\hat{V}(K) = 0.00017$, the 95% confidence interval of $N(l)$ increases approximately 6-fold and the 95% confidence interval of $F\Delta t_l$ varies approximately 3-fold. By fixing F_λ, $\rho[N(l)]$ varies from 2 to 30 as M goes from 0.1 to 1.0. By fixing M and letting F_λ vary from 0.01 to 0.4, $\rho(F)$ varies from 0 to 10.

4.4 GUIDELINES

Lai and Gallucci (1988) address how variability in the catch data and the variance of the estimates of the growth parameters combine in the results of a length cohort analysis. Very high coefficients of variation are common in cohort analysis. One of the advantages of the analytical type of sensitivity investigation is that the cause of the high variance of the estimates of $N(l)$ and $F(l)\Delta t_l$ can be traced to variation in K. By a simple examination of the equations, we therefore focused on the estimation of growth parameters. In general, the extreme sensitivity of the estimates of any LCA to variance in the input parameters is disconcerting and must be of concern whenever LCA is used to generate estimates of abundance. Perhaps LCA results should be used only as indices to reflect relative changes in abundance.

When a specific growth study has not been done and estimates of L_∞ and K are not available, Jones (1984) suggested choosing the largest fish plus 5%. Whatever is done must be done with the understanding that a unit change in L_∞ leads to a big change in $N(l)$. The choice of K is more complicated because K and L_∞ are inextricably related (Gallucci and Quinn, 1979) and because K cannot be guessed by observation. Since the ratio M/K arises naturally in the formulation of LCA, it is fortunate that M/K is more easily estimated than either M or K alone.

Similar sensitivity analyses were carried out for Pope's cohort analysis by Sims (1982, 1984). Returning briefly to the preceding chapter, Lai and Gunderson (1987) and Tyler et al. (1989) considered the consequences for yield per recruit results when aging error was present.

A final recommendation addresses the steady-state assumption, which is easily violated. Since length-frequency distributions can be examined over years, they may be used to evaluate the steady-state assumption by looking for trends or shifts in the modes of either newly recruited or fully recruited animals. Nevertheless, the greater dependence of LCA on a steady state and on the appropriateness of the von Bertalanffy growth model suggests that much less faith may be put in the results of LCA than in age cohort analysis.

To assist in the choice of a Δl that will not lead to unnecessary error, the following inequalities should be satisfied:

$$\frac{M}{K} ln\, A(l) \;<\; .3 \qquad \text{and} \qquad F(l)\Delta t_l \;<\; 1.2$$

5. LENGTH-FREQUENCY ANALYSIS

5.1 BACKGROUND

There is a sequence of papers on modal decomposition that is illuminating in the sense of how the primary methodologies relate to each other. The early work by Petersen (1892) visually separated the modes of a length-frequency distribution (lfd) at a single time, t, without the use of statistical criteria and it associated the means of each mode with an age class. Thus, modal decomposition became a form of age composition. Contemporary statistical research has considerably advanced modal separation. A logical extension from decomposition at one time t is the attempt to follow modes from time t_1 to t_2, and so on. If a mode can be followed over successive years, it increases the likelihood that an apparent mode is not an artifact and that it does indeed represent an age class.

Tanaka (1962) decomposed a length-frequency distribution into modes by noting that the natural log of a normal distribution $f(x)$ can be expressed in the form of a parabola, viz.,

$$\ln f(x) \;=\; y \;=\; ax^2 + bx + c$$

where the coefficients a, b, c contain the parameters of the $N(\mu_t, \sigma_t^2)$ for each lfd around a mean μ_t, corresponding to age t. Then σ_t would be a measure of the width of the "parabola". A modal decomposition of a length distribution relative frequency figure at one time t is illustrated in Figure 4.a. Figure 4.b illustrates the ambiguity that can occur as one tries to follow modes from one time to another. Hasselblad (1966) put this estimation problem in a firm statistical framework by deriving maximum likelihood procedures for the estimation of the $3k-1$ parameters of a mixture (or a mixed) distribution consisting of k **normally** distributed components. It is assumed that there are n observations $x_1, x_2, \ldots, x_n$ of lengths assumed to be drawn from a mixed population, $g(x)$

$$g(x) = p_1 f_1(x) + p_2 f_2(x) + \ldots + p_k f_k(x)$$

such that the proportions p_i satisfy $\sum_1^k p_i = 1$, $0 < p_i < 1$, and each $f_i(x)$ represents a normal sub-population with mean μ_i and variance σ_i^2. The problem is to estimate $\mu_1, \ldots, \mu_k; \sigma_1^2, \ldots, \sigma_k^2; p_1, \ldots, p_{k-1}$ from the sample of size n, $\{x_i\}$. For $k = 1$, the problem is classical,

$$\hat{\mu} = \bar{x} \quad \text{and} \quad \sigma^2 = \left(\frac{1}{n-1}\right)\sum_1^n (x_i - \bar{x})^2$$

For $k = 2$, Pearson (1894) solved special cases. Cassie (1954) solved the general problem graphically for arbitrary k using the cumulative distribution function.

In Hasselblad's analysis, an approximate log-likelihood function L is defined and the values of the parameters which maximize the log-likelihood function are sought. The usual procedure of trying to solve the equation below:

$$\frac{\partial L}{\partial \mu_j} = 0, \quad \frac{\partial L}{\partial \sigma_j} = 0, \text{ and } \frac{\partial L}{\partial p_j} = 0 \qquad j = 1, 2, \ldots, k$$

for the unknown $\mu's$, $\sigma's$, and $p's$ is again followed. An iterative generalized steepest descent method is one approach to obtaining the MLEs. An alternative is a Newton iteration scheme based on the second partial derivatives in the Hessian matrix. The success of the method in estimating p_j, μ_j, and σ_j and their variances depends very much upon an adequate sample size, on the separation of the modes, and on their "distinctiveness."

The next step in modal decomposition was made by MacDonald and Pitcher (1979) who extended Hasselblad's analysis with a program called NORMSEP. The contemporary version of this software package is called "MIX" and fits distributions that are assumed to be pure mixtures of normal, log-normal, gamma or exponential distributions. This is a considerable advance over the analysis of Hasselblad which was restricted to fitting only mixtures of normal distributions. Both methods compute maximum likelihood estimates.

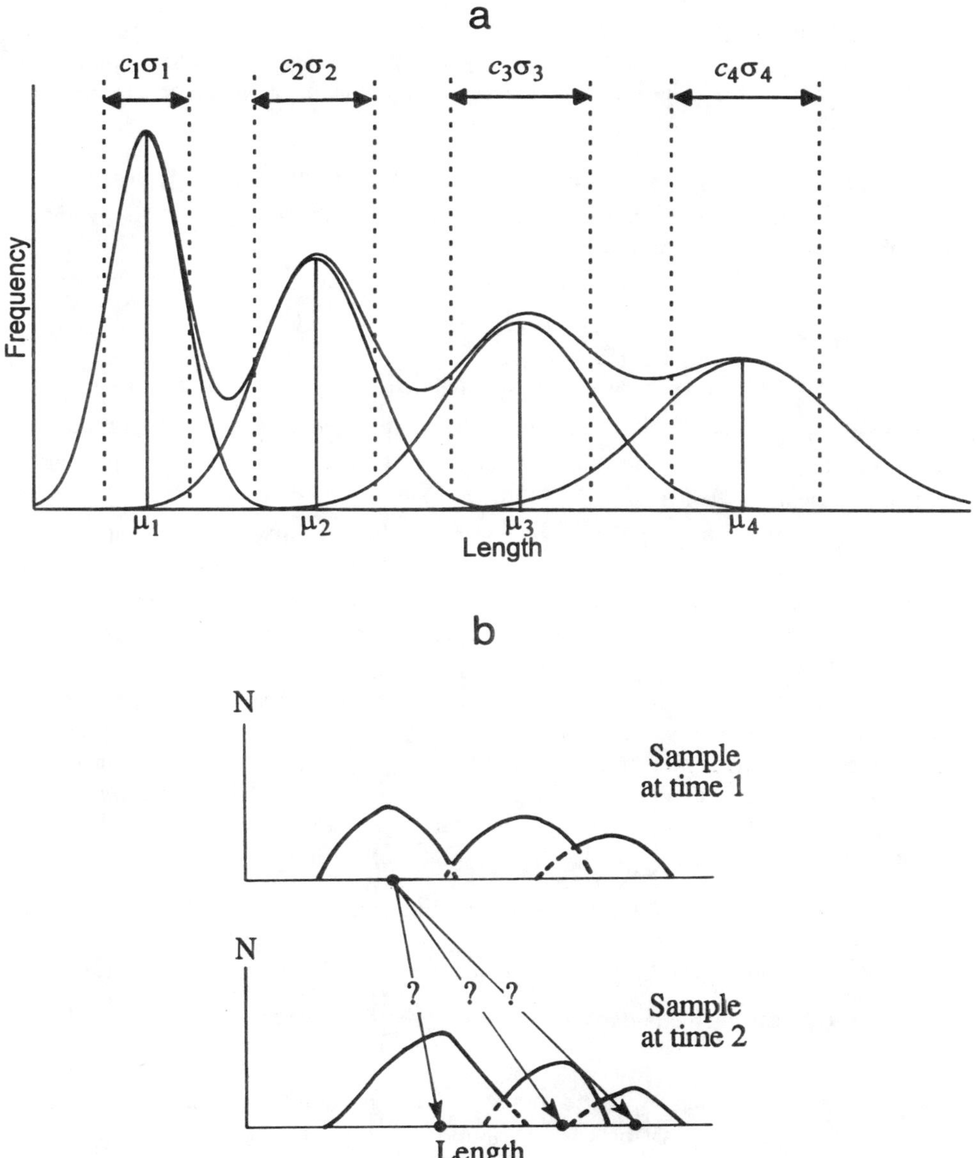

Figure 4.a. A schematic of a modal decomposition of a polymodal function showing modes around the mean μ_i's and the associated deviations $c_i\sigma_i$'s.

Figure 4.b. A schematic of the ambiguity in defining "what is a mode" and which mode(s) represent age classes. (from Pauly. 1987. *Length Based Methods in Fisheries Research.* ICLARM, Phillipines. With permission.)

MIX requires that the user play an interactive role in the fitting process by the choice of initial input values from observing the polymodal distribution and by the choice of various constraints that will reduce the dimension of the hyperspace of the fitting procedure.

5.2 STATISTICAL ANALYSIS OF MIXED PROBABILITY DISTRIBUTIONS

The following is an expanded discussion of the methodology used in the MIX program. A mixed polymodal function is again

$$g(x) \quad = \quad p_1 f_2(x) + \ldots + p_k f_k(x)$$

where $f_i(x)$ is a probability density for one of the constituent distributions with mean μ_i and variance σ_i^2, and p_i is the proportion of the total population in the ith component. The program allows the decomposition when the f_i are either normal, log-normal, exponential or gamma densities.

The objective of fitting the mixture polymodal function to data is to estimate as many of the parameters $p_1, \ldots, p_k; \mu_1, \ldots, \mu_k; \sigma_1, \ldots, \sigma_k$ as possible. Since many of the f_i components will overlap each other, there is a strong possibility that there is more complexity present than the decomposition reveals. Nevertheless, those modes which can be identified are assigned numbers 1, 2, ..., k and the maximum likelihood parameter estimates are computed by an optimization algorithm. The μ_i presumably refer to ages (see Figure 4) so if the catch begins with age t_c, then $\mu_1 = t_c$ and so on.

Because the number of variables to be fit is so large, a wide variety of constraints can be used to reduce the degrees of freedom (df) from the $3k-1$ that otherwise exist. The types of information or constraints that can be or are imposed to reduce the df include:

1. $\sum p_i = 1$ so only $(k-1)$ p-values are estimated.

2. Some of the means μ_i's are known and held fixed; μ_i's are equally spaced.

3. All σ_i's assumed equal to a constant σ or coefficients of variation are assumed equal.

4. Forcing the means μ_i to lie along an LVB growth curve. Since the biggest modes (largest p_i's) for size distributions are usually the first ones, only the first three are actually fit and the others are computed from these iteratively. The result is estimates of L_∞, K, and t_0. Reference here is made to Schnute and Fournier (1980) who introduced this approach.

The Nelder-Mead direct search optimization procedure is used to generate the MLE estimates. As usual, the user may need to specify initial starting points — a limit to the number of function evaluations and a convergence criterion as shown in the options.

Chi-square values are computed to evaluate the goodness of the fit of the model $g(x)$ with k modes to the data. The algorithm is flexible, allowing one to introduce suitable information to reduce the degrees of freedom to achieve a minimal chi-square.

5.3 ESTIMATION OF GROWTH AND MORTALITY RATES FROM SIZE DISTRIBUTIONS

Previous sections and chapters focused on the use of length-frequency distribution (lfd) data to estimate age composition, e.g., with MIX; to estimate mortality rate Z, e.g., with a catch curve; or to estimate ratios such as Z/K. The paper by Schnute and Fournier (1980) uses the lfd data. It gives an alternative estimation of the age composition, but in a single procedure that also estimates the LVB growth parameters.

The objective is precisely stated as the estimation of a set of population parameters that will lead to predicted catches that are as close as possible to the observed relative frequencies. As "close as possible" suggests a measure of closeness between the expected distribution $\left(\hat{f}_j\right)$ which are analytical functions of the population parameters and the observed frequencies $\left(f_j\right)$. The Kullback-A statistic is used to measure closeness. The estimated population parameters are the LVB parameters and the parameters of the expected $\left(f_j\right)$ distributions which define the age composition.

Another method was developed by Schnute and Fournier (1980). That analysis also begins with length composition data and uses modal decomposition concepts from MacDonald and Pitcher (1979) and Hasselblad (1966), but now in conjunction with the LVB growth model. An interesting innovation on the method is the identification of new variables to parameterize the growth model which appears to have less ambiguous interpretations than the usual set consisting of $\left(L_\infty, K, t_0\right)$. The new set is (l, L, k) where l and L correspond to the mean lengths of the first and final modes with means μ_1 and μ_M which, in turn, refer to the centers of the modal groups 1 and M, where it is assumed that there are exactly M age groups. In a similar way, the sample standard deviations, σ_l and σ_M corresponding to the 1 and M age groups, are represented by variables s and S, and these are associated with the lengths l and L in some functional way. For example, $\sigma's$ can be linear functions of $\mu's$, linear functions of ages, or possibly equal to each other: $s = S$.

Following MacDonald and Pitcher (1979) and the MIX algorithm, a fish of age a_i has probability q_{ij} that its length lies in length interval j, i.e.,

$$q_{ij} \quad = \quad \frac{1}{\sigma_i\sqrt{2\pi}}\int \exp\left[-\frac{1}{2}\left(\frac{x-\mu_i}{\sigma_i}\right)^2\right]dx$$

where the integration is over the width of the jth interval. The expected frequency of fish having length in the jth interval is

$$\hat{f}_j = n \sum_{i}^{M} p_i q_{ij}$$

where p_i is the percent of fish in age group i and n is the sample size. Finally, the parameters are fit by defining a measure of closeness A, which can be a function of the growth parameters and the standard deviation parameters, i.e., $A = A(l, L, k, s, S, p's)$. Thus, the measure of closeness of the observed $\left(f_j\right)$ and predicted $\left(\hat{f}_j\right)$ frequencies is a function of the observable growth measured in terms which are functionally related to the LVB growth parameters. Thus, minimizing the measure of closeness of the fit between the observed f_j and the predicted $\hat{f}_j$ estimates the age composition and the LVB parameters.

Statistic A is discussed by Kullback (1959), Rao (1973), MacDonald and Pitcher (1979), etc., and defined as

$$A = 2\sum_{j=1}^{N} f_j \ln\left(\frac{f_j}{\hat{f}_j}\right)$$

where N is the number of intervals and f_j, is the number of observations in the jth interval, and f_j, is the estimated number of observations in the jth interval.

The reader may recall that in the MIX program one of the options was to force the LVB curve to pass through the mean lengths μ_i per mode at age i. The method described here by Schnute and Fournier is employed by MIX.

An extension and modification of the preceding work of Schnute and Fournier (1980) involves the estimation of the age composition, the growth parameters, and the instantaneous rate of total mortality Z (Fournier and Breen, 1983). There are some differences in the methodology but it is also based upon minimizing the differences between an observed multi-modal distribution of lengths in a sample, say $\left(f_j\right)$, and the predicted distribution $\left(\hat{f}_j\right)$. The predicted $\left(\hat{f}_j\right)$ is expressed in terms of observable growth parameters functionally related to the LVB parameters (just as in the Schnute-Fournier approach) in terms of the estimated proportions in each mode p_i, which are functionally related to the mortality Z and the standard deviations s_i around the mean of each mode, which are functionally related to the age i. In particular, the proportions p_i and the standard deviations s_i are defined as

$$p_i = ce^{-iZ+e_i}$$

where age $i > t_c$ and ε_i are normally distributed error and

$$s_i = a + b\sqrt{i}$$

for age $i > t_c$. The s_i are linear functions guaranteed to cause changes in the deviations from the mean to decrease for older ages where growth slows.

The closeness of the fit between $(\hat{f}_j)$ and (f_j) is measured by a log-likelihood function modified to incorporate biological assumptions as auxiliary data. The overall objective function contains three terms:

a. $T(s, m, p) = \sum_{j=1}^{D} f_j \log\left(\frac{f_j}{\hat{f}_j}\right)$

b. $\lambda_1 \cdot$[function of LVB parameters]

c. $\lambda_2 \cdot$[function of the mortality Z]

where λ_1 and λ_2 are penalty weights used to emphasize one or the other terms in the objective function.

A user's guide and software are available for this method as Canadian Technical Report of Fisheries and Aquatic Sciences No. 1239.

Thus far, this chapter has focused on the estimation of stock parameters by decomposition of the polymodal lfd from a sample at time *t*. Little or nothing is said about samples over time, i.e., variations in polymodal distributions over time due to non-steady conditions. Nor is anything said about the logical prospects for following modes from time t to $t+1$, $t+2$, etc., each time expecting the number of animals in a given mode to shrink over time due to mortality. The next software package has such a mode following procedure as its central theme.

5.4 THE ELEFAN SOFTWARE

The ELEFAN series of programs is perhaps the best known of all the methodologies for stock assessment from size data, especially in developing countries. ELEFAN I was developed about a decade ago. Since then, additional programs have been distributed, culminating with the "Compleat ELEFAN" (Pauly, 1987) which is a synthesis and extension of the first five. We will not attempt in this handbook to explain the contents of the Compleat or any other ELEFAN since the essential contribution of the ELEFAN series is bringing together in one place methodologies with well defined computer algorithms and user's guides so that they are interfaced to make a coherent whole.

The objective of the ELEFAN I program was to use length-frequency data from successive years to estimate von Bertalanffy growth model parameters. This was done by following modes in length-frequency distributions taken over successive time periods or years. A schematic of the process is in Figure 5. A worked example is found in Sparre et al. (1989). ELEFAN II included the estimation of the parameter Z / K and of yield-per-recruit, both stock assessment quantities which we develop in earlier parts of this document.

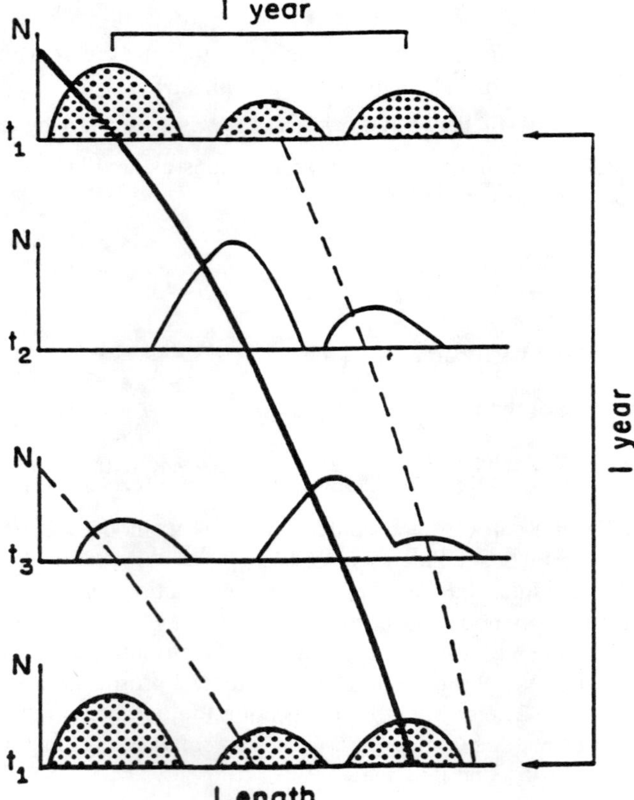

Figure 5. A schematic of mode following over successive time periods and the suggestion of a growth curve through the modes. Interpretation of modes and assignment of age are not always as straightforward. (from Pauly. 1987. *Length Based Methods in Fisheries Research*. ICLARM, Phillipines. With permission.)

ELEFAN V included the option of estimating the growth parameters from three different types of data:

1. Mark-recapture data.
2. Growth increment data from an age-at-size key.
3. Data from following of modes in length-frequency distributions over successive time periods, viz., "Modal Progression Analysis" (MPA).

The analysis of stocks using ELEFAN, especially in developing countries, led to a number of commentaries on the methodology and, ultimately, to a conference at Mazzara del Vallo and a volume entitled, "Length-Based Methods in Fisheries Research," edited by D. Pauly and G. R. Morgan (1987). Some papers in that volume and in several other literature sources are critical of the reliability and applicability of the ELEFAN programs. The primary criticisms concerned the subjectivity of the mode following procedures and algorithms. It seems that the Compleat ELEFAN includes algorithms to limit the subjectivity. A new collection of programs that includes those in the Compleat ELEFAN plus those in the FAO series is also planned. It will be called "FiSAT".

We note some reservations about the ELEFAN procedure. Our comments are closely tied to the perspective of this chapter. The ELEFAN methods do not have an underlying mathematical or statistical model. The consequence is that the estimates of the growth parameters L_∞, K, t_0, as well as other possible parameters, are not accompanied by estimates of their variance, viz., the accuracy of the estimates. Such considerations are especially important since it is known that the parameters L_∞ and K are negatively correlated (Gallucci and Quinn, 1979). At present, the ELEFAN user has no idea of variance, covariance, etc. A possible alternative is a sensitivity analysis. The alternative is to use a boot strap technique to estimate variance. The papers by Majkowski (1987), Basson et al. (1988), Rosenberg et al. (1989) and Isaac (1990) all contain comments of interest.

The absence of estimators of variance and/or of a relative sensitivity to error are serious detractors. While the preceding comments were directed primarily at the estimates of the growth parameters generated from the model, they apply equally well to the VPA estimates in ELEFAN III. It is especially worrisome because Lai and Gallucci (1988) showed very significant sensitivity in the output vectors $N(l)$ and $F(l)$ from a length-cohort analysis to mis-estimation of the parameter L_∞. This analysis is similar to that of Sims (1984) who examined the sensitivity of N and F for age-VPA.

The early parts of this chapter dealt with the different estimators of parameter Z/K. We show that there are several different estimators, some of which work better than others depending on the value of the ratio. ELEFAN II simply uses the estimators of Wetherall et al. (1987) (see earlier discussion).

It is probably true that ELEFAN algorithms have been applied to more stock assessments of different species than any other system available. There are several volumes of papers, all dealing with developing country fisheries, e.g., "Contributions to Tropical Fisheries Biology" (1986) edited by Venema, Christensen, and Pauly. For a long time ELEFAN was the only off-the-shelf package available to developing county fisheries managers. It has clearly demonstrated to new fisheries managers that they need to articulate management goals first and then analyze an array of management options, each with different costs that can be exercised to reach a goal. In some ways and at some times, this realization is more important than the actual technical stock assessment.

Compleat ELEFAN and FiSAT are available from:

ICLARM
M.C.P.O. Box 2631
0718 Makati, Metro Manila
Philippines

5.5 THE MULTIFAN SYSTEM

The preceding section discussed the ELEFAN system which has been criticized for multiple shortcomings, but was for some years the only coherent program available. At the present time, MULTIFAN or Multiple Length Frequency Analysis (Fournier et al., 1990) also contains the concept of following modes but the methodology is based on a more substantial statistical foundation. The method involves a robust likelihood-based estimation technique which facilitates hypotheses testing because it provides objective criteria for the construction of hypotheses. The method simultaneously analyzes multiple length-frequency samples taken at successive times, thus providing a version of the mode following. The method was adapted from the ideas of Hasselblad (1966) and MacDonald and Pitcher (1979) for separating a finite mixture of normal distributions. Schnute and Fournier (1980) introduced the concept that mean lengths-at-age following a von Bertalanffy curve, and Fournier and Breen (1983) introduced the steady-state concept to the method so that total mortality rate can be estimated from the fully recruited age components in the length-frequency data. However, these methods deal with only one set of length-frequency data. MULTIFAN extends this line of analysis to a time series of length-frequency data sets to estimate the population parameters over the time domain that covers the sampling schedule. Therefore, MULTIFAN is one of only a few integrated data analysis algorithms currently developed for fishery stock assessment using length-frequency data.

The principal assumptions for MULTIFAN are:

1. The lengths of fish in each age class are normally distributed with a mean length and a variance.
2. The mean length-at-age lies on or near a von Bertalanffy growth curve.
3. The standard deviations of length about the means are a simple function of the mean length at age.

Let q_{ijt} be the proportion of fish (or the probability of a fish) sampled at time t (the tth length frequency data set) that belongs to age class j and length interval i. The proportion can be expressed in terms of mean length at age i and its variance, μ_{jt} and σ_{jt}^2:

$$q_{ijt} = \frac{1}{\sqrt{2\pi}\sigma_{jt}} \int_{x_i - w/2}^{x_i + w/2} \exp\left\{ -\frac{\left(x - \mu_{jt}\right)^2}{2\sigma_{jt}^2} \right\} dx$$

As long as $\sigma_{jt} > w$, where w is the width of length intervals, q_{ijt} and can be approximated by

$$q_{ijt} = \frac{w}{\sqrt{2\pi}\sigma_{jt}} \exp\left\{ -\frac{\left(x_i - \mu_{jt}\right)^2}{2\sigma_{jt}^2} \right\}$$

The Schnute and Fournier (1980) reparameterized von Bertalanffy growth equation is used to guarantee that mean lengths lie on the LVB curve. Then

$$\mu_{jt} = m_1 + \left(m_A - m_1\right)\left(\frac{1 - \rho^{j+(m_t-1)/12}}{1 - \rho^A} \right)$$

where L_1 is the mean length of the first age class and L_N is the mean length of the last age class in the sample, ρ is the Brody $K = -\ln\rho$ growth coefficient, and $m_t - 1$ is the number of years or months after the presumed birth month of the fish in t^{th} sampling period (or t^{th} length-frequency sample).

MULTIFAN assumes that gear selectivity applies only to the first age class in the fishery. The mean length of the first age class is adjusted by a factor of $b_1(12 - m_t)/12$, i.e.,

$$\mu'_{1t} = \mu_{1t} + b_1\left(12 - m_t\right)/12$$

where b_1 is the bias parameter that shifts the mean length of the first age class from the true population value.

The standard deviation is a function of length involving two parameters, λ_1 and λ_2, viz.,

$$\sigma_{jt} = \lambda_1 \exp\left\{ \lambda_2 \left[-1 + 2\left(\frac{1 - \rho^{j+(m_t-1)/12}}{1 - \rho^A} \right) \right] \right\}$$

where λ_1 is the magnitude coefficient and λ_2 determines the effect of a length-dependent trend in standard deviation, if one exists.

Let P_{jt} be the proportion of fish sampled at time t belonging to age j. Then the proportion of fish belonging to length category i, Q_{it}, can be expressed by an inverse age-length key as

$$Q_{it} = \sum_{j=1}^{A} P_{jt} q_{ijt}$$

MULTIFAN uses the joint likelihood function

$$L = \prod_{t=i}^{T} \prod_{i=1}^{L} \left[\frac{1}{\sqrt{2\pi\left(\alpha_{it} + \frac{0.1}{L}\right)T_t}} \left(\exp\left\{ -\frac{\left(\overline{Q}_{it} - Q_{it}\right)^2}{2\left(\alpha_{it} + \frac{0.1}{L}\right)T_t^2} \right\} + 0.01 \right) \right]$$

instead of the likelihood function used in MacDonald and Pitcher (1979). Normality is assumed by $Q_{it} \sim N\left(\tilde{Q}, T_t^2\right)$, where $\tilde{Q}_{it}$ is the expected value of Q_{it} and is estimated from the model. The parameter α_{it} is $\left(1 - Q_{it}\right)Q_{it}$. This parameter plus $0.1/L$ is added to avoid the type of deviation called "Type I": a length is recorded in a region where the model predicts that there is almost no probability of observing a fish. The quantity 0.01 is added to avoid the type of deviation called "Type II": the frequency observed in a length interval is much higher or much lower than it should be considering the accuracy of the majority of the observation.

The original MULTIFAN model does not estimate the total mortality rate and recruitment. Later versions of MULTIFAN include these estimates. If the age of full recruitment is known to be the Kth age class and the total mortality rate is constant over all cohorts and time periods, then r_j is defined to be

$$r_j = \frac{i}{L} \sum_{t} \ln\left(\rho_{jt} + 0.001\right) \qquad \text{for} \quad 1 \le j \le A$$

which is used to form a new log-likelihood equation,

$$L_Z = \left[\frac{(k-1)}{10}\right] \sum_{j=k}^{A} \left(r_j - c + jz\right)^2$$

which is subtracted from the main log-likelihood function. The parameter c and Z are the parameters to be estimated. The log-likelihood function being maximized is then the sum,

$$L + L_Z$$

The partial recruitment coefficient R_j is expressed as

$$R_j = \exp\left(r_j + c - jz\right) \qquad \text{for} \quad 1 \leq j \leq k-1$$

which depends upon the estimates of Z from the MULTIFAN algorithm.

A recent application of the method to shrimp length-frequency data (Fournier et al., 1991) yielded parameter estimates judged reasonable on the basis of comparison to analyses using more extensive biological knowledge.

6. PROCESS MODELS

6.1 OVERVIEW

The category of "process" models is meant to include models where the various processes in a fishery are represented by quantitative associations that include the parameters of interest in the stock assessment. For example, size-cohort analysis which was described earlier is based upon the Baranov catch equation, which quite simply says there is a function $B(M,F)$ that relates $C(l, t)$, the catch size distribution at time t to $N(l, t)$, the population size distribution at time t or,

$$C(l, t) = B(M, F)N(l, t)$$

In contrast, a model in this section would explicitly represent the flows of biomass for each process, such as growth in length by a particular growth function, biomass loss due to fishing by a selectivity function, recruitment as a function, etc. A user would be allowed to substitute functions, e.g., to change the growth patterns or the selectivity of the gear, etc., thus permitting component by component tests of the efficacy of each process representation as well as of the whole model.

In these days of rapid computers, most such models would be formulated so that the important parameters are estimated by a nonlinear least squares optimization or a maximum likelihood procedure.

6.2 DELAY-DIFFERENCE MODEL (FOURNIER AND DOONAN)

Fournier and Doonan (1987) contributed a delay-difference model designed to estimate the parameters of exploitation. The model is quite general, incorporating multiple improvements over its closest analogy of 7 years earlier, which is found in the age-dependent literature (Deriso, 1980). We shall develop some of the background of the model and a little about the model itself because of the pedagogical value of seeing how this field is developing, and, because the model is the background basis for other work that follows.

The background assumption is that one receives three time series of data C_i, catch in year i; B_i, biomass in year i and E_i, estimated fishing effort in year i. A constant catchability coefficient q is the usual assumption in fisheries. In particular, q does not change with density or time. Then, a simple relationship holds which relates catch C_i to biomass available B_i, viz., in the absence of natural mortality i and in the presence of only fishing mortality in year i,

$$C_i = (1 - s_i) B_i$$

where s_i is a survival rate

$$s_i = e^{-q\hat{E}_t}$$

This equation with constant catchability is used with the finite difference equation for biomass in year i, B_i,

$$B_i = \rho l s_{i-1} B_{i-1} + a N_i + \rho w_{k-1} R_i \tag{32}$$

where ρ is a constant from a generalized growth model in weight w_j, the biomass per individual of age j is

$$w_j - \rho w_{j-1} = a$$

l is the survival rate from natural mortality and s_i is the survival rate from fishing in year i. N_i is the abundance in year i of fish summed over ages, j, as in

$$N_i = \sum_{j > k} N_{ij}$$

where k is the age of recruitment and, finally, R_i is recruitment in year i which is assumed to be described by the Ricker stock recruitment relationship given by

$$R_{i+k} = \alpha B_i \exp\{-\beta B_i\}$$

where α and β are parameters to be estimated. Note that the difference Equation (32) is now age-independent.

To estimate the parameters, one chooses those values for which the predicted C_i, M_i, E_i, are "closest to" their observed values C_i, M_i, E_i, where $M_i = B_i / N_i$ is the mean weight of fish in the population in year i. The maximum likelihood (ML) criterion for closeness would involve choosing parameters: B_1, N_1, l, q, α, β, a, ρ which satisfy

$$\text{Min} \left\{ \sum \left(\ln C_i - \ln \hat{C}_i \right)^2 + P \sum \left(M_i - \hat{M}_i \right)^2 \right\} \tag{33}$$

or

$$\text{Min}\left\{ \sum\left(\ln E_i - \ln \hat{E}_i\right)^2 + P\sum\left(M_i - \hat{M}_i\right)^2 \right\} \tag{34}$$

In (33), the assumptions are that there is no error in the observed fishing effort E_i and that observed catches are log-normally distributed and $M_i's$ are normally distributed. In (34), it is assumed no error exists in the observed catches C_i, and that the estimated effort is log-normally distributed and $M_i's$ are normally distributed. In (33), if C_i is the primary data and M_i auxiliary data, P is set small enough so that the model fits the observed catch data primarily and the observed mean size data secondarily. The penalty weight is in most cases determined by external considerations.

The next model is a generalization that does not require recruitment to occur in a short Δt each year, in which "q" indexes groups to which fish belong (e.g., length groups) and "f" indexes different fisheries occurring concurrently in the stock, each with its own selection and effort functions. The model and estimation scheme depend upon the first to the fourth central moments, indexed by r, of the length distribution at each fishing year i. Bringing together the above indices g, f, r, i, we let μ^r_ϕ be the rth central moment of length for fishery f in year i. The mean weight for each fishery per year M_ϕ is also computed. Naturally, the second central moment is a measure of the variance; the third, a measure of skewness; the fourth, a measure of kurtosis. Several models are proposed for the optimal estimation of the parameters. The only one given here requires the log-likelihood function be maximized

$$\begin{aligned}\text{Max}\Bigg\{ -\frac{1}{\nu_0^2}\Bigg[& P_0\sum_{fi}\left(\ln C_{fi} - \ln\hat{C}_{fi}\right)^2 + P_1\sum_{fi}\left(M_{fi} - \hat{M}_{fi}\right)^2 + P_2\sum_{fi}\left(\mu^1_{fi} - \hat{\mu}^1_{fi}\right)^2 \\ & + P_3\sum_{fi}\left(\mu^2_{fi} - \hat{\mu}^2_{fi}\right)^2 + P_4\sum_{fi}\left(\mu^3_{fi} - \hat{\mu}^3_{fi}\right)^2 + P_5\sum_{fi}\left(\mu^4_{fi} - \hat{\mu}^4_{fi}\right)^2\Bigg] \\ & + 6\sum_{fi}\ln\nu_0 + \frac{\sum_{fi}\sum_{u=0}\ln\left(P_u\right)}{2}\Bigg\}\end{aligned} \tag{35}$$

where the variances ν_i^2 are assumed to be multiples of each other so that

$$P_i = \frac{\nu_0^2}{\nu_i^2}$$

Here, a term P_0 is included which, while normally 1, allows the fitting of the catch data to have an even greater weight than otherwise by setting P_0 to more than 1. Thus, the catch data are primary, and the length and weight moments are auxiliary. The models allow checking of residuals and easy adjustment of the penalty weights P. In fact, there are a number of ways of assigning P_i-weights, which Fournier-Doonan review.

The model incorporates a difference between "availability" to the gear and "selectivity" of the gear which takes on greater importance with increased effort. The concept of "groups" is introduced as a way of maximizing homogeneity within groups, in the sense that all fish within a group experience the same natural and fishing mortality, and maximizing the heterogeneity between groups. This concept is similar to the principle of stratification in sampling theory and serves the same purpose. The only difference is that individual fish who survive proceed sequentially from group to group.

The model incorporates a general stock recruitment function and generalizes the catchability term from its normal "q" to an indexed term q_{fig} depending on the fishery, year, and group.

A generalized von Bertalanffy model is used to describe growth between length groups which are defined in terms of π_j, or mean length j. That is, the LVB model is fit to a sequence of π_j-values, $j = 1,\ 2,\ \ldots,\ m_g$.

The model is applied to several simulated data sets by Fournier and Doonan but has not been widely applied to real fishery data.

6.3 THE SIZE STRUCTURE MODEL OF SCHNUTE

Another delay difference model of the dynamics of a fishery was published by Schnute (1987). This very general size-based model is attractive since it relates the currently hypothesized model to preceding literature, most of which are age-dependent models, and well understood. The model is also an interesting framework in which to look at size structure, so we will briefly describe it for this purpose. Our description will be too brief to be useful for an application and the reader is referred to the paper by Schnute (1987) and to more recent papers with applications (Schnute et al., 1989a; Schnute et al., 1989b). The 1987 model is entirely deterministic. The 1989 presentation contains a probabilistic component. We shall see in later models how stochasticity can be introduced.

The formulation Schnute presents is in terms of weight w but other dimensions of size could also be used. He defines the dependent variables as densities, e.g., numbers per kg. If the size were length rather than weight, the dimensions would be numbers per cm. Two intervals of weight are specified to account for weight change over a time interval. V is the weight at recruitment to the fishery and V' is the weight one time unit (e.g., a year) later. Thus V and V' are boundaries and w the weight an individual takes in the interval $V,\ V'$. New recruits have weights in $[V,\ V')$, previously exploited fish have weights in $\lfloor V',\ V_\infty)$, and the entire recruited population has weights in $\lfloor V',\ V_\infty)$. It is probable that many age groups fall into either of the categories $[V,\ V')$ and $\lfloor V',\ V_\infty)$.

Numbers in the population are given by the density $N(w,\ t)$ in units of number/kg. The number of fish between weights w_1 and w_2 at time t are the integral of the density over the appropriate weight range. Thus, the number of fish in $\left[w_1,\ w_2\right)$ is

$$N_t = \int_{w_1}^{w_2} N(w,\ t)\, dw$$

Thus, the number of fish in the fishery at the start of year t is a time series

$$N_t = \int_{V}^{V_\infty} N(w,\ t)\, dw$$

and the corresponding biomass time series is

$$N_t^* = \int_{V}^{V_\infty} wN(w,\ t)w\, dw$$

R_t and R_t^* are time series of numbers and biomass in $[V,\ V')$ and P_t , $P_t^*$are numbers and biomass in $\left[V',\ V_\infty\right)$, respectively. Time series R_t , R_t^*, P_t , $P_t^*$are all defined analogously to N_t and R_t^* In similar ways, time series of densities of catch $C(w,\ t)$ and spawning stock $S(w,\ t)$, and a density index of fishing success $I(w,\ t)$, e.g., CPUE, are all defined.

The following processes are represented: growth, recruitment, survival, and capture.

Schnute's growth assumption is similar to that of Fournier and Doonan (1987), relating weight of fish the following year (w') to weight in a year (w) using two parameters W and ρ :

$$w' = W + \rho w \tag{36}$$

A specific example of this growth equation is von Bertalanffy growth, where $W = W_\infty\left(1 - e^{-k}\right)$ is the initial length and $\rho = e^{-k}$ for $k > 0$. Historically, this form of growth resulted from attempts to linearize length relationships. Walford (1946) related lengths in three successive time periods using a parameter k:

$$l_n - l_{n-1} = k\left(l_{n-1} - l_{n-2}\right), \qquad k < 1$$

He expressed length at age n as a series

$$l_n = l_1 + \left(l_2 - l_1\right) + \left(l_3 - l_2\right) + ... + \left(l_n - l_{n-1}\right) = l_1\left(\frac{1-k^n}{1-k}\right)$$

Schnute's expressions for weight (36) are derived in analogous ways, assuming that both w' and w are given indices, to get

$$w' = W\left(\frac{1-\rho^{n+1}}{1-\rho}\right), \qquad \rho \cdot w = W\left(\frac{\rho-\rho^{n+1}}{1-\rho}\right)$$

combining these two expressions leads to the growth assumption (36), again,

$$w' - \rho w = W$$

Weight at recruitment, V, can be determined by inspection of catch samples. The weight of these fish one time period later would follow from (36)

$$V' = W + \rho V$$

Schnute (1987) states four assumptions about growth, survival, and recruitment.

1. Fish growth is described by equation (36).

2. Population survival which is not size-dependent is defined by

$$\int_{w_1'}^{w_2'} N(w, t+1)\, dw = \tau_t \int_{w_1}^{w_2} N(w, t)\, dw$$

   where $N(w\ ,\ t)$ is the population size at weight w and time t, τ_t is a function of survival from natural causes (σ) and from fishing. Catches are assumed to occur prior to natural mortality:

$$\tau_t N(w, t) = \sigma\left[N(w, t) - c(w, t)\right]$$

3. Recruitment is based on a stock-recruitment function $F(\cdot)$ such as Ricker's (1975) or Schnute's (1987).

$$R_t^* = \sum_{i=k_1}^{k_2} \pi_i F\left(S_{t-i}^*\right) = \sum_{i=k_1}^{k_2} \pi_i F\left(\tau_{t-i} N_{t-i}^*\right)$$

where $\sum_{i=k_1}^{k_2} \pi_i = 1, \quad 0 \le \pi_i \le 1$. Note that if $k_1 = k_2$, then the model reduces to an age-structured one. If a stock-recruitment relationship is unavailable, constant recruitment can be assumed.

4. The index $I(w, t)$ is the ratio of catch to effort and is proportional to population size:

$$I(w, t) = \frac{c(w, t)}{E_t} = qN(w, t)$$

where E_t is effort at time t and q is a proportionality constant. This indexing is assumed to occur prior to natural mortality.

The simplest version of Schnute's model requires time series data on I_t^*, C_t^*, and X_t, the average weight of the population at the start of year t. This model consists of the following recursive equations:

$$N_t^* = \frac{I_t^* + q\theta(1-\mu(1-\sigma))C_t^*}{q(1-\theta(1-\sigma))}$$

$$S_t^* = (1-\nu(1-\sigma))N_t^* - \nu(1-\mu(1-\sigma))C_t^*$$

$$R_{t+1}^* = \sum_{i=k_1}^{k_2} \pi_i F\left(S_{t+1-i}^*\right)$$

$$N_{t+1}^* = R_{t+1}^* + \frac{X_t'}{X_t}\left(\sigma\left(N_t^* - \mu C_t^*\right) - (1-\mu)C_t^*\right)$$

$$I_{t+1}^* = q(1-\theta(1-\sigma))N_{t+1}^* - \theta(1-\mu(1-\sigma))C_{t+1}^*$$

where μ, ν, and θ are timing parameters for natural mortality, spawning, and indexing of the population. Figure 6 shows the model equations and the required inputs and parameter estimates for the case $\mu = 1$, $\nu = 1$, and $\theta = 0$, i.e., catch takes place prior to natural mortality, spawning occurs subsequent to all mortalities, and indexing reflects the population prior to all mortality.

A slightly more complicated model requires the additional time series data, Y_t and Z_t, corresponding to the average weight of fish in the intervals $[V, V')$ and $[V', V_\infty)$. Let w_t be the fraction of the exploitable biomass due to newly recruited fish:

$$w_t = \frac{R_t^*}{N_t^*} \tag{37}$$

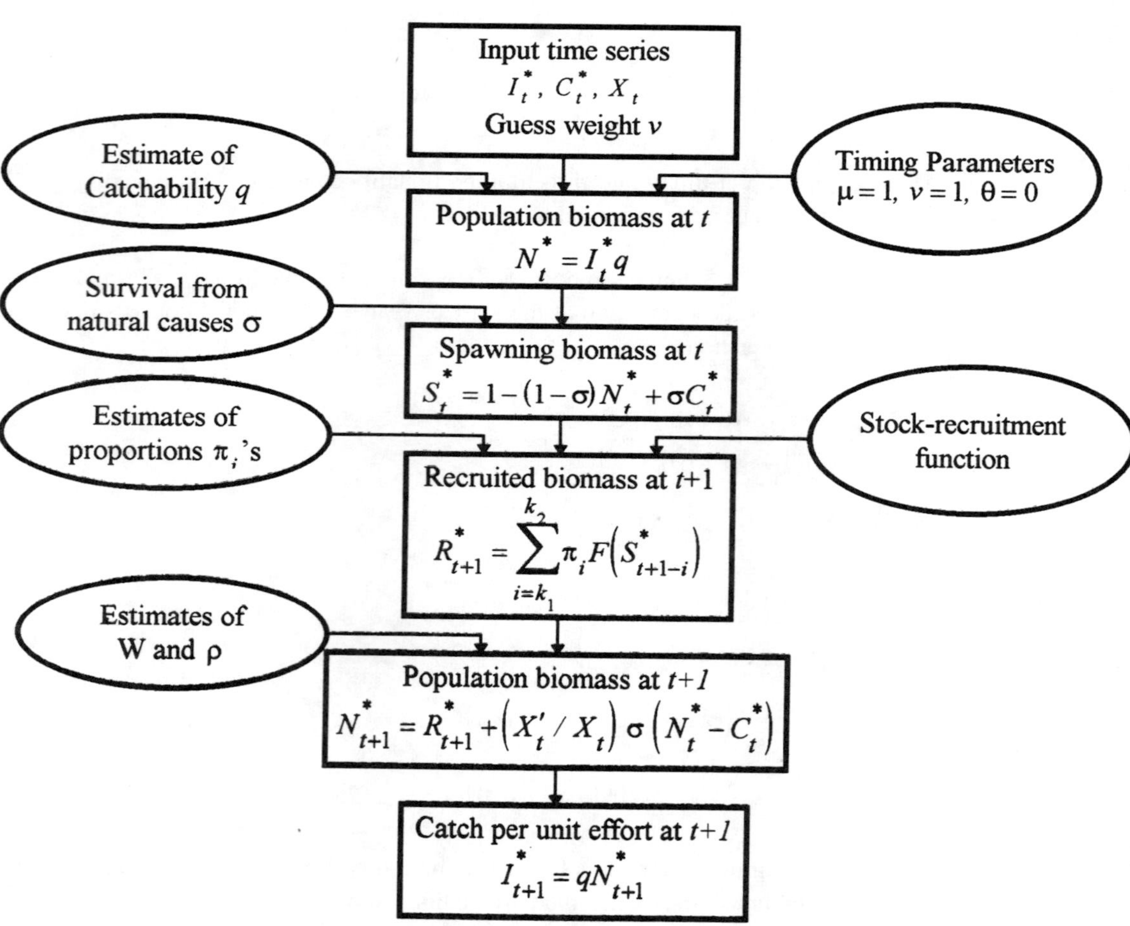

Figure 6. Simple version of Schnute's model.

Then, after some calculations using (37) and the fact that $N_t = R_t + P_t$, where P_t is the previously exploited population, we get

$$w_t = \frac{Y_t}{X_t}\left(\frac{Z_t - Y_t}{Z_t - Y_t}\right)$$

With this result, Schnute's model becomes

$$I^*_{t+1} = q\left(\frac{1-\theta(1-\sigma)}{1-w_{t+1}} R^*_{t=1} - \theta(1-\mu(1-\sigma))C^*_{t+1}\right)$$

or

$$I^*_{t+1} = q\left(\frac{1-\theta(1-\sigma)}{1-w_{t+1}} P^*_{t+1} - \theta(1-\mu(1-\sigma))C^*_{t+1}\right)$$

where P^*_{t+1}, previously exploited biomass, is given by

$$P^*_{t+1} = \frac{X'_t}{X_t}\left[\sigma\left(N^*_t - \mu C^*_t\right) - (1-\mu)C^*_{t+1}\right]$$

6.4 THE STOCK SYNTHESIS MODEL (SSM)

6.4.1 Overview

The Stage 1 version of the SSM (Methot, 1989) is based on an age-structured model whose parameters are estimated using nonlinear least squares optimization. The objective function consists of minimizing the squared deviations between the observed and estimated catch.

The model uses the maximum likelihood approach discussed by Fournier and Archibald (1982) where the log-likelihood of each observation of age or length follows a multi-nomial error structure. The likelihood approach includes a number of other log-likelihood component for: catch biomass, fishing effort, survey data, and spawner-recruitment relationships. The total log-likelihood is the sum of these components. The model is fit by maximizing this sum.

The Stage 2 model introduces size as an alternative measurement or observation. The age composition is converted into a length distribution through a length-at-age key (LA). The LA conversion is the proportion of fish at age "a" that are in the bin number "l" where the mean size-at-age is described by an LVB model. A selectivity-at-length function is applied to the presumed abundance defined by size and the LA conversion.

A Stage 3 model takes into account the effects of a size-selective fishery on the mean size-at-age of the fishable stock. This stage of the model takes into account biases in the observation process which is the fishing mortality due to gear selectivity. This aspect of the model is still under development.

The model is implemented in a FORTRAN program which runs on MS-DOS compatible microcomputers. It includes the treatment of survey abundance data, spawner-recruitment relationships, and other possible population-related processes.

6.4.2 Model Components and Parameters

The population model used in the SSM distinguishes between vulnerability V_a and availability or selectivity S_a. The dynamics of the population is described by the age-structured model

$$N_{y+1,\ a+1} = N_{ya}\left[\left(1-V_a\right)e^{-M} + V_a e^{-Z_{ya}}\right]$$

while the catch in numbers and biomass are given by

$$C_{ya} = N_{ya}V_a\left(\frac{F_{ya}}{Z_{ya}}\right)\left(1-e^{-Z_{ya}}\right)$$

$$C_y = \sum_a C_{ya}W_a$$

respectively, where:

N_{ya} = population numbers in year y and age a,
M = instantaneous rate of natural mortality,
V_a = vulnerability of age a fish to the fishery,
$F_{ya} = E_y S_a$ = fishing mortality rate in year y on age a fish,
S_a = availability (selectivity) for fish of age a,
E_y = fishing mortality in year y for fully available ages,
$Z_{ya} = M + F_{ya}$ = total mortality rate,
C_{ya} = catch of age a fish in year y
W_a = body weight at age a.

Additional parameters may be involved, for example, in "vulnerability" and "selectivity." All parameters are estimated by maximizing the total log-likelihood function, L

$$L = L1 + L2 + L3 + L4 + L5$$

where:

$L1$ = likelihood of catch biomass,
$L2$ = likelihood of fishing effort,
$L3$ = likelihood of age or length composition,

$L4$ = likelihood of survey data,
$L5$ = likelihood of spawner-recruitment relationship.

The components of the log-likelihood are expressed in mathematical notation below. Ignoring constants independent of the parameters to be estimated.

1. Log-likelihood of catch biomass:

$$L1 = \sum_y \frac{1}{2}\left(\frac{\log \hat{C}_y - \log C_y}{\sigma_C}\right)^2 - \log \sigma_C$$

where:

$C_y = \sum_a C_{ya} W_a$, observed catch biomass,
$\hat{C}_y$ = estimated catch biomass,
σ_C = standard error of $\log C_y$.

2. Log-likelihood of fishing effort:

$$L2 = \sum_y \frac{1}{2}\left(\frac{\log\left(\frac{\hat{E}_y}{Q}\right) - \log E_y}{\sigma_E}\right)^2 - \log \sigma_{E_y}$$

where:

E_y = observed fishing effort,
Q = catchability coefficient,
σ_E = standard error of $\log E_y$.

3. Log-likelihood of age or size composition:

$$L3 = \sum_y J_y \sum_{a(l)} P_{ya(l)} \log \hat{P}_{ya(l)}$$

where:

J_y = number of fish in the age (length) sample, or if the multi-nomial error structure is questionable, a weighting factor,

$P_{ya(l)}$ = observed proportion at age (length) in the sample,
$P_{ya(l)}$ = estimated proportion at age (length) in the sample.

The underlying model is age-structured. In the Stage 2 model, the population age structure is converted to the population length structure by:

$$N_{yl} = \sum_a \pi_{al} N_{ya}$$

where:

π_{al} = the proportion of age a fish in the lth size bin,
N_{ya} = the mean numbers-at-age during time period y,
N_{yl} = the mean numbers-at-length during time period y.

4. Log-likelihood of survey data:

$$L4 = \sum_y \frac{1}{2}\left(\frac{\log \hat{\mu}_y - \log \mu_y}{\sigma_{\mu_y}}\right)^2 - \log \sigma_{y\mu}$$

where:

μ_y = observed survey biomass,
$\hat{\mu}_y$ = estimated survey biomass,
σ_{μ_y} = standard error of $\log \mu_y$;

5. Log-likelihood of stock-recruitment relationship:

$$L5 = \sum_y \frac{1}{2}\left(\frac{\log \hat{R}_y - \log R_y}{\sigma_R}\right)^2 - \log \sigma_R$$

where:

R_y = recruitment estimated from a spawner-recruitment relationship,
R_y = estimated recruitment from the model,
σ_R = root mean square of $\left(\log \hat{R}_y - \log R_y\right)$.

Notice that the log-likelihoods include standard errors which will produce estimates based on confidence in the data. Methot (1990) suggests that a range of "emphasis factors" be included to produce estimates of biomass consistent with observed confidence limits. He notes, on the other hand, that the estimates should be insensitive to emphasis factors when the data is consistent with the model assumptions.

Stock synthesis uses a double logistic function to model age-specific selectivity or length-specific selectivity for Stage 1 and Stage 2 models, respectively. Size-at-age estimation in the Stage 2 model is incorporated into an additional likelihood component, *L6*:

$$L6 = \sum_a \frac{1}{2}\left(\frac{\hat{Y}_a - Y_a}{\sigma_a}\right)^2 - \log \sigma_a$$

where:

Y_a = observed mean size assigned to age *a*,

$\hat{Y}_a$ = estimated mean size assigned to age *a*,

σ_a = standard error of Y_a.

The stock synthesis model has been used with the following fisheries: Pacific whiting (Hollowed et al., 1988; Dorn and Methot, 1989), widow rockfish (Hightower and Lenarz, 1989), walleye pollock (Megrey et al., 1989), sablefish (Methot and Hightower, 1988), and northern anchovy (Methot, 1986, 1989).

6.5 CATCH-AT-SIZE ANALYSIS (CASA)

Sullivan et al. (1990) developed a stock assessment estimation methodology that makes no assumptions about the steady state of length distributions and the form of the recruitment function. This method computes the gear selectivity as a function of fish length from catch data, and describes growth with a stochastic model based on the probability of individuals moving from one length class to another. The program CASA and users' guide are available from the authors.

6.5.1 Model Description

The schematic in Figure 7 illustrates how the CASA algorithm deals with the fish in any length category in the length frequency distribution (lfd) of a population. Between $t = 1$ and $t = 2$, these fish would have two possible fates: (1) death due to natural mortality (*M*) or fishing mortality (*F*) and (2) survival and growth. The survivors will have grown by a stochastic function to enter other length categories in the lfd. The length category, under a certain length boundary, is composed, however, not only of those who grew into it, but also those who recruited to the fishery at time $t = 2$. The resultant lfd at time $t = 2$, shown in the bottom right of Figure 7, represents the exploitable population at time $t = 2$. The death due to fishing mortality constitutes the catch-at-length, i.e., lfd of catch at time $t = 1$. The selectivity to length is accounted for in the lfd of catch.

Although this schematic shows how the population lfd are formed, the estimation procedure starts with the lfd of catch C' and works backward to reconstruct the lfd of the

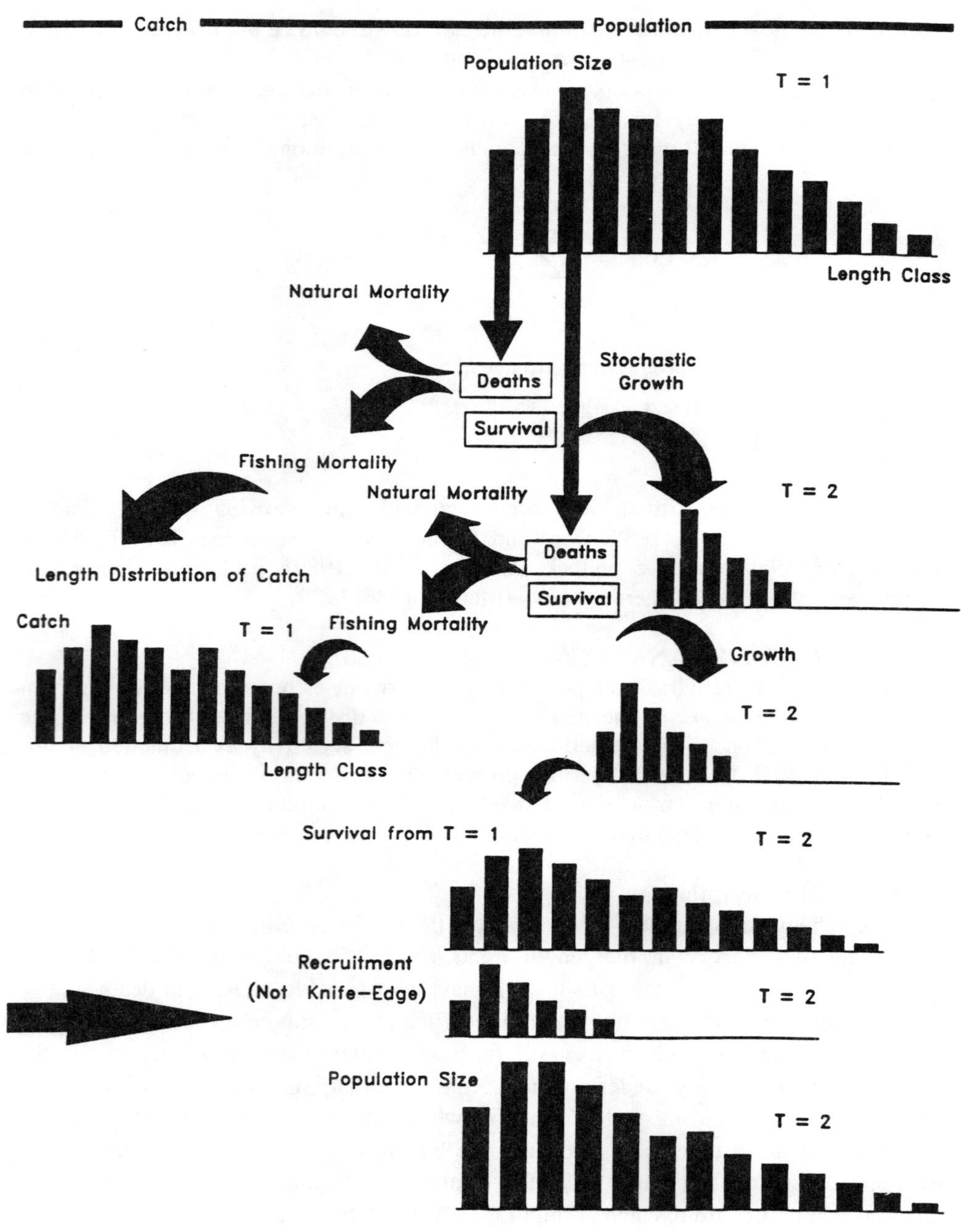

Figure 7. Schematic for CASA.

population. This is based on the fact that the population lfd is not observable, but is measurable through the catch lfd which is available from sampling catch in the fishery. The following models describe the reconstruction of the population from catch lfd.

Using the Baranov catch equation, the catch at length l and time t, $(C_{l,t})$ can be written as

$$C_{l,t} = \frac{F_{l,t}}{Z_{l,t}}\left(1 - e^{-Z_{l,t}}\right)N_{l,t}$$

where $F_{l,t}$, $Z_{l,t}$, and $N_{l,t}$ are, respectively, the instantaneous fishing mortality rate, the instantaneous total mortality rate, and the population size at length l and time t.

The length-specific fishing mortality rate $F_{l,t}$ for time t is assumed to be separable into a product of full recruitment fishing mortality rate F_t and selectivity at length

$$F_{l,t} = S_l F_t$$

The selectivity S_l represents the fraction of fish in length category l subjected to the full effect of fishing mortality. There are numerous functions that can be used to describe the selectivity.

A common selectivity function is the logistic equation which can approach the commonly found knife-edge form for large values of β_S in

$$S_l = \frac{1}{1 + \alpha_S e^{-\beta_S l}}$$

where the subscript "S" refers to selectivity.

The instantaneous total mortality rate of individuals with length l and time t is separable as the sum of the instantaneous fishing and natural mortality rates:

$$Z_{l,t} = F_{l,t} + M_{l,t} = S_l F_t + M_{l,t}$$

For each length l, the population size at time t is related to the population at a later time t by the following:

$$N_{l,t'} = N_{l,t} e^{-Z_{l,K}} \tag{38}$$

Growth combines two aspects: (1) the inherent variability seen in individual growth and (2) the general nonlinear trend frequently observed over the population as a whole. The proportion $P_{l,t}$ of surviving individuals that grow from one length class to the next length class is assumed to be represented by a probability distribution function that can be parameterized by its mean and variance. The mean represents the average growth increment that can be described by a standard and deterministic growth function. The variance describes the individual variability in growth. Under such a formulation, the mean and the variance uniquely determine the proportion of individuals moving from one length category to another.

For the von Bertalanffy growth model, the mean of the distribution can be represented by

$$l_t = L_\infty \left(1 - e^{-K(t-t_0)}\right)$$

where l_t is the length of an individual at age t, L_∞ is the theoretical asymptotic length, K is the Brody growth coefficient, and t_0 is the time correction factor for the size at birth or recruitment.

Under this model, the growth increment of a fish starting from length l during the time interval Δt is given by

$$\Delta l = l_{t+\Delta t} - l_t = (L_\infty - l)\left(1 - e^{-K\Delta t}\right)$$

If Δt is set equal to 1, the growth increment can be written as

$$\Delta l = (L_\infty - l)\left(1 - e^{-K}\right)$$

The growth is then described independently of age over the designated time unit and independently of t_0.

Applying this formula to a discrete length frequency histogram and taking into account variation in growth represented by stochasticity, we denote l' as the mid-length of class l where the fish starts to grow and Δl_i as the length increment of the ith fish in the length class l. The probability distribution of the length increment Δl is assumed to be gamma distributed with mean growth $\Delta l = \alpha_g \beta_g$ and variance $\sigma^2 = \alpha_g \beta_g{}^2$. Since $\Delta l / \beta_g$ and Δl_i are calculated from L_∞ and K, only β_g needs to be estimated. Therefore, there are three parameters to be estimated in the growth model. The proportion $\pi_{l,\,l'}$ of fish growing from the length class l to any other length class l' is found by the difference of the two cumulative gamma probabilities over the appropriate ranges:

$$\pi_{l,\,l'} = G\left(\frac{l'}{\beta_g}\right) - G\left(\frac{l}{\beta_g}\right)$$

where

$$G(x) = \int_0^x \frac{1}{\beta_g^{\alpha}\Gamma(\alpha_g)} \mu^{\alpha g-1} e^{-\mu/\beta_g} d\mu$$

Since L_∞ is a parameter in a stochastic growth equation, it may be smaller than the largest fish found in the sample. This would cause the problem of $\Delta l \leq 0$ in the length class l's where $l \geq L_\infty$. To avoid this difficulty, the following expression is used:

$$\Delta l_j = \left(1-e^{-K}\right)\Delta l_{j-1}$$

where j indexes the length class $l \geq L_\infty$

Knife-edge recruitment, as described by Ricker (1975), is not assumed in this model. Fish can enter the fishery over several length classes which may extend over several ages according to the stochasticity of growth. The probability distribution of length when recruitment occurs is described by a gamma probability density

$$g(x) = \frac{1}{\Gamma(\alpha_r)\beta_r^{\alpha_r}} x^{\alpha_r-1} e^{-x/\beta_r}$$

where x is the random variable representing length at recruitment of individual fish, α_r and β_r are parameters, and the subscript "r" refers to recruitment.

The probability of recruitment into the lth length class, with length range (l, l') is

$$P_l = G\left(\frac{l'}{\beta_r}\right) - G\left(\frac{l}{\beta_r}\right)$$

where

$$G(x) = \frac{1}{\Gamma(\alpha_r)\beta^{\alpha_r}} \int_0^x \mu^{\alpha_r-1} e^{-\mu/\beta_r} d\mu$$

The number of fish recruited into the lth length group at time t can then be computed as

$$R_{l,t} = R_t P_l$$

where

$$R_t = \sum_l R_{l,t}$$

The advantages of such a separation of variables are: (1) the reduction of the number of parameters to be estimated to the number of time periods in the catch-at-length data plus two parameters of the gamma function, i.e., α_r and β_r, and (2) recruitment estimates can be compared to estimated from standard procedures.

From the above equations, one can finally calculate the total number of fish $N_{l',t'}$ at length l' at the start of the next time period as follows:

$$N_{l',t'} = \sum_{l} \pi_{l,l'} N_{l,t} e^{-Z_{l,t}} + R_{l',t'}$$

Note again that recruitment may occur over several length classes according to the probabilistic nature of growth.

For the first period of catch-at-length data, the population size at length l can be computed as

$$N_{l,1} = N_1 P_l$$

where

$$N_1 = \sum_{l} N_{l,1}$$

and

$$P_l = \frac{N_{l,1}}{N_1} = \frac{\dfrac{C_{l,1} Z_{l,1}}{F_{l,1}\left(1-\exp\left(-Z_{l,1}\right)\right)}}{\sum_{l} \dfrac{C_{l,1} Z_{l,1}}{F_{l,1}\left(1-\exp\left(-Z_{l,1}\right)\right)}}$$

dividing the numerator and denominator of P_l by $\sum_{l} C_{l,1}$ and letting

$$Q_{l,1} = \frac{C_{l,1}}{\sum_{l} C_{l,1}}$$

we obtain:

$$N_{l,1} = N_1 \frac{\dfrac{Q_{l,1} Z_{l,1}}{F_{l,1}\left(1-\exp\left(-Z_{l,1}\right)\right)}}{\sum \left(\dfrac{Q_{l,1} Z_{l,1}}{F_{l,1}\left(1-\exp\left(-Z_{l,1}\right)\right)} \right)}$$

where $Q_{l,1}$ is the proportion of the length frequency distribution and can be obtained from the observed catch at length l and time 1. The number of parameters will then be reduced from the number of length classes to one parameter, i.e., N_1. Since the population size for other time periods can be computed recursively using (38).

6.5.2 Parameter Estimation

The nonlinear least squares technique is used to estimate the model parameters. This approach consists of minimizing the sum of squares difference between the observed and the model predicted catch-at-length l and time t; i.e.,

$$\text{Minimize} \quad \sum_{l}\sum_{t}\left(C_{l,t} - \hat{C}_{l,t}\right)^2$$

where $C_{l,t}$ is the predicted catch-at-length l and time t. The parameters to be estimated are presented in Table 3.

Table 3. Parameters to be estimated.

α_S, β_S	Parameters of the selectivity function
L_∞, k, β_g	Parameter of the growth model
α_r, β_r	Recruitment distribution parameters
R_t	Recruitment
F_t	Full recruitment fishing mortality
N_1	Initial population size

If known, some of these parameters can be fixed during the estimation procedure. Additional information on some parameters such as full recruitment fishing mortality can also be used to improve the estimates. This auxiliary information can be introduced in the objective function as follows:

$$\text{Minimize} \quad \sum_{l}\sum_{t}\left(C_{l,t} - \hat{C}_{l,t}\right)^2 + \sum_{i}\lambda_i\left(f_i(\cdot) - \hat{f}_i(\cdot)\right)^2$$

where λ_i is a weighting factor that penalizes the objective function. It can be shown that an optimal choice of λ_i, in the sense that λ_i minimizes the variances of the estimated parameters, is equal to the variance of the catch over the variance of the auxiliary information. $f(\cdot)$ is the observed information on the parameters or on a function of the parameters and $\ddot{f}_i(\cdot)$ is its predicted counterpart. Auxiliary information might include fishing effort, population estimates from surveys and hydroacoustic techniques, and/or survey data on recruitment.

6.5.3 Model Application and Analysis

For initial validation of the estimation procedure, a hypothetical data set was generated to describe the dynamics of Pacific cod (*Gadus macrocephalus*) in the eastern Bering Sea using a computer-simulated model based on the above equations. Parameter values representing the initial population size, growth, recruitment, natural and fishing mortality, and selectivity were chosen to reflect the dynamics exhibited by Pacific cod in the eastern Bering Sea. The parameter values were set equal to actual estimates arrived at by Bakkala and Low (1985), so that the model dynamics would be representative of observations from the natural system. The initial population was based on the survey estimates of population number at ages (Bakkala and Low, 1985) and transformed into 45 2-cm-wide length classes using an age-length key (Lai, 1985). The simulation depicts population numbers and catch numbers in terms of 45 2-cm-wide length classes and ten time intervals, each a year in duration. Fishing was assumed to be a logistic function. The growth increment Δl was calculated for each length class l by using the values of L_∞ and k given in Lai (1985), while β was set to equal 1, indicating a variance in growth that is equal to the mean. The recruitment distribution was set equal to the observed length distribution of age 1 individuals given for the first year (1978) of the time series in Bakkala and Low (1985). Independently distributed Gaussian error, with variance equal to 105 times the average number of individuals observed in a catch length class, was added to the observations to simulate the observation error. In this example, the constant of proportionality reflects a coefficient of variation of approximately 0.2. A summary of the parameters used in the simulation is given in Table 4.

The simulated data set is characterized by an influx of the strong 1977 year class in the first year (1978) that grows and declines, giving way in influence in subsequent years to the stronger influence of recruitment (see Figure 8). For the first estimation with the simulated data, the initial values of all parameters were set to true values (Table 5). In the second estimation, the initial values of all parameters were randomly set to ± 30% of the true values (Table 5). All parameters except natural mortality (M) were subject to estimation in these two cases.

Figure 8 compares the estimated catch values with the catch values from a simulation where the input values were varied ±30% from the (known) true values. In both estimations, the parameters estimated (Table 5) show good correspondence within known values, although some difference in the recruitment estimates can be seen over later time periods. Here, as in catch-at-age analysis (Deriso et al., 1985; Kimura, 1989), auxiliary information may prove useful in obtaining better estimates. Note however, that even with deviations of ±30%, the estimation algorithm appeared to be robust.

Table 4. Parameter values used in the simulation.

Number of age classes: 12

Number of time steps (years): 10

Number of length classes: 45

Parameter	Value	Parameter	Value	Parameter	Value
Abundance at:					
Age 2	2.14×10^8	N_0	1.36×10^9	F_0	0.42
Age 3	1.20×10^7	α_r	37.00	F_1	0.44
Age 4	1.10×10^7	β_r	0.50	F_2	0.46
Age 5	6.00×10^6	R_1	4.42×10^8	F_3	0.48
Age 6	6.00×10^6	R_2	1.32×10^8	F_4	0.47
Age 8	1.00×10^6	R_3	1.39×10^8	F_5	0.48
Age 9	1.00×10^6	R_4	8.60×10^7	F_6	0.45
Age 10	1.00×10^6	R_5	9.00×10^7	F_7	0.48
Age 11	1.00×10^6	R_6	1.65×10^8	F_8	0.40
Age 12	1.00×10^6	R_7	1.05×10^8	F_9	0.43
		R_8	1.28×10^8	L_∞	83.00
		R_9	9.70×10^7	k	0.24
		M	0.30	β	1.00
		α_S	0.10		
		β_S	30.00		

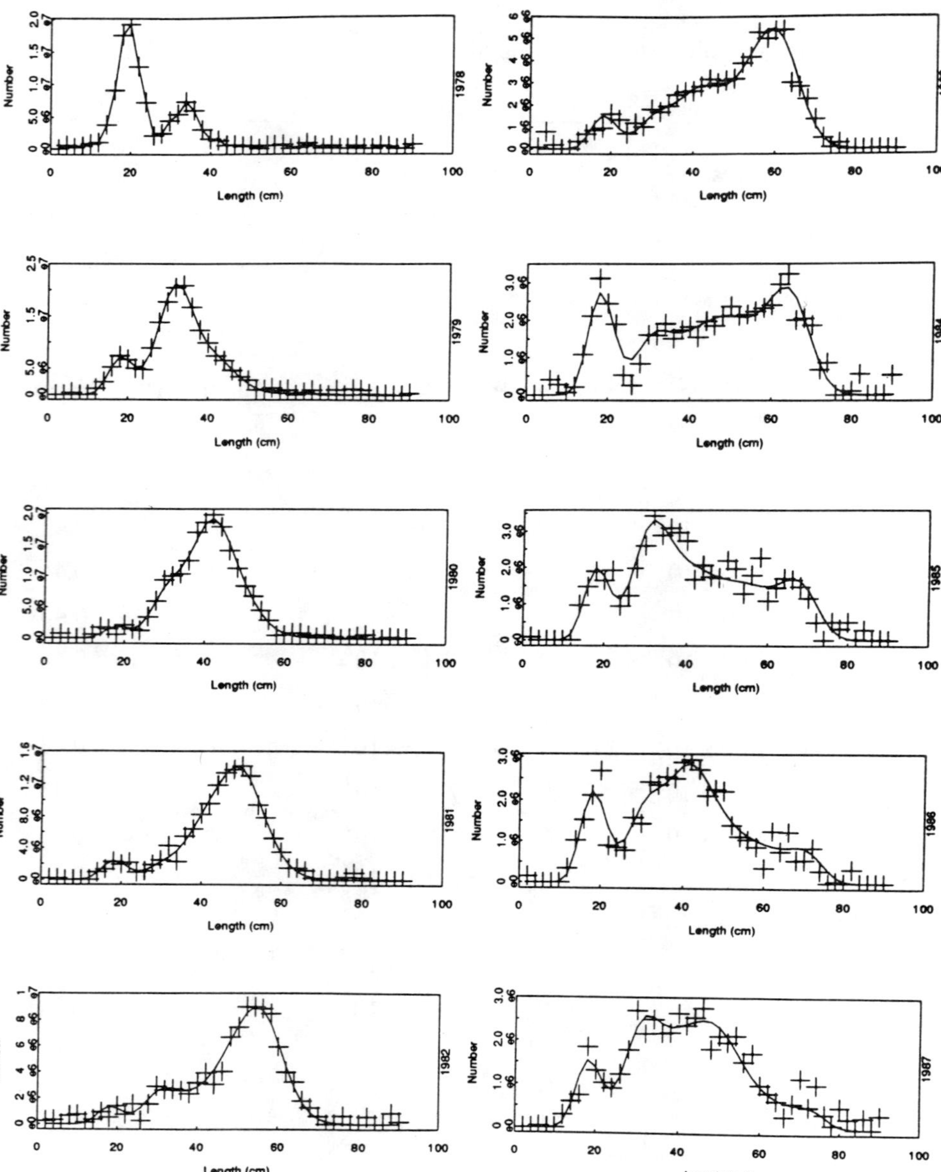

Figure 8. Simulated (+) and estimated (solid line) catch-at-length data of Pacific cod over 10 consecutive time periods. Simulated catches were generated for a Pacific cod population during 1978-1987. Catch estimates were obtained from the catch-at-length least squares algorithm, using initial parameter values which varied ±30% from the true value. (from Sullivan et al. 1990. Can. J. Fish. Aquat. Sci. 47. With permission of the Minister of Supplies and Services Canada.)

Table 5. Simulated and estimated parameter values for Pacific cod.

		Estimate 1		Estimate 2	
	True	Initial (set equal to true values)	Estimated	Initial (set equal to true values)	Estimated
N_0	1.36 x 10^9	1.36 x 10^9	1.39 x 10^9	1.00 x 10^9	1.39 x 10^9
α_r	37.00	37.00	30.40	25.00	30.40
β_r	0.50	0.50	0.56	0.70	0.56
R_1	4.42 x 10^8	4.42 x 10^8	4.13 x 10^8	5.75 x 10^8	4.12 x 10^8
R_2	1.32 x 10^8	1.32 x 10^8	1.16 x 10^8	9.30 x 10^7	1.16 x 10^8
R_3	1.39 x 10^8	1.39 x 10^8	1.50 x 10^8	1.81 x 10^8	1.50 x 10^8
R_4	8.60 x 10^7	8.60 x 10^7	8.76 x 10^7	1.06 x 10^8	8.76 x 10^7
R_5	9.00 x 10^7	9.00 x 10^7	9.85 x 10^7	6.30 x 10^7	9.85 x 10^7
R_6	1.65 x 10^8	1.65 x 10^8	1.94 x 10^8	1.15 x 10^8	1.94 x 10^8
R_7	1.05 x 10^8	1.05 x 10^8	1.40 x 10^8	1.50 x 10^8	1.40 x 10^8
R_8	1.28 x 10^8	1.28 x 10^8	1.89 x 10^8	9.00 x 10^7	1.89 x 10^8
R_9	9.70 x 10^7	9.70 x 10^7	1.42 x 10^8	1.06 x 10^8	1.42 x 10^8
α_S	0.10	0.1	0.10	0.30	0.10
β_S	30.00	30.00	32.50	20.00	32.50
F_0	0.42	0.42	0.51	0.30	0.51
F_1	0.44	0.44	0.52	0.60	0.52
F_2	0.46	0.46	0.52	0.30	0.52
F_3	0.48	0.48	0.51	0.65	0.51
F_4	0.47	0.46	0.50	0.25	0.50
F_5	0.48	0.48	0.51	0.60	0.51
F_6	0.45	0.45	0.46	0.70	0.46
F_7	0.48	0.48	0.46	0.30	0.46
F_8	0.40	0.40	0.38	0.60	0.38
F_9	0.43	0.43	0.35	0.30	0.35
L_∞	83.00	83.00	86.10	95.00	86.10
k	0.24	0.24	0.23	0.30	0.23
β	1.00	1.00	1.04	1.40	1.04

ACKNOWLEDGMENTS

The authors express appreciation to Dr. Jon Schnute, Pacific Biological Station, Nanaimo, Canada for reading an early draft from the perspective of an overview, without entering into mathematical details and to Dr. Daniel Pauly, ICLARM, the Philippines and the University of British Columbia, who read selected sections in detail. Drs. Robert Donnelly and Loveday Conquest were frequent participants in discussions. Katherine Peterson, Erin Moline, and Steven Anderson typed the many drafts. This publication was supported by the U.S. Agency for International Development under grant #DAN-4146-G-SS-5071-00, Fisheries Stock Assessment, Collaborative Research Support Program (CRSP). This publication is Contribution No. 900, University of Washington, School of Fisheries, carried out under the Management Assistance for Artisanal Fisheries (MAAF) program. Correspondence should be addressed to Vincent F. Gallucci, Center for Quantitative Science, Box No. 355230, University of Washington, Seattle, Washington 98195.

REFERENCES

Basson, M., Rosenberg, A.A. and Beddington, J.R. 1988. The accuracy and reliability of two new methods for estimating growth parameters from length-frequency data. J. Cons. Int. Explor. Mer. 44: 277-285.

Bakkala, R.G. and Low, L.L. (Eds.) 1985. Condition of groundfish resources of the eastern Bering Sea and Aleutian Islands region in 1984. NOAA Tech. Memo. NMFS F/NWG-117.

Beverton, R.J.H. and Holt, S.J. 1956. Stochastic age-frequency estimation using the von Bertalanffy growth equation. Fish. Bull. U.S. 81: 91-96.

Beverton, R.J.H. and Holt, S.J. 1957. On the dynamics of exploited fish populations. Fish Invest. Minist. Agric. Fish. Food (GB) Ser. II Salmon Freshwater Fish. 19. Fisheries Laboratory, Lowestoff.

Beverton, R.J.H. and Holt, S.J. 1959. A review of the life spans and mortality rates of fish in nature, and their relation to growth and other physiological characteristics. In *CIBA Foundation Colloquia On Aging, The Lifespans of Animals*, (Wolstenholme, G.E.W. and O'Connor, M., Eds.) London, Churchhill, Vol. 5: 142-80.

Beverton, R.J.H. and Holt, S.J. 1966. Manual of Methods for Fish Stock Assessment. Part II - Tables of Yield Functions. FAO Fisheries Technical Paper No. 38 (Rev. 1). FIb/T38 (Rev. 1).

Breen, P.A. and Fournier, D.A. 1984. A user's guide to estimating total mortality rates from length frequency data with the method of Fournier and Breen. Can. Tech. Rep. Fish. Aquat. Sci. No. 1239.

Cassie, R.M. 1954. Some uses of probability paper in the analysis of size frequency distributions. Aust. J. Mar. Freshwater Res. 5: 513-522.

Chapman, D. and Robson, D. 1960. The analysis of a catch curve. Biometrics 16: 354-368.

Chevaillier, P. and Laurec, A. 1990. Logiciels pour l'évaluation des stocks de poisson. ANALEN: Logiciel d'analyse des données de capture par classes de taille et de simulation des pêcheries multi-engins avec analyse de sensibilité. FAO Document, Technique Que sur les pêches. No. 101, Supp. 4. Rome, FAO

Deriso, R.B. 1980. Harvesting strategies and parameter estimation for an age structured model. Can. J. Fish. Aquat. Sci. 37: 268-282.

Deriso, R.B. 1987. Optimal F0.1 criteria and their relationship to maximum sustainable yield. Can. J. Fish. Aquat. Sci. 44, (Suppl. 2): 339-348.

Deriso, R.B., Quinn, T.J., II and Neal, P.R. 1985. Catch-age analysis with auxiliary information. Can J. Fish. Aquat. Sci. 42: 815-824.

Dorn, M. and Methot, R.D. 1989. Status of the Pacific whiting resource in 1989 and recommendations to management in 1990. In *Status of the Pacific coast groundfish fishery through 1989 and recommended acceptable biological catches for 1990.* Pacific Fishery Management Council, Portland, Oregon.

Ehrhardt, N.M. and Die, D.J. 1988. Size-structured yield-per-recruit simulation for the Florida gill-net fishery for Spanish mackerel. Trans. Am. Fish. Society 117(6): 591-599.

Fournier, D.A. and Archibald, C.P. 1982. A general theory for analyzing catch at age data. Can. J. Fish. Aquat. Sci. 39: 1195-1207.

Fournier, D.A. and Breen, P.A. 1983. Estimated abalone mortality rates with growth analyses. Trans. Am. Fish. Soc. 112: 403-411.

Fournier, D.A. and Doonan, I.J. 1987. A length-based stock assessment method utilizing a generalized delay-difference model. Can. J. Fish. Aquat. Sci. 44: 422-437.

Fournier, D.A., Silbert, J.A., Majkowski, J.M. and Hampton, J. 1990. MULTIFAN. Can. J. Fish. Aquat. Sci. 47: 301-317.

Fournier, D.A., Sibert, J.R. and Terceiro, M. 1991. Analysis of length frequency samples with relative abundance data for the Gulf of Maine northern shrimp (*Pandalus borealis*) by the MULTIFAN method. Can. J. Fish. Aquat. Sci. 48: 591-598.

Gallucci, V.F. and Quinn, T.J. 1979. Reparameterizing, fitting and testing a simple growth model. Trans. Am. Fish. Soc. 108: 14-25.

Gallucci, V.F., Donnelly, R. and Burr, R. 1992. A management model for a multi-species multi-mesh-size corvina fishery in the Gulf of Nicoya, Costa Rica. Presented at the Management Strategies for Exploited Fish Populations Symposium in 1992. Anchorage, AL.

Gulland, J.A. 1961. The estimation of the effect on catches of changes in gear selectivity. J. Cons. Perm. Int. Explor. Mer. 26(2): 204-214.

Gulland, J.A. 1963. Approximations to the selection Ogive, and their effect on the predicted yield. In *The Selectivity of Fishing Gear* Vol. 2. Proceedings of Joint ICNAF/ICES/FAO Special Scientific Meeting, Lisbon, 1957.

Gulland, J.A. 1965. Estimation of mortality rates. Annex to Arctic Fisheries Working Group Report, ICES, CM1965, Dec. 3 (Mimeo).

Gulland, J.A. 1983. *Fish Stock Assessment. A Manual of Basic Methods*. John Wiley and Sons, New York.

Gulland, J.A. and Boerema, L.K. 1973. Scientific advice on catch levels. Fishery Bulletin 71(2): 325-335.

Hamley, J.M. 1975. Review of gillnet selectivity. J. Fish. Res. Board Can. 32: 1943-1969.

Hampton, J. and Majkowski, J. 1987. An examination of the reliability of the ELEFAN computer programs for length-based stock assessment. In *Length based methods in fisheries research*, International Center for Living Aquatic Resources Management. (Pauly, D. and Morgan, G. R., Eds.. Makati, Metro Manila, Philippines.

Hasselblad, V. 1966. Estimation of parameters for a mixture of normal distributions. Technometrics 8: 431-444.

Hertzberg, R.C. and Gallucci, V.F. 1980. First order stochastic chemical reactions and oscillations in the variance. J. Appl. Prob. 17:1087-1093.

Hightower, J.E. and Lenarz, W.H. 1989. Status of the widow rockfish fishery. In: Status of the Pacific coast groundfish fishery through 1989 and recommended acceptable biological catches for 1990. Pacific Fishery Management Council, Portland, Oregon.

Hoenig, J.M. 1983. Estimating total mortality rate from longevity data. Ph.D. dissertation. University of Rhode Island, Kingston.

Hoenig, J.M. and Lawing, W.D. 1983. Using the *K* oldest ages in a sample to estimate the total mortality rate. I.C.E.S. CM. 1983/D. 22.

Hollowed, A.B., Methot, R.D. and Dorn, M. 1988. Status of the Pacific whiting resource in 1988 and recommendations to management in 1989. In *Status of the Pacific coast groundfish fishery through 1988 and recommended acceptable biological catches for 1989*. Pacific Fishery Management Council, Portland, OR.

Isaac, V.J. 1990. The accuracy of some length-based methods for fish population studies. ICLARM Tech. Rep. 27.

Jones, R. 1957. A much simplified version of the fish yield equation. Doc. No. P.21, presented at the Lisbon joint meeting of Int. Comm. Northwest Atl. Fish., Int. Counc. Explor. Sea, and Food Agric. Organ., United Nations.

Jones, R. 1979. An analysis of a Nephrops stock using length composition data. Rapp. P.-V. Reun. Cons. Int. Explor. Mer. 175: 259-269.

Jones, R. 1984. Assessing the effects of changes in exploitation pattern using length composition data (with notes on VPA and cohort analysis). FAO Fish. Tech. Pap. No. 256.

Jones, R. and van Zalinge, N.S. 1981. Estimates of mortality rate and population size for shrimp in Kuwait waters. Kuwait Bull. Mar. Sci. 2: 273-288.

Kimura, D.K. 1989. Variability, tuning, and simulation for the Doubleday-Deriso catch-at-age model. Can. J. Fish. Aquat. Sci. 46: 941-949.

Kullback, S. 1959. *Information Theory and Statistics*. John Wiley and Sons, New York.

Lai, H.L. 1985. Evaluation and validation of age determination of sablefish, pollock, Pacific cod and yellowfin sole; optimum sampling design using age-length key; and implications of aging variability in pollock. Ph.D. thesis, University of Washington, Seattle, WA.

Lai, H.L. and Gallucci, V.F. 1987. Effect of variability on estimates for cohort parameters using length-cohort analysis with a guide to its use and misuse. Fish. Stock Assessment, Title XII, CRSP, Tech. Rept. 2., CQS, School of Fisheries, Univ. Washington, Seattle.

Lai, H. and Gallucci, V.F. 1988. Effects of parameter variability on length-cohort analysis. J. Cons. Int. Explor. Mer. 45: 82-92.

Lai, H.L. and Gunderson, D. 1987. Effects of aging errors on estimates of growth, mortality and yield per recruit for walleye pollock (*Theragra chalcogramma*). Fish. Res. 5: 287-302.

Lai H.L., Mug V.M. and Gallucci, V.F. 1993. Management strategies in the tropical Corvina Reina (*Cynoscion albus*) in a multi-mesh size gillnet artisanal fishery. In *Proceedings of the International Symposium on Management Strategies for Exploited Fish populations*. (Kruse, G., Eggers, D.M., Marasco, R.J., Pautzke, C. and Quinn, T.J., II, Eds.) University of Alaska Sea Grant College Program Report No. 93–02. University of Alaska, Fairbanks.

MacDonald, D.D.M. and Pitcher, T.J. 1979. Age groups from size-frequency data: a versatile and efficient method of analyzing distribution mixtures. J. Fish. Res. Board Can. 36: 987-1001.

Megrey, B.A., Hollowed, A.B. and Methot, R.D. 1989. Integrated analysis of Gulf of Alaska pollock catch-at-age and research survey data using two different stock assessment procedures. Symposium on Application of Stock Assessment Techniques to Gadio.

Methot, R.D. 1986. Synthetic estimates of historical abundance and mortality for northern anchovy (*Engraulis mordax*). Southwest Fisheries Center Administrative Report: LJ 86-29.

Methot, R.D. 1989. Synthetic estimates of historical abundance and mortality for northern anchovy. In *Mathematical analysis of fish stock dynamics: review and current applications*, (Edwards, E. and Megrey, B., Eds.) Am. Fish. Soc. Symposium 6: 66-82.

Methot, R.D. 1990. Synthesis model: an adaptable framework for analysis of diverse stock assessment data. Symposium on Application of Stock Assessment Techniques to Gadio. INPFC Bull. 50: 259-277.

Methot, R.D. and Hightower, J. 1988. Status of the Washington-Oregon-California sablefish stock in 1988. In *Status of the Pacific coast groundfish fishery through 1988 and recommended acceptable biological catches for 1989*. Pacific Fishery Management Council, Portland, OR.

Mug, M., Gallucci, V.F. and Lai, H.L. 1994. Age determination of Corvina Reina (*Cynoscion albus*) in the Gulf of Nicoya, Costa Rica, based on examination and analysis of hyaline zones, morphology and microstructure of otoliths. J. Fish. Biology. 45: 177-191

Murawski, S.A. 1984. Mixed species yield-per-recruit analyses accounting for technological interactions. Can. J. Fish. Aquat. Sci. 41: 897-916.

Pauly, D. 1983a. Some simple methods for the assessment of tropical fish stocks. FAO Fish. Tech. Pap. No. 234.

Pauly, D. 1983b. Length-converted catch curves: a powerful tool for fisheries research in the tropics (Part I). Fishbyte 1(2): 9-13.

Pauly, D. 1986. On improving operation and use of the ELEFAN Programs. Part II: Improving the estimation of L_∞. Fishbyte 4(1): 18-20.

Pauly, D. 1987. A review of the ELEFAN System for analysis of length-frequency data in fish and aquatic invertebrates. In *Length based methods in fisheries research*, International Center for Living Aquatic Resources Management. (Pauly, D. and Morgan, G. R., Eds.) Makati Metro Manila, Philippines.

Pauly, D and Morgan G.R. (Eds.) 1987. Length-based methods in fisheries research. ICLARM Conference Proceedings 13, International Center for Living Aquatic Resources Management, Manila, Philippines and Kuwait Institute for Scientific Research, Safat, Kuwait: 53-74.

Pearson, K. 1894. Contributions to the mathematical theory of evolution. Phil. Trans. Roy. Soc. London A 185: 71-110.

Petersen, C.G. 1892. Fiskenbiologeske forhold i Holbock Fjord, 1890-91. Beret. Dan. Biol. St. 1890(91), 1: 24-183.

Pikitch, E.K. 1987. Use of a mixed-species yield-per-recruit model to explore the consequences of various management policies for the Oregon flatfish fishery. Can. J. Fish. Aquat. Sci. 44: 349-359.

Pope, J.G. 1972. An investigation of the accuracy of Virtual Population Analysis using cohort analysis. Res. Bull. Int. Comm. NW Atlantic Fish. 9: 65-74.

Powell, D.G. 1976. Estimation of mortality and growth parameters from the length frequency of a catch. In Report on the ICES Special Meeting on Population Assessment of Shellfish Stocks. Charlottenlund. C.M.-K:37.

Powell, D.G. 1979. Estimation of mortality and growth parameters from the length frequency of a catch. Rapp. P.-V. Reun. Cons. Int. Explor. Mer. 175: 167-169.

Quinn, T.J., II. 1987. Standardization of catch-per-unit-effort for short-term trends in catchability. Natural Resources Modeling. 1(2): 274-296

Rao, C.R. 1973. *Linear Statistical Inference and Its Applications*, 2nd ed. John Wiley and Sons, New York.

Regier, H.A. and Robson, D.S. 1966. Selectivity of gill nets, especially to lake whitefish. J. Fish. Res. Board Can. 23: 423-454.

Ricker, W.E. 1975. Computation and interpretation of biological statistics of fish populations. Bull. Fish. Res. Board Can. 191.

Robson, D.S. and Chapman, D.G. 1961. Catch curves and mortality rates. Trans. Am. Fish. Soc. 90: 181-189.

Sainsbury, K. 1984. Optimal mesh size for tropical multi-species trawl fisheries. J. Cons. Int. Explor. Mer. 41: 129-139.

Schnute, J. 1987. A general fishery model for a size-structured fish population. Can. J. Fish. Aquat. Sci. 44: 924-940.

Schnute, J. and Fournier, D. 1980. A new approach to length-frequency analysis: growth structure. Can. J. Fish. Aquat. Sci. 37: 1337-1351.

Schnute, J., Richards, L.J. and Cass, A.J. 1989a. Fish growth: investigations based on a size structured model. Can. J. Fish. Aquat. Sci. 46: 730-742.

Schnute, J., Richards, L.J. and Cass, A.J. 1989b. Fish survival and recruitment: investigations based on a size structured model. Can. J. Fish. Aquat. Sci. 46: 742-769.

Seber, G.A.F. 1973. *The Estimation of Animal Abundance and Related Parameters.* Griffin, London.

Silvestre, G.T., Soriano, M. and Pauly, D. 1991. Sigmoid selection and the Beverton-Holt yield equation. Asian Fisheries Science. 4: 85-98.

Sims, S.E. 1982. The effect of unevenly distributed catches on stock-age estimates using virtual population analysis (cohort analysis). J. Cons. Int. Explor. Mer. 40(1): 47-52.

Sims, S.E. 1984. An analysis of the effect of errors in the natural mortality rate on stock-size estimates using virtual population analysis (cohort analysis). J. Cons. Int. Explor. Mer. 41: 149-153.

Sparre, P. 1987. Computer programs for fish stock assessment. Length-based fish stock assessment (LFSA) for Apple II computers. FAO Fish. Tech. pap. (101) Suppl. 2.

Sparre, P., Ursin, E. and Venema, S.C. 1989. Introduction to tropical fish stock assessment Part I - Manual. FAO Fishing Technical Paper No. 306.1. Rome, FAO.

Ssentongo, G.W. and Larkin, P.A. 1973. Some simple methods of estimating mortality rates of exploited fish populations. J. Fish. Res. Board Can. 30: 695-698.

Sullivan, P.J., Lai, H.L. and Gallucci, V.F. 1990. A catch-at-length analysis that incorporates a stochastic model of growth. Can. J. Fish. Aquat. Sci. 47: 184-198.

Tanaka, S. 1962. A method of analyzing a polymodal frequency distribution and its application to the length distribution of the porgy, *Taius lumifrons* (T. and S.) J. Fish. Res. Board Can. 19: 1143-1159.

Tyler, A.B., Beamish, R.J. and McFarlane, G.A. 1989. Implications of age determination errors to yield estimates. Can. Spec. Publ. Fish. Aquat. Sci. 108.

Venema, S.C., Christensen, J.M. and Pauly, D. 1986. Contributions to tropical fisheries biology. FAO Fisheries Report No. 389.

Walford, C.A. 1946. A new graphic method of describing the growth of animals. Biological Bulletin 90(2): 1941-1947.

Wetherall, J.A. 1986. A new method for estimating growth and mortality parameters from length frequency data. Fishbyte 4(1): 12-14.

Wetherall, J.A., Polovina, J.J. and Ralston, S. 1987. Estimating growth and mortality in steady-state fish stocks from length-frequency data. In *Length-Based Methods in Fisheries Research*, (Pauley, D. and Morgan, G.R., Eds.) ICLARM Conf. Proc. 13: 53-74.

CHAPTER 3

AGE DETERMINATION IN FISHERIES: METHODS AND APPLICATIONS TO STOCK ASSESSMENT

Han-Lin Lai [1], Vincent F. Gallucci [2], Donald R. Gunderson [2], and Robert F. Donnelly [2]

[1] **National Marine Fisheries Service, Woods Hole, Massachusetts,**
[2] **School of Fisheries, University of Washington, Seattle**

1. INTRODUCTION

Age determination is an area of biological investigation which draws concepts and technology from many fields including physics, chemistry, geology, mathematics and statistics, ecology, and physiology. Much of the technology used in the determination of age comes from the physical sciences. Modeling and sampling designs are obtained from mathematics and statistics. Ecology and physiology enter because growth is an integrative process bringing together environmental and genetic influences. Since growth is an individual process and because there is considerable variation between individuals, the final estimation of age is a statistical statement.

Estimation of an organism's age has a long history. Paleoecological applications using the growth records stored in trees, clams, corals, and barnacles were developed long ago. Biochemical and genetic-based methods in use for fishes were sometimes developed for other organisms. Techniques, such as counting annual marks on scales and otoliths were among the first methodologies used in fisheries, while otolith micro-structural analysis represents a recent innovation.

This chapter takes a broad-based critical perspective, ideal for those interested in a

1-56670-151-1/96/$0.00+$.50

discussion of the principal methods available for estimating age. It is particularly suitable for those who must integrate age determination into a stock assessment, management, pollution or related study. The details of a particular method are found in manuals on age determination such as: Bagenal (1974), Chilton and Beamish (1982), Summerfelt and Hall (1987), Gjøsaeter et al. (1984) and Morales-Nin (1992), the last two especially for tropical fish species.

The chapter has a focus on the estimation of age of fish from tropical environments and the use of these estimates for the stock assessment of relevant fisheries, many of which are artisanal or small-scale in nature. A long-term data-base of ages for exploited species is critical for the proper management of a fishery. Most of the mathematical models for stock assessment were developed for high latitude fisheries and incorporate growth models incremented in terms of age. Even if catch-at-age data are not used in a model, the parameters of a specific growth model (e.g., K and L_∞ in the von Bertalanffy model) might be used. The extraction of age information from fish in tropical environments is especially difficult, so stock assessment models for tropical fisheries, especially those in developing countries, are usually incremented in terms of a correlate of age, typically length. While length is not a desirable variable from a mathematical/statistical modeling perspective, it is an attractive measurement from the biological point of view because it is easily taken in the field. Chapter 2 in this book addresses in detail stock assessment models based on size rather than age.

Since the stock assessment of tropical fish has largely motivated this work, many of the examples are drawn from the tropics. However, much of the theory and application of age determination originated from high latitude fisheries so the subject of age determination can only be properly treated by drawing on techniques and examples from temperate and cold water fisheries.

The preferred method of obtaining age usually involves the counting of marks on hard tissues, called "direct ageing", to obtain "chronological age". These marks are sometimes referred to as rings and are usually best viewed in bony tissue. Difficulties encountered with stock assessment and management can sometimes be attributed to errors in direct ageing methods, rather than to the sampling or modeling techniques. Most ageing methods have an element of subjectivity and depend upon individual readers. Inconsistency and non-repeatability between age readings, even for a validated ageing method, can lead to considerable variability in the results. Such problems are addressed in a later section.

When age data are not available, statistical techniques have been used to separate mixtures of length-frequency distributions, sometimes in combination with a growth model, to estimate age. Three of the best known statistical techniques based on length-frequency analysis (LFA) are MIX (MacDonald and Pitcher, 1979), NORMSEP (Hasselblad, 1966) and MULTIFAN (Fournier et al., 1990). Methods such as raising fish or tagging experiments are sometimes used to build a growth model in which the size of the fish is dependent on age. Thus, once the size of an individual is known, age can be estimated. Since these methods are not based on growth marks from hard structures, they are called "indirect ageing methods". Both direct and indirect ageing methods are discussed in this chapter.

The following chapter outline displays at a glance the variety of methods available and the auxiliary factors that may need to be addressed as part of an age determination investigation.

1. Introduction
2. Bony Tissues and Macroscopic Ageing Methods
   - 2.1 Scales
   - 2.2 Fin Rays and Spines
   - 2.3 Operculum and Headbones
   - 2.4 Pectoral Girdle
   - 2.5 Vertebrae
   - 2.6 Otoliths
     - 2.6.1 Sampling
     - 2.6.2 Extraction and Preparation
     - 2.6.3 Examination
       - 2.6.3.a *Surface Reading*
       - 2.6.3.b *Break-and-Burn Method*
       - 2.6.3.c *Thin Section*
       - 2.6.3.d *Staining Method*
3. Criteria and Validation in Age Determination
   - 3.1 Criteria for Determining Age
   - 3.2 Validation and other Possible Ageing Methods
     - 3.2.1 Direct Validation
       - 3.2.1.a *Marking the Bony Tissues*
       - 3.2.1.b *Raising Fish in Captivity*
     - 3.2.2 Indirect Validations
       - 3.2.2.a *Marginal Increment Analysis*
       - 3.2.2.b *Relative Position of Rings*
       - 3.2.2.c *Comparison of Structures, Age*
       - 3.2.2.d *Environment, Life History*
       - 3.2.2.e *Otolith Microincrement Analysis*
       - 3.2.2.f *Growth and Tagging Fish*
       - 3.2.2.g *Growth from Raised Fish*
       - 3.2.2.h *Growth and Back-Calculation*
       - 3.2.2.i *Length Frequency Analysis*
       - 3.2.2.j *Morphological Analysis*
       - 3.2.2.k *Mercury Content*
       - 3.2.2.l *Amino Acid Racimization*
       - 3.2.2.m *DNA/RNA Ratio*
       - 3.2.2.n *Lipofuscin, Age Pigment Assay*
       - 3.2.2.o *Microprobe Analysis Technique*
       - 3.2.2.p *Microradiography*
       - 3.2.2.q *Radiometrics*
4. Otolith Microincrement Analysis
   - 4.1 Sampling, Preparation, Storage
   - 4.2 Preparation for Examination
   - 4.3 Microincrement Interpretation, Analysis
   - 4.4 Validation for Daily Increment
5. Age Determination and Growth
   - 5.1 The V. B. Growth Model
   - 5.2 Estimates of Length-at-Age
   - 5.3 Fitting the V. B. Growth Model
   - 5.4 Comparison of V.B. Growth Curves
     - 5.4.1 ω-Test
     - 5.4.2 Likelihood Ratio Test
     - 5.4.3 Hotelling's T^2-Test
6. Estimation of Age Composition and Associated Sampling Designs
   - 6.1 Age-Length Key
   - 6.2 Inverse Age-Length Key
7. The Precision of Age Determination
   - 7.1 Chi-Square Test
   - 7.4 Index of Variation and Regression Analysis
   - 7.2 Average Percent Error
   - 7.5 Percent Agreement by Log-Linear Model
   - 7.3 Coefficient of Variation
   - 7.6 ANOVA with Repeated Measurements
8. Effects of Ageing Errors on Stock Assessment
9. Age Determination of Two Tropical Corvinas

The outline shows that over 25% of the chapter concerns statistical and modeling methods. This reflects the high cost and the noisy data frequently associated with age determination studies.

2. BONY TISSUES AND MACROSCOPIC AGEING METHODS

Terminology is a difficult problem in age determination. The books by Prince and Pulos (1983) and Summerfelt and Hall (1987) make an effort to collect and standardize the commonly used terminology, but confusion still exists. Our suggestion is to follow the guidelines in the above-mentioned books and, whenever possible, use the relevant terminology just as we have done. We also follow the literature tradition when necessary to convey alternative concepts or ideas. We begin by considering the use of scales for determining age and go on to the other objects that are commonly used.

2.1 SCALES

The ontogeny, growth, morphology and histology of teleost scales are described by Masterman (1924), Graham (1928), Crichton (1935), Kato (1953) and Chugunova (1963). The outer surface of a scale is formed as a series of bony ridges composed of concentric circle-like rings, often called circuli, and alternate with valley-like depressions. Multiple circuli may be formed in either a fast or a slow growing season. For example, Werder (1984) studied the scales of juvenile *Brycon melanopterus* and found that the circulus formation revealed a 2-day rhythm.

It is often assumed that the space between two circuli is correlated to the growth rate in a season. A fully developed set of wide and narrow circuli constitutes the annual growth zone. The term "annulus" refers to the zone that consists of narrow circuli although other subjective features are also used as the criteria for identifying annuli. The scales of bony fishes have been generally classified into ctenoid and cycloid types (see examples in Figure 1) and are characterized by bony ridges (also called sclerites, ridges or circuli). The morphology of cycloid and ctenoid scales is different, as are the criteria for their ageing. We start with cycloid scales which are sculptured with concentric rings or circuli, composed of individual platelets, but the inner surface is smooth (Pentilla and Dery, 1988). Pentilla (1988) used a SEM (scanning electron micrograph) to show that haddock (*Melanogrammus aeglefinus*) scales have summer platelets with curved edges and winter platelets with straight edges. The "annulus" in a pollock scale is characterized by a series of two or three narrowly spaced circuli in the anterior quadrant of the scale (Figure 1), which is often referred to as the slow growth zone. These narrowly spaced circuli are discontinuous and become irregular at the anterio-lateral quadrant, while other circuli are bent and branched. In contrast, the so-called rapid growth zone is characterized by widely spaced circuli that can be traced completely around the whole scale.

Ctenoid scales have radii or transverse grooves in the anterior quadrant of the scale. The annulus on ctenoid scales is usually described as having circuli that are "crowdy", "irregular", "bent", "segmented" or "broken" in the anterior quadrant, cutting or crossing over in the anterio-lateral quadrant (Figure 1). These narrowly spaced circuli

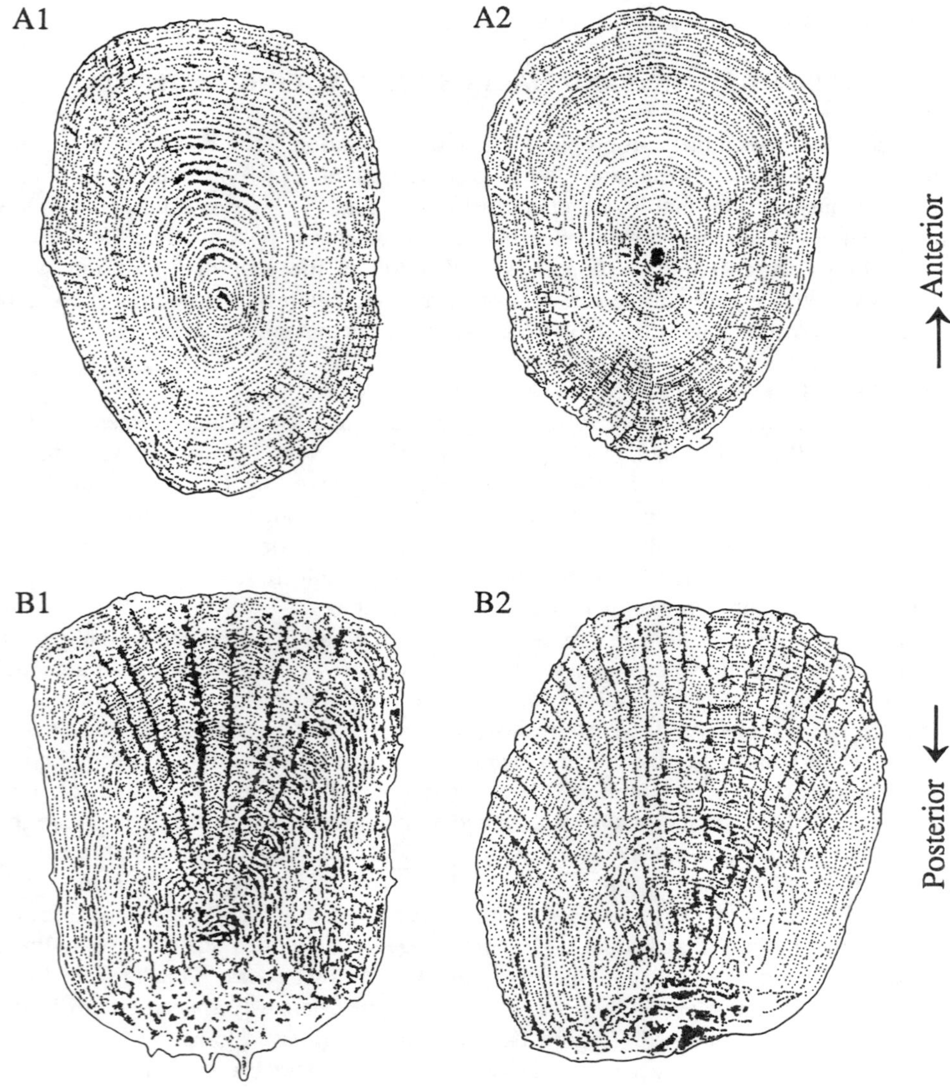

Figure 1. Clycloid and ctenoid scales from four species: A1 and A2 are cycloid scales from pacific cod and pollock, respectively; B1 and B2 are ctenoid scales from sablefish and yellowfin sole, respectively.

are segmented and bent and the radii in this same zone are branched. The cutting or crossing-over of circuli can be observed by tracing these irregular circuli in the anterio-lateral quadrant. The circulus pattern of yellowfin sole (*Pleuronectis aspera*) (Figure 2) shows the same criteria described above, except that the circuli become more irregular after the fish reaches maturity or at least when the growth rate has substantially slowed. This increased complexity with age is commonly encountered when the scales are used as the hard part for age determination.

The best location on a fish body to collect scales is determined by many factors. Some considerations for scale collection are: (1) the position below the anterior part of the first dorsal fin and above the lateral line (Chugunova, 1963) often has scales which are larger than those in other areas; (2) the positions where the scales have regular shape, growth rate and size, characteristics which are important to the back-calculation of fish size to determine age (Lander and Tanonaka, 1964; Tesch, 1968; Yeh et al., 1977); (3) the positions where the scales provide the best readability (Graham, 1928); (4) the positions with the least number of regenerated scales (Graham, 1928).

Yeh et al. (1977) and Bilton (1985) studied the variation in scales collected from different body positions and provided statistical criteria for selecting the best position to collect scales (Table 1). Regenerated scales are characterized by an obscure region in the scale center in which the first few annuli may not be visible (Figure 3). The process of regeneration is necessary to replace scales that have been lost due to environmental stress or mechanical damage. Regenerated scales may not truly represent the growth of the fish and should not be used for age determination.

Collected scales must be cleansed and preserved for later "reading". Tissue often adheres to scales and can be removed by soaking in boiling water or in a 3% to 5% potassium hydroxide solution for 1 to 2 hours, followed with a clean water rinse. The commonly used methods for scale preservation are: (1) air drying and placement in paper envelopes or booklets, (2) preservation in 50% alcohol, (3) retention in dilute formalin, or (4) freezing. It is important to use the preservation method that works the best for the species in question. For example, scales of Pacific cod (*Gadus macrocephalus*) become soft and fragile when preserved in diluted alcohol and thus must be preserved in other ways (Lai 1985).

Scales are either pressed between two glass slides or impressions are made. Impressions are made by placing the scales on a gummed card which is then covered by an acetate sheet. The two are then pressed together under pressure and heat (Koo, 1955). A blueprint of a scale impression machine can be found in Lai (1985). This technique allows for the processing of numerous scales simultaneously. Scales can then be read by viewing them with a microfiche projector or light microscope.

Many other techniques for preparing scales are also available. Staining can be a useful tool either in addition to, or separate from, the above techniques. Some staining techniques highlight the circuli; these include alizarin, ink, or DeFaures solution (Kubo and Yoshihara, 1968; Chilton, 1970). However, it seems that there is no standard staining method which is universally applicable. A polarized light microscope has also been used by many investigators (Taylor, 1916; Savage, 1919; Takemura, 1952; Miller 1955). Koo (1955) applied this technique, with some modification, to sockeye salmon

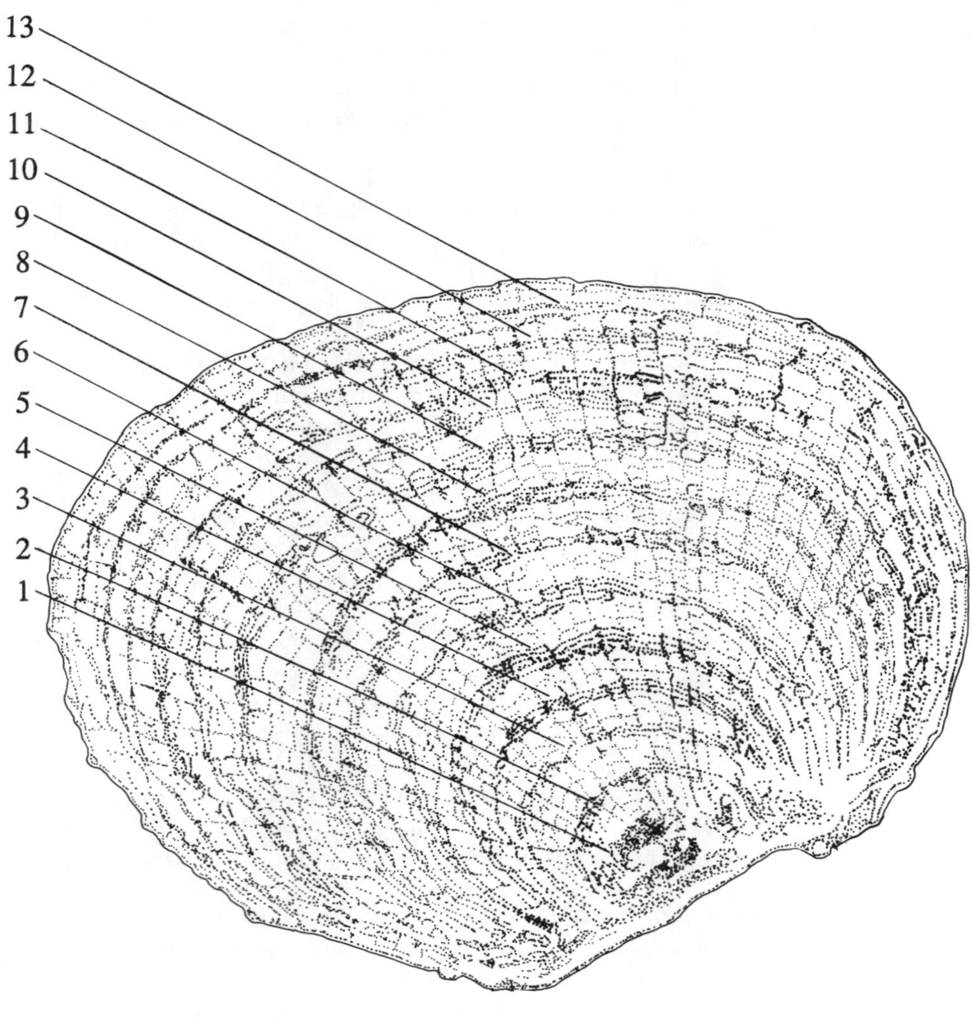

Figure 2. A ctenoid scale from a yellowfin sole. Annuli are indicated by numbers 1–13. In general, annuli become less distinct with age.

(*Oncorhynchus nerka*) where a graph is made from the topography of the scale taking into account distances and areas between circuli. The age of the fish is estimated by the peak of the curve.

Hirschhorn's digitizing technique is based on the fact that scales are formed in the very early life of a fish and that scales cover the entire body surface to protect fish from mechanical or biological injury (Small and Hirschhorn, 1987). Thus, the growth in area of scales reflects the growth of a fish body. A photograph of a scale is taken, the position of each identified annulus marked, the photograph digitized, all to measure the areas of the whole scale and the successive annuli. The length of the fish is assumed to be proportional to the square root of the scale area. The relationship between the square root of the area of successive annuli vs. age can be represented by the von Bertalanffy growth equation. Small and Hirschhorn (1987) demonstrated that this method can be used to detect false rings.

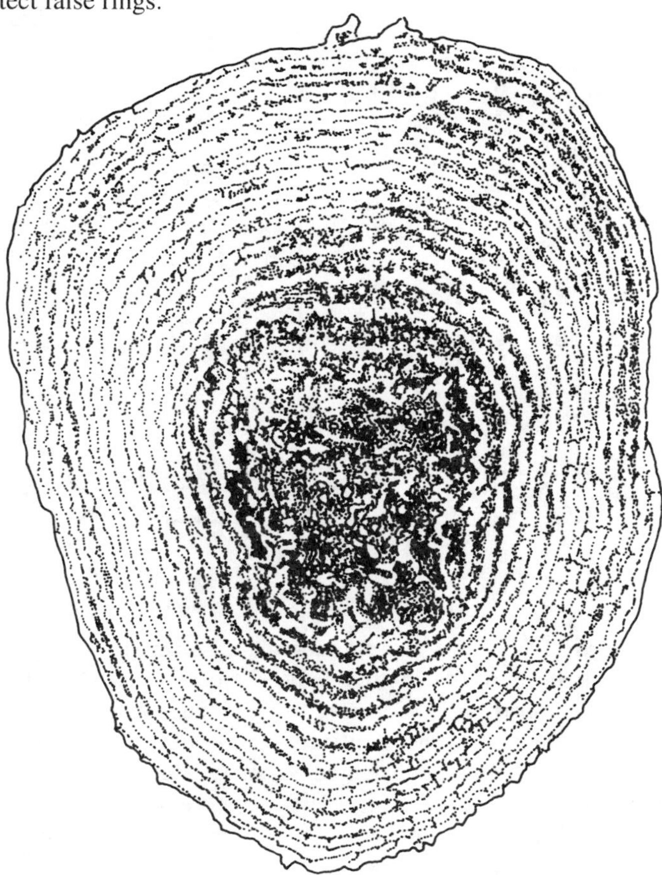

Figure 3. Regenerated cycloid scale from a pollock: note absence of discernible circuli in the center area.

Table 1. Comparison of coefficient of variation of scale-radii of scales removed from different portions of the body of the lizard fish. (from Yeh et al. 1977. Acta Ocean. Taiwanica. 7. With Permission.)

Fork Length (cm)	Item	Body Portions (see sketch below)					
		I	II	III	IV	V	VI
	Range (cm)	7.5-9.1	1.0-8.9	8.6-9.3	8.1-9.9	8.1-10.6	8.5-9.8
50.4	Mean (cm)	8.37	8.35	9.02	8.97	9.60	9.20
	C.V.	4.92	5.18	3.87	5.91	7.70	3.06
	Range (cm)	5.9-6.5	6.3-6.9	6.8-7.8	6.3-7.1	6.8-7.8	7.9-7.6
35.8	Mean (cm)	6.26	6.59	6.95	6.55	7.04	7.34
	C.V.	4.52	3.87	2.00	8.38	3.85	2.90
	Range (cm)	9.3-10.1	9.5-11.7	10.9-12.1	9.9-11.3	10.5-11.8	11.3-11.8
29.5	Mean (cm)	9.70	10.30	11.48	10.90	10.95	11.55
	C.V.	5.83	4.76	4.00	5.30	4.71	1.80

C.V.: Coefficient of variation.

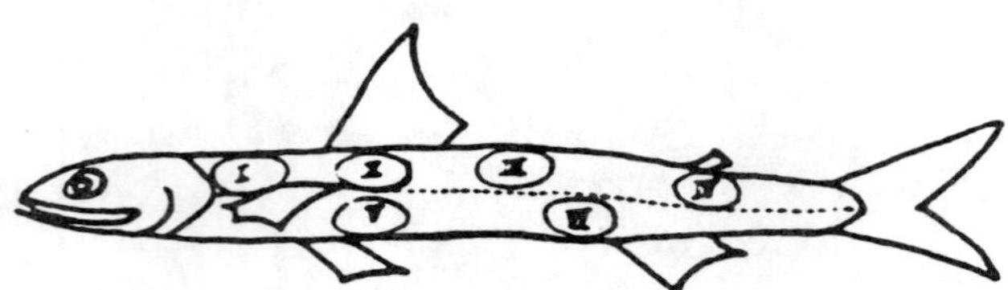

Mason (1973) tried to develop a semi-automatic machine for counting and measuring fish scale circuli. A transmitted light source was used with a rheostat control, a microscope equipped with a camera as an image detector, and a digital computer interface for analysis of the digitized photo image. van Utrencht and Schenkkan (1972) used a photometric machine to analyze age structures, including scales. A light source was used to scan the object and store the information digitally, which was then analyzed with a computer. A set of Guantimet image analyzers interfaced with a digital computer has also been used (Anon., 1980). The software automatically examines the fish scales by identifying and determining circuli spacing, determining the location of each annulus and the locations of the checks, and it identifies the edge type. Szedlmayer et al. (1991) developed a computer-based system to digitize the video image of a scale to store light intensity from a radial transect and to count circuli.

Scales are the basis for age determination of many species of fish and have many advantages that include ease of preparation, ease of preservation, ease of collection without sacrifice of the fish and ease of reading without sophisticated techniques. Together this means that scales are inexpensive structures from which age can be determined. In addition, scales are almost two dimensional in shape and increase with fish size, which is especially important for back-calculation methods (Ricker, 1969). In recent years, however, scientists have generally concluded that the scale method can substantially underestimate the age of fish, especially older fish (Beamish and McFarlane, 1987). Some authors have validated scale ages using the marginal increment technique and have then compared the results with estimates of age from back-calculations based on mean length at age from length-frequency analysis (see sections 3.2.2.g and h) (Sverlij and Arceredillo, 1991; Wassef, 1991; Lobon-Cervia et al., 1993). Readers should always be aware that age determination using scales, especially for long-lived species, may be biased.

Major increases in scale size occur prior to maturity. After maturity the growth rate of most fish is significantly reduced, especially in males, thus affecting estimates of the growth patterns when they are based on scale data (Lai, 1985). For those fish which do not have obvious growth or cease to increase in size, the annuli on the edge of scales are indistinct. For some species, especially tropical ones, none of the annuli may be visible. Lai (1985) shows that scale age is a consistent underestimator of fish age, after maturity, for some cold-water fishes. Therefore, the use of validation data is an especially important part of the use of the scales for ageing.

2.2 FIN RAYS AND SPINES

Fin rays and spines of some species of fish have been found to be useful in the determination of age. Holden and Meadows (1962) reported that the dorsal spines of the spiny dogfish (*Squalus acanthias*) are composed of several different layers (Figure 4). Beginning with the outermost, they are the enamel layer, pigment layer, outer dentine, middle dentine and inner dentine. The central cavity is also called the pulp cavity, and contains a cartilaginous rod that intrudes from the base of the fin ray. Holden and Meadows (1962) suggested that the inner dentine layer is formed by a series of overlapping concentric cones laid down by the odontoblast, which forms a layer lining the pulp cavity. The point of overlap is the enamel ridge that shows as the band of pigment on the external surface. Based on this theory, the most recently formed cone is seen at the base in a cross section of the innermost ring or in the longitudinal section of the ring. Ketchen (1975) used the second dorsal spine and the concentric cone theory to age spiny dogfish. He found a series of dark bands and ridges in the lateral view and considered each of these dark bands to be an annulus. Other investigators have also used this method to age spiny dogfish (Kaganovskaia, 1933; Chilton and Beamish, 1982; Soldat, 1982).

Beamish and McFarlane (1987) proposed another theory on the dorsal spine growth of spiny dogfish and subsequently validated the theory with oxytetracycline (OTC)

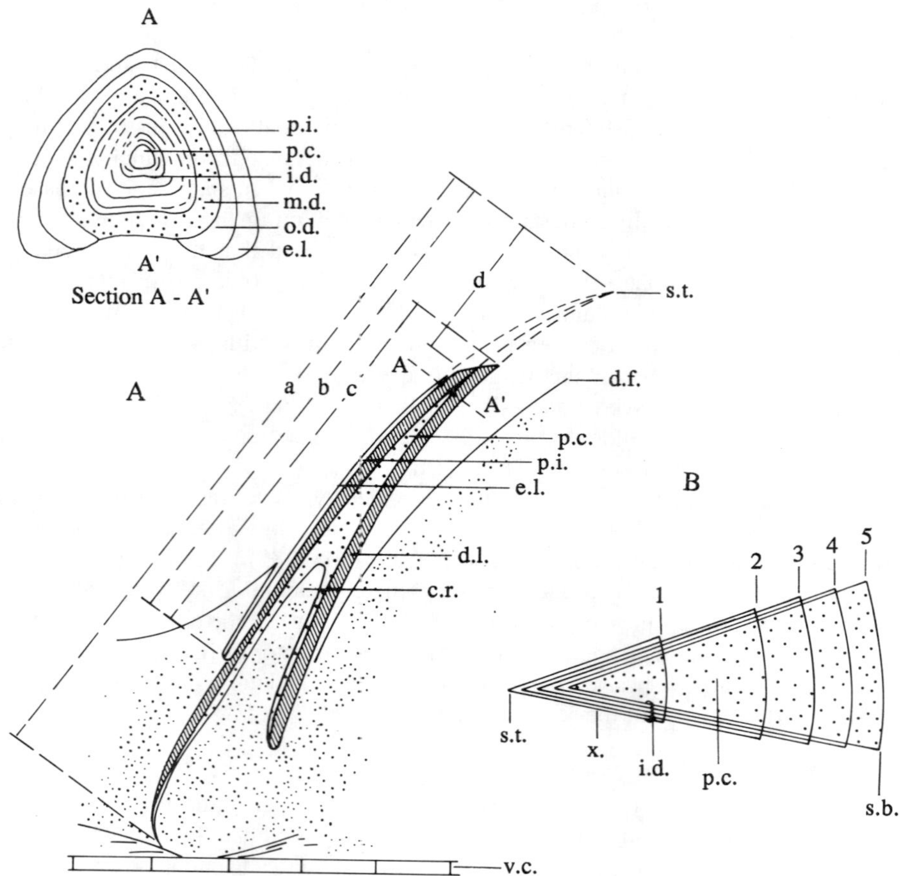

Figure 4.a. Diagram of a dorsal spine of *S. acanthius*, longitudinal and cross section. *d.f.*, dorsal fin; *d.l.*, dentine layer; *e.l.*, enamel layer; *i.d.*, inner dentine; *m.d.*, middle dentine; *c.r.*, cartilage rod; *o.d.*, outer dentine; *p.c.*, pulp cavity; *p.i.*, pigment layer; *s.t.*, spine tip; *v.c.*, nerve cord in vertebral column. Measurements: *a*, total spine length; *b*, corrected external spine length; *c*, uncorrected external spine length; *d*, distance from spine tip at which the pulp cavity becomes visible in longitudinal section.
Figure 4.b. Diagrammatic representation of inner dentine structure in the spine of a five year old fish. *i.d.*, inner dentine; *p.c.*, pulp cavity; *s.b.*, spine base; *s.t.*, spine tip; *x*, distance from spine tip at which ring number becomes maximally constant in serially cross-sectioned spines; 1, 2, 3, 4, 5, bases of successively formed cones, seen as rings in longitudinal section. (from Holden and Meadows. 1962. J. Mar. Biol. Assoc. U.K. 42. With permission of Cambridge University Press.)

marking. A longitudinal section of the spine shows three major structural components; cartilage interior, stem and mantle, arranged from inside to outside. The stem consists of three layers of dentine, inner, middle, and outer. The mantle can be scraped off with sandpaper revealing a growth pattern to the "cone bases" as described in Holden and Meadows (1962). Counts of the alternating growth zones from the OTC injected specimens showed that the growth zones were formed annually. Counts of annuli in the stem, although accurate in determining age, are not recommended because removal of the mantle is difficult and time consuming. Further, cross sections of dorsal spines reveal growth checks in the inner dentine, but counting them is difficult and not recommended (Beamish and McFarlane, 1987).

Beamish and McFarlane (1987) report that counting growth marks in the mantle of the spine is the most useful method of ageing spiny dogfish from British Columbia. The mantle is laid down on the stem dentine, originating at the enamel gland. The structure consists of an inner dentine layer, a midlayer of pigment, and an outer enamel layer. Pigment is produced by the melanophores at the inner surface of the enamel organ and is deposited between the enamel and the dentine. Dark pigmented ridges appear on the surface of the mantle and are used for age determination. Holden and Meadows (1962) and Ketchen (1975) defined the annulus to be a darkened band. Beamish and McFarlane (1987) refined this and defined the annulus as a darkened band or a ridge, or both, because the ridges are also darkened in most cases. Beamish and McFarlane (1987) confirmed the validity of this method and recommended that the mantle annuli be used for age determination of spiny dogfish.

In contrast to spiny dogfish dorsal spines, there are two major types of fin rays in bony fish: spines and soft rays. The spines consist of single rays, while the soft rays are segmented, often branched, and always bi-serial. Most teleosts have two elements in each fin ray and the most recent growth layer is located outermost on the spine (Chilton and Beamish 1982). Although the growth pattern of this type of fin ray has not been widely studied, Beamish and McFarlane (1987) validate the use of fin ray layers to age fish.

Dorsal, pectoral, pelvic and anal fin rays have all been used to determine fish ages (Kubo and Yoshihara, 1968; Beamish and Chilton, 1977; Beamish 1981). Storage and preparation of fin rays require more care than other bony tissues, such as scales and otoliths, because of the amount of tissue that adheres to these structures. Bacterial and/or fungal growth are the major concern. The fin rays must be cut off at the articulation base or information may be lost. Fin rays should be cleaned with water to remove mucus, air dried and stored in paper envelopes. It is important to keep fin rays straight during storage and not allow overlap since mechanical stress may cause damage resulting in difficulties when mounting for thin sections.

Fin rays are commonly mounted in resin and cut using an electric saw (Chilton and Beamish, 1982). A jeweler's saw is needed for small fish with fragile fin rays (Lai, 1985). Large fin rays can be cleaned, dried, cast in epoxy, and then later sectioned using a 6/0 or 7/0 blade with an industrial electric saw (Beamish, 1973). Sections, 0.4 to 0.8 mm in thickness, are the most useful for age determination (Beamish and Chilton, 1977). Sections thinner than this may not reveal the growth zones because they may become too transparent.

The methodology for using thin sections of spines involves mounting onto a glass slide and viewing microscopically with transmitted light. Beamish (1973) cemented the fin ray sections to a microscope slide with "Permount" and xylene mixed in equal amounts. Lai (1985) found that increased clarity and reduced glare was obtained when fin ray sections are immersed in water within a clear petri dish and viewed microscopically with transmitted light. It is necessary to cut sections of different thickness and at different distances from the base of the fin ray to connect annuli: usually three sections of about 0.5 mm in thickness are sufficient (Beamish 1973).

The uses of dorsal fin ray sections for age determination have been validated for several North Pacific species, e.g., Pacific cod (*Gadus macrocephalus*) and walleye pollock (*Theragra chalcogramma*) (Beamish and McFarlane, 1983; Lai, 1985; Lai et al., 1987). Chur et al. (1986) applied thin sections of dorsal fin rays to tropical skipjack *Katsuwonus pelamis*. Fin ray sections have also been used to determine the age of many other species, especially marine pelagic fish (Prince and Pulos 1983).

Fin rays should be used with caution because the core of spines or fin rays of older fish may undergo absorption and become vascularized, obscuring or even eliminating the first few growth zones (Casselman, 1983; Lai, 1985). In older fish, the distal translucent zones may be so close together that they appear to coalesce and are difficult to differentiate, resulting in an underestimate of fish age.

Annual growth zones of fin rays are identified by a narrow translucent (light) and a wide dark zone that correspond to slow winter growth and fast summer growth, respectively. Readers should be aware that different species have different appearances (Prince and Pulos, 1983). Fin rays of some North Pacific species have small, translucent, oblong or crescent-shaped centers (Beamish, 1973 and 1981).

Deelder and Williemse (1966, 1973) developed a method to use solid fin rays. The fin ray was cut transversely and the cut surface polished and air dried, attached via modeling clay to a slide, and placed upside-down under the microscope. A dark screen was put between the fin ray and a sidelight to control the quantity of incident light. Alternating dark and light bands could then be seen and counted if the screen was properly adjusted. Ketchen (1975) applied this method to age spiny dogfish.

A major advantage of using fin rays for age determination is that fish need not be sacrificed when a fin ray is removed. A fin ray can be taken at the time a fish is tagged and released (Mills and Beamish, 1980) and an additional fin ray can be taken when the fish is recaptured. An age is assigned to the fish which can then be verified by comparing the age determined from the fin ray to the time between tagging and recapture.

However, there are also disadvantages to using fin rays for age determination. Interpretation of the first annulus is difficult, especially when the section is taken at some distance from the articulation base (Chilton and Beamish, 1982). The outside annuli are distorted when an oblique cut is made (Chilton and Beamish, 1982) and when the distal end is peeled away. Fin rays are difficult to handle, prepare and preserve and are time-consuming to use.

2.3 OPERCULUM AND OTHER HEADBONES

Some head bones, especially those with a flat overall structure, have been used for age determination of fish. The operculum, the supra-occipital crest, the interopercle, the cleithrum and the bronchiostegals are examples (LeCren, 1947; Menon, 1950; Bardach, 1955; Palmen, 1956; Cooper, 1967; Fagade, 1974; Casselman, 1974, 1978, 1979; Lai, 1985; Eslava de Gonzalez, 1991; Babaluk et al., 1993; Banda, 1992). Because the methods of extraction, preservation, preparation and viewing of headbones are similar, the operculum will be used as an illustration. The operculum is removed with a scalpel by cutting the membranes between the preoperculum and the operculum and between the operculum and the branchiostegal rays. The adhering tissue can be removed by hand with hot water. Boiling water must be avoided because it may cause the structure to distort after it cools. For preservation, the operculum is air dried and kept in Kraft paper envelopes. Any stress or bending of the operculum during storage should be avoided because it may cause mechanical damage, adding to reading difficulties.

Annuli on the operculum and other headbones can be viewed directly or with the help of low power magnification. The broad layer appears dark and the narrow layer appears transparent under transmitted light. In contrast, the broad layer appears white-opaque and the narrow layer dark under reflected light with a dark background. Casselman (1983) indicated that the use of fluorescent light can enhance optical zonation as compared to incandescent light. Also, incident light with the bone viewed against a dark background is better than transmitted light. Because of the flatness of the structure, Casselman (1978) recommended that opercula be used for the determination of age and to obtain growth information. Lai (1985), however, encountered difficulties when using head bones for age determination. Annuli may be obscured because chips form at the inner side of these bones and form an array radially from center to margin. The disadvantages of using head bones include: (1) removal of any head bones sacrifices the fish, (2) false checks are frequently confused with, and difficult to differentiate from, annuli, and (3) preparation is labor intensive and time consuming (Lai 1985).

2.4 PECTORAL GIRDLES

Pectoral girdle bones used for age determination include the cleithrum, pectoral radii, scapula and coracoid (Figure 5). On the cleithrum, growth zones can be viewed on the surface of the lower and elongated section, although the thickness and spongy ridge-like radii obscure the appearance of the first few growth zones. Based on the criteria of Casselman (1974), yearly growth is characterized by wide concave zones separated by a sharp ridge.

Some scapula show clear growth patterns on the edge using a binocular stereomicroscope and transmitted light, but the central part of the scapula is obscured by a spongy radial ridge (Figure 5). The coracoid (Figure 5) is a dermal bone connected to the scapula and a triangular component adjacent to the scapula. This triangular component consists of two thin transparent bony layers with cartilage filling the space between. Under a binocular microscope and transmitted light, alternating dark (opaque) and light (translucent) growth zones are visible Lai et al. (1987).

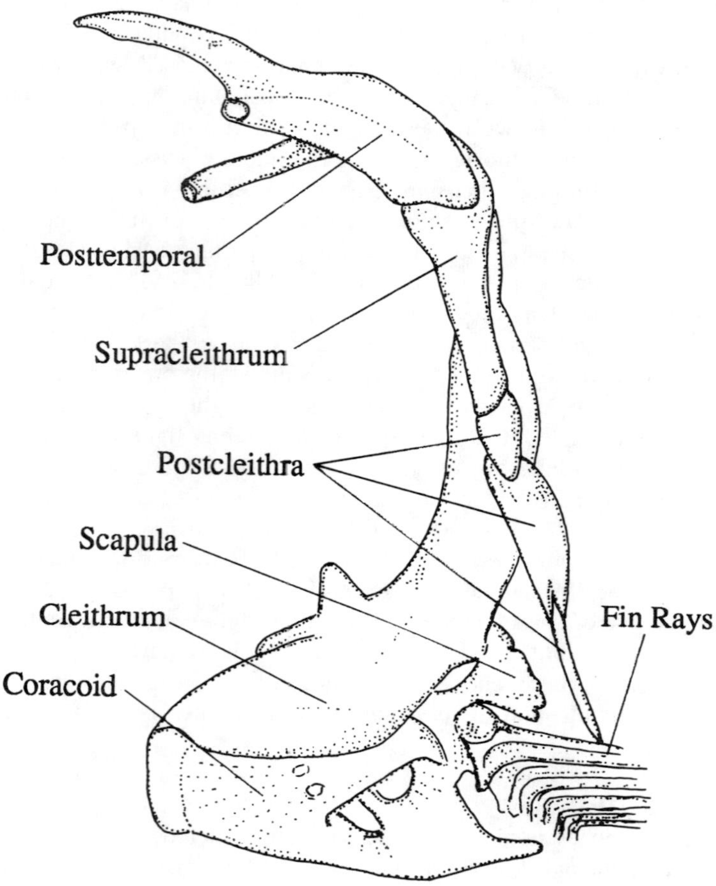

Figure 5. One half of the pectoral girdle, showing the relative locations of the coracoid, scapula, radii, cleithrum and fin rays. (from Bigelow. 1963. *Fishes of the Western North Atlantic*. With permission.)

As with otoliths, a major disadvantage of using pectoral girdle bones for age determination is the necessity of sacrificing the individual. Other problems exist, primarily with establishing criteria for the interpretation of rings and for maintaining these criteria over time.

2.5 VERTEBRAE

The centrum of vertebrae have been used for ageing elasmobranchs (sharks, rays and skates), tunas, and other demersal fishes (Lai and Liu, 1974, 1977; Stevens, 1975; Berry et al., 1977; Farber and Lee, 1981; Thorson and Lacy, 1982; Cailliet et al., 1983; Lee et al., 1983; Prince and Pulos, 1983; Ferreira and Vooren, 1991; Abdel-Aziz, 1992; Armstrong et al., 1992; Vassilopoulou and Papaconstantinou, 1992; Simpfendorfer, 1993). Vertebrae are collected by cutting through the flesh of the fish; a time-consuming process and impractical if sampling commercial landings. Therefore, only caudal or nearby vertebrae are commonly collected if these bones are necessary for age determination (Farber and Lee 1981).

Vertebrae are either cleaned immediately or kept frozen in plastic bags until cleaned. The accepted cleaning procedure involves placing the vertebrae in boiling water to remove the adhering tissue, soaking in 3% potassium hydroxide solution for final removal of oil and remaining tissue, rinsing in tap water and air drying (Lai and Liu, 1974, 1977). Annuli on the centrum of vertebrae appear as ridges and each ridge is associated with a transparent band. The collected measurements can be used for back-calculation and marginal increment determination.

Three methods have been developed for age determination using vertebrae. The whole vertebra can be embedded in wax with the articulative fossae mounted upward and then viewed through a magnifying glass where the articulative fossae can be examined (Chugunova 1963). Vertebrae can be sectioned, which involves sawing or grinding lengthwise in the dorso-ventral direction (Lai and Liu, 1974, 1977). Half of the vertebra is fixed in modeling clay with the concave centrum facing upward. Berry et al. (1977) and Abdel-Aziz (1992) reported that staining can enhance the viewing of annuli for better results. Details on staining solutions, techniques and interpretation of marks are in their articles. Berry et al. (1977) preferred reflected light from a blue-filtered, high-intensity, incandescent bulb for viewing stained vertebrae. When viewing the vertebrae, the centrum can be dried or covered with clear liquid (e.g., cedar oil, clove oil or water). Other staining solutions and techniques are discussed in Prince and Pulos (1983). Longitudinal thin sections of vertebrae can be obtained by grinding and polishing until they become transparent (van Utrencht and Schenkkan 1972). The section is mounted with CEDAX after dehydration in a alcohol and xylol mixture. A series of light and dark zones becomes apparent when viewed with transmitted light under a microscope. The dark zones are not uniform in density and length. Dark zones represent summer growth, while the light zones represent the winter growth. Electron microprobe and radiometric techniques can also be used on vertebrae (Jones and Geen 1977, Welden et al. 1987, Cailliet and Radtke 1987).

Although vertebrae have been used as an effective tool for age determination of elasmobranchs, tunas, and other pelagic fishes, there are disadvantages. The collection of vertebrae requires: (1) that the fish be sacrificed and (2) that fish be dissected. Berry et al. (1977) describe storage and preparation methods to account for the oily nature of vertebrae, which may otherwise obscure the identification of annuli.

2.6 OTOLITHS

In general, teleosts have three pairs of otoliths present in three optic sacs, one pair in each of the utriculus, the sacculus and the lagena (Figure 6) The pair in the utriculus are the lapillae, those in the sacculus are the sagittae, and the asteriscus are in the lagenae (Lagler et al. 1962). On occasion, one or more pairs of otoliths may be degenerated or absent: for example, the New Zealand elephant fish (*Callorhynchus milii*) lacks a lapillus but has well-developed cristae which serve the balance function (Gauldie et al. 1987) and butterfly tuna (*Gasterochisma melampus*) have lost the sagitta and asteriscus (Gauldie and Radtke 1990). To add confusion, other terminologies have been used to name the otolith pairs; see Secor et al. (1992; Table 3) for a short summary. Generally the sagittae are the largest pair of otoliths in fish and have been widely used for age determination. Therefore, when the term "otolith" is used, people are usually referring to the "sagittae".

Otoliths are composed of both inorganic and organic materials. The inorganic material, calcium carbonate ($CaCO_3$), is deposited on an organic matrix (Degens et al., 1969). The organic material in otoliths is thought to be primarily a collagen fiber matrix of protein, which is generally in the form of otolin. The majority of the calcium comes from the surrounding water via the branchial pathway (Simkiss. 1974; Mugiya et al.. 1981; Campana. 1983) with a little from food (Ichii and Mugiya, 1983). The chemical deposition rates of inorganic and organic materials reflect fast and slow growing seasons within a year and environmental and physiological changes during the life history of the fish (Kalish, 1989, 1991). The alternating deposition of inorganic and organic materials result in different appearances that depend upon the viewing methods, and are the basis for the identification of "annual rings" (Figure 7) (Christensen, 1964; Mina, 1968; Blacker, 1974; Williams and Bedford, 1974). Environmental or physiological disturbances occasionally occur which can change the deposition rates of different materials in otoliths, creating "sub-annual" rings or checks called "false checks".

The calcium carbonate ($CaCO_3$) is in the form of calcite, aragonite, vaterite or calcium carbonate monohydrate (Fitch, 1951; Kitano et al., 1980; Mulligan and Gauldie, 1989). These different crystal forms of $CaCO_3$ are found in different portions of the otolith. Different portions of the otolith may have different metabolic activities and pH-values related to the microscopic differences within the sac containing the otolith (Mulligan and Gauldie, 1989; Gauldie, 1990; Gauldie and Nelson, 1990). Furthermore, the differences in crystal growth, size and jointing are related to the ring locations (Davies et al., 1988; Gauldie, 1990). Studies on deposition rates of organic materials and $CaCO_3$ have been related to seasons, and especially to temperature (Fitch, 1951; Irie, 1960; Simkiss, 1974; Hickling, 1983; Radtke et al., 1985; Casselman, 1990; Gaudie et al., 1990). In general, $CaCO_3$ is thought to deposit at a faster rate during the

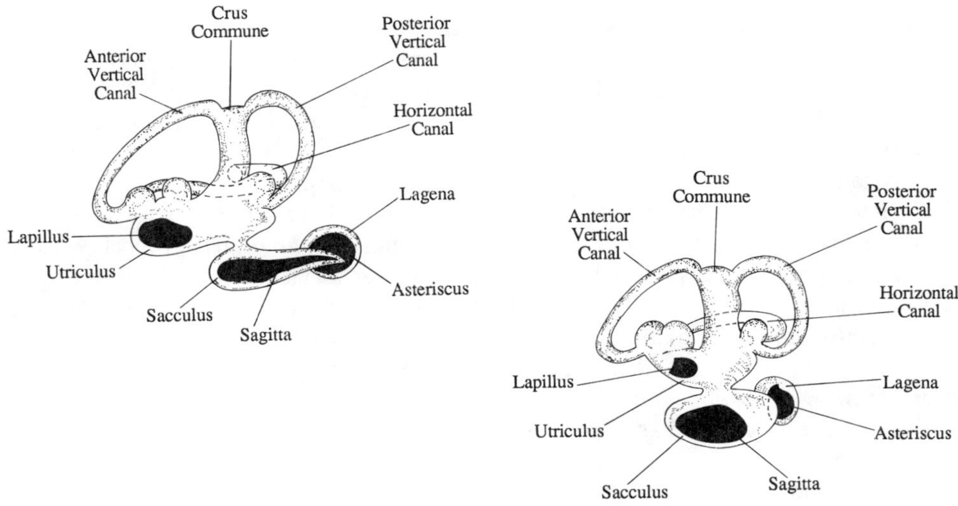

Figure 6. Location of otoliths in the inner ear of teleosts: cyprinoid and salmonid types. (from Lagler et al. 1962. *Ichthyology*. John Wiley & Sons, New York. With permission.)

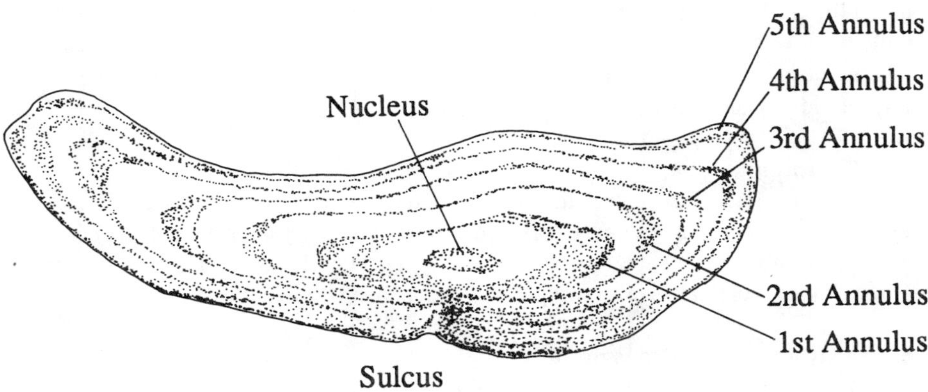

Figure 7. Cross section of the otolith of a pollock showing annuli.growing season when the calcium metabolic rate is high. Stress-induced hyaline-edge effects have been found in otoliths of juvenile marine fishes (Campana, 1983).

growing season when the calcium metabolic rate is high. Stress-induced hyaline-edge effects have been found in otoliths of juveniles marine fishes (Campana, 1983).

In addition to the calcium and protein ratios, Radtke et al. (1990) and Townsend et al. (1992) found that the strontium/calcium ratio in the aragonite crystal form is related to environmental conditions during the time of growth and applied this knowledge to determination of the age. Welleman (1990) has summarized the recent literature on biochemical, physiological and environmental control of otolith growth with a substantial accompanying list of references. The chemistry of otoliths is probably one of the most important areas of study for the development of new ageing methodologies. Radiometric analysis, electron probe and, in some extensions, microstructural analysis using electron microscopes are examples of new techniques.

The otolith microincrement technique has undergone a spurt of development since Pannella's (1971) work. The applications have reached beyond the field of age determination. The technique has proved useful in the study of environmental stresses that affect the physiology of fish in different life history stages. This subject is discussed further in section 4,"Otolith Microincrement Analysis".

A major disadvantage of using otoliths for age determination is the need to sacrifice the individual to obtain otoliths. Other problems include establishing criteria for the interpretation of rings and the maintenance of these criteria over time.

The role of age determination in fishery science is not restricted to the estimation of the growth parameters of the von Bertalanffy growth model. However, the importance of age data in fishery management should not be underestimated since these data are a primary input into computer-intensive stock assessment models such as CAGEAN (Deriso et al., 1985) and the Stock Synthesis model (Methot, 1990). While other computer-intensive stock assessment models exist which do not require age data (e.g., CASA, Sullivan et al., 1990), the availability of age data fitted to a growth model is a major plus for accurate stock assessment. See Chapter 4 on stock assessment for more detail.

We now consider some factors central to the collection, preparation and interpretation of otoliths for age determination.

2.6.1 Sampling

Sampling design is central to any age determination precisely because age determination is labor intensive and expensive. That this is a major area of research today will become obvious in Section 6, devoted to age-length keys. For the moment, we assume that a statistically valid sample of fish has been collected.

2.6.2 Extraction and Preparation

Removing the otolith from the skull of a fish usually involves cutting the fish across the occipital portion at a right angle to the line associating the joint of the operculum and the lateral line. But there are many other methods to extract otoliths, one of which is to use a sharp knife to cut open the upper surface of the mouth cavity. Once obtained, the otoliths should be cleaned before storage.

Otolith preservation (storage) methods differ for different life history stages, species, otolith compositions and lengths of storage. Examples of different storage media include 50% ethanol, glycerin-water mixture (50/50), fresh water and thymol, seawater with thymol, glycerol, xylol, and 95% isopropanol with anisole. For long-term preservation, otoliths are usually kept refrigerated to prevent contamination with bacteria and fungi. Thymol or anisole is often added to the storage medium to prevent mold or bacterial infection. Some workers have kept air-dried otoliths in envelopes (Mosher, 1954; Kao and Liu, 1972) but this method may lead to opaqueness for some species (LaLanne, 1975).

Otolith dissolution in storage media can be a problem during preservation. For example, Lai (1985) found that the otoliths of yellowfin sole could not be stored in an alcohol solution or in any weakly acidic solution because the otoliths would dissolve over time. Thus, otoliths of yellowfin sole must be stored dry. Prior to reading, they are put into distilled water or 50% ethanol to increase the visibility of the annuli. After reading, the otoliths are washed in tap water and stored dry for later use. Although formalin has been used, with and without a marble chip or sodium carbonate buffer, it generally is not recommended because formalin causes protein fixation and, occasionally, otolith dissolution.

2.6.3 Examination of Otoliths

A binocular stereomicroscope is the standard equipment for reading otoliths. Other equipment includes electric diamond saw, fiber optic light, petri dishes, modeling clay, alcohol burner and brush, depending on the technique used. It is recommended that a detailed description of the equipment and methods be given for every study to avoid future confusion. Although scientists agree that the alternation of opaque and translucent zones can be interpreted as fish age (Williams and Bedford, 1974), differences in terminology are a problem (Prince and Pulos, 1983; Summerfelt and Hall, 1987), especially when inconsistent equipment and methods have been used.

Commonly used terminologies, such as opaque, white, cloudy, translucent, hyaline and dark zones (or bands, rings) are not adequate for defining ageing criteria. The confusion involved can be illustrated with the following example: Molander (1947), Dannevig (1956) and Irie (1960) suggested that organic materials are present in the "opaque" zone and aragonite crystals in the "hyaline" zone. However, Christensen (1964) and Blacker (1974) reported that optical "dark" parts, after burning, corresponded to "hyaline" zones because of the higher organic content. Confusion such as this has prompted many scientists to recognize the importance of standardization and redefinition of the growth zones present in otoliths. Casselman (1983) recommended the use of terms corresponding to the structural appearance or light properties. The contemporary definitions can be found in Prince and Pulos (1983) and Summerfelt and Hall (1987).

Another difficulty in age determination is caused by "false" or "check" rings (Jensen, 1965). False rings, or checks, may be due to changes in environmental and physiological conditions, e.g., spawning, food supply, environmental stress (Campana, 1983). Many researchers use the width of the "translucent" zone between two rings, or the irregularity and discontinuity of rings, as the criteria for identifying false rings. It should be noted that there are different growth rates in different portions of otoliths, which can be confused for rings. For example, Gauldie and Nelson (1990) reported that the sagittal otoliths of teleosts rest in a groove in the base of the otic cleft, resulting in restriction of growth along part of the ventral edge of the otolith. The rate of crystal deposition is low at the sulcus surface, rising to a maximum at the growing edge of the otolith. This finding suggests that growth due to crystalline deposition between the sulcus and otolith edge is maintained by a pH gradient resulting from high to low levels of metabolic activity. In addition, the reader's subjectivity may play a major role in identifying false rings from "annual" rings. Only a carefully designed validation study can solve the problems of false rings.

The literature contains references to a number of methods for the preparation and examination of otoliths. Brothers (1987) has a good summary of methods that have been widely used, but the actual selection of a method for a specific species usually involves some trial and error. In general, otolith preparation and examination involves one or more possible procedures. Otoliths can be viewed whole or in part. Readability is enhanced by grinding or cutting (then polishing), use of the so called break and burn method, staining or acid etching (normally with 1% HCl) parts of the exposed surface, or immersion in clearing agents. The following sections describe these and other techniques.

2.6.3.a *Surface Reading*

The so-called "surface reading" is the part of the whole otolith that is viewed. The otolith is immersed in water, glycerin, or 50% alcohol in a dark-background petri dish and is viewed with reflected light. The liquid acts to enhance light filtering and to reduce glare reflected from the otolith. The annulus appears dark when viewed under a dissecting microscope because the dark background is seen through the translucent zone. Some thick otoliths must be ground to allow better light transmission (Kao and Liu 1972, Mug 1993, Mug et al 1994). Vero et al. (1986) described a simple technique for surface grinding the eel sagittae. However, a concave otolith should be ground with care because the information at the edge of otolith may be lost after grinding (Mug 1993). Etching otoliths in an HCl solution to dissolve $CaCO_3$, or staining otoliths with dyes, (see below) can increase the contrast of growth zones.

The surface reading method has been criticized because of different growth rates in different parts of the otolith. Otoliths of many species do not grow evenly in all dimensions. Kimura et al. (1979) showed that otoliths of old yellowtail rockfish (*Sebastes flavidus*) do not show growth zones on the distal surface (external side related to the body of the fish); instead the growth zones are shown on the proximal surface (internal side). Many studies (e.g., Beamish 1979, Chilton and Beamish 1982, Lai 1985, Lai et al. 1987 and Withall and Wankowski 1988) have found that fish ages obtained from surface readings are consistently lower than those from sectioned otoliths. Williams and Bedford (1974) also reported difficulties in the identification of annual rings at the edge of otoliths of older fish because the growth rate is so diminished that rings cannot be distinguished. The surface reading technique may not be valid, especially for older fish, and thin section, break-and-burn, or another cross-section technique should be considered.

2.6.3.b *Break-and-Burn Method*

This technique was first developed by Christensen (1964). Recent authors have considered it to be more precise than the surface reading technique because transversely broken otoliths show growth zones on the proximal surface (Bennett et al., 1982; Chilton and Beamish, 1982; Beamish et al., 1983). The otolith is broken (by hand) or cut transversely (with a saw) across the nucleus or focus. The broken surface is burned with an alcohol lamp flame until the color becomes dark brown. After burning, the zones with higher protein concentration turn dark, which is used as the criterion for age determination. Care should be exercised because damage can be caused by inadequate heat (see Chilton and Beamish, 1982, for guidelines). The broken-and-burned otolith is then put in soft modeling clay. The burned surface is painted with cedar- or vegetable-oil and examined under a stereomicroscope with reflected light. Lai (1985) preferred to hold the otolith in soft modeling clay and immerse it in water. The advantage of this technique is that it provides homogeneous light absorbency and reduces the glare from a burned surface painted with oil.

Another way to deal with a broken otolith is to bake it (Dannevig, 1956; Pentilla, and Derry 1988). Otoliths are baked 3–6 minutes in a radiant heat oven at 275°C resulting in a caramel color. Visibility of the annuli is enhanced since the "hyaline (i.e., protein-rich)" zones turn brown in contrast to the white "opaque" (i.e., $CaCO_3$-rich), zones. The advantage of baking over burning is that the otoliths remain unchanged for several years whereas burned otoliths fade.

2.6.3.c *Thin Section.*

A low speed electric saw is used to cut a thin section through the center of an otolith mounted in epoxy, resin or wax (Nichy, 1977; Pentilla and Derry, 1988). The optimum thickness of the section should range from 0.4 to 0.5 mm. The thin sections are mounted on glass slides and viewed under a compound microscope with transmitted light. The annuli are identified by reference to surface readings. The orientation of the cut surface may be important because of the asymmetrical growth of an otolith (Gauldie and Nelson, 1990).

2.6.3.d *Staining Method*

Staining techniques have also been applied to otoliths to increase the contrast between growth zones. Albrechtsen (1968) used a dye prepared by adding 0.05 g of methyl violet B to 30 ml distilled water and 1 ml 39% HCl (hydrochloric acid) solution. The dye solution must be used within a few hours and is applied to the sectioned otolith with a soft brush. The solution is neutralized by the $CaCO_3$ in the otolith within 20–40, seconds and becomes a light violet color. Pannella (1980) indicated that the method is not satisfactory for tropical fish. Bouain and Siam (1988) developed a method to stain transversely sectioned otoliths. In their procedure, the cut surface is polished, immersed in acid fuchsin for 12–14 hours, drained, transferred to Amido-Schwartz 5% for 1–3 minutes, and dried with filter paper. They interpreted the darkly stained deep blue as the zone of growth cessation and the light pink part as the growth zone. They suggested that acid fuchsin enables acetic acid to dissolve the junction between the "winter" rings and the "summer" rings. Thus the organic material of winter zones would be stained by the Amido-Schwartz solution. Richter and McDermott (1990) found that the appearance of stained rings resembled the results from the burning technique, but found that darker staining provides better contrast. The success of a staining method is highly dependent upon the appropriateness for the biochemistry of the otolith.

3. CRITERIA AND VALIDATION IN AGE DETERMINATION

Age determination is essential to the proper management of a fishery, but it is a very detailed operation with many considerations, these include: (1) selection of structures for ageing, (2) establishing ageing criteria, (3) maintaining consistent criteria between and within age readers over time, and (4) validation of the ageing method. Although these problems may become tedious and annoying, they are important and worthy of repeated evaluation to assure the quality and usefulness of ageing data. We now discuss the problems in two categories: ageing criteria and validation. A quantitative discussion on maintaining consistent age criteria is given later.

3.1 CRITERIA FOR DETERMINING AGE

Age determination is important because it is fundamental to many areas of fisheries research as well as to fisheries management. However, the process called age reading (that is, the counting or reading of the annuli) is dependent on human vision and subjective judgment. Vision and judgment may be distorted under different circumstances, e.g., different microscopes, different light sources and treatment of age structures, and they may be complicated by the formation of "false checks" which are formed when fish are exposed to environmental and physiological stresses. Once ageing criteria are established, guesswork is removed and consistency becomes the rule. There are many procedures that can be applied to establish ageing criteria and to maintain quality control during subsequent readings.

If particular bony structures have not been defined *a priori* for estimating age, a sample of fish is collected to obtain different bony parts which might be used. Ideally, these samples should be taken from a wide range of fish sizes which, presumably, are related to age. Some effort may be needed to collect samples at specific life stages where special biological processes can cause the deposition of a false check on the bony tissues. The readings are begun with the smallest fish, ideally one year or less in age. Photos, written records and even specimens must be kept of the characteristics of each check that is identified as an annulus. This is important, especially for young ages, because such characteristics frequently remain over the life span of a fish. At this juncture, bony tissues which prove unsuitable can be eliminated from the samples. Also, records should be kept for specific characteristics of checks that may be worthy of future reader-training and workshops.

It is a good idea to carry out a comparative study using various bony tissues and ageing methods before deciding on an optimal ageing method. Caution should be exercised if there are systematic differences in age among the bony tissues. For example, researchers have often found that ages determined from scales are consistently younger than from other tissues such as otoliths and fin rays. Scales should be used only as a last resort in this case. If several bony tissues are possible candidates, then the choice depends on: (1) the technique to be applied and the readers' experience, (2) ease of collection and preparation of the tissue(s), (3) precision of the age readers and (4) equipment and budget considerations.

The relative position of each identified annulus can be measured to assure that the annuli are deposited at a regular basis. This is done by pooling all samples from fish of the same age and measuring the radii of the annuli and length of the centrum. The radii are then plotted against the length of the centrum showing the distributions of radii from young to older fish.

The next stage in developing criteria is personnel training and regular workshops to reconcile subjectivity and differences. Unfortunately, this stage is almost always neglected and may result in future problems. This is especially true if there is a high rate of turnover among age readers.

The final step is validation. Various methods of validation have been used, but researchers should only rely on a method that can provide validation for the entire life history of the fish. Several validation methods are summarized in the next section.

3.2 VALIDATION AND OTHER POSSIBLE AGEING METHODS

Validation is used to establish and maintain ageing criteria for annuli in the bony tissues of fish. Two issues must be solved: (1) the interpretation of major discernible checks (or rings) in the bony tissues as to whether they are cyclic, periodic or, most importantly, annual; and (2) the possibility of missing annuli due to environmental and physiological stresses, especially after sexual maturity is achieved. These issues have caused serious problems in the past (Beamish and McFarlane, 1983, 1987), leading to erroneous stock assessment and management decisions (Lai and Gunderson, 1987).

False checks are frequently found in the bony tissues within each year of life. Many of these checks may be due to preparation, storage, or equipment used in the laboratory. The false checks may bias the counts of annuli if environmental and physiological factors dominate the factors that produce annuli. However, Gauldie (1988) has found a case where checks on the otolith may be due to the shape of the otolith and are only secondarily related to the age and growth of fish. False checks are subject to variations that exist within and between individuals in the population.

Missing annuli can result from two causes: (1) otolith morphology and (2) environmental and physiological stress. Williams and Bedford (1974) reported that annuli at the edge of otoliths are hard to distinguish, or may not be formed, due to a slow rate of calcium deposition. Kimura et al. (1979) and Chilton and Beamish (1982) reported that age reading should only be done on otolith sections that reveal all growth parts. Chilton and Beamish (1982) and Beamish and McFarlane (1987) concluded that annuli may not be formed in fish scales after fish become sexually mature. Missing annuli in older fish underscore that validation should be carried out for the entire life-history of the fish species. Using an inappropriate validation method may be as damaging to the acquisition of good data as forgetting validation.

The methods of validation are diverse, ranging from the use of supplemental information, such as marginal increment and mode tracing on length-frequency distributions, to applying highly complex bioassay procedures. Many of these methods have also been used as independent methods of age determination for various organisms besides fishes. Brothers (1987) classified the validation methods into direct and indirect. Hales and Belk (1992) have applied several different validation techniques to otoliths under controlled circumstances in a reservoir.

3.2.1 Direct Validation

Direct validation is determined from the time lag between two observed events so that the periodicity of ring-formation can be confirmed. Direct validation is the most accurate type of validation; however, the methods should be applied to the entire life history to avoid extrapolation into unobserved life history stages.

3.2.1.a *Marking the Bony Tissues*

This method is widely used in different life stages and age structures (Kobayashi et al., 1964; Odense and Logan, 1974; Hettler, 1984; Schmitt, 1984; Tsukamoto, 1985; Villavicencio de Muck, 1989). The number of checks or annuli between a chemical tag and the margin is compared with the known elapsed time. Beamish et al. (1983) indicated that the number of otolith annuli found after oxytetracycline (OTC) marking in sablefish was consistent with the age readings of readers who were either told or not told the time of marking.

This method may be subject to the following criticisms: (a) the effect of OTC on the growth rate of fish can be significant; (b) the number of OTC marked and recaptured fish is usually too low to cover all age classes in the population; (c) the time-lag and effective duration of OTC marks is not clear; (d) the tagged fish must cover the whole life span of the species and recaptures must be spread over a sufficiently long time period and (e) adequate sample size of recovered marked fish, which has not been addressed in the literature. The criteria (a)–(c) are of major concern because of substantial efforts in laboratory experimentation. The toxicity of OTC was determined by Beamish and McFarlane (1983). Dosages over 0.57 mg/kg body weight were found to be fatal to sablefish. A dosage of 0.25 mg/kg was recommended for sablefish and spiny dogfish. The OTC markers in otoliths, scales and vertebrae are clearly visible. Other chemicals, such as acetozolamide (Brothers, 1987), strontium and calcein (Wilson et al., 1987) have also been used. However, Beckman and Wilson (1990) used calcein and reported that there were inconsistencies in the incorporation of fluorescent calcein marks in otoliths between laboratory-reared and wild fish.

Application of OTC injection or immersion has been extended to otolith microincrement analysis for larvae and juveniles. Application of OTC can be through feeding or intraperitoneal injection (Wild and Foreman, 1980). Mass marking techniques have been described in Hettler (1984) and Schmitt (1984).

3.2.1.b *Raising Fish in Captivity with Sequential Sacrifice*

Lai (1985) applied this method to validate the age determination of sablefish up to four years old. The young-of-the-year sablefish were captured in the wild and raised in captivity over 4 years. These artificially reared fish showed four annuli in their otoliths, determined by both surface and break-and-burn techniques. Although the characteristics of the first four annuli on the surface of whole otoliths are the same for both raised and wild specimens, Lai (1985) reported that the broken-and-burned otoliths of the artificially reared sablefish consisted of many checks, which may be due to feeding rhythm and changes in food items. Also, the growth rate of sablefish in captivity is higher than in the wild. The comparison of growth rates between captive and wild fish is not appropriate. Heidinger and Clodfelter (1987) also used the pond-reared known-age smallmouth bass (*Micropterus dolomieu*), striped bass (*Morone saxatilis*) and walleye (*Stizostedion vitreum*) to validate the scale ages of wild stocks. The best way of using this validation technique is sequential sacrifice over the entire life history. However, the difficulties and budget involved in raising fish are sometimes major concerns. Nonetheless, this method, incorporating OTC injection or stress manipulation, has been widely used in the validation of otolith microincrement analysis.

3.2.2 Indirect Validations

The indirect methods do not necessarily rely on ring counts; rather they depend on auxiliary information, such as the seasonal growth patterns of bony tissues, otolith shapes and growth rates, or the coincidence of growth patterns in bony tissues with environmental and physiological stresses to confirm the periodicity and synchronicity of ring-formation. Otolith growth can be in terms of otolith mass or otolith microchemistry. Application of highly technical methodologies and equipment to otolith microchemical analysis is an area of rapid development which will continue as technology advances. Although studies of microchemistry have been carried out primarily on otoliths, these methods can also be applied to other bony tissues such as scales, vertebrae and the cleithrum.

3.2.2.a *Marginal Increment Analysis*

Time of annulus formation can be determined by two methods. Estimation of the probability of the appearance of an annulus at the edge of a hard part (Ogata 1956, LaLanne 1975, Maeda 1982) and measurement of the relative marginal increment. An example for red snapper (*Lutjanus sanguineus*) from the Arafura Sea and the northwest shelf of Australia (Lai and Liu 1977) shows that measuring the centrum-radii (R) and the annulus (r_i) from sectioned vertebra, the marginal increment (M.I.) is estimated as M.I. $= (R - r_i)/(r_i - r_{i-1})$, where r_i and r_{i-1} are annulus-radii of the ultimate and penultimate annuli, respectively. The monthly frequency distribution and mean of M.I. are plotted in Figure 8. Because the smallest mean M.I. occurs in the period from October to December, the annulus formation time was determined to be in that period.

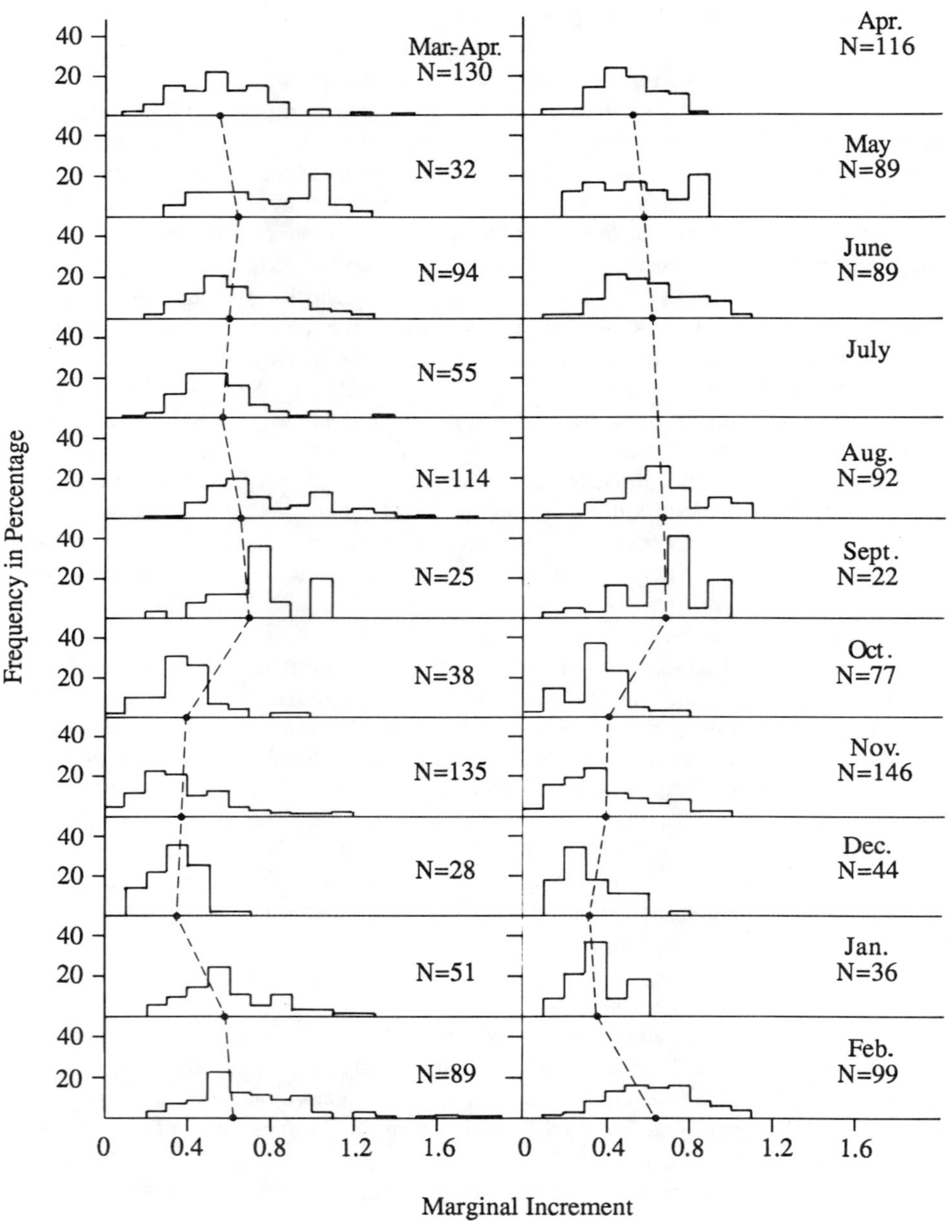

Figure 8. Monthly fluctuation of marginal increment from sectioned vertebrae of red snapper. (from Lai and Liu. 1979. Acta Ocean. Taiwanica. 6. With permission.)

Fowler (1990) used the time series of frequency distributions of $(R - r_i)$ to verify the time of annulus formation. A Kolmogorov-Smirnov non-parametric test was also used to compare the means of $(R - r_i)$ from three sampling occasions. Samuel et al. (1987) applied the marginal increment analysis to tropical fish from the Arabian Gulf.

3.2.2.b *Relative Position of Rings*

As in marginal increment analysis, consistency in the readings suggests patterns that make sense and are correct. Figure 9 shows regression lines of ring radius vs. scale radius for *n* 3-year old fish. The regression lines show clear separations (Yeh et al., 1977). Figure 10 shows another ring-radius method, where a frequency distribution is made of ring-radii measurements. The idea is similar to traditional mode separation methods, used in length-frequency analysis (Morales-Nin 1992).

3.2.2.c *Comparison of Different Structures and Age Readings*

In this method, readings from different structures and age readers are compared. Kimura et al. (1979), Lai (1985) and Lai et al. (1987) have all used this method for age validation. The comparisons are carried out by statistical hypothesis tests summarized later in this chapter. The method is based on the "modal" assumption of age readings; i.e., the true age is assigned with the highest probability (Richards et al. 1992). The validity of this assumption should be carefully evaluated since other validation procedures may be required if the assumption is unreasonable.

Baillon and Kulbicki (1988) used the otolith surface and cross-section methods to estimate the age of *Diagramma piatum.* The growth curves derived with these two methods are comparable. However, the analysis of life history traits (size at first reproduction) indicates that these methods are likely to underestimate the growth rate of the fish.

3.2.2.d *Correlation with Environmental and Life History Events*

Mustafa and Ansari (1983) found that the decline of feeding intensity in *Gudusia chapra* in tropical reservoirs of India during the first year of life and the spawning stress in the second and subsequent years resulted in the formation of what appeared to be annuli on the scales. Seasonal temperature change is commonly believed to be a major factor in annuli formation. Garcia-Arteaga and Reshetnikov (1985) found that tropical bar jack *Caranx ruber* formed two markers annually on the posterior hyoid bones. One marker formed during the period of February to May is the annual marker, while the second one formed in November is due to the feeding cycle. Prutko (1987) discussed some possible relationships between different growth zones on otoliths and environmental/physiological events in the life of a fish. Schramm (1989) estimated the marginal increments (translucent zones distal to opaque zones) formed at 22°C or higher in the otoliths of hatchery-spawned bluegills (*Lepomis macrochirus*). When bluegills are exposed to a cycle of warm water they form a second opaque zone and they form the marginal increment of an opaque zone when returned to cool water. Opaque zones can also be induced by changes in photoperiod and feeding rate.

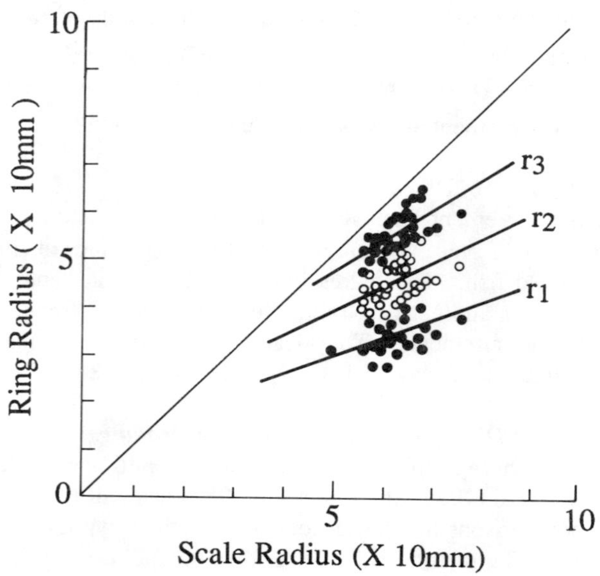

Figure 9. Relationships of ring-radius (r_i) vs. scale-radius (R) of 3-year-old lizard fish. (from Yeh et al. 1977. Acta Ocean. Taiwanica. 7. With permission.)

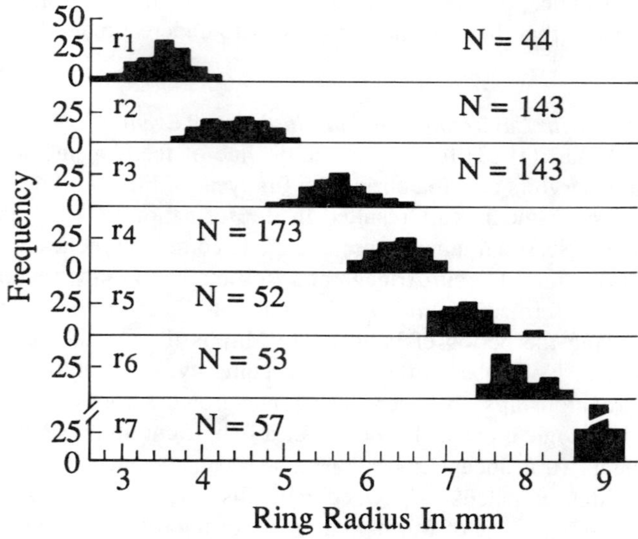

Figure 10. Ring-radius frequency distribution. N denotes the number of specimens. (from Yeh et al. 1977. Acta Ocean. Taiwanica. 7. With permission.)

3.2.2.e *Otolith Microincrement Analysis*

This method has been shown to be useful in estimating the age of larvae, juveniles and adults. Prince et al. (1991) applied this method to Atlantic blue marlin (*Makaira nigricaus*) ranging in age from 21 to 495 days (1.4 years) and found it could not be applied with confidence to individuals older than 1.4 years. Baillon and Kulbicki (1988) reported that otolith microincrement analysis can only be applied to those species with a small zone of readable otolith and not to species and/or individuals with large seasonal growth variation. Morales-Nin (1988) used otolith microstructure analysis to validate the annual growth rings in the tropical fish, *Lethrinus nebulosus,* and found there was consistency and close correlation between the mean length-at-age determined by otolith readings and back-calculation.

Applications of otolith microincrement analysis to age determination of adult fish has been developed by Ralston (1976, 1981, 1985), Ralston and Miyamoto (1981, 1983), Morales-Nin (1989), Morales-Nin and Ralston (1990), Mug (1993) and Mug et al. (1994). The general procedure for this method is given in Morales-Nin (1992). Mug et al. (1994) applied this method to validate the age of the tropical corvina reina (*Cynoscion albus*) in individuals up to 17 years of age. Figure 11 shows the micro-increments as seen under a scanning electron microscope (SEM). The preparation of an otolith for surface reading is time-consuming, regardless of the age of the fish. The effort required to read the age of older fish increases with fish age. Mug et al. (1994) found comparable age estimates from application of otolith microincrement measurements with the SEM and measurements from otolith surface readings.

The application of sophisticated statistical methods in otolith microincrement analysis for investigation of environmental response is applicable since microincrement analysis and time series analysis have been used to study the seasonal periodicity of ring formations correlated with the growth season of bivalve and gastropod shells and cephalopod statoliths (Adlerstein, 1987; Lønne and Gray, 1988). In the area of tree ring analysis, the climatic response model (Cook, 1985; Clendenen et al., 1978) and the application of the Kalman filter method in climatic response models (van Deusen, 1987; Visser and Molenaar, 1987) have been developed. Section 4 of this chapter is devoted to microincrement analysis.

3.2.2.f *Comparison of Growth Rate by Tagging/Marking Fish*

In this technique fish are collected in the wild for tagging or marking and then are released back into the wild. Fish are measured at both the time of initial capture and subsequent recapture. The estimated growth rate, between the times of release and recapture, is used to estimate the age of the tagged fish or to compare elapsed time with that estimated from age determination. There are several assumptions associated with this technique: (1) growth rate, mortality rate and activity of the fish are not influenced by tagging and marking, (2) recaptures do not take place a short time after release, and the release period is long enough to estimate the growth rate throughout the life span, and (3) the recaptures cover the whole life history of the fish. Problems encountered are: (1) it is frequently found that the length at recapture is smaller than that at release due to the stress of tagging or measurement error (Lai, 1985), and (2) some

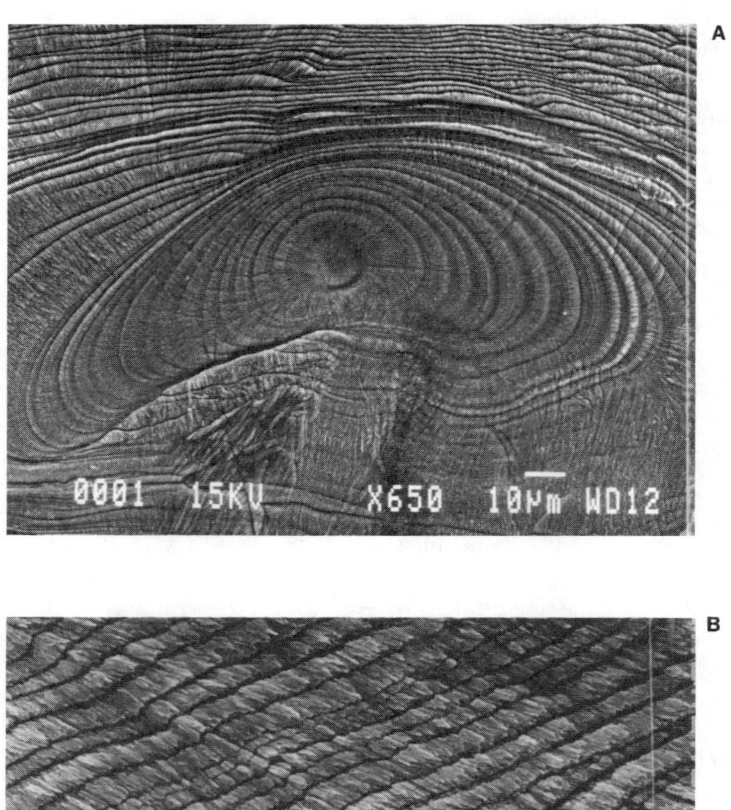

Figure 11. Scanning electronmicroscope photographs of the microstructure of the otolith of *Cynoscion albus*. (A) Nucleus: note the well-defined early microincrements and the irregularity of later depositions. (B) Microincrements about 3 mm from the nucleus: note the difference in the regularity of increment widths near and away from the nucleus, (from Mug et al., 1994. J. Fish. Biol. 45. With Permission.)

capture/recapture methods may introduce bias because of gear selectivity and site-selection.

The techniques related to tagging and marking are diverse. The selection of a method is important because it can affect the physiology of the fish. McFarlane et al. (1990) has provided further reviews of various tagging/marking methods.

Kirkwood (1983) proposed a maximum likelihood method to estimate the parameters of a von Bertalanffy growth curve using growth data from a tagging experiment associated with ageing data. He indicated that the use of age as a dependent variable in a nonlinear regression to estimate the von Bertalanffy growth curve is not appropriate. The growth curve and growth rate determined from the age-structure are compared to those derived from tagged fish. Examples of the use of tagging/marking methods on the studies of growth for tropical and subtropical fish include Olsen (1954), Randall (1961) and Joseph and Calkins (1969). Also see Francis (1990) for cautionary statements.

Frequently, the growth study from tagging and marking assumes that tagging or marking does not affect the growth of the fish and thus the estimates of the growth parameters are unbiased. This assumption is violated in many studies and Xiao (1994) developed a model that allows the effects of tagging on fish growth to be estimated.

3.2.2.g *Comparing Growth from Raised Fish*

Meyer (1878) was probably the first scientist who raised fish to estimate age and growth. A growth model is needed for the cultivated animals in which the length of fish is expressed as a function of age, i.e., age is the independent variable and length is the dependent variable. When the model is used to estimate age from the length of fish, the model must be rearranged such that length becomes the independent variable and age becomes the dependent variable. This statistical procedure is called calibration. Although many studies show that calibration based on linear regression is valid (Scheffe 1973, Williams 1969, Hunter and Lamboy 1981, Rosenberg and Pope 1987), the validity of calibration based on non-linear regression models is yet to be investigated.

The use of artificially raised fish in age determination can be questioned because culture conditions are almost always different from natural conditions. Also, techniques necessary for the long-term cultivation of fish are frequently difficult, especially for off-shore species, and the methods can be time-consuming and costly.

Some studies have introduced radioactive elements, e.g., carbon-14 and calcium-45, in foods and water of the pen-raised fish. The radioactive elements are incorporated into the bony tissues (Irie, 1960; Ottaway and Simkiss, 1977; Ottaway, 1978) and the deposition rate obtained from these reared fish is then compared to those from the wild.

3.2.2.h *Estimation of Age by the Method of Back-Calculation*

The method of back-calculation has been used to estimate lengths at earlier ages, i.e., to reconstruct the growth history of a fish. It has also been used to estimate the mean scale sizes for specific body sizes, as well as a variety of other relationships that may exist between an appropriate hard tissue and body size. For the purpose of validation, the back-calculated lengths at various ages are compared to the estimated mean length of age groups from a sample to investigate their discrepancies. If the discrepancy is greater than the growth from or to the adjacent ages, the ageing results would be doubtful.

The back-calculation method is based upon the assumption of a proportional relationship between a detectable growth dimension of a hard tissue (e.g., scale, otolith, etc.) and the body size of the fish. The growth dimensions of marks on the tissue, usually the radius of each annulus, are then measured. These two procedures imply that an empirical growth axis on the hard tissue should be determined and that all measurements are taken along this axis (see Figure 12.a).

A linear relationship is generally used for back-calculation; however, nonlinearity may occur. Linear back-calculation assumes that increments in fish lengths are directly proportional to increments between the annuli on the hard tissue (Figure 12.a). Many methods have been developed to accommodate this assumption, e.g., Lea (1910), Lee (1920), (also see Tesch, 1968; Jearld, 1983 for summaries). Francis (1990) formalized the methods for obtaining a back-calculation formula (BCF), as follows. Let L and S represent the fish body length and "scale" radius ("scale" is a representative hard part here), respectively; let L_c and S_c be the corresponding measurements at the time of capture; and let L_i and S_i be the corresponding measurements at the time of formation of the ith scale radius as $i = 1, 2, 3, \ldots, n$, from a sample of size n. A BCF uses the S_i and the S_c and L_c measurements to estimate corresponding L_i - values. For a given sample, a BCF is a family of curves. A particular curve depends upon the values of S_c and L_c. A back-calculation "line" is a linear BCF (two examples are in Figure 12.b), which is defined by the relationship assumed to exist between S and L.

The S - L relationship is developed by defining functions $f(L)$, the mean scale radius for a fish of length L, and $g(S)$, the mean body length for a fish with scale radius S. Assumption of a linear relationship between scale and body growth increments implies that $f(L)$ and $g(S)$ are represented by the two linear regression lines,

$$f(L) = \mathrm{a} + \mathrm{b}L \qquad \text{and} \qquad g(S) = \mathrm{c} + \mathrm{d}S \; ,$$

based on regressing S on L and L on S, respectively. Under Francis' scale proportional hypothesis (SPH), $f(L)$ becomes

$$f(L_i) = (S_i/S_c)f(L_c) \; , \qquad \text{(i)}$$

leading to a linear BCF, L_i. His body proportional hypothesis (BPH) also leads to a linear BCF,

$$L_i = [g(S_i)/g(S_c)]L_c \; . \qquad \text{(ii)}$$

Both BCFs follow directly from substitution of the appropriate $f(L)$ and $g(S)$

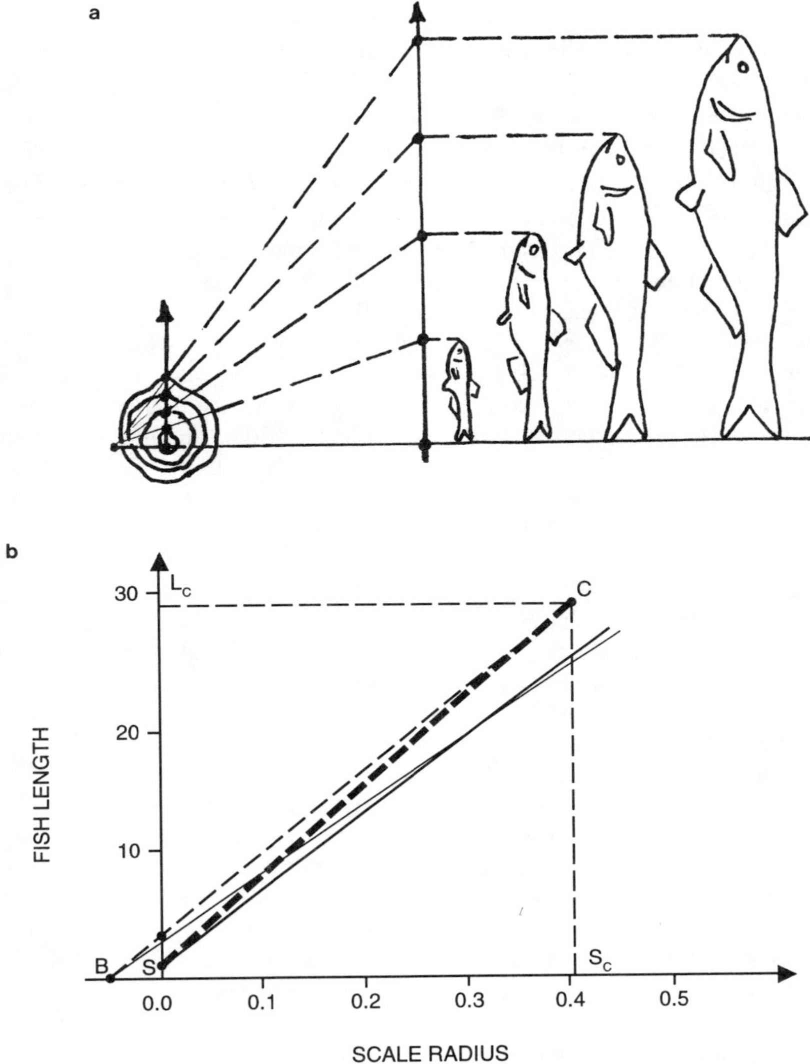

Figure 12.a. Schematic illustrating direct proportionality between scale increments and corresponding length increments. (from Tesch. 1971. In *Methods for the assessment of fish production in fresh waters*. With permission.)

Figure 12.b. The back-calculation method illustrating the differences between the SPH (bold lines) and BPH (fine lines). Continuous lines are regression lines: *S* on *L* (bold) and *L* on *S* (fine). Broken lines are corresponding back-calculation lines: the SPH (*c-s* = bold) and the BPH (*c-b* = fine) based on joining points of intersection of regression lines with axes and point *c*. Point $c = (S_c, L_c)$ represents the scale and length data for this fish at time of capture. (from Francis. 1990. J. Fish. Biol. 36. With permission.)

functions into (i) and (ii) with the constants a, b, c and d estimated from the linear regressions. The example in Figure 12.b shows back-calculation lines under the SPH and the BPH. The choice of hypothesis specifies whether the regression is S on L or L on S. Back-calculation lines, whether based on $f(L)$ (SPH) or $g(S)$ (BPH), must pass through the point (S_c, L_c). The theory underlying both hypotheses should be understood before proceeding with a back-calculation, since it is possible to assume one hypothesis and to incorrectly use the calculating formula for the other hypothesis.

As noted above, the relationship between fish length and the radius on the hard tissue may be non-linear (Lai and Liu, 1974; Lee et al., 1983; Wright et al., 1990; Chauvelon and Bach, 1993) or it may not be possible to find (Radtke, 1983; Wilson and Dean, 1983). In addition, Lee's phenomenon (Lee, 1920) refers to the tendency of back-calculated lengths at annulus formation to be smaller than observed. Possible causes include: (1) use of incorrect formulas, (2) non-random sampling, (3) selective natural mortality and (4) fishing mortalities due to gear selectivity (Tesch, 1968; Kubo and Yoshihara, 1968). Sneed (1950) describes an anti-Lee's phenomenon resulting in back-calculated lengths greater than observed in live fish.

3.2.2.i *Length Frequency Analysis* (*LFA*)

Mode-progression analysis has frequently been applied to the estimation of age and growth. The samples obtained for this technique should be from the same population. Other theoretical methods have been developed based on the fact that variation in fish length in each age group depends upon presumed theoretical probability density distributions, which are frequently normally distributed. Cassie's (1954) probability paper method, Hasselblad's (1966) NORMSEP, and Macdonald and Pitcher's (1979) MIX model and software are commonly used statistical methods for separating mixed probability distributions (Figure 13.a). The separation into age groups and the corresponding mean length-at-age obtained from these methods can be compared with that obtained from direct ageing methods (Lai et al. 1987). The age composition obtained from the age-structure (either by an age-length key or by random sampling) is constructed from several consecutive year classes. The year-class strength can then be related to a survey or to other biological evidence believed to produce a strong year class. Lai et al. (1987) applied this method to dismiss the scale method as invalid for the age determination of Pacific cod. Because of the difficulty and expense of age determination, estimating the age distribution and growth from length distributions has become very attractive. In particular when bony tissues, such as scales, do not reveal distinct patterns the LFA of length distributions may be the only recourse (Watson and Balon 1985). Length-frequency analysis may be broadly classified into two groups: statistical decomposition, e.g., MIX and mode progression methods, e.g., ELEFAN (Pauly and David 1981 and Sparre and Venema 1992). Figures 13.a and 13.b represent the basic ideas of each method. The former is more statistically intense, providing more statistical detail and the latter is more ad hoc, providing the von Bertalanffy parameters, L_∞ and K. Both have been used for tropical species, e.g., Olsen (1954), Sarojini (1957), Bennett (1961), Pantulu (1962), Longhurst (1965), Fryer and Iles (1972), LeGuen and

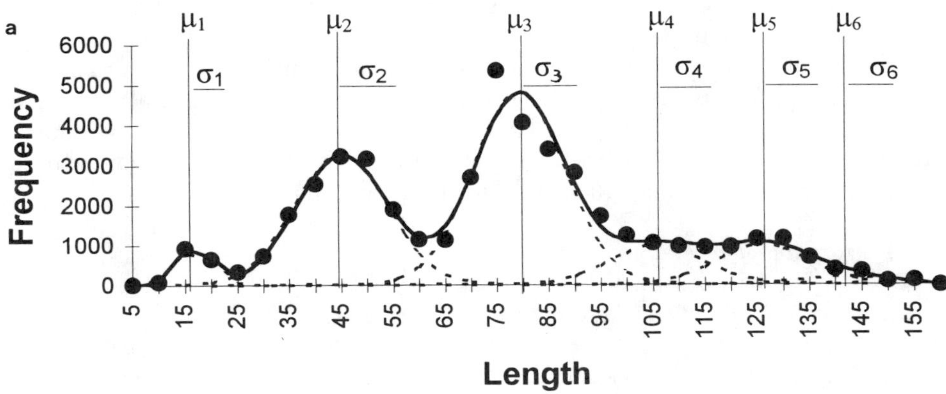

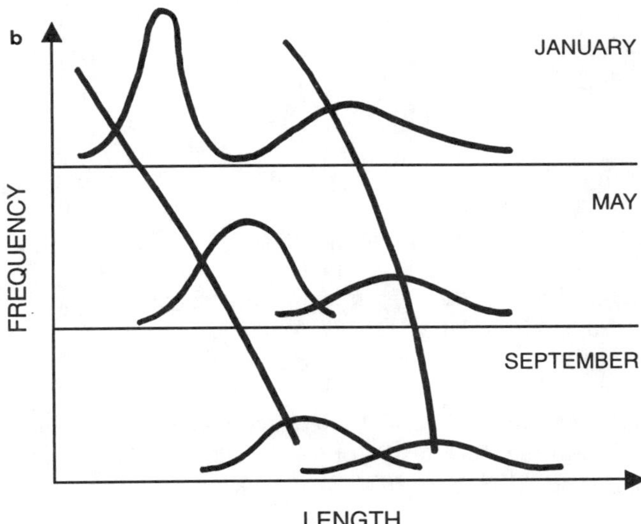

Figure 13. Modal decomposition: **a.** A schematic of a polymodal function showing modes around means (μ_i) and the associated standard deviations (σ_i). **b.** A schematic of the mode following method of fitting a growth curve based upon a time series of length frequency data. (from Sparre and Venema. 1992. FAO Fisheries Technical Paper 306/1. With permission.)

Sakagawa (1973), Stevens (1975) and Isaac (1990).

The disadvantages of LFA are: (1) the modes representing age groups in the length-frequency distribution may not be separable for older fish when they reach their approximate maximum size, (2) individual variation, especially between sexes and cohorts, may distort modes, (3) sampling biases, such as a small number of younger and older fish, gear and site selectivity, makes modal separation difficult, and (4) many fish, especially in the tropics, have a lunar or semi-lunar spawning and juvenile recruitment periodicity (Johannes 1978). Thus, non-annual peaks in recruitment may appear in the length-frequency distributions (Randall, 1961; Feddern, 1965).

3.2.2.j *Morphological Analysis*

Boehlert (1985) constructed a multivariate model with age as the dependent variable and otolith dried weight, otolith length, otolith width, the respective square and cubic terms, and the interaction terms of otolith weight/length and otolith length/width of males and females as independent variables. The differences between the ages from such multivariate models and ages from break-and-burn otolith ages were not significant. This kind of model depends upon having ages from any direct method and treating them as dependent variates. If the ageing method is not validated, the results can be ambiguous. The differences of otolith morphometrics due to temporal and geological variations may not support a universal application for this method. Lai (1985) found that otoliths of sablefish belonging to the 1977 year class have a special shape which was then quantified by discriminate analysis.

3.2.2.k *Mercury Content*

The accumulation of mercury and other heavy elements in the body of aquatic organisms is often noted. Forrester et al. (1972) reported that mercury content is as high as 1.96 ppm in dogfish from the Strait of Georgia. Ketchen (1975) assumed that if the average annual net uptake per fish is constant, the mercury content existing in dogfish could be used as an indicator of age. According to the relationship of mercury content and body length calculated by Forrester et al. (1972), Ketchen back-calculated the age from length and projected the relationship of age against mercury content. Unfortunately, the confidence limit of the fitted curve was too broad to allow the technique to be useful. Another drawback is that the mercury content in embryos and juveniles of spiny dogfish was not the result of being subjected to the mercury concentration in seawater. Thus, Ketchen (1975) assumed that the mercury burden came from the parents.

3.2.2.l *Amino Acid Racimization Dating*

The amino acid racimization dating technique was originally developed for ageing fossils such as bones, shells and wood (Bada and Schroeder, 1975; Schroeder and Bada, 1976; Williams and Smith, 1977; Masters and Bada, 1978). The method is based on the principle that all amino acids (except glycine) exist in two different isomeric forms dextro (D)- and levo (L)- rotary under polarized light. In the natural environment the two isomers are in equal abundance, but a living organism maintains a disequilibrium condition because the organism utilizes the L-form only. After an organism dies or becomes metabolically stable, the biochemical reactions to maintain disequilibrium of the two isomeric protein forms cease and the reaction called "racimization" begins. That is, L-amino acids are converted to D-amino acids until an equilibrium state has been reached and the determination of the D/L ratio is a possible way to age the animal.

3.2.2.m *RNA/DNA Ratio*

This technique is based on the ratio of RNA to DNA and has been used to assess the relative growth rate of organisms. The amount of DNA in any given species is essentially constant and unaffected by environmental differences or nutritional conditions (Lehninger, 1975). However, various forms of RNA change as the rates of protein synthesis fluctuate during growth (Buckley, 1984; Bulow, 1987). Buckley and Lough (1987) reported that the RNA/DNA ratio is a sensitive indicator of recent (2–3 days) food shortage. This fine time resolution is valuable in inferring short-term changes in prey availability. The application of this technique has been fruitful for larval and juvenile stages of fishes (Buckley, 1979; Buckley and Lough, 1987; Clemmesen, 1994; Malloy and Targett, 1994) and American lobster, *Homarus americanus* (Juinio et al., 1992). A review of this subject is given by Bulow (1987).

3.2.2.n *Lipofuscin and Age Pigment Assay*

Oguri (1986) applied the concentration of lipofuscin in internal cells to the age determination of anglerfish (*Lophius litulon*) and leopard shark (*Triakis scyllia*). Hill and Radtke (1987) used the measurement of lipofuscin and gerontological pigment in the brain tissue of damselfish (*Dascyllus albisella*) to define physiological age. From assays of the metabolically accumulated cellular pigment, Hill and Radtke (1987) found that the concentration of lipofuscin is directly correlated to otolith microincrement count.

3.2.2.o *Microprobe Analysis Technique*

Several microprobe-related techniques have been used in age determination of fish. These include: electron, proton and ion microprobes. Controversy exists over the validity of application of the electron microprobe (Edmonds et al., 1989; Kalish, 1989). The controversies are probably due to variation in sample preparation, equipment and experimental design (Gunn et al., 1992). Readers should refer to Radtke and Shafer (1992) for further details. The ion and proton microprobe techniques appear to be even more promising because they are more sensitive to rare elements such as strontium.

Townsend et al. (1989) and Radtke et al. (1990) applied microprobe techniques to otoliths of herring (*Clupea harengus*). Gauldie et al. (1986) applied the proton microprobe to otoliths of chinook salmon, Edmonds et al. (1989) used the electron microprobe to analyze trace element in otoliths of pink snapper (*Chrysophrys auratus*) for stock identification, Cailliet and Radtke (1987) used electron microprobe technique to determine calcium and phosphorus levels across the surface of thin-sectioned shark vertebra.

Elemental analysis using wave-length dispersive electron microprobe techniques on otoliths (Kalish, 1989, 1991; Radtke et al., 1990) has also been explored and Radtke et al. (1990) found that Strontium/Calcium concentration rations are correlated with water temperature for fish reared in captivity. Kalish (1989) found that the relationship was weak for Australian salmon *Arripis trutta* reared in a laboratory under different temperature conditions and found that there was a significant correlation between otolith chemistry and fish age that can be reasonably explained by growth rates and reproduction. Thus, Kalish (1989) suggested that seasonal and age-reared variation in otolith strontium content is largely due to changes in the proportions of free Ca and strontium present in blood plasma and that this in turn is a function of the quantity and type of proteins in plasma.

3.2.2.p *Microradiography*

This technique is closely related to electron and ion microprobe techniques. It is based on the concept that the absorption of X-rays by the atoms of a particular element is proportional to its atomic number (Brain, 1966). The rate of absorption depends on the types and relative proportions of the atoms present in the molecules. Therefore, different tissue components can be distinguished by differential absorption due to the composition and density of the elements.

Casselman (1974) used this method to examine the elemental composition in relation to the cleithrum zonation of pike (*Esox lucius*). The results related directly to differential calcification of the zones and confirmed that the translucent zone has a higher calcium content and is totally inorganic. Calcium and the total mineral content of both zones increases with age and a decreasing growth rate. Calcium content is relatively low in opaque zones and in the older and slower growing parts of the cleithra. Cailliet et al. (1983) applied the method to shark vertebral centra for age determination.

The method is subject to error due to the correlation of changes in the degree of mineralization (radiopacity) with changes in structure (optical density) or in stainability (Perrin and Myrick, 1980). Hohn (1980) used sections of 150 mm thickness for the dolphin, *Tursiops truncatus*, 600 mm thickness for the dugong, *Dugong dugon*, and 50 mm thickness for the seal, *Phocoena phocoema*.

3.2.2.q *Radiometrics*

The idea of using isotopes and radiometric analysis in the age determination of fish may come from fossil dating. Irie's (1960) study using ^{14}C is a classic application of this analysis. Other isotopes may also be useful for age determination. In addition to the classic papers by Irie (1955, 1957, 1960), recent studies on this subject are summarized in Radtke and Shafer (1992). Scientists in other fields have also applied the idea to other aquatic animals, for example, *Nautilus pompilus* (Cochran et al., 1981), hermatypic corals (Moore and Krisnaswami, 1972; Dodge and Thompson, 1974), bivalves (Turekian et al., 1975; Turekian and Cochran, 1981; Turekian et al., 1979), spider crab, *Maja squinato* (Le Foll et al., 1989) and European lobster *Homarus gammarus* (Le Foll et al., 1989).

Various isotope pairs, e.g., ^{228}Th / ^{228}Ra, ^{210}Po / ^{210}Pb, and ^{210}Pb / ^{226}Ra, have been used for the purpose of age determination. The use of these isotope pairs is based on the theory of radioactive decay of certain isotopes incorporated into otolith composition. A summary of the decay theory of radioactive elements, necessary assumptions and laboratory procedures to apply the technique is given in Kastelle et al. (1994), Smith et al. (1991) and Fenton and Short (1992).

Examples and special techniques using radiometric analysis of otoliths are: splitnose rockfish, *Sebastes diploproa* (Bennett et al., 1982), Atlantic redfish, *Sebaste mentella* (Campana et al., 1990), blue grenadier, *Macruronus novaezelandiae* (Fenton et al. 1990), orange roughy, *Hoplostethus atlanticus* (Fenton et al., 1991), and sablefish (Kastelle et al., 1994). Welden et al. (1987) applied the method to the vertebrae of elasmobranchs in California. The technique is applicable to species with life spans ranging from 2 to 120 years.

The assumptions of radiometric analysis that apply to age determination are: (1) the calcium deposition rate is constant over the specific stages of life history (e.g., pre- or post-mature), (2) the uptake rate of isotope pairs is constant and parallel to that of calcium, (3) the initial ratio of the isotope pair must be close to zero and (4) dissolution or resorption of calcium and daughter isotopes does not occur in otoliths. These assumptions imply that the otolith mass growth model should be assumed and estimated *a priori* and are very important when radiometric analysis is done with a whole otolith (Bennett et al., 1982; Fenton et al., 1991). Bennett et al. (1982) has shown that the calcium deposition rate is constant up to the maturity of splitnose rockfish. They found that the ages predicted from radiometric analysis of whole otoliths are in agreement with those measured by the otolith section method. Campana et al. (1990), Smith et al. (1991) and Kastelle et al. (1994) analyzed the daughter isotope pairs in the center of otoliths. Fenton and Short (1992) favored this approach because it reduces reliance on use of the otolith mass growth model and because the above assumptions are more easily attained.

4. OTOLITH MICROINCREMENT ANALYSIS

The use of otolith microincrements for age determination did not attract great attention until Pannella (1971) noted that there were approximately 360 lines between annuli of some fish otoliths. These lines were visible when viewed microscopically, thus they became known variously as "microincrements", "microstructures" or more often "daily increments". Pannella (1971) further noted that microincrements in adult tropical fish otoliths followed a 14-day cycle that was interpreted as relating to a lunar behavior pattern.

Brothers (1979) reemphasized the importance of microincrement analysis for tropical fishes where counting "annuli" in otoliths often does not work satisfactorily and where seasonal growth patterns must be studied. Gjøsaeter et al. (1984) and Morales-Nin (1992) have described microincrement analysis applied to tropical species, while Stevenson and Campana (1992) present an overview of the technique applied to fishes in general. Other reviews of microincrement investigations include Brothers (1979, 1982, 1987), Struhsaker and Uchiyama (1976), Pannella (1980), Campana and Neilson

(1985), Geffen (1987) and Neilson (1992). Microincrement analysis is considered to be highly technical and sensitive to sources of error at every stage of the process, from sampling to viewing. Neilson's (1992) paper should be consulted prior to any serious microincrement analysis work.

Applications of otolith microincrement analysis to bony fishes expanded during the 1980's and will likely continue to expand and be applied in other fields outside of fisheries, such as dendrochronology (see, e.g., Rhoads and Lutz, 1980; Cook, 1985; van Deusen, 1987; Visser and Molenaar, 1987).

4.1 SAMPLING, EXTRACTION, PREPARATION AND STORAGE OF OTOLITHS

The effects of gear selectivity on different phases of population dynamics and fishery management for adult fishes are well documented in the literature (e.g., Ralston 1982, 1990; Lokkeborg 1990). Fish larvae and juveniles collected with plankton nets are subject to the same selection biases (Somerton and Kobayashi, 1989). Failure to sample all segments of a population may result in biased estimates of growth rates and mortality (e.g., Lo et al., 1989; Morse 1989). Taggart and Leggett (1984) and Weinstein and Davis (1980) address gear efficiency for fish and larvae.

The fish sampled for microincrement studies must be properly preserved because the otoliths of larvae and juveniles may degrade or dissolve and the fish may shrink, eliminating accurate length estimates for comparisons to age. Butler (1992) summarized a list of existing preservative media and the percent of size shrinkage for different fishes. Radtke (1989) reported that immediate fixation in ethanol resulted in minimal shrinkage for larvae of Atlantic cod (*Gadus morhua*).

Methods for the extraction of otoliths from fish are diverse. Uchiyama et al. (1986) used a method that "squashed" the larvae by placing samples between a glass slide and a cover glass for direct observation. Secor et al. (1992) developed a series of otolith extracting methods for striped bass based on the size of the fish size and otolith location. Similar protocols can be developed for other species according to their characteristics such as fish size, otolith size and otolith location. Microscopic methods are critical for fish with tiny otoliths. Immersion oil, glycerin or xyline can help clear the hard parts of a fish skull and make otoliths easier to locate within the skull. However, specimens should not be extensively immersed in these oils because over-clearing and the loss of microincrements can occur (Struhsaker and Uchiyama, 1976).

Extracted otoliths must be completely cleaned prior to preservation and storage because adhering tissues may cause bacterial infection or reduce the resolution during examination. Otoliths may be immersed in 10% sodium hypochlorite or 3% potassium hydroxide solution for few minutes and the adhering tissues teased away with fine forceps and dissecting needles. The cleaned otoliths should also be rinsed thoroughly with tap water and either air dried or dried in an incubator before being put into a preservative. Because of the small size of some otoliths, care should be exercised during handling (Secor et al., 1992).

Otoliths from larger fish may be stored dry or in liquid preservatives such as ethanol, mineral oil or glycerin. Careful selection of the storage medium is essential and is dependent on the species. Struhsaker and Uchiyama (1976) note that long-term storage in immersion oil or glycerin can cause over-clearing, loss of microincrement resolution and degradation of otoliths. Brothers (1987) recommended that otoliths be extracted from freshly sampled fish and that fixation and storage be in 95% ethanol, but that freezing was the safest method. Any fixation solution containing formalin should be avoided because the pH value may change during the time of preservation, resulting in totally degraded otoliths of small fish. It is also important to check and adjust the pH value during the time of preservation.

Otoliths of larvae are very difficult to handle and are thus best preserved as slide-mounted specimens. Mounting media include: Permount, epoxy, polyester or casting resin, eupanal, acrylic adhesive, Canada balsam, thermoplastic glues and nail glues (Secor et al., 1992). Brothers (1987) also noted that small otoliths can break after the mounting medium sets up. Fine nylon monofilament or glass fibers may be used to avoid this problem.

4.2 PREPARATION FOR EXAMINATION

An extremely flat surface that facilitates viewing of the microincrements is the first objective. Small otoliths can sometimes be viewed directly under a dissecting microscope. Zhang et al. (1991) report on staining otoliths for microincrement examination. The size of large otoliths is a problem. Preparation procedures include: grinding, thin-sectioning and acetate-peel replicas. The location of the otolith core (center) should be frequently monitored during the course of preparation. Concave otoliths are especially difficult because the margin on the concave side is easily lost. Lost information near the core and/or the margin is obviously detrimental to microincrement analysis.

Thin-section and acetate-peel replica procedures are described in Pannella (1980), where it is noted that dorso-ventral transverse sections perpendicular to the growth surface are less desirable because they may show a discontinuous sequence of microincrements. Ashford et al. (1993) reported a method for the processing of large numbers of otoliths to be viewed with SEM and light microscopes. The method of preparation may have advantages over traditional ones since sample size could be greatly increased. However, the method ignores edge problems and a resulting section may miss the core of the otolith.

Grinding and polishing methods can be generally classified into two types. The first embeds the otolith within a mold to form a plastic block which is then mounted on a glass slide for grinding or polishing. The plastic blocks can be epoxy resin, casting resin or euparal. The hardness of the block is adjusted by changing the percent of hardening component used. The media for attaching the embedded otoliths to a glass slide was described in the last section. Air bubbles in the embedding and attaching media during hardening should be avoided. Placement of the embedded blocks and/or the mounted glass slides on a hot plate or incubator reduces air bubbles and facilitates hardening. Otoliths embedded in plastic blocks or mounted on glass slides are ready for long-term storage.

The second method is described by Brothers (1987) who prefers grinding and polishing by finger for small otoliths mounted on glass slides. This method has also been used by Wild and Foreman (1980), Wilson and Larkin (1980), and Campana and Neilson (1982). However, Karakiri and Westernhagen (1988) suggest that the method should be avoided to prevent edge over-grinding effects caused by forces exerted in a non parallel direction to the grinding plane. Methods of grinding or polishing otoliths mounted on glass slides include the use of stones, sandpaper, lapping wheels, polishing cloth, glass plates, and metallurgy grinding/polishing instruments. The grinding/polishing media are diverse: aluminum slurry, carborundum powder, carbide or aluminum oxide and diamond paste (see Secor et al., 1992, for additional details). Successively finer grit grades of sandpaper or grinding/polishing media are normally used. A drawback to finger grinding/polishing is that mounted otoliths can only be ground or polished from one side. Although solutions can be applied to remove mounted otoliths, remounting involves so many uncertainties that it is not recommended. Otoliths embedded in a plastic block can be ground and polished from both sides and can be sectioned in any desired direction. The block is attached to a glass slide with thermoplastic glue which can be softened with heat so the block can be turned over for polishing the other side. An Isomet saw can be used for initial thinning if a block is thick. The grinding and polishing procedures for otoliths embedded in blocks are the same as that of otoliths mounted on glass slides. Neilson and Geen (1981), Karakiri and Westernhagen (1988) and Volk et al. (1990) have developed apparatus which facilitate grinding and polishing and the simultaneous processing of a large number of otoliths. A note of caution: care should be exercised to align the otolith cores in one plane to avoid off-center grinding/polishing.

Otoliths are next etched by a dilute HCl or EDTA solution to improve the visibility of the increments. This is required if the otoliths are to be examined under the SEM. The acetate replica technique is frequently cited in the literature but Brothers (1987) has criticized this technique because the replicas may be poor reproductions of the surface structures. The ground surfaces should be at right angles to all the increments or some growth increments may not be carried over onto the replicas.

Brothers (1987) also compared the advantages of using the SEM and the light microscope. The light microscope is generally simpler and easier to use with greater tolerance for imperfections in the grinding procedures. The resolution of the light microscope may not be sufficient to distinguish microincrement widths of less than 1mm which leads to the use of the SEM. Preparation for SEM use involves sectioning, grinding, polishing, and coating the section with a 100–200 Å layer of gold/palladium alloy. Results are also affected by many variables independent of the sample, including tilt, accelerating voltage, and spot size. Morales-Nin (1988) compared light microscopy and SEM results from the otoliths of 14 tropical and subtropical fishes and found light microscopy to be inadequate for the detection of microincrements.

To some extent, sample preparation for SEM analysis is machine dependent so we do not go into detail here. SEM user guides are available for each machine and are usually quite detailed.

4.3 MICROINCREMENT INTERPRETATION AND ANALYSIS

Partially calcified primordia, exocysted by the inner ear cells, are thought to be the initial formation sites of otoliths. Geffen (1987) reported that the fusion of primordia may occur before hatching; Mugiya et al. (1981) and Mugiya (1984) found diel variation in calcium deposits; Degens et al. (1969) and Ross and Pote (1984) reported that the organic matrix is a template for the crystallization of $CaCO_3$ once incremental growth starts. Dunkelberger et al. (1980) found differences in organic materials between the crystalline and proteinaceous zones, which suggests that the organic matrix deposition is probably subject to diel variation.

Some authors refer to the periodic microincrements as daily increments (Mugiya et al., 1981; Mugiya, 1984) but environmental and physiological variables such as temperature (Umezawa and Tsukamoto, 1991; Suthers and Sundby, 1993; Wright and Huntingford, 1993), feeding (Jenkins et al., 1993) and endogenous circadian rhythms may disturb the deposition of increments. For example, De Vries et al. (1990) found subdaily microincrements in the larvae of king (*Scomberomorus cavalla*) and Spanish mackerel (*S. maculatus*). Campana and Neilson (1985) undertook an in-depth review of the literature and noted three important laboratory conditions that cause problems in interpretation: (1) non constant temperature in the laboratory, (2) otolith preparation and (3) the resolving power of the microscope. Gjøsaeter et al. (1984) published a list of all fish species in which the periodicity of the microincrement pattern has been studied.

The day when the first daily increment is deposited appears to be dependent on the species. Some species such as sockeye salmon and Mummichog (*Fundulus heteroclitus*) deposit the first daily increment before hatching (Marshall and Parker, 1982; Radtke and Dean, 1982). English sole (*Parophrys vetulus*) deposit the first daily increment after hatching (Rosenberg, 1982). Radtke (1989) determined that Atlantic cod deposited their first daily ring one day after hatching.

The hatching or birth date analysis is a promising tool for studying recruitment processes (Campana and Jones, 1992). When a back-calculated hatching date distribution reveals a discrepancy, daily-specific mortality induced by environmental factors may be the cause, as well as sampling biases due to gear selectivity and time of day. Methot (1983) has related monthly differences in survival rates to environmental factors and to the overall effect of year-class strength. Hatching date analysis was used to estimate life-history parameters and the recruitment date of: herring (Fossum and Moksness, 1991; Moksness and Fossum, 1992; Moksness 1992), eel (*Anguilla japonica*; Umezawa and Tsukamoto, 1990) and sardine (*Sardina pilchardus*; Alvarez and Morales-Nin, 1992). The ability to describe and predict the effects of environmental factors on year-class strength, potential yields and future recruitment is a potential boon to fishery management.

Checks (or discontinuities), as distinct from daily increments, can be found in otolith growth sequences if fish have encountered stress or perturbations in their environment. These checks may occur at random, be correlated with disturbances or they may delimit incremental patterns of weekly or fortnightly periodicity (Pannella, 1971, 1980; Campana, 1984). These authors relate checks to sexual maturation or the lunar cycle. Changes in otolith microstructure have also been found to be correlated with yolk sac absorption and larval metamorphosis, (e.g., lesser sandeel (*Ammodytes marinus*; Wright, 1993) and Atlantic cod, (Radtke, 1989)). Campana and Neilson (1985) discuss other causes that interrupt otolith growth. These activities have stimulated the development of a new field of study on natural microstructural features and natural chemical fingerprints specifically related to habitats, water bodies, climatic events, seasons, etc. (Brothers, 1990).

The obvious application of otolith microincrement analysis is to the estimation of growth rate. Different growth models have been used to describe larval and juvenile growth. Models can be purely empirical, such as polynomial regression (e.g., Wilson and Larkin, 1982; West and Larkin, 1987), power or exponential curves (e.g., Campana and Hurley, 1989; Tzeng and Yu, 1988) or linear regression (e.g., Geffen, 1982; Leak and Houde, 1987; Victor, 1987; Moksness and Fossum, 1992; Wexler, 1993). These models should be used only for a specific life history stage. Extrapolation of models outside of the observation range may not be appropriate. Empirical models are easily modified to include environmental factors (Campana and Hurley, 1989). Traditional back-calculation methods (Francis, 1990) can also be used. Secor and Dean (1992) used Fraser-Lee's formula for larval striped bass microincrements. The Gompertz model has been used to describe growth in larval and juvenile stages (Methot and Kramer, 1979; Lough et al., 1982; Smith and Kostlan, 1991). The von Bertalanffy growth model has also been used (Ralston, 1976; Wild and Foreman, 1980; Laroche et al., 1982; Young et al., 1988) to describe growth in larval and juvenile stages.

Growth at successive times or ages can be a state space condition. That is, the width of microincrements at adjacent time steps may be autocorrelated and amenable to time series analysis. Gutierrez and Morales-Nin (1986) used the cross-correlation function to relate daily increment and water temperature. State space methods like the Kalman filter as found in Box and Jenkins (1976) will find ready applications to microincrement data in the future.

Age composition information obtained from otolith microincrement analysis can be used to estimate mortality or survival rates of particular life history stages of larvae and juveniles (Methot, 1983; Owen et al., 1989; Pepin, 1991). If sample size is sufficiently large to follow a daily cohort, a simple negative exponential model can be use to estimate the mortality rate of the cohort. However, if sample size is too small for the negative exponential, catch-curve analysis (Chapman and Robson, 1960) can be used. See Campana and Jones (1992) for a review of this subject.

The age of larvae and juveniles has also been estimated by multivariate statistical analysis. Radtke and Hourigan (1990) and Radtke et al. (1993) used multivariate analysis on *Nototheniops nudifrons* and *Pleuragramma antarcticum* from the Antarctic. Ralston (1982) developed a model-based method to estimate the age of adults utilizing otolith microincrements. Ralston and Miyamoto (1983) and Morales-Nin and Ralston (1990) have expanded this method to other fishes. Mug (1993) and Mug et al. (1994) applied this method to determine the age of an adult corvina reina (*Cynoscion albus*) from the Gulf of Nicoya, Costa Rica, which was neither juvenile nor short-lived.

The study of crystalline deposition has contributed to a better understanding of microstructural patterns. Gauldie and Radtke (1990) found that microincrement widths in otoliths showed short-term changes in step with temperature, length and volume changes, especially in the anabolic phase of metabolism. Check lines occurred in the areas of otoliths corresponding to metabolic conditions that are usually associated with low pH values, suggesting that the metabolic basis driving the pH gradient causes the deposition of check rings, not age. Gauldie and Nelson (1988) found that teleost otoliths composed of twinned aragonite crystals were laid down as a series of protein-rich and protein-deficient bands consisting of bundles of crystals orthogonal to the orientation of the protein bands. The cycle of protein-rich and deficient mineral was correlated to diurnal rhythms by the amount of haemotoxylin-eosin staining protein vesicles within the kinocilia cells of the macula. The results suggested that twinning leads to rapid growth of otoliths and that crystal increments are laid down under neural control. This would lead to obligatory daily growth rings under normal physiological conditions (see also: Gauldie and Nelson (1988, 1990), Gauldie and Radtke (1990) and Gauldie et al. (1990).

4.4 VALIDATION FOR DAILY INCREMENTS

Validation of microincrements as daily increments is a central part of their use to estimate age. Commonly used validation methods have been summarized by Geffen (1987, 1992). These methods are usually applied to laboratory reared fish and are subject to criticism when the results are used to compare wild and laboratory stocks. As stated previously, there are many factors that can reduce the reliability of microincrement readings. Therefore, validation procedures must be designed to insure: (1) that increment deposition is constant and predictable over time, (2) that the increment number reflects the age of a specimen, (3) that increment deposition is independent of larval growth rate and feeding and (4) that increment deposition is controlled by the same natural phenomenon throughout the population in question. This calls for direct validation on wild stocks. Gutierrez and Morales-Nin (1986) thought that time series analysis might be useful in correlating the increment width and daily environmental effects. For example, the method has been successfully used to validate the deposition of daily increments in gastropod shells (Adlerstein 1987 and Lønne and Gray 1988).

Nevertheless, the most useful and direct validation is via oxytetracycline injection or immersion. Geffen (1992) gives examples of these techniques. Other chemicals, such as calcein, alizarin complexone and acetazolimide can be substituted for oxytetracycline. There are advantages and disadvantages to the use of these chemicals. The criteria for selecting suitable chemicals, the method of getting the chemical into the fish (injection or immersion), and the dose are related to the size of fish, mortality after treatment, uptake rate of the chemicals, and duration of chemical bounding with otolith materials. Other methods of validation are via known-age larvae and introduction of stress that may cause identifiable patterns such as via feeding or temperature regimes (Geffen, 1992).

5. AGE DETERMINATION AND GROWTH

The estimation of age is normally based on the use of: (1) age-length data, (2) back-calculation techniques applied to bony tissue, (3) direct counts of annual or microincrement lines or (4) tagging data. The data for (1) are described in section 3.2.2.i; the data for (2) are described in section 3.2.2.h; and the data for (3) are described in section 3.2.2.f. But, in a more general sense, all of the 18 methods described in section 3.2.2 may lead to age data which are suitable for model building. Although there are exceptions, most age determination investigations will build a growth model which may then be used for a particular application. This section of the chapter describes the transition from data based on one of the 18 methods to the selection and application of a mathematical model, via the steps of hypothesis testing and statistical parameter estimation.

The von Bertalanffy growth model is the most widely used of the many possible growth models since it appears to fit so many populations (Beverton and Holt, 1957). Although this section focuses on the von Bertalanffy (LVB) growth model, the methods of fitting and testing for other growth models are similar.

5.1 THE VON BERTALANFFY GROWTH MODEL

For a complete derivation of the LVB growth model, see von Bertalanffy (1938), Beverton and Holt (1957) and Gallucci and Quinn (1979). The LVB growth model is written as:

$$l_t = L_\infty \left(1 - e^{-K(t - t_0)} \right) \tag{1}$$

where l_t is the mean length of a fish at age t, L_∞ is the mean asymptotic length from age-length data, K is the Brody growth coefficient which describes the curvature of a growth curve, and t_0 is the age at zero length, which acts as an adjustment factor moving the curve to the left or right. More versatile models still based upon the von Bertalanffy model can be found in Chapman (1960), Schnute (1981) and many other sources.

5.2 ESTIMATES OF LENGTH-AT-AGE

Assume that n fish have been sampled for age determination and classified into "A" age classes of n_t fish, for t = 1, ..., A. The simplest analysis for the age data would be to compute the mean length-at-age, $\bar{l}_t$, and variance s_t^2 for each age-class t. This computational procedure is based on the assumption that the samples for age determination are collected using a simple random sampling design so that the samples fall into an age-class at random. In the next section we discuss the use of age length keys, age composition, sampling and costs. The age-length key method should not be directly applied to estimate mean-length at age because of bias due to nonrandom sampling. A weighted sample mean method should be used, in which the probability density of the length-frequency distribution is the weighting factor.

Back-calculation is another way to estimate mean length-at-age. The method has traditionally been applied to scales. The basic assumption is that the growth in fish length is linearly proportional to the increase in the radius of a scale. Different computational methods have been proposed to predict the mean length-at-age of the fish (at annulus formation) given radius measurements of annuli taken from the scale. An assumption of the validity of linear proportionality for bony tissues other than scales should be made with care.

5.3 FITTING THE VON BERTALANFFY (LVB) GROWTH MODEL

The traditional method of fitting the LVB growth model is the Walford plot (Gulland 1969, Ricker 1975), which applies the linear property and related linearized approximations of the discrete form of the model. Other studies have shown that a non-linear regression technique is superior to the Walford plot by producing less variable estimates (Gallucci and Quinn, 1979; Sundberg, 1984; Vaughan and Kanciruk, 1982). That is, statistical goodness of fit tests of the data to the growth curve show improved fits from nonlinear regression compared to those based on linear regression and a Walford plot (Allen, 1976).

Since fish do not grow at a constant rate throughout their life history nor are rates among individuals the same, a stochastic version of the model would, in principle, be preferred. One type of stochastic LVB model that can be fitted to the data is written as:

$$\bar{l}_t = L_\infty\left(1 - e^{-K\left(t - t_0\right)}\right) + e_t$$

or

$$l_{t_i} = L_\infty\left(1 - e^{-K\left(t_i - t_0\right)}\right) + e_{t_i}$$

where e is the random error and $i = 1, 2, \ldots, n$ is for n individuals. These equations are based on an assumed additive random error. If the random error appears to be multiplicative, e.g., lognormally distributed random error, a log-transformation of mean length at age is necessary. Other forms of stochastic models are also possible (Sainsbury, 1980) but will not be explored here.

Commonly used methods for parameter estimation are nonlinear least squares (Gallucci and Quinn, 1979) or maximum likelihood (Kimura, 1979). The assumption made about the structure of the random error is an essential difference between the two methods. Maximum likelihood estimation requires an assumption of normality on the probability density distribution of random error, e, while the least squares method does not. If it is true that $e \sim N\left(0, \sigma^2\right)$ the maximum likelihood method reduces to a least squares method (Kimura, 1977). In the least squares method (the LS method) the numerical values of the parameters L_∞ and K are those which minimize the sum of squares of the objective function for the stochastic models:

$$\sum_t \left(\bar{l}_t - E(\bar{l}_t)\right)^2 \quad \text{or} \quad \sum_{t_i} \left(l_{t_i} - E\left(l_{t_i}\right)\right)^2 ,$$

where $E(\bar{l}_t)$ and $E\left(l_{t_i}\right)$ are the expected mean and individual lengths-at-age, respectively. These two objective functions are based on the assumption of a constant variance over ages,

$$e_t \sim \left(0, \sigma^2\right)$$

or over individuals,

$$e_{t_i} \sim \left(0, \sigma^2\right).$$

If the distribution of fish in each age class in the sample is not representative of that in the population, weighted least squares must be used. Biased samples occur due to circumstances such as gear selectivity, differential age class availability or sampling design. For example: if the variance $\left(\sigma_t^{\,2}\right)$ of $\bar{l}_t$ changes with age t and sample size n_t, the weighted least square function (Kimura, 1979) is:

$$\sum \left(\frac{n_t}{\sigma_t^{\,2}}\right) \left(\bar{l}_t - E(l_t)\right)^2 .$$

In fact, any value may be used as a weighing factor W_t or w_{t_i} to minimize the sum of squares,

$$\sum_t W_t \left(\bar{l}_t - E(\bar{l}_t)\right)^2 \quad \text{or} \quad \sum_{t_i} W_{t_i} \left(l_{t_i} - E\left(l_{t_i}\right)\right)^2 ,$$

where W_t or W_{t_i} indicate the relative importance of $\bar{l}_t$ or of l_{t_i} in constructing a representative growth curve.

Commonly used nonlinear least squares techniques are numerical iteration methods based on the Gauss-Newton algorithm (e.g., Marquardt, 1963). A byproduct of these techniques is an estimate of the large-sample covariance matrix, also known as the inverse of the "information matrix" evaluated at the convergence value of the LS estimate (Ratkowsky, 1983; Bard, 1974).

Schnute (1981) proposed using Nelder and Mead's (1965) simplex method, a direct search method, to fit the parameters K and L_∞. This algorithm, however, will not estimate the variance and covariance of the parameters. Many statistical software packages (e.g., SPSS, SAS and BMDP) also contain nonlinear regression programs which calculate asymptotic standard errors and correlation coefficients for the estimated parameters. For interactive programs that can be used to estimate growth curves, several spreadsheets (e.g., MS-EXCEL and LOTUS) contain nonlinear optimization routines suitable for exploratory data analysis, but variance estimation is not usually part of these software.

5.4 COMPARISON OF VON BERTALANFFY GROWTH CURVES

Comparison of the parameters from two of the same growth curves with different parameters provides a mechanism for the comparison of populations in different areas, times, and environmental conditions. Growth curves can be compared using univariate or multivariate tests based on t or χ^2 test statistics (Kingsley, 1979; Gallucci and Quinn, 1979; Misra, 1980, 1986) test statistics. The multivariate Hotelling's T^2-test has been recommended by Kingsley (1979) and Bernard (1981). Cerrato (1990) estimated and compared the LVB parameters using the Hotelling's T^2- and likelihood ratio tests and concluded that a likelihood ratio test should be more reliable than the Hotelling's T^2-test.

5.4.1 ω-test

Gallucci and Quinn (1979) suggested reparameterizating the LVB growth model as follows:

$$l_t = \frac{\omega}{K}\left(1 - e^{-K(t-t_0)}\right)$$

where $\omega = K L_\infty$, which is the slope of the growth curve at $t = 0$, i.e., the growth rate at the birth of an animal. The test of equality of ω between two growth curves can be a χ^2 test of homogeneity for large sample sizes or a t-test for a small sample sizes. The test is based upon the model-based negative correlation between K and L_∞. That is, the mathematical model requires that K increase when L_∞ decreases, and vise versa. Thus, tests of comparison are confounded.

5.4.2 Likelihood Ratio Test

Kimura (1980) suggested a likelihood ratio test where the maximum likelihood estimator can be obtained by maximizing the likelihood function under the assumption that the random error term $e_t \sim N(0, \sigma^2)$. The likelihood function is:

$$L = -\frac{n}{2}ln\left(2\pi\sigma^2\right) - \frac{1}{2\sigma^2}S\left(L_\infty, k, t_0\right) \tag{2}$$

where $S\left(L_\infty, k, t_0\right)$ is the sum of squares for the growth model. In order to find $\sigma^2 = \frac{1}{n}S\left(L_\infty, k, t_0\right)$ set $\frac{\partial L}{\partial \sigma^2} = 0$. That is, the maximum likelihood method is reduced to a least squares method when $e_t \sim N\left(0, \sigma^2\right)$. The likelihood ratio test procedure to compare two growth curves is:

(i) Fit each of the empirical growth curves with the separate parameter sets,

$$\left(L_{\infty_1}, K_1, t_{0_1}\right) \text{ and } \left(L_{\infty_2}, K_2, t_{0_2}\right)$$

and compute the likelihood values L_1 and L_2 using equation (2).

(ii) If the test is on one of the three parameters, e.g., H_0: $L_{\infty_1} = L_{\infty_2} = L_\infty$, then the growth curve is fit to the combination of the two data sets with $\left(L_\infty, K_1, K_2, t_{0_1}, t_{0_2}\right)$. The likelihood value L is computed next.

(iii) The likelihood ratio statistic is $T = L - (L_1 + L_2)$. The statistic $(-2T)$ has an asymptotic chi-square distribution (Dixon, 1983) with degrees of freedom given by the difference in the number of estimated parameters between the two hypothetical parameter sets. In this case the degree of freedom is 1.

(iv) If the hypothesis test involves more than two growth parameters, e.g., H_0: $L_{\infty_1} = L_{\infty_2} = L_\infty$; $K_1 = K_2 = K$; $t_{0_1} = t_{0_2} = t_0$, then the growth curve is fit to the combination of two data sets with (L_∞, K, t_0) and the likelihood value L is computed using the procedure in (iii). This method can be used to compare more than two growth curves. Kirkwood (1983) has extended the likelihood method for use with tagging data.

5.4.3 Hotelling's T^2-test

The procedure of Cerrato (1990) is used instead of Bernard (1981) because Cerrato's procedure relaxes some of the assumptions. Bernard (1981) has been corrected in the literature. The procedure of Cerrato (1990) involves the computation of a covariance matrix for the parameters to be estimated. The procedure is summarized as follows:

(i) Fit each individual curve with separate parameter sets,

$$\theta_1 = \left(L_{\infty_1}, K_1, t_{0_1}\right) \text{ and } \theta_2 = \left(L_{\infty_2}, K_2, t_{0_2}\right)$$

Use a least squares technique to estimate the growth parameters by minimizing $S(\theta_1)$ and $S(\theta_2)$. Statistical software such as SAS, SPSS and BMDP provide estimates of the covariance matrices U_1 and U_2, respectively, for the parameters in θ_1 and θ_2.

(ii) Define $d = \theta_1 - \theta_2$. The hypothesis to be tested is H_0: $d = 0$. The test statistic is:

$$T^2 = d'U^{-1}d$$

where d' is the transpose of d and $U = U_1 + U_2$ Assuming that the growth parameters are from a multivariate normal distribution, the test statistic T^2 is approximately distributed as a Hotelling's T^2-distribution. The degrees of freedom for T^2 are three for the growth parameters and f^*, which is found from:

$$\frac{1}{f^*} = \frac{1}{f_1}\left(\frac{d'U^{-1}U_1U^{-1}d}{d'U^{-1}d}\right)^2 + \frac{1}{f_2}\left(\frac{d'U^{-1}U_2U^{-1}d}{d'U^{-1}d}\right)^2$$

where $f_1 = n_1 - 3$ and $f_2 = n_2 - 3$ and are the degrees of freedom of the two growth curves being compared. The critical value for Hotelling's T^2-distribution is calculated from an F distribution:

$$T^2_\alpha(p, f^*) = \frac{pf^*}{f^* - p + 1} F^\alpha_{p, f^* - p + 1}$$

where $p = 3$ is the number of growth parameters being compared.

More could be written about fitting, testing and comparing growth models in general, and the LVB model in particular, but these methods will serve to introduce the reader to the principle issues and methods.

6. ESTIMATION OF AGE COMPOSITION AND SAMPLING DESIGNS

The age composition of a stock is the principle input into an age-dependent stock assessment model. Age determination data such as we discussed earlier are central to estimating the age composition. Data for age determination are usually collected from scientific sampling of the stock. The classic sampling techniques, simple random sampling, stratified random sampling and cluster sampling can be used but the most cost-efficient is the double sampling technique called the "age-length key" (ALK) method.

The ALK method is based on two assumptions: (1) the samples for age determination are drawn from a sufficiently large number of samples to estimate a length-frequency distribution of the population, i.e., samples for age determination are an unbiased subsample of the length composition of the entire population; and (2) there are no discrepancies beyond sampling differences in growth rate, gear selectivity or age composition among the samples and between the samples and the stock under consideration. If age determination is not feasible and no age data are available for the population from which length samples are collected, the "inverse age-length key" (IALK) method and length-frequency analysis (LFA) can be applied. The ALK and IALK are both discussed below.

6.1 AGE–LENGTH KEY

Fredrikssen (1933) developed a double sampling technique to estimate the age composition of a fish population. Tanaka (1953) used double sampling theory to derive the age-length key method now widely used in fisheries. The first stage is to sample a large number of fish for a length frequency distribution (called the length sample); the second stage is to subsample a small number of fish from each length stratum for age determination (called the age subsample). Age determination is the costly part of the process. After determining the age of the fish in the age subsamples, the data are organized as a contingency table with age vs. length, commonly called an age-length key (ALK). Lai (1993) uses an example from Ricker (1969, 1975), which was also used by Jinn et al. (1987), to illustrate the estimation procedures and optimal sampling designs.

Fisheries scientists frequently use two different subsampling schemes to obtain age-length data: (1) fixed age subsampling, where the size of the subsample for age in each length stratum (say, n_i for the ith stratum) is constant, i.e., $n_i = n / L$, where $n = \sum_i n_i$ is the total number of age subsamples and L is the number of length strata; (2) proportional age subsampling, in which the size of each age subsample is proportional to the number of individuals in each length stratum, i.e., $n_i = nl_i$ where l_i is the proportion of fish in the ith length stratum. Although these two subsampling schemes are frequently used, other types of length-based subsampling schemes can be designed to account for the increasing difficulty and cost of aging older fish (see Jinn et al., 1987; Lai, 1993). The fixed and proportional age subsamples are special cases of the length-based optimal sampling design (Lai, 1993). The following summary is based on the derivations in Lai (1993).

Let N = total number of fish sampled;

N_i = number of fish in the ith length stratum, $i = 1, \ldots, L$;

l_i = proportion of fish in the ith length stratum, $\left(l_i = N_i / N\right)$

$$n = \sum_{i=1}^{L} n_i$$

n_i = number of subsamples taken randomly from the ith length stratum to be aged;
n_{ij} = number of fish from n_i assigned to the jth age class;
q_{ij} = the proportion of fish from the ith length stratum belonging to the jth age-class $\left(\hat{q}_{ij} = n_{ij}/n_i\right)$;
A = number of age classes;
L = number of length strata;
p_j = proportion of the population in the jth age class.

The proportion of fish at age j is estimated as $\hat{p}_j = \sum_{i=1}^{L} \hat{l}_i \hat{q}_{ij}$. The variance estimate of $\hat{p}_j$ is approximated by:

$$\mathrm{Var}\left(\hat{p}_j\right) \cong \sum_{j=1}^{L} \left[\frac{\hat{l}_i^2 \hat{q}_{ij}\left(1-\hat{q}_{ij}\right)}{n_i} + \frac{\hat{l}_i\left(\hat{q}_{ij} - \hat{p}_j\right)^2}{N} \right].$$

The covariance of $\hat{p}_j$ and $\hat{p}_k$ is approximated by

$$\mathrm{Cov}\left(\hat{p}_j, \hat{p}_k\right) \cong \sum_{i=1}^{L} \frac{-\hat{l}_i^2 \hat{q}_{ij}\hat{q}_{ik}}{n_i} + \sum_{i=1}^{L} \frac{\hat{l}_i \hat{q}_{ij}\hat{q}_{ik}}{N} - \frac{\hat{p}_j \hat{p}_k}{N}.$$

A quadratic loss function is used to infer the precision of the estimated age composition, $\hat{p}' = \left(\hat{p}_1, \ldots, \hat{p}_A\right)$:

$$\begin{aligned}
\lambda(\hat{\boldsymbol{p}}, \boldsymbol{p}) &= (\hat{\boldsymbol{p}} - \boldsymbol{p})' \boldsymbol{W} (\hat{\boldsymbol{p}} - \boldsymbol{p}) \\
&= \sum_{j=1}^{A} w_{jj} \mathrm{Var}\left(\hat{p}_j\right) + \sum_{j \neq k}^{A} w_{jk} \mathrm{Cov}\left(\hat{p}_j, \hat{p}_k\right) \\
&= \sum_{i=1}^{L} \left(\frac{a_i - u_i}{n_i} \right) + \frac{b + v - m}{N}
\end{aligned}$$

where
$$a_i = \sum_{j=1}^{A} w_{jj} \hat{l}_i^2 \hat{q}_{ij} \left(1 - \hat{q}_{ij}\right),$$

$$u_i = \sum_{j \neq k}^{A} w_{jk} \hat{l}_i^2 \hat{q}_{ij} \hat{q}_{ik}$$

$$b = \sum_{i=1}^{L} \sum_{j=1}^{A} w_{jj} \hat{l}_i \left(\hat{q}_{ij} - \hat{p}_j\right)^2$$

$$v = \sum_{i=1}^{L} \sum_{j \neq k}^{A} w_{jk} \hat{l}_i \hat{q}_{ij} \hat{q}_{ik}$$

$$m = \sum_{j \neq k}^{A} w_{jk} \hat{p}_j \hat{p}_k$$

The elements in the matrix of weights W, w_{jk}, can be zero or any positive real number, depending on the precision of the estimate from the corresponding age-classes compared with other age classes. Usually, a diagonal matrix, i.e., $w_{jk} = 0$ for $j \neq k$, is sufficient for the optimal sampling design of ALK.

For an optimum allocation of n and N, we need a cost function where, for simplicity, a linear function of n and N is assumed. Let c_1 be the unit cost of observing the length of a fish, and c_{2i} be the unit cost of ageing a fish in the ith length stratum. The total cost for an optimal sampling design can be written as:

$$C = c_1 N + \lambda \sum_{i=1}^{L} c_{2i} n_i \quad .$$

For an optimum sampling design in an ALK, we can either find optimum values of n and N (denoted, n^* and N^*) to minimize λ for a given and fixed total cost C or, find n^* and N^* to minimize the total cost C with a specified level of λ. In either case, the Cauchy-Schwarz inequality method is used (Cochran, 1977; Kendall and Stuart, 1977). To apply this method, first write the product of λ and C and then apply the Cauchy-Schwarz inequality to obtain the optimal age-subsampling ratio between age subsamples and length samples in the ith length stratum. The solution for this subsampling ratio is:

$$r_i^* = \sqrt{\frac{c_1 \left(a_i - u_i\right)}{c_{2i} \left(b + v + m\right)^2}}$$

The optimum subsampling ratio is the same for either choice, but the optimum set of n^* and N^* is dependent on the constraints specified in each case.

For the first choice: the total cost is given and r^* is substituted into the cost function to obtain the optimum size of the length sample:

$$N^* = \frac{C}{c_1 + \sum_{i=1}^{L} c_{2i} r^*}$$

and, subsequently, the optimum size of the age subsample is:

$$n^* = r_i^* N^*$$

The minimum loss in precision is:

$$\min \lambda = \sum_{i=1}^{L} \left(\frac{a_i - u_i}{n_i^*} \right) + \frac{b + v - m}{N^*}$$

For the second choice, a desired λ is preset, the optimum N^*, n_i^* and min C are:

$$N^* = \frac{1}{\lambda} \left[\sum_{i=1}^{L} \left(\frac{a_i - u_i}{r_i^*} \right) + (b + v - m) \right]$$

$$n_i^* = r_i^* N^*$$

$$\min C = c_1 N^* + \lambda \sum_{i=1}^{L} c_{2i} n_i^*$$

An example of this method is given in Lai (1993)

The results of the above derivations can be extended to the traditional fixed and proportional age subsampling schemes. The per unit cost of ageing a fish is length-independent, i.e., $c_{2i} = c_2$ for all i's. The optimal combination of (N^*, n_i^*) follows from substituting $n_i = n/L$ for fixed age subsampling or $n_i = nl_i$ for the proportional age subsampling into the loss function and applying the Cauchy-Schwarz inequality. The results are summarized in the table on the following page.

An ALK requires a large random sample of fish from which a length-stratified subsample is collected for aging. Most fishery data are collected either from surveys where fish are collected from different tows, or samples are collected from commercial

Choice	Fixed age subsampling	Proportional age subsampling
	$r^* = \sqrt{\dfrac{c_1 \sum_{i=1}^{L} L(a_i - u_i)}{c_2(b+v-m)}}$	$r^* = \sqrt{\dfrac{c_1 \sum_{i=1}^{L} (a_i - u_i)/\hat{l}_i}{c_2(b+v-m)}}$
fixed C (or first)	$N^* = c_1 + c_2 r^*$ $n^* = r^* N^*$ $\min.\lambda = \dfrac{\sum_{i=1}^{L}(a_i - u_i)}{n^*} + \dfrac{b+v-m}{N^*}$	$N^* = c_1 + c_2 r^*$ $n^* = r^* N^*$ $\min.\lambda = \dfrac{\sum_{i=1}^{L}(a_i - u_i)/\hat{l}_i}{n^*} + \dfrac{b+v-m}{N^*}$
fixed λ (or second)	$N^* = \dfrac{1}{\lambda}\left[\dfrac{\sum_{i=1}^{L}(a_i - u_i)}{r^*} + (b+v-m)\right]$ $n^* = r^* N^*$ $\min.C = c_1 N^* + c_2 n^*$	$N^* = \dfrac{1}{\lambda}\left[\dfrac{\sum_{i=1}^{L}(a_i - u_i)/\hat{l}_i}{r^*} + (b+v-m)\right]$ $n^* = r^* N^*$ $\min.C = c_1 N^* + c_2 n^*$

catches in which fish from different vessel-trips are sampled. Each vessel's contribution is called a "cluster". Pooling data over clusters is necessary because of the cost of data gathering.

Fishery data are frequently stratified into time-area, fisheries (or gears), and sex strata (Kimura 1989). The question is how to make the optimal sampling design of an ALK apply in these situations. The following factors are considered: (1) the possible need for stratification; (2) ALK sampling within a stratum; and (3) combined-strata estimation. Westrheim and Ricker (1978) showed that an ALK obtained from a population at a time interval could not be universally applied to length-frequency data sets from the other populations, or other time intervals, if growth and survival rates are different among populations and among time intervals. Therefore, the factors that may result in differences in growth and survival rates must be evaluated and the stratification scheme should account for these factors.

The special publication on sampling (Doubleday and Rivard, 1983) has a good collection of sampling designs. Quinn et al. (1983) and Kimura (1989) adopted the strategy in which length-frequency data collected from clustered sampling units (e.g., tows or vessel-trips) within a stratum are pooled, then a length-stratified subsample is collected for ageing. Southward (1963) evaluated an older method in which a set of length and age data are collected from each landing of a vessel-trip. Because this method was not developed from a probability sampling design, the within-vessel variability in fish lengths is assumed to be less than between-vessel variability, but

Southward (1963) showed that this assumption was not valid and that the estimated variances of age composition are too large to have confidence in the technique.

The length-frequency data pooled over clusters should be a representative sampling of that stratum. Therefore, the weighting factor of each sample should be included in the pooling. Ignoring the weighting factor will bias the estimated age composition (Kimura 1989). Quinn et al. (1983) described a sampling-rate method in which a fixed proportion of halibut were sampled from landings of more than a 1000 pounds and then age data were subsampled from the pooled length samples from these landings. All length samples are self-weighted and can be pooled directly.

Quinn et al. (1983) evaluated the methods of combined-strata estimation and found that the "project-and-add" method (i.e., total catch-at-age is estimated for each stratum and then the estimates are added over strata) produces unbiased estimators if all strata are sampled. The project-and-add method uses the concept of a stratified random sampling technique (Cochran, 1977). Therefore, Cochran's rules (Cochran, 1977) of optimal allocation for stratified random sampling can be applied. In a given stratum, take larger length and age samples if: (1) the stratum is larger; (2) the stratum is internally more variable and, (3) sampling is cheaper in the stratum. The first two rules are the basis of the Neyman allocation (Cochran, 1977). The sampling-rate method proposed by Quinn et al. (1983) is built upon the first rule and can easily incorporate other rules into a sampling program.

It is clear that the number of strata and variability of each stratum should be evaluated before designing an ALK sampling plan. The total cost should be allocated into various strata according to Cochran's rules. Once the total cost for each stratum is determined, an optimal sampling design for an ALK can be applied. The sampling rate method of Quinn et al. (1983) can be used to collect the optimal length sample size from the clusters. After pooling the length samples, a length-stratified subsample is collected for aging.

There are no clear rules to determine the number of length strata needed for an ALK. Tanaka (1953) partially evaluated this issue and suggested that the number of length strata should be based on the width of the length interval. The width must be kept as small as possible, but not so small that the number of individuals in each stratum becomes too small to sample.

6.2 INVERSE AGE-LENGTH KEY

There are some cases where the only quantitative data available are length-frequency data and perhaps some anecdotal ageing data (e.g., age-length data from another time or another geographic area). In general, two possible approaches are possible: (1) use an LFA method or (2) use an IALK method. We begin with a description of the IALK method.

An important advance was made when it was found that age composition among samples could be nonconstant (Clark, 1981; Bartoo and Parker, 1983). These authors imposed the "inverse age-length key" where: $\sum_{j=1}^{L} q_{ij} = 1$ was applied, rather than the ordinary age-length-key in which: $\sum_{i=1}^{A} q_{ij} = 1$, where q_{ij} is the probability that an age i fish belongs to the jth length category (a note of caution, the q_{ij} defined in this section is actually equivalent to that in the analysis of mixture distributions, e.g., Macdonald and Pitcher (1979). In this IALK formulation, the length frequency composition is the linear transform of the age composition. In linear algebra terms:

$$[l_j] = [q_{ij}]'[p_i], \quad \text{or} \quad \boldsymbol{l} = \boldsymbol{Q'p} ,$$

where l_j and p_i have the same specification as in the age-length key and the prime symbol indicates the transform operator in matrix theory. The matrix $\boldsymbol{Q}$ is called the inverse ALK. Bartoo and Parker (1983) solved this system of equations by ordinary least squares, while Clark (1981) used a restricted least squares technique in which each proportion is restricted to a nonnegative value of p_i where $\sum p_i = 1$. Both of these approaches assume that $[q_{ij}]$ are known. Neither Bartoo and Parker (1983) nor Clark (1981) suggests a procedure to estimate the variance.

Another development for estimating the age distributions is the use of the EM (estimation and maximization) algorithm (Dempster et al., 1977). The EM algorithm was applied to the estimation of age distribution by Kimura and Chikuni (1987) and Hoenig and Heisey (1987). Kimura and Chikuni (1987) treated the observed $[q_{ij}]$ as mixtures of empirical distributions of independent length samples, while Hoenig and Heisey (1987) applied a log-linear model. Both methods have the same basic attributes: (i) age and length samples may be collected from the two populations; (ii) growth rate and gear selectivity may be different for the two populations, and (iii) the two populations may have different age compositions.

If the age and length of fish are estimated for the two populations (k = 1,2), the two samples from each of these populations can be classified into a A x L x 2 contingency table. If a random sample is collected from k = 2 and is aged and classified entirely by length and age categories it is called "complete data". In contrast, if the other sample, k = 1, is collected with a length measurement and is classified by length category with unknown age it is called "incomplete data". Then the problem is to estimate the age composition p for the sample k = 1.

To apply the EM algorithm, prior information about q_{ij} is needed. This can be obtained from the complete data, k = 2 (i.e., from q_{ij2} by assuming that the probability that a fish of age i falls into length category j remains constant from sample to sample. Let $q_{ij2} = P(j \mid i)_{k=2}$ be the probability of length j being read as age i in the sample k = 2 (Hoenig and Heisey, 1987), viz., the inverse age-length key. Kimura and Chikuni (1987) called it an iterative age-length key. Also, let p_{i2} be the probability that an animal from sample $k = 2$ has an age i, while p_{i1} be the unknown age distribution to be estimated for the sample $k = 1$.

For the sample $k = 2$, the probability that an animal lies in cell i, j is given as:

$$P(i, j \mid k = 2) = q_{ij2} p_{i2}$$

In this application of the EM algorithm, the joint distribution is taken of the complete and incomplete data.

Instead of taking the joint distributions of the complete and incomplete data, Kimura and Chikuni (1987) started with the EM algorithm to find the $(\hat{p}_{ij})$ that maximized the likelihood function. The algorithm can be summarized as follows:

(i) Initially set $\hat{p}_{i2} = 1/A$ where A is the number of age classes.

(ii) Use the available age-length data to find an empirical length at age distribution $\hat{q}_{ij2}$, then compute $\hat{q}_{ij1} = \hat{p}_{i2}\hat{q}_{ij2} / l_j$.

(iii) Calculate $\hat{p}_{i1} = \sum_j l_j \hat{q}_{ij1}$.

(iv) If the maximum absolute deviation over ages between current and previous $\hat{p}_{i1}$ is less than a small constant, the optimum estimates are reached; otherwise repeat (ii)–(iv).

The asymptotic variance of $\{\hat{p}_{i1}\}$ can be estimated by the inverse of the information matrix obtained from the maximum likelihood methods.

Although the EM-algorithm is a state-of-the-art statistical method, the IALK method is nevertheless still subject to sampling errors and uncertainties in the population. It is too early to conclude whether IALK or LFA methods should be used preferentially in the estimation of age composition.

7. THE PRECISION OF AGE DETERMINATION

Estimates of the chronological age of fish depend upon the assumption that annual growth rings or annuli can be identified and counted in the age-structures. It is a commonly held belief that fish living in temperate or higher latitudes are easier to age than tropical species. However, the literature shows that errors in age determination prevail throughout all latitudes. The appearance of an annulus in young or juvenile temperate latitude fish may be very consistent and easy to interpret, but interpretation of an apparent annulus in a mature fish is more difficult. Thus, age determination of fish is confounded by factors of human bias and fish physiology. The problems of validity and variation in age determination have been associated with age determination since the beginning of such studies. Fisheries scientists working on age determination have agreed to use **precision** to refer to the degree of repeatability of estimates Thus, precision is related to the variability between readers, between readings, between age structures, or between laboratories, while **accuracy** relates to the degree of bias from the true age. This section focuses on the **precision** of age determination.

Many statistics have been used to compare the precision of age determination (e.g., Campbell and Babaluk, 1979; Kimura et al., 1979). Since the difficulty of age determination increases with age, it becomes a general requirement that the statistic reflect this difficulty. The importance of this requirement has been illustrated by commonly used percent agreement statistics (Beamish and Fournier, 1981). They point out that if 95% of age readings by two readers agree within ± 1 yr for Pacific cod, this would be very poor precision since most commercial samples are comprised of only a few age classes. However, if 95% of the age readings in the case of spiny dogfish are within ± 5 yr, this can be an indication of good precision since spiny dogfish may reach 60 years of age and be exposed to the fishery during 30 of those age classes. Thus, the ideal estimator of precision is a statistic which is a function of age and can take into account the sample size. Six means of comparing the precision of age determination are: (1) the chi-square test, (2) average percent error, (3) coefficient of variation, (4) index of variation and regression analysis, (5) percent agreement by the log-linear model, and (6) ANOVA with repeated measurements.

7.1 CHI-SQUARE TEST

Suppose we have a sample of N fish aged by two readers (or two age structures, or two readings by the same reader, etc.). The frequencies of these N fish are classified into k age classes as shown in the following table.

	Age Class				
Reader	1	2	...	k	Total
1	n_{11}	n_{12}	...	n_{1k}	$n_{1\cdot}$
2	n_{21}	n_{22}	...	n_{2k}	$n_{2\cdot}$
Total	$n_{\cdot 1}$	$n_{\cdot 2}$	...	$n_{\cdot k}$	$n_{\cdot\cdot}$

Then the chi-square test statistic is:

$$\chi^2 = \sum_{i=1}^{2}\sum_{j=1}^{k} \frac{\left(n_{ij} - \frac{n_{i\cdot}\, n_{\cdot j}}{n_{..}}\right)^2}{\frac{n_{i\cdot}\, n_{\cdot j}}{n_{..}}}$$

with $(2-1)(k-1)$ degrees of freedom, $n_{..} = 2N$, and $n_{1.} = n_{2.} = N$ for this example of two readers. This method can be expanded to m readers. The method does not reflect the difficulty of determining age as the number of age classes increase and is not valid when the number in any cell, $n_{ij} < 5$.

7.2 AVERAGE PERCENT ERROR

Any statistic used to measure the relative precision of age readings must be able to reflect the difficulty of determining age. To deal with this phenomenon, Beamish and Fournier (1981) proposed the average percent error technique. Suppose that N fish are aged R times (i.e., R replicates) and let X_{ij} be the ith age reading for the jth fish, the average age based on R readings for the jth fish is

$$X_j = \frac{1}{R}\sum_{i=1}^{R} X_{ij} \ .$$

Then, the average error (AE_j) in determining the age of the jth fish is a fraction of the average of the age estimates. That is,

$$AE_j = \frac{1}{R}\sum_{i=1}^{R} \frac{\left|X_{ij} - X_j\right|}{X_j}$$

The index of the average percent error is defined as

$$APE = \frac{1}{N}\left(\sum_{j=1}^{N} AE_j\right) \times 100\% \ .$$

The problems of statistical inference about the average percent error, the probability distribution, and test statistics remain unaddressed.

7.3 COEFFICIENT OF VARIATION

Chang (1982) criticized the index of average percent error because the range of fish year classes available to be aged increases in proportion to the average age of fish in the fishery, i.e., the standard deviation is proportional to the mean. With the same notation as in the previous section, the coefficient of variation of the age of jth fish

$$\mathrm{V}_j = \frac{1}{X_j}\sqrt{\frac{\left(X_{ij} - X_j\right)^2}{R(R-1)}} .$$

The percent error contributed by each observation to the average age-class may be estimated by the index of precision (D_j) as:

$$\mathrm{D}_j = {}^{\mathrm{V}_j}\!\big/_{\sqrt{R}} .$$

Chang (1982) proposed that the average V and D for all fish aged is:

$$\mathrm{V} = {}^{\mathrm{V}_j}\!\big/_{N} \quad \text{and} \quad \mathrm{D} = {}^{\mathrm{D}_j}\!\big/_{N} .$$

Examples from Beamish and Fournier (1981) are given in Chang (1982). It can be seen that there are benefits to be obtained by the use of precision vs. the indices V and D above. In contrast, the benefits of using V and D include: (1) Variance is a better estimator than the absolute differences, as it is statistically unbiased and consistent. The estimated mean and variance converge to the population mean and variance as sample size increases; the coefficient of variation shares these properties. (2) The index of precision can be used to show the percent error contributed by each observation to the estimate of average age for the jth fish. (3) The probability density function for V can be derived and tested as described in Sokal and Rohlf (1969) and Zar (1974), while the average percent error cannot. Therefore, for statistical purposes, one might use V and D instead of the average percent error.

7.4 INDEX OF VARIATION AND REGRESSION ANALYSIS

As previously stated, age determination is labor-intensive, time-consuming, and costly. There are difficulties associated with the average percent error, and the coefficient of variation (V and D) tests. All of these methods require at least three readings for each reader (or ageing method), and the sample size must be large to test for excessive numbers of age classes in a fish population. Also, these statistics do not describe systematic differences between two readings from different bony tissues (e.g., otoliths and scales) (Lai, 1985). To overcome these problems, Lai (1985) developed the regression analysis and index of variation (IV).

Campbell and Babaluk (1979) compared the regression line of the two age readings against the 45° line through the origin. The full regression model, H_0: $Y = b_0 + b_1X + \varepsilon$, was fit to the data. The 45° diagonal line is a reduced model, H_0: $Y = X + \varepsilon$, or $b_0 = 0$ and $b_1 = 1$. The generalized F-test (Draper and Smith 1981, Weisberg 1980) was used to test for systematic differences between the two age readings. If H_0 is rejected, there is a significant systematic error.

This test only shows the existence of a systematic difference between the two readings or, in other words, there are balances between positive and negative deviations of the two readings for all age classes. Nevertheless, this does not necessarily imply high precision since the statistical test does not provide a description of the deviation of each of the two readings. High precision between two age readings must show a very "small" random error, i.e., the scatter plot of the two readings is not widely dispersed over the 45° diagonal line. In order to examine the degree of precision, an index of variation (IV) must be calculated,

$$\text{IV} = 100\% \times \sqrt{(Y - X)^2 \Big/ \frac{N-1}{2}\left(\overline{X}^2 + \overline{Y}^2\right)}$$

where, Y: is the tested age as a dependent variable,
X: is the reference age as an independent variable,
N: is the sample size,
$\overline{X}$ and $\overline{Y}$: are the mean ages of X and Y.

The statistic, IV, is the residual mean square from fitting $Y = X$ to the data, divided by the mean of the two estimated mean ages, $\overline{X}$ and $\overline{Y}$. Therefore, IV indicates the degree of variation of the two age readings being compared and measures the relative precision of the two age readings by accounting for the distribution of ages in the sample.

If "X" represents the known (true) age of fish, the reading accuracy can be measured by the coefficient of variation (c.v.) as

$$\text{C.V.} = \sqrt{(Y - X)^2 / (N-1)\overline{X}^2} \ .$$

Lai (1985) shows an example for Pacific cod where there are no systematic differences within the readers using coracoids, otoliths and dorsal-fin rays, but there are in the pectoral-fin rays and scales. The values of IV are 14%, 13%, 15%, respectively, for the coracoids, otoliths, and dorsal-fin rays. Therefore, from age validation and tests of precision, Lai et al. (1987) suggested that the dorsal fin ray method is the most suitable ageing method for Pacific cod.

7.5 TEST OF PERCENT AGREEMENT BY LOG-LINEAR MODEL

A statistical test for percent agreement using a normal approximation (Snedecor and Cochran, 1967; Fleiss, 1977) can be employed, but it is considered inappropriate because age classes may act as an important factor in the measure of agreement. Therefore, Lai (1985) and Lai et al. (1987) developed a log-linear model method to test the repeatability of the ageing methods and the age class. For this statistical test, two readings on each fish by different ageing methods are usually required, but independent samples for each of the different ageing methods will suffice.

Assume that three factors are involved: ageing method (M), age class (A) and repeatability (R). The terminology of Fienberg (1981) was used. The aged fish are cross-classified into cells denoted by the factors M, A and R. A multi-dimensional contingency table is constructed where y_{ijk} is the observed cell frequency for the ith ageing method, the jth row of the age class and the kth column of repeatability; m_{ijk} is the expected value of the y_{ijk} and $q_{ijk} = \log (m_{ijk})$.

The general log-linear model for the three dimensional contingency table is

$$q_{ijk} = \mu + \lambda_i^{(M)} + \lambda_j^{(A)} + \lambda_k^{(R)} + \lambda_{ik}^{(MR)} + \lambda_{jk}^{(AR)} + \lambda_{ij}^{(MA)} + \lambda_{ijk}^{(MAR)}$$

where all effects sum to zero over any subscript (as is usual for an ANOVA model). The brackets above the terms in q_{ijk} indicate the interaction terms represented.

The goodness-of-fit test to the model is

$$G^2 = 2\sum_i \sum_j \sum_k y_{ijk} \log \frac{y_{ijk}}{m'_{ijk}}$$

where m'_{ijk} is the expected value of the cell frequencies. The G^2 test statistic can be partitioned to determine which interactions to include. Investigation of the effects of ageing method and age class repeatability indicated that percent agreement decreased with increasing age. For example, in Lai et al. (1987) age determination using coracoids and dorsal-fin rays had a positive effect on agreement, i.e., agreement between these methods was higher than the average of the other methods, which all had negative effects (i.e., agreement was lower than average).

7.6 ANOVA WITH REPEATED MEASUREMENTS

Suppose that at least two age readings are carried out by different ageing methods (or by different readers) for a total of N fish. Comparison of these data can be done with an analysis of variance using a factorial design with repeated measures on the subjects. That is, a single subject which is repeatedly measured by various methods. The ageing methods or readers are termed “the within-subject factor” or “the trial factor”.

The ANOVA model

$$x_{ijn} = m + p_n + M_i + (Mp)_{in} + R_j + (MRp)_{ijn} + \varepsilon_{ijn}$$

where x_{ijn} is the observed age of the nth fish by the ith ageing method and on the jth reading. In Lai et al. (1987) i = 1, 2, 3, 4 ageing methods, j = 1, 2 first and second readings, n = 1, 2, ..., N number of fish; m is the grand mean, p the effect between individual fish and the M the effect of the ageing method. The test statistic Q (Snedecor and Cochran 1967) was used to test the differences between the mean ages between readings and between ageing methods.

Observations made to study variability in the ageing of Pacific cod (Lai, 1987) are presented to illustrate the types of results that can be inferred. The choice of the ageing method was significant for all cod older than age 3. There was no significant difference (5% level) between readings except for ages 5–6, which probably resulted from differences between age readings from otoliths and pectoral-fin rays. Mean square error (MSE) for the ageing method effect increased with age and was the dominant component of the within-subject variation for all age categories. Therefore, variability in age determination was mainly due to the ageing method, rather than inconsistent interpretations by readers.

In further analysis using the Q-statistic, the mean ages of the two readings were not significantly different, except for age group 5–6, using the otolith and pectoral-fin ray method. Significant differences between ageing methods were found in all age categories except the youngest. Age readings from dorsal-fin rays and pectoral-fin rays were not significantly different for fish younger than age 6. Age readings from otoliths and pectoral-fin rays were not significantly different for fish older than age 7. Otolith readings indicated older readings than other methods for fish younger than age 6, but gave younger age estimates than dorsal-fin ray readings for fish older than age 7. Scale readings gave consistently younger ages than the other methods.

8. EFFECTS OF AGEING ERRORS ON STOCK ASSESSMENT

Gulland (1955, 1958) reported that estimation of survival rate or mortality rate was significantly affected by over- and under-estimation of age, but if the error was randomly over and under, even if for as much as half of the sample, the estimate of survival rate was essentially unaffected.

Ricker (1969, 1975) showed that: (i) if the error in age reading is random and the magnitude remains the same at all ages, then all mean length-at-age estimates will decrease; (ii) if the error in age reading is random and the magnitude increases with age, then all mean length-at-age estimates will again decrease, but the magnitude may be reduced compared to (i); and (iii) if negative error exceeds positive error, and this difference increases with age, then survival rate estimates will vary from a small over-estimate to a large under-estimate.

Brander (1974) examined the error in age determination of Atlantic cod and found that the reassigned age-length key gave very small differences in estimated abundance. The survival rate estimate was increased from 0.1% to 0.2% if the proportion of misclassification was increased by 2.5% between successive ages.

Le Cren (1974) assessed the effect of errors in ageing on the estimated production of fish populations. Again, unbiased errors in ageing, especially for older fish, had little effect on production estimates. Estimates of juvenile production required data based on interpretation of the first annulus. Simulated experiments where errors arise from a doubling or halving of the true age resulted in a halving or doubling of the production estimates.

Mortera and Levi (1983) examined the consequences of errors in age determination on cohort analysis. They investigated different ages, and misclassification rates between successive and preceding age reading. The results are summarized in Table 2.

Barlow (1984) simulated a misclassification matrix with the probability of random ageing error evenly distributed along the diagonal and applied this matrix to a simulated age composition with known instantaneous total mortality rate Z. The results indicated that Z is subject to substantial bias, even when the degree of ageing error is considered small. Errors in age determination have been widely documented and fisheries scientists have become increasingly concerned about the problems that arise from these errors (e.g., Linfield, 1974; Beamish, 1979; Beamish and Chilton, 1982; Beamish and McFarlane, 1983, 1987; Boehlert and Yoklavich, 1984; Lai, 1985; Clark et al., 1986; Lai and Yeh, 1986; Lai et al., 1987; Kimura, 1989).

Error in age determination can be classified into three types (Lai, 1985): (i) normally distributed error around the reference age, (ii) skewed error around the reference age, but with zero mean difference, and (iii) biases due to systematic under- or over-ageing of samples. Lai (1985) and Lai and Gunderson (1987) carried out Monte Carlo simulations to investigate the consequences of these types of errors in age determination.

Table 2. Classification of consequences of errors in age determination.

Age	Fishing Mortality	Abundance
Ages 1+2 to age 1	Overestimated in early ages but underestimated in older	Underestimated
Ages 1+2 to age 1 and ages 3+4 to age 2	Same as above	Same as above
Ages 6 onward grouped	Underestimated in all ages	Overestimated
2, 5,10, 29% misclassification between a given age and the preceding one	Overestimated at given ages	Underestimated
2, 5,10, 20% misclassification between a given age and the successive one	Underestimated	Overestimated

The simulation included estimation of LVB parameters. Estimates of K, L_∞ and survival changed little (less than 10%) when type (i) error was simulated. Type (ii) error led to a significant bias (>10%) in K. It was shown that increases in the variability of ageing error substantially affected the stock assessment.

9. AGE DETERMINATION OF TWO TROPICAL CORVINAS

This last section is selected as a case history that synthesizes many of the preceding concepts and some of the methods that have been referred to in this chapter. The estimation of age is normally based on the use of data from: (1) an age-length analysis, (2) a back-calculation or other statistical technique applied to bony tissue or (3) a tagging or related marking experiment. The data for (1) and (2) are obtained directly from an age determination investigation, whereas the data for (3) are based upon a mark and recapture study, which is an investigation of another type. Marking studies are sometimes used to validate age determination results based upon (1) or (2) above.

These accepted methods present special problems for tropical situations where the physiological and environmental events that lead to definable marks on hard parts of the fish do not occur. As a consequence, age data for tropical stocks are often either unavailable or unreliable, despite the central position that such data play in the usual stock assessment models. As noted earlier, this has led to the use of size-based methods of stock assessment, as presented in Chapter 2 of this book. When the only data available are the length distributions of the catch over some time period, various methods of analysis have been applied, including: length cohort analysis (LCA) and CASA, for stock assessment and MIX for a modal analysis, from which age information can be inferred. As noted earlier, such methods are inherently less accurate than methods which use age data, due in part, to the variability in the number of sizes that can be associated with any one age, as seen in Figure 13.b.

In many cases, especially in tropical environments, a mixed situation may occur. High technology methods such as the SEM may not be practical for the examination of microincrements due to their expense and time consuming nature. Mark-recapture methods cannot be applied for validation because of the undeveloped nature of the fishery and/or the environment. However, it may still be possible to carry out an examination of hard tissue, such as otoliths, with a light microscope and length-frequency modal analysis (for LFA, see 3.2.2.i). In particular, if the fish is a species where the reading of otoliths with a light microscope reveals what appear to be annual marks for at least some years and where modes can be identified in the LFA so that they can be used for validation, the results would be an improvement over those based upon the use of length frequency data alone. Although such a procedure would not suffice as a growth analysis, it would be a valuable component in a stock assessment and would not be very costly in terms of either technology or personnel training. The entire analysis could be improved further by the use of computer simulations to reduce errors in reading hard parts and in the identification of modes.

This is the situation behind studies leading to estimates of the von Bertalanffy growth parameters of three species of corvina in the Gulf of Nicoya, Costa Rica. The species, coliamarilla (*Cynoscion stolzmanni*), aguada (*C. squamipinnas*) and reina (*C. albus*) are the dominant corvinas in the fishery. The estimates for the first two species were based on the use of otoliths from which surface readings provided estimates of the growth parameters and where validation was based on a modal analysis, using the program MIX (Lai and Campos, 1989). The third species, reina, is the largest and most economically valuable of the three. It was the object of a different study by Mug (1993) and Mug et al. (1994) where the primary data were from a SEM microincrement analysis of the otoliths and the validation data were from surface readings of the otoliths, with a light microscope. Microincrement analysis is described in 3.2.2.e above. The following describes the otolith surface reading-modal methods, as carried out as part of a preliminary investigation, and compares the estimates to earlier estimates of the parameters based exclusively upon length frequency analysis.

Otoliths were selected for analysis after otoliths, scales, dorsal-fin rays, cleithra, gill-covers and vertebrae were collected from 30 specimens of each species. Examination of these hard tissue structures as possible sources for an age determination study using the light microscope led to the conclusion that only otoliths showed identifiable growth marks.

Otoliths from all three species were therefore collected, along with associated data on size. After capture, the fish were transported to the laboratory, the otoliths were removed, cleansed and preserved in anhydrous glycerin. Otoliths of coliamarilla and aguada were prepared as described in the section on surface readings. They were read by putting them into a dark-background petri dish (using rubber glue to attach a black color velvet on the bottom of the petri dish) filled with glycerin and were viewed with a dissecting binocular microscope using reflected fiber optic light. Fiber optic light is preferred because light intensity and light angle are easily controlled.

The growth of young individuals was so rapid that the naked eye or a magnifying glass was adequate for otoliths with less than three annuli. In general, annual growth marks were identified as a series of concentric translucent zones that appeared dark under reflected light on a dark background. These zones would commonly be assumed to be formed in winter and are called "winter zones" or "annuli", terms which usually have little meaning for tropical fish.

Each otolith was read twice, independently. A log-linear model (Fienberg, 1981; Lai et al., 1987) was used to test for an association between the repeatability (R_i) of the two readings and the age of fish (A_j). A repeatability index measured the deviation between the two age readings from the same otolith, i.e., R_i = (age from the second reading) – (age from the first reading).

To test for a systematic difference between the two readings, a linear regression analysis was used to evaluate the possibility of a relationship between the first and second readings, i.e., to test the null hypothesis, H_0: $Y = X$, against the H_a: $Y = a + bX$, where Y is the second reading and X is the first reading. If H_0 is not rejected, the two readings produce the same age, except for nonsystematic differences.

To validate the estimates of age, the length frequency distributions of the two species were analyzed with the MIX program (Macdonald and Green 1985), based on

A

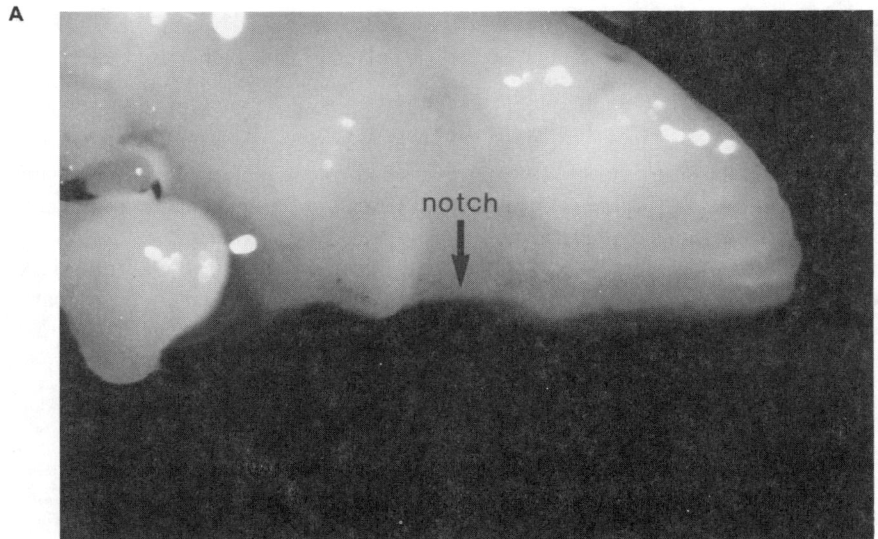

B

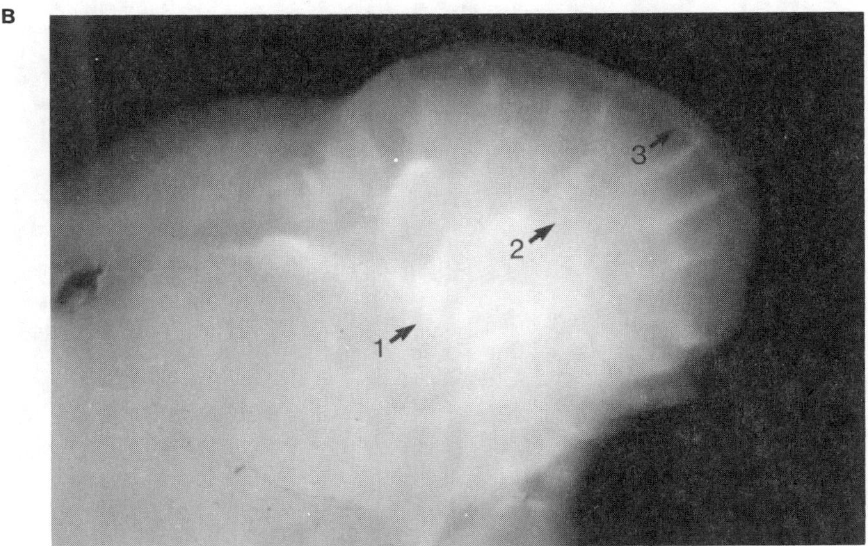

Figure 14. Photographs of an otolith from a 3-year old *Cynoscion stolzmanni* showing: (A) the characteristic notch associated with the first annual mark and (B) the difficulty of reading at margin as the otolith becomes thinner.

informed length frequency analysis (LFA). The mean lengths-at-age estimated from age determination and from MIX were compared. For coliamarilla, the first annual mark is identified by the dark zone associated with a distinguishing notch on the posterior lobe of the concave surface of the otolith (Figure 14.a). However, this notch may become obscure as the fish ages. The second and third annual marks are more complicated since there are checks associated with them. This problem becomes more serious when the annual mark is close to the otolith margin, which is frequently transparent since it is very thin in this area (Figure 14.b). In most cases, the annual marks and checks were more easily distinguished after the otoliths were soaked in glycerin.

For aguada, the appearance of annual growth rings was determined from dark bands under reflected light against a dark-background (Figure 15). The first two marks were frequently associated with a deep notch. The transparent appearance at the margin of the otoliths again presented a problem for the identification of annual growth marks. This was more of a problem for 1 and 2 year old fish since growth appeared fast at these stages and the annual growth marks were wider than for older fish.

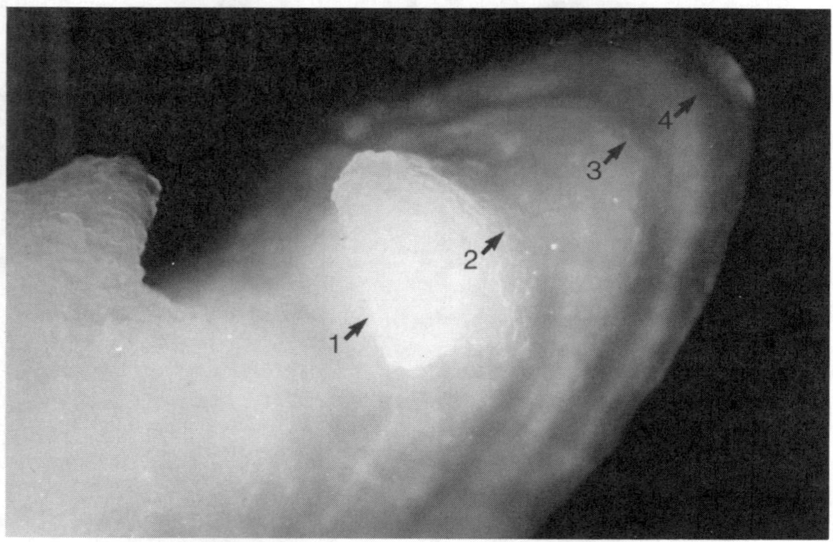

Figure 15. Photograph of an otolith from a 4 year old *Cynoscion squamipinnas*. Note the notch associated with the first 2 years of growth.

Validation of age readings based on LFA rests upon the assumption that the length-frequency distributions are a mixture of several normal distributions and that each distribution has a mean length-at-age and a variance. The estimated proportions, mean lengths, and their variances are computed by the MIX program (Macdonald and Green 1985). Examination of the estimated standard deviation and the estimated mean lengths-at-age for each age class clearly showed that no statistical difference existed between the ages predicted by the two methods. This suggests that the surface reading method used was accurate. The results of the two methods of age determination for aguada and coliamarilla are in Table 3.

Table 3. Mean lengths-at-age and standard deviation (S.D.) of *Cynoscion stolzmanni* and *Cynoscion squamipinnas* estimated by otolith age determination and MIX.

A.	*Cynoscion stolzmanni* (coliamarilla)									
	Age	I	II	III	IV	V	VI	VII	VIII	IX
I.	From MIX									
	Mean length (cm)	49.54	57.35	64.78	71.22	85.41 *				
	S.D.	4.54	0.61	2.21	1.99	3.92				
II.	From otoliths									
	Mean length (cm)	46.91	55.77	68.06	72.95	82.63	98.60	84.90	—	88.90
	S.D.	3.94	4.30	3.99	1.02	6.18	—	2.97	—	—
	N	114	26	11	4	4	1	2	—	1

B.	*Cynoscion squamipinnas* (aguada)							
	Age	I	II	III	IV	V	VI	VII
I.	From MIX							
	Mean length (cm)	35.22	41.99	44.95	49.51	52.21	55.78 *	
	S.D.	—	3.29	2.65	1.94	1.48	2.27	2.86
II.	From otoliths							
	Mean length (cm)	27.13	35.67	41.88	46.72	48.94	51.10	
	S.D.	5.03	2.31	3.78	3.62	2.77	2.97	
	N	14	30	26	33	14	2	

* Includes the ages equal to and greater than the corresponding age category.

The von Bertalanffy growth parameters and their standard deviations were estimated from the mean lengths-at-age in Table 4. Both species grew to about 49% of L_∞ in their first year of life, which is a very rapid growth. If the growth curves were fitted, K was 0.506 for coliamarilla and 0.647 for aquada, while the values of L_∞ remained nearly the same, independent of horizontal translation along the age axis. The length distributions at age covered a wide range, which suggests continuous or semi-continuous recruitment and spawning during the year, which is expected in a tropical environment. Growth rate in the first year is therefore especially variable, and important, given its role in determining the ages of sexual maturity and recruitment to the fishery, both of which are essential aspects of the estimation of year class strength and yield potential.

The use of length frequency data for validation of direct age readings is not the preferred method. Preferable methods include use of known age specimens in an

Table 4. Estimated parameters of the von Bertalanffy growth curves for *Cynoscion stolzmanni* and *Cynoscion squamipinnas* from the otolith surface readings.

	Cynoscion stolzmanni		*Cynoscion squamipinnas*	
Parameters	Estimate	S.D.	Estimate	S.D.
L_∞	96.682	7.271	55.651	0.915
K	0.318	0.146	0.370	0.026
t_0	– 0.946	0.935	– 0.797	0.106

independent experiment such as oxytetracycline tagging. The analysis of length frequency distributions was selected as an alternative to other more expensive and essentially impossible experiments for this remote tropical environment.

The von Bertalanffy growth parameters of aguada have also been estimated by Stevenson (1981a, b) using the method of NORMSEP (Hasselblad, 1966; Young and Skillman, 1975) based on length frequency distributions calculated from assessment survey data and by Madrigal (1985) using ELEFAN (Pauly and David, 1981) with length frequency distributions calculated from data gathered from the fishery. Table 5 clearly shows that the estimates of L_∞ and K from ELEFAN and NORMSEP analyses are different from those estimated by the otolith method. The differences probably result

Table 5. Growth parameter estimates and standard deviations (if known) of *Cynoscion squamipinnas* in various studies.

	L_∞(cm)	K	t_0	Remarks
Stevenson (1981a, b)	60.00	0.26	-	Head Gulf (76–77), NORMSEP
	60.00	0.65	-	Mid Gulf (76–77), NORMSEP
Madrigal (1985)	75.5	0.4	– 0.1607	1979 survey, ELEFAN
	72.5	0.4	– 0.1621	1982 survey, ELEFAN
Present study	55.651	0.37	– 0.797	Otolith analysis and MIX
	(0.915)	(0.026)	(0.106)	Standard deviations

from the misinterpretation of modes in the length frequency distributions. Other possible explanations for the observed differences between the estimates include changes in population structure due to recruitment fluctuation, gear selectivity, growth variation, and harvesting pressure. While not very much is known about these possible explanations, none of the available evidence indicated a dramatic shift in the fishery over the time periods involved. Unfortunately, none of the three species appear in the list of 13 sciaenid species analyzed by Isaac (1990) using length-based methods.

9.1 CONCLUDING COMMENTS

The results from the age determination of the two tropical corvinas, based on surface reading of their otoliths and validation with LFA via MIX (Lai and Campos 1989), combined with the age determination work of Mug et al. (1994), based on the microincrements of the otoliths and validation with surface readings, are a fitting end to this chapter. The technological sophistication available for ageing has changed significantly, even over the 5 years separating these two reports. Statistical methods of analysis have advanced from the early fitting of apparent modes to constrained optimization methods which provide estimates of age and asymptotic variance.

Compared to other treatments of age determination, this chapter is distinguished by its extensive coverage of the technological and statistical aspects of collecting age data for stock assessment and the consequences of errors in age estimation. Chapter 2 on size-based methods and models deals with stock assessment methodology when size is a surrogate independent variable for age. Later chapters on sampling designs present other methods for the efficient collection of a data base for stock assessment.

ACKNOWLEDGMENTS

The authors express appreciation to Shayne MacLellan of Pacific Biological Station, Nanaimo, British Columbia, for reading multiple drafts of this chapter. Prof. T. Pietsch assisted with fish identifications and drawings of anatomical parts, and T. Tsai assisted with literature searches; both of the University of Washington. The U.S. Agency for International Development under grant No. DAN-4146-G-SS-5071-00, provided support. This publication is Contribution No. 898, University of Washington, School of Fisheries, Seattle WA 98195, carried out under the Management Assistance for Artisanal Fisheries (MAAF) Program. Address correspondence to Vincent Gallucci, Center for Quantitative Science, Box No. 355230 University of Washington, Seattle, Washington 98195.

REFERENCES

Abdel-Aziz, S.H. 1992. The use of vertebral rings of the brown ray (*Raja miraletus*) Linneaus 1758 off Egyptian Mediterranean coast for estimation of age and growth. Cybium 16: 121-132.

Adlerstein, S. 1987. Age and growth of *Concholepas concholepas* (Brugiere 1789) using microgrowth structure and spectral analysis. M.S. Thesis, University of Washington, Seattle.

Albrechtsen, K. 1968. A dyeing technique for otolith age reading. J. Cons. Int. Explor. Mer. 32: 278-280.

Allen, R.L. 1976. Method for comparing fish growth curves. N. Z. J. Mar. Freshwater Res. 10: 687-692.

Alvarez, F. and Morales-Nin, B. 1992. An attempt to determine growth and hatching dates of juvenile sardine (*Sardina pilchardus*) in the western Mediterranean Sea. Mar. Biol. 114: 199-203

Anon. 1980. Development of a software program for automatic image analysis of fish scales from haddock. Final Report. Cambridge Instrument Company. Cambridge, MA.

Armstrong, M.P., Musick, J.A. and Colvocoresses, J.A. 1992. Age, growth and reproduction of the goosefish, *Lophius americanus* (*Pisces Lophiiformes*). U.S. Fish. Bull. 90: 217-230.

Ashford, J.R., Robinson, K. and White, M.G. 1993. A method for preparing large numbers of otolith sections for viewing by scanning electronic microscope. ICES J. Mar. Sci. 50: 227-229.

Bada, J.L. and Schroeder, R. A. 1975. Amino acid racemization reactions and their chemical implications. Naturwiss. 61: 71.

Babaluk, J.A., Craig, J.F. and Campbell, J.S. 1993. Age and growth estimation of walleye (*Strizonstedion vitreum*) using opercula. Can. Manus. Rep. Fish. Aquat. Sci. 2183.

Bagenal, T.B. (Ed.) 1974. *The Ageing of Fish*. Unwin Brothers Ltd., Surrey, England

Baillon, N. and Kulbicki, M. 1988. Ageing of adult tropical reef fish by otoliths: A comparison of three methods on *Diagramma pictum*. In *Proc. of 6th International Coral Reef Symposium*. (Choat, J.H. et al., Eds.) Vol. 2: 341-346.

Banda, M.C. 1992. Age and growth parameters of Chambo (*Oreochomis spp.*) and the southeast arm of Lake Malawi, as determined from opercula bones. FAO FI/DP/MLW/86/013-Field-Doc-20.

Bard, Y. 1974. *Nonlinear Parameter Estimation*. Academic Press, New York.

Bardach, J.E. 1955. The opercula bone of the yellow perch (*Perca flavescens*) as a tool for age and growth studies. Copeia 1955: 107-109.

Barlow, J. 1984. Mortality estimation: biased results from unbiased ages. Can. J. Fish. Aquat. Sci. 41: 1843-1847.

Bartoo, N.W. and Parker, K.R. 1983. Stochastic age-frequency estimation using the von Bertalanffy growth equation. Fish. Bull. U.S. 81: 91-96.

Beamish, R.J. 1973. Determination of age and growth of populations of the white sucker (*Catostomus commersoni*) exhibiting a wide range in size at maturity. J. Fish. Res. Bd. Can. 30: 607-616.

Beamish, R.J. 1979. Differences in the age of Pacific hake (*Merluccius productus*) using whole otoliths and sections of otoliths. J. Fish. Res. Bd. Can. 36: 141-151.

Beamish, R.J. 1979b. New information on the longevity of Pacific ocean perch (*Sebastes alutus*). J. Fish. Res. Bd. Can. 36: 1395-1400.

Beamish, R.J. 1981. Use of fin ray sections to age walleye pollock, Pacific cod, and albacore, and importance of this method. Trans. Am. Fish. Soc. 110: 287-299.

Beamish, R.J., and Chilton, D.E. 1977. Age determination of lingcod (*Ophiodon elongatus*) using dorsal fin rays and scales. J. Fish. Res. Bd. Can. 34: 1305-1313.

Beamish, R.J. and Chilton, D.E. 1982. Preliminary evaluation of a method to determine the age of sablefish (*Anoplopoma fimbria*). Can. J. Fish. Aquat. Sci. 39: 277-287.

Beamish, R.J. and Fournier, A. 1981. A method for comparing the precision of a set of age determinations. Can. J. Fish. Aquat. Sci. 38: 982-983.

Beamish, R.J. and McFarlane, G.A. 1983. Validation of age determination estimates: The forgotten requirement. In *Proc. of International Workshop on Age Determination of Oceanic Pelagic Fishes: Tunas, Billfishes and Sharks*. (Prince, E.D. and Pulos, L.M., Eds.) NOAA Tech. Rep. NMFS. 29-33

Beamish, R.J. and McFarlane, G.A. 1983. The forgotten requirement for age validation in fisheries biology. Trans. Am. Fish. Soc. 112: 735-743.

Beamish, R.J. and McFarlane, G.A. 1987. Current trends in age determination methodology. In *The Age and Growth of Fish*. (Summerfelt, R.C. and Hall, G.E., Eds.) Iowa State University Press, Ames, IA.

Beamish, R.J., McFarlane, G.A. and Chilton, D.E. 1983. Use of oxytetracycline and other methods to validate a method of age determination for sablefish.. In *Proceedings of the International Sablefish Symposium*. (Melteff, B.R., Ed.) 95-118. Alaska Sea Grant Report. 83-8.

Beckman, D.W. and Wilson, C.A. 1990. Variability in incorporation of calcein as a fluorescent marker in fish otoliths. Am. Fish. Soc. Symp. 7: 547-549.

Bennett, J.T., Boehlert, G.W. and Turekian, K.K. 1982. Confirmation of longevity in *Sebastes diploproa* (*Pisces*: *Scorpaenidae*) using Pb-210/Ra-226 measurements in otoliths. Mar. Biol. 71: 209-215.

Bennett, S. 1961. Further observations on the fisher and biology of "chooday" (*Sardinella spp.*) of Madapam area. Indian J. Fish. 8: 157-168.

Bernard, D.R. 1981. Multivariate analysis as a means of comparing growth in fish. Can. J. Fish. Aquat. Sci. 38: 233-236.

Berry, F.H., Lee, D.W. and Bertolino, A.R. 1977. Progress in Atlantic bluefin tuna ageing attempts. ICCAT Collect. Sci. Pap. Madrid 6: 305-317.

Beverton, R.J.H. and Holt, S.J. 1957. On the dynamics of exploited fish population. U.K. Min. Agric. Fish., Fish. Invest., Ser. 2. 19.

Bigelow, H.B. (Ed.) 1963. *Fishes of the Western North Atlantic*. Sears Foundation for Marine Research. Yale University, New Haven, CT.

Bilton, H.T. 1985. Variation in scales sampled from different body areas of adult pink salmon. Can. Tech. Rep. Fish. Aquat. Sci. No. 1360.

Blacker, R. W. 1974. Recent advances in otolith studies. In *Sea Fisheries Research.* (Jones, F.R.H., Ed.) Wiley, New York. 67-90.

Boehlert, G.W. 1985. Using objective criteria and multiple regression models for age determination in fishes. U.S. Fish. Bull. 83: 103-117.

Boehlert, G.W. and Yoklavich, M.M. 1984. Variability in age estimates in *Sebastes* as a function of methodology, different readers, and different laboratories. (unpublished manuscript)

Bouain, A. and Siam, Y. 1988. A new Technique for staining fish otoliths for age determination. J. Fish. Biol. 32: 977-978.

Box, G.E.P. and Jenkins, G.M. 1976. *Time Series Analysis, Forecasting and Control.* Hoden-Day, San Francisco.

Brain, E.B. 1966. *The Preparation of Decalcification Sections.* Chas. C Thomas, Springfield, IL.

Brander, K. 1974. The effect of age reading errors on the statistical reliability of marine fishery modeling. In *Ageing of Fish.* (Bagenel, T.B., Ed.) Unwin Bros. Surrey, England. 181-191.

Brothers, E.B. 1979. Age and growth studies on tropical fishes. In *Stock Assessment for Tropical Small-Scale Fisheries* (Roedel, P.M. and Saila, S.B., Eds.) International Center for Marine Resources Development. University of Rhode Island. 119-136

Brothers, E.B. 1982. Aging reef fishes. In *The Biological Bases for Reef Fishery Management* (Huntsman, G.R., Nicholson, W. and Fox, W.W., Eds.) National Oceanic and Atmospheric Administration. Washington D.C. 1-23.

Brothers, E.B. 1987. Methodological approaches to the examination of otoliths in aging studies. In *The Age and Growth of Fish.* (Summerfelt, R.C. and Hall, G.E., Eds.) Iowa State University Press. Ames, IA.

Brothers, E.B. 1990. Otolith marking. Am. Fish. Soc. Symp. 7: 183-202.

Buckley, L.J. 1979. Relationships between RNA-DNA ratio, prey density, and growth rate in Atlantic cod (*Gadus morhua*) larvae. J.Fish. Res. Bd. Can. 36: 1497-1502.

Buckley, L.J. 1984. RNA-DNA ratio: an index of larval fish growth in the sea. J. Fish. Res. Bd. Can. 80: 291-298.

Buckley, L.J. and Lough, R.G. 1987. Recent growth, biochemical composition, and prey field of larval haddock (*Melanogrammus aeglefinus*) and Atlantic cod (*Gadus morhua*) on Georges Bank. Can. J. Fish. Aquat. Sci. 44: 14-25.

Bulow, F.J. 1987. RNA-DNA ratios as indicators of growth in fish: a review. In *The Age and Growth of Fish.* (Summerfelt, R.C. and Hall, G.E., Eds.) Iowa State University Press. Ames, IA.

Butler, J.L. 1992. Collection and preservation of material for otolith analysis. In *Otolith Microstructure Examination and Analysis* (Stevenson, D.K. and Campana, S.E., Eds.) 13-17.

Cailliet, G., Martin, L.K., Kusher, D., Wolf, P., and Welden, B.A. 1983. Techniques for enhancing vertebral bands in age estimation of California elasmobranchs. In *Proc. of International Workshop on Age Determination of Oceanic Pelagic Fishes: Tunas, Billfishes and Sharks*. (Prince, E.D. and Pulos, L.M., Eds.) NOAA Tech. Rep. NMFS-8.

Cailliet, G.M. and Radtke, R.L. 1987. A progress report on the electron microprobe analysis technique for age determination and verification in elasmobranchs. In *The Age and Growth of Fish*. (Summerfelt, R.C. and Hall, G.E., Eds). Iowa State University Press. Ames, IA.

Campana, S.E. 1983a. Calcium deposition and check formation during periods of stress in coho salmon (*Oncorhychus kisutch*). Comp. Biochem. Physiol. 75A: 215-220.

Campana, S.E. 1983. Feeding periodicity and the production of daily growth increments in otoliths of steelhead trout (*Salmo gairdneri*) and starry flounder (*Platichtys stellatus*). Can. J. Zool. 61: 1591-1597.

Campana, S.E. 1984. Interactive effects of age and environmental modifiers on the production of daily growth increments in otoliths of plainfin midshipman (*Porichthys notatus*). Fish. Bull. U.S. 82: 165-177.

Campana, S.E. and Hurley, P.C.F. 1989. An age- and temperature-mediated growth model for cod (*Gadus morhua*) and haddock (*Melanogrammus aeglefinus*) larvae in the Gulf of Maine. Can. J. Fish. Aquat. Sci. 46: 603-613.

Campana, S.E. and Jones, C.M. 1992. Analysis of otolith microstructure data. In *Otolith Microstructure Examination and Analysis* (Stevenson, D.K. and Campana, S.E., Eds.) Can. Spec. Publ. Fish. Aquat. Sci. Ser. 117: 73-100.

Campana, S.E. and Neilson, J.D. 1982. Daily growth increments in otolith of starry flounder (*Platichthys stellatus*) and the influence of some environmental variables in their production. Can. J. Fish. Aquat. Sci. 39: 937-942.

Campana, S.E. and Neilson, J.D. 1985. Microstructure of fish otoliths. Can. J. Fish. Aquat. Sci. 42: 1014-1032.

Campana, S.E., Zwanenburg, K.C.T. and Smith, J.N. 1990. Pb-210/Ra-226 determination of longevity in redfish. Can. J. Fish. Aquat. Sci. 47: 163-165.

Campbell, J.S. and Babaluk, J.A. 1979. Age determination of walleye (*Stizostedium vitreum vitreum*) (Mitchell), based on the examination of eight different structures. Fish. Ms. Tech. Rep. 849.

Casselman, J.M. 1974. Analysis of hard tissue of pike (*Esox lucius*) (Bagenal, L.T.B., Ed). Ageing of Fish. Aquat. Sci. 60.

Casselman, J.M. 1978. Calcified tissue and body growth of northern pike (*Esox lucius, Linnaeus*). Ph.D. Thesis. University of Toronto.

Casselman, J.M. 1979. The esocid cleithrum as an indicator calcified structure In *Proc. 10th Warm Water Workshop* (Duke, J. and Gravle, Y., Eds.) Spec. Publ. N.E. Am. Fish. Soc. 249-171.

Casselman, J.M. 1983. Age and growth assessment of fish from their calcified structures: techniques and tools. U.S. Dept. of Commerce, NOAA Tech. Rept. NMFS 8: 1-17.

Casselman, J.M. 1990. Growth and relative size of calicified structures of fish. Trans. Am. Fish. Soc. 119: 673-688.

Cassie, R.M. 1954. Some uses of probability paper in the analysis of size frequency distributions. Aust. J. Mar. Freshw. Res. 5: 513-522.

Cerrato, R.M. 1990. Interpretable statistical tests for growth comparisons using parameters in the von Bertalanffy equation. Can. J. Fish. Aquat. Sci. 47: 1416-1426.

Chang, W.Y.B. 1982. A statistical method for evaluating the reproductivity of age determination. Can. J. Fish. Aquat. Sci. 39: 1208-1210.

Chapman, D.G. 1960. Statistical problems in dynamics of exploited fisheries populations. Proc. 4th Berkeley Symp. on Math. Stat. and Prob. Berkeley, CA. 153-168.

Chapman, D.G. and Robson, D.S. 1960. The analysis of catch curves. Biometrics. 16: 354-368.

Chauvelon, P. and Bach, P. 1993. Modeling the otolith shape as an ellipse, an attempt for back-calculation purpose. ICES J. Mar. Sci. 50: 121-128.

Chilton, D.E. 1970. Preparation and mounting of Pacific cod scales for age determination. Fish. Res. Bd. Can. MS. Rept. 1086.

Chilton, D.E. and Beamish, R.J. 1982. Age determination methods for fishes studies by the groundfish programs at Pacific Biological Station. Can. Spec. Publ. Fish. Aquat. Sci. 60.

Christensen, J.C. 1964. Burning of otoliths, a technique for age determination of soles and other fish. J. Cos. 29: 73-81.

Chugunova, N.I. 1963. Age and growth studies in fish. Office of Tech. Serv. Washington, D.C. (Russian version 1959).

Chur, V.N., Ouchinnikov, V.V. and Korolevich, L.I. 1986. Some problems of age and growth of skipjack (*Katsuwonces pelamis*) from the eastern Atlantic Ocean. In *Proc. of ICCAT Conf. on the International Skipjake Year Program*. (Symons, P.E.K., Miyake, P.M. and Sakagawa, G.T., Eds.).

Clark, W.E., Mandapat, R.R. and Field, L.J. 1986. Comparative studies of yellowtail rockfish age readings. WDF. Tech. Rep. 97.

Clark, W.G. 1981. Restricted least-squares estimates of age composition from length composition. Can. J. Fish. Aquat. Sci. 38: 297-307.

Clemmesen, C. 1994. The effect of food availability, age or size on the RNA/DNA ratio of individually measured herring larvae: laboratory calibration. Mar. Biol. 118: 377-382.

Clendenen, G., Gallucci, V.F. and Gara, R.I. 1978. On the spectral analysis of cyclical tussock moth epidemics and corresponding climatic indices, with a critical discussion of the underlying hypotheses. In *Time Series and Ecological Processes*. (Shugart, H.H., Jr., Ed.) SIAM, Philadelphia.

Cochran, J.K., Rye, D.M. and Landman, N.H. 1981. Growth rate and habitat of *Nautilus pompilius* inferred from radioactive and stable isotope studies. Paleobiol. 7: 469-480.

Cochran, W.G. 1977. *Sampling Techniques.* 3rd Edition. John Wiley & Sons, New York.

Cook, E.R. 1985. The use and limitations of dendrochronology in studying effects of air pollution on forests. In *Proceedings of NATO advanced research workshop: Effects of acidic deposition on forests, wetlands, and agricultural ecosystems.* 12-17 May 1985. Toronto, Canada.

Cooper, R.A. 1967. Age and growth of the tautog, (*Tautoga onitis*) (L.) from Rhode Island. Trans. Am. Fish. Soc. 96: 134-142.

Crichton, M.I. 1935. Scale absorption in salmon and sea trout. Scotland Fishery Board. Salmon Fisheries. No. 4.

Dannevig, E.H. 1956. Chemical composition of the zones in cod otoliths. J. Cons. 21: 156-159.

Davies, N.M., Gauldie, R.W., Crane, S.A. and Thompson, R.K. 1988. Otolith ultrastructure of smooth oreo (*Pseudocyttus maculatus*) and black oreo (*Allocyttus sp.*) species. U.S. Fish. Bull. 86: 499-516.

Deelder, C.L. and Williemse, J.J. 1966. Doe het zelf. Leer ook de leefijd kennen van Uw partner: De Zoetwatervis. De Sportvisser. 14(5): 141-145.

Deelder, C.L. and Williemse, J.J. 1973. Age determination in freshwater teleosts, based on annular structures in fin rays. Aquaculture. 1: 365-371.

Degens, E.T., Deuser, W.G. and Haedrich, R.L. 1969. Molecular structure and composition of fish otoliths. Mar. Biol. 2: 105-113.

Dempster, A.P., Laird, N.M. and Rubin, D.B. 1977. Maximum likelihood estimation from incomplete data via the EM algorithm. J. Roy. Stat. Soc. B 39: 1-39.

Deriso, R.B., Quinn, T.J. II and Neal, P.R. 1985. Catch-at-age analysis with auxiliary information. Can. J. Fish. Aquat. Sci. 42: 815-824.

De Vries, D.A., Grimes, C.B., Lang, K.L. and White, D.B. 1990. Age and growth of king and spanish mackerel larvae and juveniles from the Gulf of Mexico and U.S. Atlantic Bight. Environ. Biol. Fish. 29: 135-143.

Dixon, W.J. (Ed.). 1983. *BMDP Statistical Software.* University of California Press. Berkeley, CA.

Dodge, R.E. and Thompson, J. 1974. The natural radiochemical and growth records in contemporary hermatypic corals from the Atlantic and the Carribbean. Earth Planet Science Letter. 23: 313-322.

Doubleday, W.G. and Rivard, D. (Eds.). 1983. Sampling commercial catches of marine fish and invertebrates. Can. Spec. Publ. Fish. Aquat. Sci. 66.

Draper, N.R. and Smith, H. 1981. *Applied Regression Analysis.* John Wiley and Sons. New York.

Dunkelberger, D.G., Dean, J.M. and Watanabe, N. 1980. The ultrastructure of the otolithic membrane and otolith in the juvenile mummichog (*Fundulus heteroclitus*). J. Morphol. 163: 367-377.

Edmonds, J.S., Moran, M.J., Caputi, N. and Morita, M. 1989. Trace element analysis of fish sagittae as an aid to stock identification: pink snapper (*Chryosphys auratus*) in Western Australian waters. Can. J. Fish. Aquat. Sci. 46: 50-54.

Eslava de Gonzalez, N. 1991. Comparison of the use of scales and cleithra for determining age and growth of chere-chere (*Haemulon steindachneri*) Jordan and Gilbert 1882. Mem. Soc. Cienc. Nat. la Salle 51: 97-107.

Fagade, S.O. 1974. Age determination in *Tilapia melanotheron* (Ruppell) in the Lagos Lagoon, Lagos, Nigeria. In *Ageing of Fish.* (Bagenal, T.B., Ed.) Unwin Bros. Surrey, England. 71-77.

Farber, M.I. and Lee, D.W. 1981. Ageing eastern Atlantic bluefin tuna (*Thunnus thynnus*) using tagging data, candal vertebrae and otoliths. ICCAT. Collect. Vol. Sci. Papp. Madrid 15: 288-301.

Feddern, H.A. 1965. The spawning growth and general behavior of the bluehead wrasse (*Thalassoma bifascatum*) (Pisces Labridae) Bull. Mar. Sci. 15: 896-941.

Fenton, G.E., Ritz, D.A. and Short, S.A. 1990. Pb-210/Ra-226 disequilibria in otoliths of blue grenadier (*Macruronus novaezelandiae*) problems associated with radiometric ageing. Aust. J. Mar. Freshwat. Res. 41(4): 467-473.

Fenton, G.E., Short, S.A. and Ritz, D.A. 1991. Age determination of orange roughy (*Hoplostethus atlanticus*) (Pisces: *Trachichthyidae*) using Pb-210/Ra-226 disequilibria. Mar. Biol. 109: 197-202.

Fenton, G.E. and Short, S.A. 1992. Fish age validation by radiometric analysis of otoliths. Aust. J. Mar. Freshw. Res. 43: 923-933.

Ferreira, B.P. and Vooren, C.M. 1991. Age, growth and structure of vertebrae in the school shark (*Galeorhinus galeus, Linnaeus* 1750) from southern Brazil. U.S. Fish. Bull. 89: 19-32.

Fienberg, S.E. 1981. *The Analysis of Cross-Classified Categorical Data.* MIT Press. Cambridge, MA.

Fitch, J.E. 1951. Age composition of the southern California catch of Pacific mackerel 1939-41 through 1950-51. Calif. Fish. Game, Fish. Bull. 83: 7-73.

Fleiss, J.L. 1977. *Statistical Methods for Rates and Proportions.* John-Wiley and Sons, New York.

Forrester, C.R., Ketchen, K.S. and Wong, C.C. 1972. Mercury content of spiny dogfish (*Squalus acanthias*) in the Strait of Georgia, British Columbia. J. Fish. Res. Bd. Can. 29: 1487-1490.

Fossum, P. and Moksness, E. 1991. Estimation of daily growth rate and birthdate distribution in Norwegian spring spawning herring (*Clupea harengus* L.), from daily increment studies of 2–4 month old juveniles. Alaska Sea Grant Program Rep. 91-01: 37-52.

Fournier, D.A., Sibert, J.R., Majkowski, J. and Hampton, J. 1990. MULTIFAN a likelihood-based method for estimating growth parameters and age composition from multiple length frequency data sets illustrated using data for southern bluefin tuna (Thunnus maccoyii). Can. J. Fish. Aquat. Sci. 47: 301-317.

Fowler, A.J. 1990. Validation of annual growth increments in the otoliths of small, tropical coral reef fish. Mar. Ecol. Prog. Ser. 64: 25-38.

Francis, R.I.C.C. 1990. Back-calculation of fish length: a critical review. J. Fish. Biol. 36: 883-902.

Fredrikssen, A. 1933. On the calculation of age distribution within a stock of cod by means of relative few age determinations as a key to measurements on large scale. Rapp. P.-V. Reun. Cons. Int. Explor. Mer. 86: 1-14.

Fryer, G. and Iles, T.D. 1972. *The Cichlid Fishes of the Great Lakes of Africa.* Oliver and Boyd, Edinburgh.

Gallucci, V.F. and Quinn, T.J. 1979. Reparameterizing, fitting, and testing a simple growth model. Trans. Am. Fish. Soc. 108: 14-25.

Garcia-Arteaga, J.P. and Reshetnikov, Y.S. 1985. Age and Growth of the barjack (*Caranx ruber*) off the coast of Cuba. J. Ichthyol. 25: 120-131.

Gauldie, R.W. 1988. Function, form and time-keeping properties of fish otoliths. Comp. Biochem. Physiol.. 91A: 395-402.

Gauldie, R.W. 1990. Phase differences between check ring locations in the orange roughy otolith (*Hopolostethus atlanticus*). Can. J. Fish. Aquat. Sci. 47: 760-765.

Gauldie, R.W., Mulligan, K. and Thompson, R.K. 1987. The otoliths of a chimaera, the New Zealand elephant fish (*Callorhynchus milii*). N.Z. Mar. Freshwat. Res. 21: 275-280.

Gauldie, R.W., Fournier, D.A, Dunlop, D.E. and Coote, G. 1986. Atomic emission and proton microprobe studies of the ion content of otoliths of chinook salmon aimed at recovering the temperature life history of individuals. Comp. Biochem. Physiol. A 84: 607-615.

Gauldie, R.W., Coote, G., West, I.F. and Radtke, R.L. 1990. The influence of temperature on the fluorine and calcium composition of fish scales. Tissue Cell; 22: 645-654.

Gauldie, R.W., Davies, N.M., Coote, G. and Vickridge, I. 1990. The relationship between organic material and check rings in fish otoliths. Comp. Biochem. Physiol. 97A: 461-474.

Gauldie, R.W. and Nelson, D.G.A. 1988. Argonite twinning and neuroprotein secretion are the cause of daily growth ring in fish otoliths. Comp. Biochem. Physiol. 90A: 501-509.

Gauldie, R.W. and Nelson, D.G.A. 1990. Otolith growth in fishes. Comp. Biochem. Physiol. 97A: 119-135.

Gauldie, R.W. and Nelson, D.G.A. 1990. Interactions between crystal ultrastructure and microincrement layers in fish otoliths. Comp. Biochem. Physiol. 97A: 449-459.

Gauldie, R.W. and Radtke, R.L. 1990. Using the physical dimensions of the semicircular canal as a probe to evaluate inner ear function in fishes. Comp. Biochem. Physiol. 96A: 199-203.

Gauldie, R.W. and Radtke, R.L. 1990. Microincrementation: facultative and obligatory precipitation of otolith crystal. Comp. Biochem. Physiol. 97A: 137-144.

Geffen, A.J. 1982. Otolith ring deposition in relation to growth rate in herring (*Clupea harengus*) and turbot (*Scophthalmus maximus*) larvae. Mar. Biol. 71: 317-326.

Geffen, A.J. 1987. Methods of validating daily increment deposition in otoliths of larval fish. In *The Age and Growth of Fish.* (Summerfelt, R.C. and Hall, G.E., Eds). Iowa State University Press, Ames, IA.

Geffen, A.J. 1992. Validation of otolith increment deposition rate. In *Otolith Microstructure Examination and Analysis*. (Stevenson, D.K. and Campana, S.E., Eds). Can. Spec. Publ. Fish. Aquat. Sci. Ser. 117: 101-113.

Gjøsaeter, J., Dayaratne, P., Bergstad, O.A., Gjøsaeter, H., Sousa, M.I. and Beck, I.M. 1984. Ageing tropical fish by growth rings in the otoliths. FAO Fish. Cir. No. 776.

Graham, M. 1928. Studies of age-determination in fish. Part II. A survey of the literature. Fish. Inv. Ser. II. 11(3): 1-50.

Gulland, J.A. 1955. Estimation of growth and mortality in commercial fish population. Fish. Invest. Lond. Ser.II, XVIII(9).

Gulland, J.A. 1958. Age determination of cod by fin rays and otoliths. Spec. Publ. Int. Comm. N.W. Atlantic Fish. 1: 179-190.

Gulland, J.A. 1969. Manual of methods for fish stock assessment. Part I. Fish population analysis. FAO Manu. Fish. Sci, No. 4, FRs/M4.

Gunn, J.S., Harrowfield, I.R., Proctor, C.H. and Thresher, R.E. 1992. Electron probe microanalysis of calcified tissues in fish-evaluation of techniques for studying age and stock discrimination. J. Exp. Mar. Biol. Ecol. 158: 1-38.

Gutierrez, E. and Morales-Nin, B. 1986. Time series analysis of daily growth in *Dicentrarchus labrax* L. otoliths. J. Exp. Mar. Biol. Ecol. 103: 163-179.

Hales, L.S. Jr. and Belk, M.C. 1992. Validation of otolith annuli of bluegills in a southeastern thermal reservoir. Trans. Am. Fish. Soc. 121: 823-830.

Hasselblad, V. 1966. Estimation of parameters for a mixture of normal distributions. Technometrics 8: 431-442.

Heidinger, R.C. and K. Clodfelter. 1987. Validity of the otolith for determining age and growth of walleye, striped bass and smallmouth bass in power plant cooling ponds. In *The Age and Growth of Fish*. (Summerfelt, R.C. and Hall, G.E., Eds). Iowa State University Press. Ames. 261-251.

Hettler, W.F. 1984. Marking otoliths by immersion of marine fish larvae in tetracycline. Trans. Am. Fish. Soc. 113: 370-373.

Hickling, C.F. 1983. The natural history of the hake: age determination and growth rate. Fishery Invest. London. Ser. 2. 13: 120 p.

Hill, K. and Radtke, R.L. 1987. Otolith increment deposition and lipofuscin accumulation in the damselfish (*Dascyllus albisella*) as related to growth. Sec. Intern. Symp. on Indo-Pacific Mar. Biol. Agana (Guam), 23-28 June 1986. 41(2).

Hoenig, J.M. and Heisey, D.M. 1987. Using a log-linear model with the EM algorithm to correct estimates of stock composition and to convert length to age. Trans. Am. Fish. Soc. 116: 232-243.

Hohn, A.A. 1980. Analysis of growth layers in the teeth of *Tursiops truncatus* using light microscopy, microradiography, and SEM. In *Age Determination of Toothed Whales and Sirenians*. (Perrin, W. E. and Myrick, A.C. Jr., Eds.) IWC Sp. Issue 3.

Holden, M.J. and Meadows, P.S. 1962. The structure of the spine of the spurdogfish (*Squalus acanthias* L.) and its use for age determination. J. Mar. Biol. Assoc. U.K. 42: 179-197.

Hunter, W.G. and Lamboy, W.F. 1981. A Bayesian analysis of the linear calibration problem. Technometrics. 23: 323-350.

Ichii, T. and Mugiya, Y. 1983. Comparative aspects of calcium dynamics in calcified tissues in the goldfish (*Carassius auratus*). Bull. Jpn. Soc. Sci. Fish. 49: 1039-1044.

Irie, T. 1955. The crystal structure of the otolith of a marine teleost (*Pseudoaciaena*. J.) Fac. Fish. Anim. Husb. Hiroshima Univ. 1: 1-8.

Irie, T. 1957. On the forming season of annual rings in the otoliths of several marine teleosts. J. Fac. Fish. Anim. Husb. Hiroshima Univ. 3: 311-317.

Irie, T. 1960. The growth of the fish otolith. J. Fac. Fish. Anim. Husb. Hiroshima Univ. 3: 203-221.

Isaac, V.J. 1990. The accuracy of some length-based methods for fish population studies. ICLARM Tech. Rep. 27.

Jearld, A. Jr. 1983. Age determination. In *Fishery Techniques*. (Nielson, L.A. and Johnson, D.L., Eds.) AFS, Bethesda, MA.

Jenkins, G.P., Shaw, M. and Stewart, B.D. 1993. Spatial variation in food-limited growth of juvenile greenback flounder (*Rhombosolea tapirina*) evidence from otolith daily increments and otolith scaling. Can. J. Fish. Aquat. Sci. 50: 2558-2567.

Jensen, A.C. 1965. A standard terminology and Notation for otolith readers. ICNAF Res. Bull. 2: 5-7.

Jinn, J.H., Sedransk, J. and Smith, P. 1987. Optimum two-phase stratified sampling for estimation of the age composition of a fish population. Biometrics 43: 343-353.

Johannes, R.E. 1978. Reproductive strategies of coastal marine fishes in the tropics. Environ. Biol. Fish. 3: 65-84.

Jones, B.C. and Geen, G.H. 1977. Age determination of an elasmobranch (*Squalus acanthias*) by X-ray spectrometry. J. Fish. Res. Bd. Can. 34: 44-48.

Joseph, J. and Calkins, T.P. 1969. Population dynamics of the skipjack tuna (*Katsuwonns plelamis*) of the eastern Pacific Ocean. IATTC. Bull. 13: 1-273.

Juinio, M.A.R., Cobb, J.S., Bengston, D. and Johnson, M. 1992. Changes in nucleic acids over the molt cycle in relation to food availability and temperature in *Homarus americanus* postlarvae. Mar. Biol. 114(1): 1-10.

Kaganovskaia, S. 1933. A method of determining the age and the composition of the catches of the spiny dogfish (*Squalus acanthias* L.). Vestn. Dal'n. Fil. Akai. Nauk. SSSR 1933. No. 1-3: 139-141. (Transl. Fish. Res. Bd. Can. Transl. Ser. No. 281, 1960).

Kalish, J.M. 1989. Use of otolith microchemistry to distinguish the progeny of sympatric anadromous and non-anadromous salmonids. Fish. Bull. U.S. 88: 657-666.

Kalish, J.M. 1991. Oxygen and carbon stable isotopes in the otoliths of wild and laboratory-reared Australian salmon (*Arripis trutta*). Mar. Biol. 110: 37-47.

Kao, C.L and Liu, H.C. 1972. Age determination of golden thread (*Nemiptera virtus*). Acta. Oceanog. Taiwanica.

Karakiri, M. and von Westerhagen, H. 1988. Apparatus for grinding otoliths of larval and juvenile fish for microstructure analysis. Mar. Ecol. Prog. Ser. 49: 195-198.

Kastelle, C.R., Kimura, D.F., Nevissi, A.E. and Gunderson, D.R. 1994. Using Pb-210/Ra-226 disequilibria for sablefish (*Anaplopama fimbria*), age validation. Fish. Bull. 92: 292-301.

Kato, K. 1953. Cacion oxalate and other calcium salts in fish scale. Sci. Rept. Saituma Univ. Ser. B. 1: 51-58.

Kendall, M. and Stuart, A. 1977. *The Advanced Theory of Statistics*, Vol. 1. *Distribution Theory*. 4th edition. Macmillan, New York.

Ketchen, K.S. 1975. Age and growth of dogfish *Squalus acanthias* in British Columbia waters. J. Fish. Res. Board. Can. 32: 43-59.

Kimura, D.K. 1977. Statistical assessment of the age-length key. J. Fish. Res. Bd. Can. 34: 317-324.

Kimura, D.K. 1979. Likelihood methods for the von Bertalanffy growth curve. Fish. Bull. 77: 765-775.

Kimura, D.K. 1980. Likelihood methods for the von Bertalanffy growth curve. Fish. Bull., U.S. 97: 765-776.

Kimura, M. 1980. Variability in techniques of counting destinal growth layer groups in a tooth of a known-age dolphin (*Tursiops truncatus*). In *Age Determination of Toothed Whales and Sirenians*. (Perrin, W.E. and Myrick, A.C., Jr., Eds.) International Whaling Commision (IWC) Sp. Issue 3.

Kimura, D.K. 1989. Variability in estimating catch-in-numbers-at-age and its impacts on cohort analysis. In *Effects of Ocean Variability on Recruitment and an Evaluation of Parameters Used in Stock Assessment Models* (Beamish, J.R. and McFarlane, G.A., Eds.) Can. Spec. Publ. Fish. Aquat. Sci. 108: 57-66

Kimura, D.K. and Chikuni, S. 1987. Mixtures of empirical distributions: an iterative application of the age-length key. Biometrics 43: 23-35.

Kimura, D.K., Manapat, R.R. and Oxford, S.L. 1979. Method, validity, and variability in the age determination of yellowtail rockfish (*Sebastes flavidus*) using otoliths. J. Fish. Res. Bd. Can. 36: 377-383.

Kingsley, M.C.S. 1979. Fitting the von Bertalanffy growth equation to polar bear age-weight data. Can. J. Zool. 57: 1020-1025.

Kirkwood, G.P. 1983. Estimation of von Bertalanffy growth curve parameters using both length increment and age-length data. Can. J. Fish. Aquat. Sci. 40: 1405-1411.

Kitano, Y., Kanamori, N. and Yoshioka, S. 1980. Argonite to calcite transformation in corals in aquatic environment. In *The Mechanisms of Biomineralization in Animals and Plants. Proc. 3rd Int. Biomineralization Symp.* (Omori, O.M. and Wataler, N., Eds.) Tokai University Press. Toyko. 269-278.

Kobayashi, S., Yuki, R., Furui, T. and Kosugiyama, T. 1964. Calcification in fish and shell-fish. I. Tetracycline labeling pattern on scale, centrum and otolith in young goldfish. Bull. Jpn. Soc. Sci. Fish. 30: 6-13.

Koo, T.S.Y. 1955. Biology of red salmon (*Oncorhynchus nerka*) (Walbaum). University of Washington, Seattle. 1-64.

Kubo, I., and Yoshihara, T. 1968. *Fishery Biology*. Kyoritsu Shuppan. Tokyo.

Lagler, K.F., Bardach, J.E. and Miller, R.R. 1962. *Ichthyology*. John Wiley and Sons, New York.

Lai, H.L. 1985. Evaluation and validation of age determination for sablefish, pollock, Pacific cod and yellowfin sole; optimum sampling design using age-length key; and implications of ageing variability in pollock. Ph.D. thesis. University of Washington, Seattle.

Lai, H.L. 1987. Optimum allocation for estimating age composition using age-length key. Fish. Bull. U.S. 85: 179-185.

Lai, H.L. and Campos, J. 1989. Age determination and growth for two corvinas, *Cynoscion stolzmanni* and *C. squamipinnas* in the Gulf of Nicoya. CRSP Working Paper No. 67. MAAF School of Fisheries, University of Washington, Seattle.

Lai, H.L. and Gunderson, D.R. 1987. Effects of ageing errors on estimates of growth, mortality and yield per recruit for walleye pollock (*Theragra chalcogramma*). Fish. Res. 5: 287-302.

Lai, H.L. and Liu, H.C. 1974. Age determination of red snapper (*Lutjanus sanguinius*) in the South China Sea. J. Fish. Soc. Taiwan 3: 39-57.

Lai, H.L. and Liu, H.C. 1979. Age determination and growth of red snapper, (*Lutjanus sanguinius*) in the north Australian waters. Acta Ocean. Taiwanica 10: 160-170.

Lai, H.L. and Yeh, S.Y. 1986. Age determination of walleye pollock (*Theragra chalcogramma*) using four age structures. Bull. Int. N. Pac. Fish. Comm. 45: 66-98.

Lai, H.L., Gunderson, D.R. and Low, L.L. 1987. Age determination of Pacific cod (*Gadus macrocephalus*) using five ageing methods. U.S. Fish. Bull. 85: 713-723.

Lai, H.L. 1993. Optimal sampling design for using the age-length-key to estimate age composition of a fish population. Fish. Bull. 92: 382-388.

LaLanne, J.J. 1975. Age determination of walleye pollock (*Theragra chalcogramma*) from otoliths. Northwest and Alaska Fish. Center, Proc. Rept. Sept. 1975.

Lander, R.H. and Tanonaka, G.K. 1964. Marine growth of western Alaska sockeye salmon (*Oncorhychus nerka*) (Walbaum). INPFC. Bull. 14: 1-27.

Laroche, J.L., Richardson, S.L. and Rosenberg, A.A. 1982. Age and growth of a pleuronectid (*Parophrys vertulus*) during the pelagic larval period in Oregon coastal waters. Fish. Bull. U.S. 80: 93-104.

Lea, E. 1910. On the methods used in herring investigations. Publs Circonst. Cons. Perm. Explor. Mer. No. 53.

Leak, J.C. and Houde, E.D. 1987. Cohort growth and survival of bay anchovy (*Anchoa mitchilli*) larvae in Biscayne Bay, Florida. Mar. Ecol. Prog. Ser. 37: 109-122.

LeCren, E.D. 1947. The determination of the age and growth of perch (*Perca fluviiatilis*) from the opercular bone. J. Anim. Ecol. 16: 188-204.

LeCren, E.D. 1974. The effects of errors in ageing in production studies. In *Ageing of Fish*. (Bagenel, T.B., Ed.) Unwin Bros. Surrey. 221-224

Lee, D.W., Prince, E.D. and Crow, M.E. 1983. Interpretation of growth bands on vertebrae and otoliths of Atlantic bluefin tuna, *Thunnus thynnus*. In *Proc. International Workshop on Age Determination of Oceanic Pelagic Fishes: Tunas, Billfishes and Sharks*. NOAA Tech. Rep., NMFS 8.

Lee, R.M. 1920. A review of the methods of age determination in fishes by means of scales. Fishery Invest. Lond. Ser. 2,4,2.

Le Foll, D., Brichett, E., Reyss, J.L., Lalou, C. and Latrouite, C. 1989. Age determination of the spider crab (*Maja squinado*) and the European lobster (*Homarus gammarus*) by super (228)Th/ super(228)Ra chronology: Possible extension to other crustaceans. Can. J. Fish. Aquat. Sci. 46: 720-724.

Lehninger, A.L. 1975. *Biochemistry: The Molecular Basis of Cell Structure and Function*. Worth Publishers. New York.

LeGuen, J.C. and Sakagawa, G.T. 1973. Apparent growth of yellowfin tuna from the eastern Atlantic Ocean. Fish. Bull. U.S. 71: 175-187.

Linfield, R.S.J. 1974. The errors likely in ageing roach (*Rutilus rutilus*) (L.) with special reference to stunted populations. In *Ageing of Fish*. (Bagenel, T.B., Ed.) Unwin Bros. Surrey. 167-172.

Lo, N.C.H., Hunter, J.R. and Hewitt, R.P. 1989. Precision and bias of estimates of larval mortality. Fish. Bull. U.S. 87: 399-416.

Lobon-Cervia, J., Utrilla, C.G., Querol, E. and Puig, M.A. 1993. Population ecology of pike-cichlid (*Crenicichla lepidota*) in two streams of the Brazilian Pampa subject to a severe drought. J. Fish. Biol. 43: 537-557.

Lokkeborg, S. 1990. Reduced catch of under-sized cod (*Gadus morhua*) in longline by using artificial bait. Can. J. Fish. Aquat. Sci. 47: 1112-1115.

Longhurst, A.T. 1965. The biology of Western African polynamid fishes. J. Cons. Int. Explor. Mer. 30: 58-74.

Lønne, O.J. and Gray, J.S. 1988. Influence of tides on microgrowth bands in *Cerastoderma edule* from Norway. Mar. Ecol. Prog. Ser. 42: 1-7.

Lough, R.G., Pennington, M., Bolz, G.R. and Rosenberg, A.A. 1982. Age and growth of larval Atlantic herring, *Clupea harengus*, in the Gulf of Maine-Georges Bank region based on otolith growth increments. Fish. Bull. U.S. 80: 187-200.

Macdonald, P.D.M. and Pitcher, T.J. 1979. Age-groups from size-frequency data: a versatile and efficient method of analyzing distribution mixtures. J. Fish. Res. Bd Can. 36: 987-1001.

Macdonald, P.D.M. and Green, P.E.J. 1985. User's guide to program MIX: An interactive program for fitting mixtures of distributions. Ichthus Data System, Ontario, Canada.

Madrigal, A.E. 1985. Dinamica pesquera de tres especies de sciaenidae (corvinas) en el Golfo de Nicoya, Costa Rica. M.S. thesis, University of Costa Rica.

Maeda, R. 1982. Ageing of sablefish (*Anoplopoma fimbria*) from otoliths and scales: an evaluation of readability, reliability, and derived growth patterns. M.S. Thesis, Humboldt State University, Humboldt, CA.

Malloy, K.D. and Targett, T.E. 1994. The use of RNA:DNA ratios to predict growth limitation of juvenile summer flounder (*Paralichtys dentatus*) from Delaware and North Carolina estuaries. Mar. Biol. 118: 367-376.

Marquardt, D.W. 1963. An algorithm for least squares estimation of nonlinear parameters. J. Soc. Ind. Appl. Math. 2: 431-441.

Marshall, S.L. and Parker, S.S. 1982. Pattern identification in the microstructure of sockeye salmon (*Oncorhychus nerka*) otoliths. Can. J. Fish. Aquat. Sci. 39: 542-547.

Mason, J.E. 1973. A semi-automatic machine for counting and measuring circuli on fish scales. In *Ageing of Fish.* (Bagenel, T.B., Ed.) Unwin Bros. Surrey. 221-224

Masterman, A.T. 1924. On the scales of certain freshwater fish in relation to age determination. Fish. Invest. Minis, Agric. Fish., Ser. I, 1: 1-16.

Masters, P.M. and Bada, J.L. 1978. Amino acid racemization dating of bone and shell. In *Archaeological Chemistry*, Vol. II. (Carter, G. F., Ed.)

McFarlane, G.A., Wydoski, R.S. and Prince, E.D. 1990. Historical review of the development of external tags and marks. AFS Symp. 7: 9-29.

Menon, M.D. 1950. The use of bones, other than otoliths, in determining the age and growth-rate of fishes. J. Cons. Int. Expl. Mer 16: 311-335.

Methot, R.D. 1983. Seasonal variation in survival of larval northern anchovy (*Engraulis mordax*) estimated from age distribution of juveniles. Fish. Bull. U.S. 81: 424-423.

Methot, R. 1990. Synthesis model: an adaptable framework for analysis of diverse stock assessment data. INPFC Bull. 50: 259-277.

Methot, R.D. and Kramer, D. 1979. Growth of northern anchovy (*Engraulis mordax*) larvae in the sea. Fish. Bull. U.S. 77: 413-423.

Meyer, H.A. 1878. Beobachtungen über des Wachsthum des Herings im westlichen Theile er Ostsee. Jahresb. Komm. Deutsch. Meere, Kiel, 4-6 Jahrg. 227-252.

Miller, D.J. 1955. Studies relating to the validity of the scale method for age determination of the northern anchovy (*Engraulis mordax*). Calif. Fish. Game, Fish. Bull. 101: 7-34.

Mills, K.H. and Beamish, R.J. 1980. Comparison of fin ray and scale age determinations for lake whitefish (*Coregonus clupeaformis*) and their implications for estimates of growth and annual survival. Can. J. Fish. Aquat. Sci. 37: 534-544.

Mina, M.V. 1968. A note on a problem in the visual qualitative evaluation of otolith zones. J. Cons. Perm. Int. Explor. Mer. 32: 93-97.

Misra, R.K. 1980. Statistical comparisons of several growth curves of the von Bertalanffy type. Can. J. Fish. Aquat. Sci. 37: 920-926.

Misra, R.K. 1986. Fitting and comparing several growth curves of the generalized von Bertalanffy type. Can. J. Fish. Aquat. Sci. 43: 1656-1659.

Moksness, E. 1992. Otolith microstructure: a new method in recruitment studies and management of herring (*Clupea spp.*). Ph.D. thesis, University of Bergen, Bergen, Norway.

Moksness, E., and Fossum, P. 1992. Daily growth rate and hatching-date distribution of Norwegian spring-spawning herring (*Clupea harengus* L.). ICES J. Mar. Sci. 49: 217-221.

Molander, A.R. 1947. Observations on the growth of the plaice and on the formation of annual rings in its otoliths. Svenska. Hydrog.-Biol. Komma. Skr. NY. Ser. Biol. 2: 3-11.

Moore, W.S. and Krishnaswami, S. 1972. Coral growth rates using Ra-226 and Pb-210. Earth Sci. Lett. 15: 187-190.

Morales-Nin, B. 1988. Age Determination in a tropical fish (*Lethrinus nebulosis*) (Forskal 1775) (*Teleosti: Lethrinidae*) by means of otolith interpretation. Invest. Pesq. 52: 237-244.

Morales-Nin, B. 1988. Caution in the use of daily increments for ageing tropical fishes. Fish Byte. 6: 5-6.

Morales-Nin, B. 1989. Growth determination of tropical marine fishes by means of otolith interpretation and length frequency analysis. Aquatic Living Resources, Paris. 2: 241-253.

Morales-Nin, B. 1992. Determination of growth in bony fish from otolith microstructure. FAO Fish. Resh. Rep. 322.

Morales-Nin, B. and Ralston, S. 1990. Age and growth of *Lutjanus kasmira* (Forsskal) in Hawaiian water. J. Fish Biol. 36: 191-203.

Morse, W.W. 1989. Catchability, growth, and mortality of larval fishes. Fish. Bull. U.S. 87: 417-446.

Mortera, J. and Levi, D. 1983. Bias in age reading and consequences on age/length key, on growth curve and in virtual population analysis. Annex D. FAO Fish. Rep. 257, FIPL/R257: 73-79.

Mosher, K.H. 1954. Use of otoliths for determining the age of several fishes from the Bering Sea. J. Cons. 19: 337-344.

Mug, V.M. 1993. Age determination of corvina reina (*Cynoscion albus*) in the Gulf of Nicoya based on otolith surface readings and microincrement analysis. Masters Thesis, Oregon State University, Corvallis, OR.

Mug, V.M., Gallucci, V.F. and Lai, H. L. 1994. Age determination of corvina reina (*Cynoscion albus*) in the Gulf of Nicoya, Costa Rica, based on examination and analysis of hyline zones, morphology and microstructure of otoliths. J. Fish. Biol. 45: 177-191.

Mugiya, Y. 1984. Diurnal rhythm in otolith formation in the rainbow trout (*Salmo gairdneri*) seasonal reversal of the rhythm in relation to plasma calcium concentrations. Comp. Biochem. Physiol. 78A: 289-293.

Mugiya, Y., Watabe, N., Yamada, J., Dean, J.M., Dukelberger, D.G. and Shimuzu, M. 1981. Diurnal rhythm in otolith formation in the gold fish (*Carassius auratus*). Comp. Biochem. Physiol. 68A: 659-662.

Mulligan, K.P. and Gauldie, R.W. 1989. The biological significance of the variation in crystalline morph and habit of otoconia in elasmobranchs. Copeia 1989: 856-871.

Mustafa, S. and Ansari, A.R. 1983. Growth of teleostean fish (*Gudusia chapra*) in tropical reservoirs of India. Z. Angen Zool. 70: 57-72.

Neilson, J.D. and Geen, G.H. 1981. Method for preparing otoliths for microstructure examination. Prog. Fish-Cult. 43: 90-91.

Neilson, J.D. 1992. Sources of error in otolith microstructure examination. In Otolith microstructure Examination and Analysis. (Stevenson, D.K. and Campana, S.E., Eds.) Can. Spec. Publ. Fish. Aquat. Sci. Ser. 117: 115-125.

Nelder, J.A. and Mead, R. 1965. A simplex method for function minimization. Comp. J. 7: 308-313.

Nichy, F. 1977. Thin sectioning fish ear bones. Sea Tech. 2: 27.

Odense, P.H. and Logan, V.H. 1974. Marking Atlantic salmon (*Salmo salar*) with oxytetracycline. J. Fish. Res. Bd. Can. 31: 348-350.

Ogata, T. 1956. Studies on fisheries and biology of important fish: Alaska Pollock. Bull. Jap. Sea Reg. Fish. Res. Lab. 4: 93-139.

Oguri, M. 1986. On the yellowish brown pigment granules in the interrenal cells of anguler fish and leopard shark. Bull. Jap. Soc. Sci. Fish. 52: 801-804.

Olsen, A.M. 1954. The biology, migration, and growth rate of the school shark (*Galeorhinus anstralis*) (Maclay) (Coercharhanidae) in South-eastern Australian waters. Aust. J. Mar. Freshw. Res. 5: 353-410.

Ottaway, E.M. 1978. Rhythmic growth activity in fish scales. J. Fish. Biol. 12: 615-623.

Ottaway, E.M. and K. Simkiss. 1977. "Instantaneous" growth rates of fish scales and their use in studies of fish populations. I. Zoology. (Lond.) 181: 407-419.

Owen, R.W., Lo, N.C.H., Butler, J.L., Theilacker, G.H., Alvarino, A., Hunter, J.R., and Watanabe, Y. 1989. Spawning and survival patterns of larval northern anchovy (*Engraulis mordax*) in contrasting environments-a site-intensive study. Fish. Bull. U.S. 87: 673-688.

Palmen, A.T. 1956. A comparison of otoliths and interopercular bones as age indicators of English sole. Wash. Dept. Fish. Fish. Res. Pap. 1: 5-20.

Pannella, G. 1971. Fish otoliths: daily growth layers and periodical patterns. Science 173: 1124-1127.

Pannella, G. 1980. Methods of preparing fish sagittae for the study of growth patterns. In *Skeletal Growth of Aquatic Organisms: Biological Records of Environmental Change*. (Rhoades, D.C. and Lutz, R.A., Eds.) Plenum Press. New York. 619-624.

Pantulu, V.R. 1962. On the use of the pectoral spines for the determination of age and growth of *Pangasius pangasius* (Han. Buch.). J. Cons. Perm. Int. Explor. Mer. 27: 192-216,

Pauly, D. and David, N. 1981. A BASIC program for the objective extraction of growth parameters from length-frequency data. Meeresforschung. 28: 205-211.

Pentilla, J. 1988. Haddock (*Melanogrammas aeglifinus*). In *Age Determination Methods for Northwest Atlantic Species*. (Pentilla, J. and Dery, L.M., Eds.) NOAA Tech. Rep. NMFS 72: 23-30.

Pentilla, J. and Dery, L.M. (Eds.) 1988. Age determination methods for Northwest Atlantic species. NOAA Tech. Rep. NMFS. 72.

Pepin, P. 1991. Effect of temperature and size on development, mortality and survival rates of the pelagic early life history stages of marine fish. Can. J. Fish. Aquat. Sci. 48: 503-518.

Perrin, W.E. and Myrick, A.C. Jr., (Eds). 1980. Age Determination of Toothed Whales. IWC Sp. Issue 3.

Prince, E.D. and Pulos, L.M., (Eds). 1983. Proceeding of the international workshop on age determination of oceanic pelagic fishes: tunas, billfishes, and sharks, NOAA Tech. Rep. 8.

Prince, E.D., Lee, D.W., Zweifel, J.R. and Brothers, E.B. 1991. Estimating age and growth of young Atlantic blue merlin (*Makaira nigricans*) from otolith micro structure. Fish. Bull U.S. 89: 441-459.

Prutko, V.G. 1987. Age and growth of *Diaphus suborbitalis* (Myctophidae) from the equatorial part of the Indian Ocean. J. Ichthyol. 27: 14-23.

Quinn, T.J. II, Best, E.A., Bujsterveld, L. and McGregor, I.R. 1983. Sampling Pacific halibut (*Hippoglossis stenolepis*) landings for age composition: history, evaluation and estimation. Int. Pac. Halibut Comm. Sci. Rep. No. 68.

Radtke, R.L. 1983. Codfish otoliths: information storage structures in the propagation of cod (*Gadus morhua* L.). Flodevigen Rapp. 1: 273-298.

Radtke, R.L. 1989. Larval fish age, growth, and body shrinkage: information available from otoliths. Can. J. Fish. Aquat. Sci. 46: 1884-1894.

Radtke, R.L. and Dean, J.M. 1982. Increment formation in the otoliths of embryos, larvae, and juveniles of the mumichog (*Fundulus heteroclitus*). U.S. Fish. Bull. 80: 41-55.

Radtke, R.L., Fine, M.C. and Bell, J. 1985. Somatic and otolith growth in the oyster toadfish (*Opsanus tau* L.). J. Exp. Mar. Biol. Ecol. 90: 259-275.

Radtke, R.L. and Hourigan, T.F. 1990. Age and growth of antarctic fish, *Nototheniopes nudifrons*. Fish. Bull. U.S. 88: 557-571.

Radtke, R.L. and Shafer, D.J. 1992. Environmental sensitivity of fish otolith microchemistry. Aust. J. Mar. Freshw. Res. 43: 935-951.

Radtke, R.L., Hubold, S., Folsom, S.D. and Lenz, P.H. 1993. Otolith structure and chemical analyses: the key to resolving age and growth of the antarctic silverfish (*Pleuragramma antarcticum*). Antarc. Sci. 5: 51-61.

Radtke, R.L., Townsend, D.W., Folsom, S.D. and Morrison, M.A. 1990. Strontium/calcium concentration ratios in otoliths of herring larvae as indicators of environmental histories. Environ. Biol. Fish. 27: 51-61.

Ralston, S. 1976. Age determination of a tropical reef butterflyfish utilizing daily growth rings of otoliths. U.S. Fish. Bull. 74: 990-994.

Ralston, S. 1981. A study of Hawaiian deepsea headline fishery with special reference to the population dynamics of Opakapaka (*Pristipomoides filamentosus*) (Pisces: Lutjanidae). Ph. D. Dissertation, University of Washington, Seattle WA.

Ralston, S. 1982. Influence of hook size in the Hawaiian deep-sea handline fishery. Can. J. Fish.Aquat. Sci. 39: 1297-1302.

Ralston, S. 1985. A novel approach to ageing tropical fish. ICLARM Newsletter. 8(1): 14-15.

Ralston, S. 1990. Size selection of snappers (Lutjanitae) by hook and line gear. Can. J. Fish. Aquat. Sci. 47: 696-700.

Ralston, S. and Miyamoto, G.T. 1981. Estimation of the age of a tropical reef fish using the density of daily growth increments. In *The Reef and Man, Proceedings of the 4th International Coral reef Symposium.* Vol. 1. Marine Science Center, University of the Phillippines, Queson City.

Ralston, S. and Miyamoto, G.T. 1983. Analyzing the width of daily otolith increments to age the Hawaiian snapper (*Pristipomoides filamentosus*). Fish. Bull. U.S. 81: 523-535.

Randall, J.E. 1961. A contribution to the biology of the convict surgeon fish of Hawaiian Islands (*Acanthurus triostegus sandvicensis*). Pac. Sci. 15: 215-272.

Ratkowsky, D.A. 1983. *Nonlinear regression modeling.* Marcel Dekker, Inc. New York.

Rhoads, D.C. and Lutz, R.A. (Eds.) 1980. *Skeletal Growth of Aquatic Organisms, Biological Records of Environmental Change.* Plenum Press, New York.

Richards, L.J., Schnute, J.T., Kronlund, A.R. and Beamish, R.J. 1992. Statistical models for the analysis of ageing error. Can. J. Fish. Aquat. Sci. 49: 1801-1815.

Richter, H. and McDermott, J.G. 1990. The staining of fish otoliths for age determination. J. Fish. Biol. 36: 773-779.

Ricker, W.E. 1969. Effects of size-selective mortality and sampling bias on estimates of growth, mortality, production and yield. J. Fish. Res. Bd. Can. 26: 479-541.

Ricker, W.E. 1975. Computation and interpretation of biological statistics of fish populations. Fish. Res. Board. Can. Bull. 191.

Rosenberg, A.A. 1982. Growth of juvenile English sole (*Parophrys vetulus*) in estuarine and open coastal nursery grounds. Fish. Bull. U.S. 80: 245-252.

Rosenberg, A.A. and Pope, J.G. 1987. Comments on age-length vs. length-age relationship. In *Length-Based Methods in Fisheries Research.* (Pauly, D. and Morgan, G.R., Eds.) ICLARM Conf. Proc. No.13. ICLARM, Manila, Philippines and Kuwait Institute for Scientific Research, Safat, Kuwait.

Ross, M.D. and Pote, K.G. 1984. Some properties of Otoconia. Philos. Trans. R. Soc. Lond. 304: 445-452.

Sainsbury, K.S. 1980. Effect of individual variability on the von Bertalanffy growth equation. Can. J. Fish Aquat. Sci. 37: 241-247.

Samuel, M., Mathews, C.P. and Bawazeer, A.S. 1987. Age and validation of age from otoliths for warm water fishes from the Arabian Gulf. In *Age and Growth in Fish.* (Summerfelt, R.C. and Hall, G.E., Eds.) Iowa State University Press, Ames. 253-265.

Sarojini, K.K. 1957. Biology of *Mugil parsia* Hamilton with notes on its fishery in Bengal. Indian J. Fish. 4: 160-2-7.

Savage, R.E. 1919. Report on age determination from scales of young herring, with special reference to the use of polarized light. Fish. Inv. Minis. Agric. Fish. Ser. II. 4: 1-27.

Scheffe, H. 1973. A statistical theory of calibration. Ann. Stat. 1: 1-37.

Schmitt, P.D. 1984. Marking growth increments in otoliths of larval and juvenile fish by immersion in tetracycline to examine the rate of increment formation. Fish. Bull. U.S. 82: 237-242.

Schnute, J. 1981. A versatile growth model with statistically stable parameters. Can. J. Fish. Aquat. Sci. 38: 1128-1140.

Schramm, H.L. Jr. 1989. Formation of annuli in otoliths of bluegills. Trans. Am. Fish. Soc. 118: 546-555

Schroeder, R. A. and Bada, J. L. 1976. A review of the geochemical applications of the amino acid racemization reaction. Earth Sci. Rev. 12: 347.

Secor, D.H. and Dean, J.M. 1992. Comparison of otolith-based back-calculation methods to determine individual growth histories of larval striped bass (*Morone saxatilis*). Can. J. Fish. Aquat. Sci. 49: 1439-1454.

Secor, D.H., Dean, J.M. and Laban, E.H. 1992. Otolith removal and preparation for microstructural examination. In *Otolith Microstructure Examination and Analysis*. (Stevenson, D.K. and Campana, S.E., Eds.) Can. Spec. Publ. Fish. Aquat. Sci. 117: 19-57.

Simkiss, K. 1974. Calcium metabolism in fish in relation to ageing. In *The Ageing of Fish*. (Bagenal, T.B., Ed.) Unwin Bros. Surrey.

Simpfendorfer, C.A. 1993. Age and growth of the Austrian sharpnose shark (*Rhizopriondon taylori*) from north Queensland, Australia. Environ. Biol. Fish. 36: 233-241.

Small, G.J. and Hirschhorn, G. 1987. Computer-assisted age and growth pattern recognition of fish scales using a digitizing tablet. In *The Age and Growth of Fish*. (Summerfelt, R.C. and Hall, G.E., Eds.) Iowa State University Press, Ames, IA.

Smith, M.K. and Kostlan, E. 1991. Estimates of age growth of ehu (*Etelis carbunculus*) in four regions of the Pacific from density of daily increments in otoliths. Fish. Bull. U.S. 889: 461-472.

Smith, J.M., Nelson, R. and Campana, S.E. 1991. The use of Pb-210/Ra-226 and Th-228/Ra-228 disequilibria in the ageing of otoliths of marine fish. In *Radionuclides in the study of marine processes*. (Kershaw, P.J. and Woodhead, D.S., Eds.) Elsevier Applied Science London. 350-359.

Snedecor, G.W. and Cochran, W.A. 1967. *Statistical Methods*. 6th ed. Iowa State University Press. Ames, IA.

Sneed, K.E. 1950. A Method for calculating the growth of channel catfish (*Ictalurus lacustris punctatus*). Trans. Am. Fish. Soc. 80: 174-183.

Sokal, R.R. and Rohlf, F.J. 1969. *Biometry*. Freeman, San Francisco.

Soldat, V.T. 1982. Age and size of spiny dogfish (*Squalus acanthias*) in the Northwest Atlantic. NAFO Sci. Counc. Stud. 3: 47-52.

Somerton, D.A. and Kobayashi, D.R. 1989. A method of correcting catches of fish larvae for the size selection of plankton nets. Fish. Bull. U.S. 87: 447-455.

Southward, G.M. 1963. Sampling landings of halibut for age composition. Int. Pac. Halibut Comm. Sci. Rep. No. 58.

Sparre, P. and Venema, S.C. 1992. Introduction to tropical fish stock assessment. FAO Fish. Tech. Pap. 306/1 FAO, Rome, Italy.

Stevens, J.D. 1975. Vertebral rings as a means of age determination in the blue shark (*Prionace glanca* L.). J. Mar. Biol. Assoc. U.K. 55: 657-665.

Stevenson, D.K. 1981a. Assessment survey: Costa Rica. *In Small-Scale Fisheries in Central America: Acquiring Information for Decision Making*. (Sutinen, J. and Pollnac, R., Eds.) University of Rhode Island, Kingston. 45-65.

Stevenson, D.K. 1981b. Assessment of fisheries resources, Gulf of Nicoya, Costa Rica. In *Small-scale fisheries in Central America: Acquiring Information for Decision Making*. (Sutinen, J. and Pollnac, R., Eds.) University of Rhode Island, Kingston. 187-203.

Stevenson, D.K. and Campana, S.E. 1992. Otolith microstructure examination and analysis. Can. Spec. Publ. Fish. Aquat. Sci. Ser. 117.

Struhsaker, P., and Uchiyama, J.H. 1976. Age and growth of the nehy (*Stolephorus purpurens*) (Pisces: Engraulidae) from the Hawaiian Islands as indicated by daily growth increments of sagittae. Fish. Bull. U. S. 74: 9-17.

Sullivan, R.J., Lai, H.L. and Gallucci, V.F. 1990. A catch-at-length analysis that incorporates a stochastic model of growth. Can. J. fish. Aquat Sci. 47: 184-198.

Summerfelt, R.C. and Hall, G.E., (Eds.) 1987. *Age and Growth of fish*. Iowa State University, Ames, IA.

Sundberg, P. 1984. A Monte Carlo study of three methods for estimating the parameters in the von Bertalanffy growth equation. J. Cons. Int. Explor. Mer. 41: 248-258.

Suthers, I.M. and Sundby, S. 1993. Dispersal and growth of pelagic juvenile Arcto-Norwegian cod (*Gadus morhua*) inferred from otolith microstructure and water temperature. ICES Mar. Sci. 50: 261-270.

Sverlij, S.B. and Arceredillo, J.P.M. 1991. Growth of the argentine silverside (*Odontesthes-bonariensis*) (Pisces Artheriniformes) in Florida reservoir, San Luis Argentina. Rev. Hydrobiol. Trop. 24: 183-196.

Szedlmayer, S.M., Szedlmayer, M.M. and Sieracki, M.E. 1991. Automated enumeration by computers of age -0 weakfish (*Cynoscion regalis*) scale circuli. Fish. Bull. U.S. 89: 337-340.

Taggart, J.V. and Leggett, W. 1984. Comparison of final ages assigned to a common set of Pacific Ocean perch otoliths. WDF Tech. Rep. 81.

Takemura, Y. 1952. Observation of the scale of the "hokke" (*Plenrogrammus azanus*) by the use of polarized light. Bull. Jap. Soc. Fish. Sci. 17: 369-370.

Tanaka, S. 1953. Precision of age-composition of fish estimated by double sampling method using the length for stratification. Bull. Jap. Soc. Sci. Fish. 19: 657-670.

Taylor, H.F. 1916. The structure and growth of the scales of the squetaque and the pigfish as indicated of life history. Bull. U. S. Bur. Fish. 34: 289-330.

Tesch, F.W. 1968. Age and growth. In *Methods for Assessment of Fish Production in Fresh Water. I.B.P. Handbook 3*. (Ricker, W.E., Ed.) Blackwell, London.

Tesch, F.W. 1971. Age and growth. In *Methods for the assessment of fish production in fresh waters*. (Ricker, W.E., Ed). Blackwell Scientific Publications. Oxford, U.K.

Thorson, T.B. and Lacy, E.J., Jr. 1982. Age, growth rate and longevity of *Carcharhinus leucas* estimated from tagging and vertebral rings. Copeia 1982: 110-116.

Townsend, D.W., Radtke, R.L., Morrison, M.A. and Folsom, S.D. 1989. Recruitment implications of larval herring overwintering distributions in the Gulf of Maine, inferred using a new otolith technique. Mar. Ecol. Prog. Ser. 55: 1-13.

Townsend, D.W., Radtke, R.L., Morrison, M.A. and Folsom, S.D. 1992. Strontium:calcium ratios in juvenile herring (*Clupea harengus* L.) otoliths as a function of temperature. J. Exp. Mar. Biol. Ecol. 160: 131-140.

Tsukamoto, K. 1985. Mass marking of ayu eggs and larvae by tetracycline-tagging of otoliths. Bull. Jap. Soc. Sci. Fish. 51: 903-911.

Turekian, K.K. and Cochran, J.K. 1981. Growth rate of the viscomyid clam from the Galapagos spreading center. Science 214: 909-911.

Turekian, K.K., Cochran, J.K., Kharkar, D.P., Cerrato, R.M., Vaisnys, J.R., Sanders, H.L., Grassle, J.F. and Allen, J.A. 1975. Slow growth rate of a deep-sea clam determined by Ra-228 chronology. Proc. Natl. Acad. Sci. U.S.A. 72: 2829-2832.

Turekian, K.K., Cochran, J.K. and Nosaki, Y. 1979. Growth rate of a clam from the Galapagos Rise hot spring using natural radionuclide ratios. Nature (Lond.). 280: 385-387.

Tzeng, W.N. and Yu, S.Y. 1988. Daily growth increments in otoliths of milkfish (*Chanos chanos*) larvae. J. Fish. Biol. 32: 495-504.

Uchiyama, J.H., Burch, R.K. and Kraul, S.A., Jr. 1986. Growth of dolphins, (*Coryphaena hippurus* and *C. equiselis*) in Hawaiian waters as determined by daily increments on otoliths. Fish. Bull. U.S. 84: 186-191.

Umezawa, A. and Tsukamoto, K. 1990. Age and birth date of the class eel (*Anguilla japonica*) collected in Taiwan. Bull. Jap. Soc. Sci. Fish. 56: 1199-1202.

Umezawa, A. and Tsukamoto, K. 1991. Factors influencing otolith increment formation in Japanese eel (*Anguilla japonica*) (T and S elvers). J. Fish. Biol. 39: 211-223.

van Deusen, P.C. 1987. Some applications of the Kalman filter to tree ring analysis. In *Proceedings of the International Symposium on Ecological Aspects of Tree Ring Analysis*. 17-21 Aug. 1986. Palisades N.Y.

van Utrencht, W.L. and Schenkkan, E.J. 1972. On the analysis of the periodicity in the growth of scales, vertebrae and other hard structures in a teleost. Aquaculture 1: 293-316.

Vassilopoulou, V. and Papaconstantinou, C. 1992. Age, growth and mortality of the red porgy (*Pagrus pagrus*) in the eastern Mediterranean Sea (Dodecaneses, Greece). Vie Milieu 42: 51-55.

Vaughan, D.S. and Kanciruk, P. 1982. An empirical of estimation procedures for the von Bertalanffy growth equation. J. Cons. Int. Explor. Mer. 40: 211-219.

Vero, M., Paulovits, G. and Biro, P. 1986. An improved grinding technique for examining fish otoliths for age and growth studies with special consideration of the eel (*Anguilla anguilla*). L. Aquacult. Fish. Manage. 17: 207-212.

Victor, B.C. 1987. Growth, dispersal, and identification of planktonic labrid and pomacentrid reef-fish larvae in the eastern Pacific Ocean. Mar. Biol. 95: 145-152.

Villavicencio de Muck, Z. 1989. Tetracycline labeling for age and growth studies in fish, with emphasis on the peruvian auchoveta. The Peruvian Upwelling Ecosystem: Dynamics and Interactions. ICLARM Conf. Proc. 18: 174-178.

Visser, H. and Molenaar, J. 1987. Kalman filter analysis in dendroclimatology. In *Proceedings of the International Symposium on Ecological Aspects of Tree Ring Analysis*. 17-21 Aug. 1986. Palisades, N.Y.

Volk, E., Schroder, S.L. and Fresh, K.L. 1990. Inducement of unique otolith banding patterns as practical means to mass-Meirk juvenile Pacific salmon. Am. Fish. Soc. Symp. 7: 203-215.

von Bertalanffy, L. 1938. A quantitative theory of organic growth. Hum. Biol. 10: 181-213.

Wassef, E.A. 1991. Comparative growth studies on *Lethrinus-lentjan* Lacepede 1802 and *Lethrinus-mahsena* Forsskal 1775 pisces Lethrinidae in the Red Sea. Fish. Res. 11: 75-92.

Watson, D.J. and Balon, E.K. 1985. Determination of age and growth in stream fishes of northern Borneo. Environ. Biol. Fish. 13: 59-70

Weinstein, M.P. and Davis, R.W. 1980. Collection efficiency of seine and rotenone samples from tidal creeks, Cape Fear River, North Carolina. Estuaries 3: 98-105.

Weisberg, S. 1980. *Applied Linear Regression*. John-Wiley and Sons, New York.

Welden, B.A, Cailliet, G.M. and Flegal, A.R. 1987. Comparison of radiometric with vertebral band age estimates in four California elasmobranchs. In (Summerfelt, R.C. and Hall, G.E., Eds.) Int. Symp. on Age and Growth in Fish. Iowa State University, Ames, IA.

Welleman, H.C. 1990. Annual check formation in otoliths as a starting point for ageing by means of computer analysis. C.M. 1999/G:29. ICES.23

Werder, U. 1984. Age determination by scale analysis in juvenile Matrincha (*Brycon* cf. *melanopterus* Mueller and Troschel, *Teleostei*: *Characoidei*)., a tropical characin from the central Amazon. Anim. Res. Dev. 19: 48-66.

West, C.J. and Larkin, P.A. 1987. Evidence for size-selective mortality of juvenile sockeye salmon (*Oncorhynchus nerka*) in Babine Lake, British Columbia. Can. J. Fish. Aquat. Sci. 44: 712-721.

Westrheim, S.J. and Ricker, W.E. 1978. Bias in using an age-length key to estimate age-frequency distributions. J. Fish. Res. Bd. Can. 35: 184-189.

Wexler, J.B. 1993. Validation of daily growth increments and estimation of growth rates of larval and early-juvenile black skipjack (*Euthynnuis lineatus*) using otoliths. Inter-Am. Trop. Tuna Comm. 20.

Wild, A. and Foreman, T.J. 1980. The relationship between otolith increments and time for yellowfin and skipjack tuna marked with tetracycline. Inter-Am. Trop. Tuna Comm. 17: 509-560.

Williams, E.J. 1969. A note on regression methods in calibration. Technometrics. 11: 189-192.

Williams, H.M., and Smith, G.G. 1977. A critical evaluation of the application of amino acid racemization to geochronology and geothermometry. Origins of Life, 8: 91.

Williams, T., and Bedford, B.C. 1974. The use of otoliths for age determination. In *Ageing of Fish*. (Bagenal, T. B. Ed.) Unwin Bros. Surrey.

Wilson, C.A., and Dean, J. M. 1983. The potential use of sagittae for estimating age of Atlantic swordfish (*Xiphias gladius*). In *Age determination of oceanic pelagic fishes: tunas, billfishes and sharks*. (Prince E.D. and Pulos L.M., Eds.) NOAA Tech. Rep. NMFS 8: 151-156.

Wilson, C.A., Beckman, D.W. and Dean, J. M. 1987. Calcein as a fluorescent marker of otoliths of larval and juvenile fish. Trans. Am. Fish. Soc. 116: 668-670.

Wilson, K.H., and Larkin, P A.,. 1980. Daily growth rings in the otoliths of juvenile sockeye salmon (*Oncorhynchus nerka*). Can. J. Fish. Aquat. Sci. 37: 1495-1498.

Wilson, K.H. and Larkin, P.A. 1982. Relationship between thickness of daily growth increments in sagittae and change in body weight of sockeye salmon (*Oncorhynchus nerka*) fry. Can. J. Fish. Aquat. Sci. 39: 1335-1339.

Withall, A.F. and Wankowski, J.W. 1988. Estimates of age and growth of ocean perch (*Helicolenus percoides*) Richardson in southeastern Australian waters. Aust. J. Mar. Freshwat. Res. 39: 441-459.

Wright, P.J. 1993. Otolith microstructure of the lesser sandeel, *Ammodytes marinus*. J. Mar. Biol. Assoc. U.K. 73: 245-248

Wright, P.J. and Huntingford, F.A. 1993. Daily growth increments in the otoliths of the three-spined stickleback (*Gasterosteus aculeatus* L.). J. Fish. Biol. 42: 65-77.

Wright, P.J., Metcalf, N.B. and Thorpe, J.E. 1990. Otolith and somatic growth rate in Atlantic salmon parr (*Salmo salar*. L.). Evidence against coupling. J. Fish. Biol. 36: 241-249.

Xiao, Y. 1994. Growth models with corrections for the retardative effects of tagging. Can. J. Fish. Aquat. Sci. 51(2): 263-267.

Yeh, S.Y., Lai, H.L. and Liu, H.C. 1977. Age and growth of lizard fish (*Saurida tumbil*) (Block) in the East China Sea and the Gulf of Tonkin. Acta Ocean. Taiwanica 7: 134-145.

Young, M.Y. and Skillman, R.A. 1975. A computer program for analysis of polymodal frequency distribution (NORMSEP), FORTRAN IV. Fish. Bull. U.S. 73: 681-710.

Young, J.W., Bulman, C.M., Blaber, S.J.M. and Wayte, S.E. 1988. Age and growth of the lanternfish (*Lampanyctodes hectoris*) (Myctophidae) from eastern Tasmania, Australia. Mar. Biol. 99: 569-576.

Zar, J.H. 1974. *Biostatistical Analysis*. Prentice-Hall, Inc., Englewood Cliffs, NJ.

Zhang, Z., Runham, N.W. and Pitcher, T.J. 1991. A new technique for preparing fish otoliths for examination of daily growth increments. J. Fish. Biol. 38: 313-315.

CHAPTER 4

SAMPLING METHODS FOR STOCK ASSESSMENT FOR SMALL-SCALE FISHERIES IN DEVELOPING COUNTRIES

Loveday Conquest[1], Robert Burr[2], Robert Donnelly[1], Juan Chavarria[3], and Vincent Gallucci[1]

[1]School of Fisheries, University of Washington, [2]Nursing Research Office, University of Washington, [3]Department of Statistics, University of Costa Rica

1. INTRODUCTION

Tropical marine artisanal fisheries present many interesting management problems, not the least of which is the difficulty of establishing resource monitoring procedures to track long term biological trends. Near-shore tropical fisheries are often very productive and are important in the economies of many developing nations. Artisanal fisheries, as the name implies, are dominated by many individual fishermen, each generating relatively small amounts of fishing effort. Artisanal fishermen are often able to compete effectively in shallow-water areas viewed as inaccessible or uneconomical by the industrial fleets. The highly diverse fishing activity of many individuals over a large area creates logistical problems for those charged with managing such activity. Management and ecological issues presented by multi-species artisanal fisheries are distinct from the North American and European fisheries so well represented in the scientific literature. The traditional focus of temperate fisheries management on

1-56670-151-1/96/$0.00+$.50

monitoring catch and effort needs to be broadened in the tropical artisanal fishery to encompass the multi-species, multi-gear, and spatial-temporal interactions of these complex fisheries systems.

In spite of the high biomass production typical of tropical ecosystems, they can easily be overexploited, with serious negative social and biological consequences. Effective management requires control of fishing effort and information on stock productivity and composition. This may be difficult to achieve in a multi-species artisanal fishery. As a result, examples of extensive data collection in tropical marine artisanal fisheries are scarce.

The case study in this chapter is the Gulf of Nicoya, Costa Rica. It is representative of shallow, muddy, tropical estuarine embayments ringed by mangroves. The fishery in the Gulf of Nicoya presents a rich complex of features that are also typical of many tropical artisanal fisheries worldwide. The Gulf of Nicoya produces about 50% of the total catch on the Pacific coast of Costa Rica. The fishery is characterized by mixed gears and numerous species (more than 100) and is generally comprised of small, owner-operated vessels made of locally available materials. The artisanal fishery provides the economic basis for a large number of families that live in villages along the coastal areas.

Historically, the fishery in the Gulf of Nicoya has been dominated by two basic types of capture gear: gillnet and hook. In recent years larger trawl vessels have begun to participate, but the relative effort is low. The gillnets used represent efficient, selective designs and are constructed from nylon line. Gillnets are usually deployed at night across the current, then allowed to drift with the tide. Mesh size, net length, and net height vary considerably. The hook fishery generally uses two methods of deployment: longline sets and standard hook-and-line from small vessels.

The total catch of the artisanal fleet in the Gulf of Nicoya has experienced a slow-but-steady geometrical growth since the early 1950s, and some evidence indicates that the fishery is entering a mature phase, with declining CPUE (catch-per-unit-effort) in the most profitable commercial catch categories. The oceanographic environment varies significantly over the year in response to, among other things, meteorological (a wet-dry seasonal cycle) conditions. Summarizing data by month probably provides the maximum temporal integration possible for managers needing timely and accurate information about the system.

Sampling of the fishery for statistical information can be viewed as an economic optimization, trading off precision and bias of the developed information for cost and timeliness. The cost of the precision needs to be carefully considered with respect to the intended usage of the developed statistical estimates. There are limits to what even a complete census of the sampling frame can communicate about the future of the fishery and the appropriate management adaptations. In realistic settings, where a fishery is tracked over a long baseline, limitations include inaccessibility of sites, inaccuracy in data collection, and high natural variability. Sampling theory prescribes methods to optimize precision with respect to a fixed cost or to minimize cost for a given required precision.

Traditional sampling theory, as developed in the first half of this century and contained in the classic work of Cochran (1977), serves as the basis for most of the sampling plans supporting catch and stock assessment in the temperate and sub-Arctic industrial fisheries. Yet, persistent anecdotal evidence from professional managers involved in international fisheries development hints at limited utility for these approaches in small-scale, artisanal, multi-species settings. The international collaboration in the Gulf of Nicoya offers an opportunity to evaluate a number of sampling models in a complex setting typical of many other artisanal fisheries around the world.

Because sampling theory is well developed and because there exist decades of applied experience, we began our sampling evaluations with traditional approaches. Section 2 contains an overview of some well-known approaches to sampling, including simple random sampling, stratified random sampling, systematic sampling, cluster sampling, adaptive cluster sampling, and transect sampling. Section 3 discusses considerations for implementing sampling programs for artisanal fisheries in general and introduces a case history using fisheries data from the Gulf of Nicoya. Section 4 details the analysis of Gulf of Nicoya catch slip data for catch from that particular fishery.

This chapter puts sampling problems and designs into a perspective that is useful for viewing artisanal, developing country fisheries as well as for viewing selected problems of larger, developed country fisheries. Among those we are trying to reach are students and persons responsible for the overview and integration of a stock assessment-management project. We expect such persons to be reasonably knowledgeable about the basics of statistics and thus able to read this chapter without our duplicating either an elementary statistics text or other manuals aimed more at the field technician such as the FAO manual: *Guidelines for Statistical Monitoring*: *Practical guidelines for statistical monitoring of fisheries in manpower limited situations* (Caddy and Bazigos, 1985).

2. OVERVIEW OF SAMPLING METHODS

The purpose of sampling is to estimate parameters of the population (e.g., mean, variance, population abundance) from a small fraction of the total population. Probability samples, which have an element of "planned randomness" in the sampling procedure, result in estimators whose properties can be assessed using results of probability theory. What this means is that in addition to estimates of desired quantities, one also can obtain variances for those estimates and confidence intervals around the estimates. Table 1 contains a summary list of formulas for estimating means, totals, or proportions according to various sampling schemes, explained below. It also contains formulas for approximate sample sizes required to estimate the parameter of interest with a bound "B" of 2.0 standard errors around the unknown parameter (these would be approximate 95% confidence limits). We have used notation consistent with Scheaffer et al. (1990), an excellent introductory text on survey sampling. Other reference texts for survey sampling include Cochran (1977), Kish (1965), Hansen et al. (1953) and Thompson (1992).

Table 1. Sampling approaches, Estimators, and Variances.

1. Simple Random Sampling

	Estimate of	Variance of estimate	Sample size n required to estimate parameter with a bound B of 2.0 standard errors (approximate 95% confidence limits)
Population Mean μ	$\hat{\mu} = \bar{y} = \frac{\sum_{i=1}^{n} y_i}{n}$	$\hat{V}(\bar{y}) = \frac{s^2}{n}\left(\frac{N-n}{N}\right)$; $s^2 = \frac{\sum_{i=1}^{n}(y_i - \bar{y})^2}{n-1}$	$\frac{N\sigma^2}{(N-1)\frac{B^2}{4}+\sigma^2}$; σ^2 = population variance (or best guess thereof)
Population Total τ	$\hat{\tau} = N\bar{y}$	$\hat{V}(\hat{\tau}) = N^2\frac{s^2}{n}\left(\frac{N-n}{N}\right)$	$\frac{N\sigma^2}{(N-1)\frac{B^2}{4N^2}+\sigma^2}$
Population Proportion p	$\hat{p} = \frac{\sum_{i=1}^{n} y_i}{n}$ where y_i = 0 or 1	$\hat{V}(\hat{p}) = \frac{\hat{p}\hat{q}}{n-1}\left(\frac{N-n}{N}\right)$; $\hat{q} = 1-\hat{p}$	$\frac{Npq}{(N-1)\frac{B^2}{4}+pq}$; q = 1 - p
Population Ratio $R = \frac{y}{x}$	$r = \frac{\sum_{i=1}^{n} y_i}{\sum_{i=1}^{n} x_i}$	$\hat{V}(r) = \left(\frac{N-n}{nN}\right)\frac{1}{\bar{x}^2}\frac{\sum_{i=1}^{n}(y_i - rx_i)^2}{n-1}$	$\frac{N\hat{\sigma}^2}{N\frac{B^2\bar{x}^2}{4}+\hat{\sigma}^2}$; $\hat{\sigma}^2 = \frac{\sum_{i=1}^{n'}(y_i - rx_i)^2}{n'-1}$ $\hat{\sigma}$, $\bar{x}^2$ from preliminary sample of size n'

2.a. Stratified Random Sampling

L = number of strata

N_i = number of sampling units, stratum i

$N = \sum_{i=1}^{L} N_i$ = number of sampling units in population

w_i = fraction of observations allocated to stratum i

σ_i^2 = variance, stratum i (or best guess thereof)

p_i = population proportion, stratum i

	Estimate of	Variance of estimate	Sample size n required to estimate parameter with a bound B of 2.0 standard errors (approximate 95% confidence limits)
Population Mean μ	$\hat{\mu}_{ST} = \bar{y}_{ST} = \frac{1}{N}\sum_{i=1}^{L} N_i \bar{y}_i$	$\hat{V}(\bar{y}_{ST}) = \frac{1}{N^2}\sum_{i=1}^{L} N_i^2\left(\frac{N_i - n_i}{N_i}\right)\frac{s_i^2}{n_i}$	$\dfrac{\sum_{i=1}^{L} N_i^2 \sigma_i^2 / w_i}{N^2 \frac{B^2}{4} + \sum_{i=1}^{L} N_i \sigma_i^2}$
Population Total τ	$\hat{\tau}_{ST} = \sum_{i=1}^{L} N_i \bar{y}_i$	$\hat{V}(\hat{\tau}_{ST}) = \sum_{i=1}^{L} N_i^2\left(\frac{N_i - n_i}{N_i}\right)\frac{s_i^2}{n_i}$	$\dfrac{\sum_{i=1}^{L} N_i^2 \sigma_i^2 / w_i}{\frac{B^2}{4} + \sum_{i=1}^{L} N_i \sigma_i^2}$
Population Proportion p	$\hat{p}_{ST} = \frac{1}{N}\sum_{i=1}^{L} N_i \hat{p}_i$	$\hat{V}(\hat{p}_{ST}) = \frac{1}{N^2}\sum_{i=1}^{L} N_i^2\left(\frac{N_i - n_i}{N_i}\right)\frac{\hat{p}_i \hat{q}_i}{n_i - 1}$; $\hat{q}_i = 1 - \hat{p}_i$	$\dfrac{\sum_{i=1}^{L} N_i^2 \, p_i q_i / w_i}{N^2 \frac{B^2}{4} + \sum_{i=1}^{L} N_i p_i q_i}$; $q_i = 1 - p_i$

2.b. Stratified Random Sampling with Cost Allocation

Allocation minimizing cost for fixed value of $V(\bar{y}_{ST})$ or $V(\hat{\tau}_{ST})$ or minimizing $V(\bar{y}_{ST})$ or $V(\hat{\tau}_{ST})$ for fixed cost:

Overall $n = \dfrac{\left(\sum_{i=1}^{L} \dfrac{N_i \sigma_i}{\sqrt{c_i}}\right) \times \left(\sum_{i=1}^{L} N_i \sigma_i \sqrt{c_i}\right)}{N^2 D + \sum_{i=1}^{L} N_i \sigma_i^2}$

c_i = cost of a single observation from i^{th} stratum

$$D = \begin{cases} \dfrac{B^2}{4} & \text{for estimating } \mu \\ \dfrac{B^2}{4N^2} & \text{for estimating } \tau \end{cases}$$

Sample size, i^{th} stratum, $n_i = n \times \dfrac{\dfrac{N_i \sigma_i}{\sqrt{c_i}}}{\sum_{k=1}^{L} \dfrac{N_k \sigma_k}{\sqrt{c_k}}}$

Sample size under equal costs, $n = \dfrac{\left(\sum_{i=1}^{L} N_i \sigma_i\right)^2}{N^2 D + \sum_{i=1}^{L} N_i \sigma_i^2}$, $n_i = n \times \dfrac{N_i \sigma_i}{\sum_{k=1}^{L} N_k \sigma_k}$

Allocation minimizing cost for fixed value of $V(\hat{p}_{ST})$ or minimizing $V(\hat{p}_{ST})$ for fixed cost:

$$n = \frac{\left(\sum_{i=1}^{L} N_i \sqrt{\frac{p_i q_i}{c_i}}\right) \times \left(\sum_{i=1}^{L} N_i \sqrt{p_i q_i c_i}\right)}{N^2 \frac{B^2}{4} + \sum_{i=1}^{L} N_i p_i q_i} \qquad n_i = n \times \frac{N_i \sqrt{\frac{p_i q_i}{c_i}}}{\sum_{k=1}^{L} N_k \sqrt{\frac{p_k q_k}{c_k}}}$$

3. Systematic Sampling

	Estimate of	Variance of estimate	Sample size n required to estimate parameter with a bound B of 2.0 standard errors (approximate 95% confidence limits)
Population Mean μ	$\hat{\mu}_{SY} = \bar{y}_{SY} = \frac{\sum_{i=1}^{n} y_i}{n}$	$* \hat{V}(\bar{y}_{SY}) = \frac{s^2}{n}\left(\frac{N-n}{N}\right)$ $** \hat{V}_d(\bar{y}_{SY}) = \frac{N-n}{Nn}\frac{1}{2(n-1)}\sum_{i=1}^{n-1}(y_{i+1}-y_i)^2$	$\frac{N\sigma^2}{(N-1)\frac{B^2}{4}+\sigma^2}$
Population Total τ	$\hat{\tau}_{SY} = N\bar{y}_{SY}$	$* \hat{V}(\hat{\tau}_{SY}) = N^2\frac{s^2}{n}\left(\frac{N-n}{N}\right)$ $** \hat{V}_d(\hat{\tau}_{SY}) = \frac{N(N-n)}{n}\frac{1}{2(n-1)}\sum_{i=1}^{n-1}(y_{i+1}-y_i)^2$	$\frac{N\sigma^2}{(N-1)\frac{B^2}{4N^2}+\sigma^2}$
Population Proportion p	$\hat{p}_{SY} = \frac{\sum_{i=1}^{n} y_i}{n}$	$* \hat{V}(\hat{p}_{SY}) = \frac{\hat{p}_{SY}\hat{q}_{SY}}{n-1}\left(\frac{N-n}{N}\right)$; $\hat{q}_{SY} = 1-\hat{p}_{SY}$ $** \hat{V}_d(\hat{p}_{SY}) = \frac{N-n}{Nn}\frac{1}{2(n-1)}\sum_{i=1}^{n-1}(y_{i+1}-y_i)^2$	$\frac{Npq}{(N-1)\frac{B^2}{4}+pq}$

(* Assumes a randomly ordered population. ** Assumes a nonrandomly ordered population.)

4. Cluster Sampling

N = number of clusters in the population

n = number of clusters in a simple random sample

m_i = number of elements in cluster i, $i = 1, \ldots, N$

$\overline{m} = \frac{1}{n}\sum_{i=1}^{n} m_i$ = average cluster size for the sample

$\overline{M} = M/N$ = average cluster size for the population

y_i = total of all observations in cluster i

a_i = total number of elements in cluster i possesing a characteristic of interest

$M = \sum_{i=1}^{N} m_i$ = number of elements in population

	Estimate of	Variance of estimate	Sample size n required to estimate parameter with a bound B of 2.0 standard errors (approximate 95% confidence limits)
Population Mean μ	$\hat{\mu}_{CL} = \overline{y}_{CL} = \frac{\sum_{i=1}^{n} y_i}{\sum_{i=1}^{n} m_i}$	$\hat{V}(\overline{y}_{CL}) = \left(\frac{N-n}{Nn\overline{M}^2}\right)\frac{\sum_{i=1}^{n}(y_i - \overline{y}_{CL}m_i)^2}{n-1}$	$\frac{Ns_c^2}{N\frac{B^2\overline{M}^2}{4} + s_c^2}$; $* \; s_c^2 = \frac{\sum_{i=1}^{n'}(y_i - \overline{y}m_i)^2}{n'-1}$
Population Total τ	$\hat{\tau}_{CL} = M\frac{\sum_{i=1}^{n} y_i}{\sum_{i=1}^{n} m_i}$	$\hat{V}(\hat{\tau}_{CL}) = N^2\left(\frac{N-n}{Nn}\right)\frac{\sum_{i=1}^{n}(y_i - \overline{y}_{CL}m_i)^2}{n-1}$	$\frac{Ns_c^2}{\frac{B^2}{4N} + s_c^2}$
	or use $\hat{M} = N\overline{y}_t = \frac{N}{n}\sum_{i=1}^{n} y_i$ for M unknown	or $N^2\left[\frac{N-n}{Nn}\right]\frac{\sum_{i=1}^{n}(y_i - \overline{y}_t)^2}{n-1}$ for M unknown	$\frac{Ns_t^2}{\frac{B^2}{4N} + s_t^2}$; $* \; s_t^2 = \sum_{i=1}^{n'}(y_i - \overline{y}_t)^2$
Population Proportion p	$\hat{p}_{CL} = \frac{\sum_{i=1}^{n} a_i}{\sum_{i=1}^{n} m_i}$	$\hat{V}(\hat{p}_{CL}) = \left(\frac{N-n}{Nn\overline{M}^2}\right)\frac{\sum_{i=1}^{n}(a_i - \hat{p}_{CL}m_i)^2}{n-1}$	$\frac{Ns_c^2}{N\frac{B^2\overline{M}^2}{4} + s_c^2}$; $* \; s_c^2 = \frac{\sum_{i=1}^{n'}(a_i - \hat{p}_{CL}m_i)^2}{n'-1}$

(* From preliminary sample of size n'.)

5. Adaptive Cluster Sampling

Ψ_i = network including unit i ; m_i = number of units in that network to which unit i belongs.

$t_i = \frac{1}{m_i}\sum_{j\in\Psi_i} y_j$ = average of the observations in the network that includes i^{th} unit of original sample.

n = number of units in original sample.

	Estimate of	Variance of estimate	Sample size n required to estimate parameter with a bound B of 2.0 standard errors (approximate 95% confidence limits)
Population Mean μ	$\hat{\mu} = \frac{1}{n}\sum_{i=1}^{n} t_i$	** $\hat{V}(\hat{\mu}) = \frac{N-n}{Nn(n-1)}\sum_{i=1}^{n}(t_i-\hat{\mu})^2$	* $\frac{N\gamma^2}{(N-1)\frac{B^2}{4}+\gamma^2}$
Population Total τ	$\hat{\tau} = N\hat{\mu}$	** $\hat{V}(\hat{\tau}) = \frac{N(N-n)}{n(n-1)}\sum_{i=1}^{n}(t_i-\hat{\mu})^2$	* $\frac{N^3\gamma^2}{(N-1)\frac{B^2}{4}+\gamma^2 N^2}$
Population Proportion p	$\hat{p} = \frac{1}{n}\sum_{i=1}^{n} t_i$ y_i , contained in t_i , are either 0 or 1	** $\hat{V}(\hat{p}) = \frac{N-n}{Nn(n-1)}\sum_{i=1}^{n}(t_i-\hat{p})^2$	* $\frac{N\gamma^2}{(N-1)\frac{B^2}{4}+\gamma^2}$

(* γ^2 = best guess of $\frac{\sum_{i=1}^{N}(t_i-\mu)^2}{N}$, or from preliminary sample. ** Assumes sampling without replacement.)

2.1 SIMPLE RANDOM SAMPLING

The basic sampling methodology behind all probabilistic inference of point estimates is simple random sampling (SRS). SRS means that out of the entire population of N elements (e.g., fish catch slips, or the fish themselves in a day's catch), a group of n elements is selected according to certain probability rules, and the desired information is recorded. For example, this information could be the total biomass in kilograms (kg), or the length, weight, and species. The probability sampling rule that is followed in simple random sampling is this: that any group of n elements drawn from the population has the same chance of being selected as any other group. This is what is meant by the phrase, sampled "at random". For selecting a sample of size n out of a total population of N elements, the probability attached to drawing any particular group of n elements is

$$\binom{N}{n} = \frac{N!}{n!\,(N-n)!}.$$

For example, there are $\binom{10}{3} = \frac{10!}{3!\ 7!} = 120$ ways of selecting a group of 3 elements out of a population consisting of 10 elements. Many statistics texts contain tables of random numbers for this purpose; many statistical packages contain computer programs which can also be used to generate samples of random numbers.

We have used fish catch slips (papers that fishermen are required to complete for reduced fuel costs when the catch is landed; see Section 3.2 for further details) as the sampling units to obtain information on fish catch per boat trip and have employed simple random sampling to investigate the sampling distributions of mean catch estimators for varying sample sizes (see Section 4.2.3, "Results of Subsampling Experiments"). Table 1.1 summarizes the sampling estimators used for a simple random sampling design. These are the basis for almost all of the other formulas given for other sampling designs in Table 1.

2.2 STRATIFIED RANDOM SAMPLING

A stratified random sample can be obtained by partitioning the sampling elements, in our case catch slips, into disjoint (non-overlapping) groups called strata on the basis of additional information, and then drawing a simple random or systematic sample from each defined stratum. If the responses within each stratum are more likely to be homogeneous (i.e., less variable than responses between strata), stratification will produce smaller standard errors for parameter estimates than would be obtained from a simple random sample of the entire population. Also, by stratifying the elements of the population into convenient groupings, the cost per observation in the survey may be reduced. Furthermore, it may be desirable to obtain estimates of population parameters for certain subgroups of the population. Potential categories for stratification include a unit of time (e.g., month), location (e.g., landing site), gear type, boat type, and so on. For instance, if it is desired to obtain information on catch per boat for each landing site for each month, this suggests stratifying by both landing site and month before beginning the sampling process. The number of samples allocated to each stratum can be optimized with respect to a number of criteria and constraints to capitalize on things such as differences in total landings and predicted standard deviations among strata. For example, strata with lower presumed variation can be allocated fewer sampling resources. Also, strata with larger numbers of elements require larger numbers of units to be sampled. Finally, strata with higher unit sampling costs will result in fewer units being sampled, if one objective is to keep sampling costs to a minimum. Tables 1.2.a and 1.2.b summarize the formulas for stratified random sampling designs when the costs of executing the data collection are not considered and then, where the cost is an extra constraint on choosing the optimal sampling design.

A well established rule for allocation of effort in stratified sampling is the one which, in estimating a population mean or total, minimizes overall cost for a fixed value of the variance of that estimate, or equivalently, for a fixed value on the bound of the final error around that estimate at the 95% confidence level. Similarly one could fix the overall sampling cost and then allocate sampling effort that minimizes the variance of the final estimate. Both approaches lead to the same equations for n, the overall sample size, and n_i, the sample size in the ith stratum. These are given in Table 1.2.b under "Allocation minimizing cost for fixed value of $V(\bar{y}_{ST})$, or minimizing $V(\bar{y}_{ST})$ or $V(\hat{\tau}_{ST})$ for fixed cost."

As an illustrative example, suppose there are 3 strata (e.g., 3 different landing sites) of sizes 100, 500, and 1,000 elements. Furthermore, suppose that the respective stratum standard deviations are 5.0, 10.0, and 15.0 (e.g., kg). Assume that the cost of sampling a given element is \$5 U.S. (or the equivalent in local currency) in the first stratum, \$8 in the second stratum, and \$12 in the third. Finally, in estimating, say, a sample mean (e.g., mean catch per boat in kg), we desire a bound on the error to be equal to ± 3.0 (e.g., 3.0 kg in the case of estimating mean catch per boat). Then, following the equations in Table 1, the total sample size n for the 3 strata is as follows:

$$n = \frac{\left(\left(100\frac{5}{\sqrt{5}}\right)+\left(500\frac{10}{\sqrt{8}}\right)+\left(1000\frac{15}{\sqrt{12}}\right)\right)\left((100\cdot 5\cdot\sqrt{5})+(500\cdot 10\cdot\sqrt{8})+(1000\cdot 15\cdot\sqrt{12})\right)}{\left((100+500+1000)^2\frac{3^2}{4}\right)+(100\cdot 25+500\cdot 100+1000\cdot 225)}$$

= 70.38, or ≈ 71 sampling units. The sample size n_i for each stratum is then:

$$n_1 = 71\times\frac{\left(100\cdot\frac{5}{\sqrt{5}}\right)}{\left(100\cdot\frac{5}{\sqrt{5}}\right)+\left(500\cdot\frac{10}{\sqrt{8}}\right)+\left(1000\cdot\frac{15}{\sqrt{12}}\right)}$$

$= 2.51$, or ≈ 3 sampling units

$$n_2 = 71 \times \frac{\left(500 \cdot \frac{10}{\sqrt{8}}\right)}{\left(100 \cdot \frac{5}{\sqrt{5}}\right) + \left(500 \cdot \frac{10}{\sqrt{8}}\right) + \left(1000 \cdot \frac{15}{\sqrt{12}}\right)}$$

$= 19.85$, or ≈ 20 sampling units

$$n_3 = 71 \times \frac{\left(1000 \cdot \frac{15}{\sqrt{12}}\right)}{\left(100 \cdot \frac{5}{\sqrt{5}}\right) + \left(500 \cdot \frac{10}{\sqrt{8}}\right) + \left(10{,}000 \cdot \frac{15}{\sqrt{12}}\right)}$$

$= 48.63$, or ≈ 49 sampling units

Note that $n_1 + n_2 + n_3 = 3 + 20 + 49 = 72$ sampling units, the difference from the previously calculated n of 71 being due to rounding error. The increasing number of sampling units in the strata (from 3 to 20 to 49) is a function of the increasing variability, the increasing stratum sizes, and the increasing sampling costs. For equal costs per sampling unit, the equations simplify considerably (all the $\sqrt{c_i}$ terms drop out) and the sample sizes remain a function of the strata sizes, strata variability, and the desired bounds on the estimate. If there is more homogeneity of response within strata than between strata, stratification can result in a substantial decrease in the number of samples to be collected for a fixed cost or for a fixed variance level. The stratified random sampling examples used in this chapter are based on month as the defined stratum. The one exception is January 1987 where we stratified the data within that month by landing site. Other ways of stratifying the data are conceptually possible, including fishing time, boat size, or horsepower. Other criteria for stratification were deemed unusable for this data set primarily due to the inconsistencies of reporting (for example, the number of slips that contained information on boat size would have severely limited the total sample size).

2.3 SYSTEMATIC SAMPLING

Systematic sampling can be popular with field workers because of its ease of execution. If a good sampling frame is not available (i.e., a list of all the elements in the population), then simple random samples or stratified random samples may be difficult to obtain and can be subject to selection errors. In systematic sampling the first element is selected randomly (i.e., via a table of random numbers or computer-generated random numbers), and then every succeeding k^{th} element is selected. This means that the elements in the population must appear in some sort of order (e.g., catch slips stacked in piles) so that the idea of "sampling every k^{th} element" makes sense. Table 1.3 presents the formulas for systematic sampling designs.

For example, systematic sampling was identified as procedurally convenient for the Gulf of Nicoya vessel catch slips. The reams of vessel landing reports provide the units to be sampled, and systematic samples of the catch slips could be drawn with minimal handling costs. In research conducted at the University of Costa Rica, a variety of systematic sampling protocols were performed on available catch slip data. A representative analysis is presented in Section 4.2.3.

Note that the first set of variance estimates listed in Table 1.3 assumes that the elements of the population are randomly ordered. If this is not the case, then the variance estimates for parameters being estimated may underestimate the true variance. If there are any periodicities in the data, systematic sampling can be very sensitive to this. For example, when sampling every tenth element of the population, we might accidentally hit the "high points" for the response variable we are investigating. Then the final estimate of the population mean or total will tend to be too high and the associated variance estimate too low. In order to check for periodicities in a data set, it is necessary to look at the autocorrelation function, which is the ordinary computed statistical correlation between elements which are "one apart" (i.e., next to each other, also known as "lag 1"), two apart (lag 2), three apart (lag 3), and so on. If the autocorrelation for any particular lag is high, then the effect is to underestimate the estimated variance for the population mean or total. If, however, the autocorrelation for any particular lag is low or zero, then systematic sampling performs as well as simple random sampling. Autocorrelation results for Gulf of Nicoya catch data appear in Section 4.2.3.

2.4 CLUSTER SAMPLING

Cluster sampling is a technique that may prove to be cost-effective for artisanal fisheries. The individual elements are formed into logical groups (examples are given below), and a simple random sample of groups is drawn from the new (and much smaller) frame of group elements, called clusters. Cluster sampling is a good choice when it is easier to inventory clusters than individual elements of the whole population—for example, when the cost of obtaining a sampling frame listing all the elements in the population is prohibitively high, or when the cost of obtaining observations increases as the distance separating the elements increases. Cluster sampling may be the best strategy when the sampling will be carried out over a geographic area and travel costs become significant, frequent conditions for artisanal fisheries.

Clusters may be arbitrarily chosen to minimize costs, but are often naturally defined in spatially distributed artisanal fisheries where a number of small communities may exploit a constrained geographic region. For example, it may be impossible to obtain a list of all the boats in a large geographic area; it may be easier to sample clusters (groups) of boats which could contain all different types of vessels. One can easily envision that the cost of sampling could rise rapidly as the distance one might have to travel between boats rises; it would be much easier to sample all the boats in a group.

In this case one would want the boats in a cluster to be of as many different types as possible in order to represent the various types in the population. Clustering may occur naturally when the target animals, the capture technology, and the fishermen themselves are highly localized. In the case of surveying sparse or aggregated populations where the clusters only reveal themselves during the sampling process itself, adaptive sampling techniques (Section 2.5) may be employed.

Cluster sampling takes advantage of natural partitions in the population in much the same way as stratified sampling. The difference between stratified sampling and cluster sampling is that with the former, the elements within strata should be as alike as possible and the strata themselves should be as different as possible; while in cluster sampling, one wants the elements within each cluster to be as heterogeneous as possible, but one cluster should look very much like another in terms of representing the variety in the population. But where every stratum must be visited in stratified sampling, only a sampling of the total number of clusters must be visited in order to execute cluster sampling (but the entire cluster must be enumerated once it is chosen). Optimum cluster size is influenced by the degree of local heterogeneity and the economic overhead of traveling to and establishing each cluster base. If there is a high degree of local homogeneity, then it is generally better to sample a relatively larger number of smaller clusters, as the information gained from each additional within-cluster element is slight due to their great similarity. Sampling a few large clusters may be the best strategy when the responses are well mixed.

The formulas for estimating a population mean, total, or population proportion are given in Table 1.4; the notation is complicated due to the fact that cluster sampling may be thought of as simple random sampling where each sampling unit (a group or cluster) contains a number of elements (e.g., boats, or fish). Note that the estimates actually take the form of ratio estimators (as in the first part of Table 1 under Simple Random Sampling), with the cluster size and total of responses from that cluster providing the additional information in the denominator of the ratio estimator.

2.5 ADAPTIVE CLUSTER SAMPLING

Adaptive cluster sampling involves a procedure for selecting sampling units where the selection procedure actually depends upon the response values of the units being selected. It is useful when sampling aggregated populations, such as fish, ore deposits, or disease clusters. Thus, having identified units with (say) fish in them suggests that neighboring units are also likely to have fish. These neighboring units will form a network (defined below) that can be added to the sample to increase the efficiency of estimators. Thompson (1992) contains several chapters addressing various aspects of adaptive sampling; much of this material is also contained in Thompson (1990), Thompson (1991a) and Thompson (1991b). Some basic results are highlighted here (see Table 1.5).

In addition to the units in the original sample, units in "networks" associated with the original sample are also included in the estimation process. Thompson defines a network as "a set of observed units with a given linkage pattern" such that selection of any unit in the network leads to inclusion of every other unit in the network: the sampling universe can be partitioned into mutually exclusive networks. A unit which adds no further units to its network is simply a network of size one. For example, if a given sampling area is found to have fish in it, then all geographically defined nearest neighbors are added to the sample. Since adaptive designs depend on actually observing the response values of a unit to see whether to add other units to the sample, traditional parameter estimates will be biased unless appropriate modifications are introduced. Thompson (1990) displays an example where the sampling units are squares on a grid; the objective is to estimate the number of objects in a region if a random sample of 10 squares is chosen. If any one of these squares has at least one object in it, adjacent neighboring units (units "north", "south", "east", and "west" of the original unit) are added to the sample. If any of the added squares has at least one object in it, further adjacent neighboring units are added ; the process continues until no more units are added to the original sample. This process takes advantage of the phenomenon that for populations distributed in a patchy manner, finding a unit with objects in it means that more units with objects in them are likely to be nearby.

In comparing the estimators for adaptive sampling to those for simple random sampling (Tables 1.1 and 1.5), note that the y-responses enter the adaptive sampling estimators through the networks to which they belong. Similarly, the associated variances are based on the squared deviations of the network averages around the estimated parameter.

2.6 TRANSECT SAMPLING

Stock assessment problems may also occur that do not involve sampling fish on the deck of a boat or at a landing site. Examples of situations include sampling the number of fishing boats via water-level or aerial overflights or sampling stationary or non-stationary organisms such as flounders, coral reefs or bivalves. In these cases, it may be useful to use the methodology known as line transect theory. Zoologists have historically used an informal line transect approach to estimate intertidal abundance, species zonation, etc. and hydroacoustic estimates of abundance of fish are also based on variants of line transect theory.

As a starting point, we imagine a sampler moving along a straight line, flying, sailing or swimming, recording the number of observations. Sometimes an observed sample is essentially stationary relative to the motion of the sampler, e.g., an aerial transect sampler of porpoise schools; sometimes the relative motion is about equal, e.g., a diver swimming a transect "flushes" or disturbs flatfish buried in the sediment which suddenly swim away. In each case, the sampler records an observation when a porpoise school is sighted or a fish swims away.

In this chapter, a great deal of attention is placed upon variations of random sampling where the samples originate from *a priori* defined designs, all of which are characterized by having a probabilistic or random component. Line transect methodology is another variant on this general case. We will demonstrate how line transect problems might be carried out without going far into statistical intricacies. Interesting questions will be discussed: e.g., how to choose a sighting model, how to place transects, and how to test if a sighting model is appropriate.

Some transect experiments involve counting animals sighted within a strip of prespecified width placed symmetrically about a transect line. This is commonly called a strip or belt transect. However, when animals within the strip are overlooked, which is almost inevitable, the result is an underestimate of abundance. References on this point include Seber (1973), Anderson and Pospahala (1970), Eklund and Atwood (1962). In addition, for any distance from the transect line which is nontrivial, it is usually very difficult to determine whether an animal is within the strip or not. Such problems are alleviated in the line transect methodology, in which the strip or belt transect becomes a special case.

Figure 1 shows the transect line "*l*" and a perpendicular distance y from the line to the sighted animal. The perpendicular distances y are based on a sighting model g(y) to correct for animals that are present but which were overlooked. Table 2 contains the formulas pertaining to the relationship between the observed counts n and N, the estimate of the total population. The formula for the variance follows from using the delta approximation (Seber, 1973, pp. 4–6) for the product of two random variables. A selection of sighting models is shown in Table 3. Whereas the estimator for abundance is the same in all cases, estimates for the variance may differ according to the sighting model (e.g., Quinn and Gallucci, 1980). The following example demonstrates how to use a sighting model to obtain an estimate of abundance.

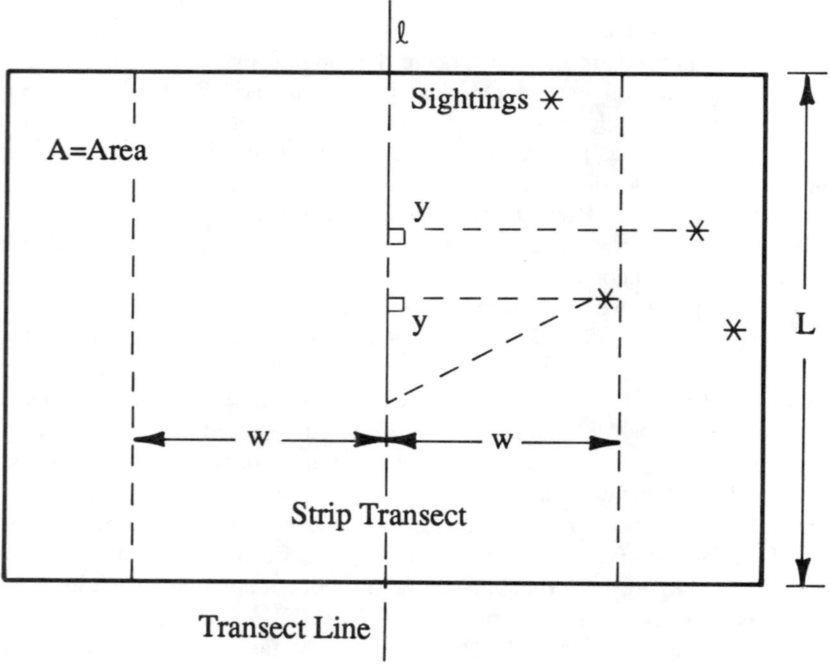

Figure 1. Schematic for line transect estimation

Table 2. General transect formulas and assumptions.

$$\hat{N} = \left(\frac{A}{2L}\frac{1}{\hat{c}}\right)n = \left[\frac{A}{2L}\hat{f}(0)\right]n$$

$$\text{var}\left(\hat{N}\right) = \hat{N}\left[(cv)^2\left(\hat{c}^{-1}|n\right)+(cv)^2(n)\right]$$

$$= \left(\frac{A}{2L}\right)^2\left[E^2(n)\,\text{var}\left(\hat{c}^{-1}\right)+E^2\left(\hat{c}^{-1}\right)\text{var}(n)\right]$$

where,

N = unknown population size

A = area over which sampling occurs

L = transect length

n = number of animals sighted

y_i = perpendicular distance to i^{th} animal

g(y) = sighting model to correct for animals present but overlooked

p = probability of sighting from the line

$c = \int_0^{\infty} g(y)dy$ = effective width on one side of transect

f(y) = histogram of observations of y_i values; $c = \frac{1}{f(0)}$

c.v. = coefficient of variation

Assumptions:

- Animals are distributed randomly and independently
- Sightings are independent
- No animal is counted more that once
- No error is made in the measurement of y_i
- No change in animal behavior occurs during the survey
- $g(0) = 1$

Table 3: Selected sighting models.

Model Name	Model Formula g(y)		Reference
Generalized Exponential (GEM)	$\exp\left[-\frac{1}{\beta}\left(\frac{y}{a}\right)^{\beta}\right]$	$0 \le y < \infty$	Quinn and Gallucci, 1980
Gates, Exponential, GEM with $\beta = 1$	$\exp[-\lambda y]$	$0 \le y < \infty$	Gates et al., 1968
Power Law	$\begin{cases} 1-\left(\frac{y}{w}\right)^{a} \\ 0 \end{cases}$	$0 \le y < \infty$ $y > w$	Eberhardt, 1968 Seber, 1973
Right-Triangle (Power Law, a = 1)	$\begin{cases} 1-\frac{1}{b}y \\ 0 \end{cases}$	$0 \le y \le w$ $y > w$	Gates, 1969
Strip	$\begin{cases} 1 \\ 0 \end{cases}$	$0 \le y \le w$ $y \ge w$	Seber, 1973

Note: y denotes the perpendicular distance to an animal.

Example 2.6.1

A survey of flatfish was done in the subtidal area with the diver pushing a T-shaped device to disturb fish lying slightly below the sediment surface. The diver observed 502 "flushed" fish. The total transect length was 150 m. The total area through which the transect(s) is/are placed is 1500 m^2. The farthest observation was 4 m from the transect line. Suppose that the histogram of the observations suggests a right triangle sighting model,

$$g(y) = \begin{cases} 1-\frac{1}{b}\cdot y & 0 \le y \le 4 \text{ m} \\ 0 & y > 4 \text{ m} \end{cases}$$

where b is a constant. The objective is to estimate abundance, N, and var(N).

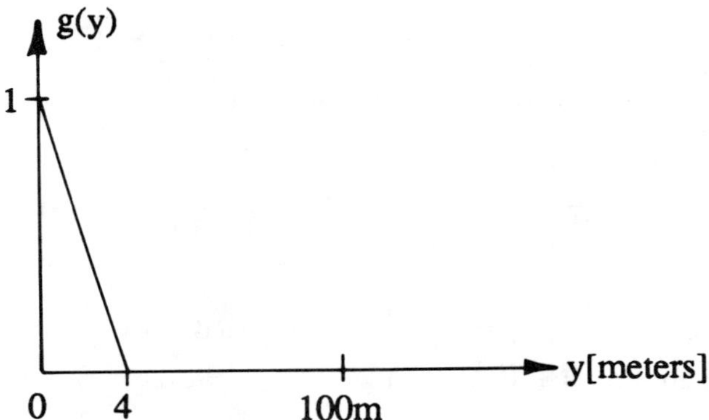

Figure 2. Triangular sighting model.

(i) Given L = 150 m, A = 1500 m^2, n = 502 fish. The schematic in Figure 2 represents the situation where the 4 m perpendicular distance represents the furthest distance of a sighting. The right triangle model is a special case of the power law model where a = 1 (Table 3).

(ii) $g(y) = 1 - \frac{1}{b}y$ where b^{-1} is the slope of the line in Figure 2; b = +4.0. We now solve for c, the effective transect width,

$$c = \int_0^\infty g(y)\,dy = \int_0^\infty \left(1 - \frac{1}{b}y\right) dy = y\Big|_0^4 - \frac{1}{2b}y^2\Big|_0^4 = 2.0 \text{ m}$$

The effective width c, is now used in the formula for $\hat{N}$, the estimated total abundance (Table 2).

$$\hat{N} = \frac{A}{2L\hat{c}}n = \frac{1500}{2(150)(2)}(502) = (2.50)(502) = 1255 \text{ animals}$$

(iii) Now use the delta method (Seber, 1973) to obtain the variance for $\hat{N}$

$$\text{var}(\hat{N}) = \left(\frac{A}{2L}\right)^2\left[E^2(n)\,\text{var}\left(\frac{1}{c}\right) + E^2\left(\frac{1}{c}\right)\text{var}(n)\right]$$

when, $E(n) \cong \hat{N}\hat{P}$, $E\left(\frac{1}{c}\right) = \frac{1}{2.0}$, $var(n) \cong \hat{N}\hat{P}\hat{Q}$ and $var\left(\frac{1}{c}\right) = 0$, because c is a known constant. Therefore the estimated variance of $\hat{N}$, $v\hat{a}r(\hat{N})$, is:

$$v\hat{a}r(\hat{N}) = \left(\frac{A}{2L}\right)^2\left(\frac{1}{c}\right)^2 \hat{N}\hat{P}\hat{Q} = (5.0)^2\left(\frac{1}{2.0}\right)^2(1255)(0.40)(0.60) = 1883$$

where $\hat{P} = \frac{2Lc}{A} = \frac{2(150)(2.0)}{1500} = 0.4$ and $\hat{Q} = 1 - \hat{P} = 0.6$. Thus the standard error of $\hat{N} = \sqrt{1883} = 43$.

(iv) Thus, from a right triangle sighting model and an observation of 502 animals, we estimate an abundance of 1255 fish with a standard error of 43.
If the Gates model, $g(y) = e^{-\lambda y}$, had been used, we would estimate λ from

$$\hat{\lambda} = \frac{(n-1)}{\sum y_i} \cong \frac{(n-1)}{\sum n_i m_i}$$

where the n_i are the number of sightings in the ith distance cell along y and m_i is the midpoint of that cell (Gates et al., 1968).

(v) Then, the effective transect width c is found from

$$c = \int_0^\infty g(y)\,dy = \int_0^\infty e^{-\lambda y}\,dy = -\frac{1}{\lambda}\left(e^{-\lambda y}\right)\Big|_0^\infty = \frac{1}{\lambda}$$

Then the estimated abundance $\hat{N}$ and its estimated variance $v\hat{a}r(\hat{N})$ are as follows:

$$\hat{N} = \frac{A}{2L}\hat{\lambda} n \quad \text{and} \quad v\hat{a}r(\hat{N}) = \frac{n}{\hat{P}^2}\left[\frac{n}{n-2} + \hat{Q}\right]$$

where

$$\hat{P} = \frac{2Lc}{A} \quad \text{and} \quad \hat{Q} = 1 - \hat{P}$$

Example 2.6.2

Four line transects are used to estimate the abundance of fishing boats in a gulf as an independent estimate of fishing effort. Each transect has a length of 1000 m and the gulf has a fishable area of $10^8\,m^2$. The data from the four transects are given below in intervals of 100 m from the survey vessel over a distance of 400 m on each side of the transect. The Gates sighting model is chosen, based on a sketch of how the observations decrease with y.

Transect i	Distance (m)				$n_{i.}$
	0 – 100	100 – 200	200 – 300	300 – 400	
1	19	7	3	1	30
2	14	5	2	1	22
3	26	10	4	1	41
4	12	5	2	1	20
Total = $n_{.j}$	71	27	11	4	$n_{..}$ = 113

There are at least two methods to estimate the abundance and its properties. One is to treat each transect independently and to **average** the result to find a **mean** total estimate, $\hat{\bar{N}}_m$, and the other is to **pool** all the transects into one long transect, $\hat{N}_p$. A different estimator of variance would be used in each case. In either method of estimation, the histogram of sightings is plotted as in Figure 3.

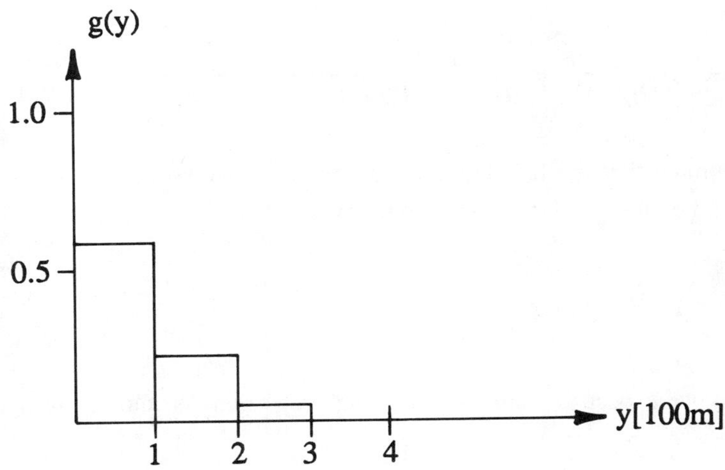

Figure 3. Histogram of data for Gates sighting model.

	Distance $\times 10^2$ m			
	0 – 1	1 – 2	2 – 3	3 – 4
Observed Relative Frequencies	0.63	0.24	0.10	0.03
Predicted Relative Frequencies	0.63	0.23	0.09	0.03

This figure is the basis for choosing the Gates model. But the question arises, does this exponential model reasonably fit the observations? To determine this, we estimate predicted relative frequencies. But we first need an estimate of λ for the Gates model.

The observed relative frequencies are found from $\frac{n_{\cdot j}}{n_{\cdot\cdot}}$. The predicted relative frequencies follow from the use of the Gates model which follows below.

Adapting the $\hat{\lambda}$ for the Gates model to the pooled data, we have

$$\hat{\lambda} = \frac{(n_{\cdot\cdot} - 1)}{\sum_{1}^{4} m_j n_{\cdot j}}$$

where $n_{\cdot\cdot}$ is the observations summed over i and j, $n_{\cdot j}$ is the number of observations summed over i, and m_j is the midpoint of the jth column. Thus, the respective values of m_j are 50, 150, 250, and 350; the respective n_j are 71, 27, 11, and 4; and $n_{\cdot\cdot} = 113 = 71 + 27 + 11 + 4$. So

$$\hat{\lambda} = \frac{113-1}{[(50)(71)+(150)(27)+(250)(11)+(350)(4)]} = \frac{112}{11750} = 0.01$$

Assuming that the histogram of observations follows $f(y) = \lambda e^{-\lambda y}$ we expect to find the following proportions in each interval [a, b]:

$$\int_a^b f(y)\,dy = \int_a^b \lambda e^{-\lambda y}\,dy = e^{-\lambda y}\Big|_a^b = e^{-\lambda a} - e^{-\lambda b}$$

Thus we expect to sight the following proportions of animals over the respective intervals:

Interval	Proportion of animals sighted	Expected number of animals sighted [proportion x 113]
[0,100],	$e^{-0.01(0)} - e^{-0.01(100)} = 1 - 0.37 = 0.63$	71.2
[100,200],	$e^{-0.01(100)} - e^{-0.01(200)} = 0.37 - 0.14 = 0.23$	26.0
[200,300],	$e^{-0.01(200)} - e^{-0.01(300)} = 0.14 - 0.05 = 0.09$	10.2
[300,400],	$e^{-0.01(300)} - e^{-0.01(400)} = 0.05 - 0.02 = 0.03$	3.4

We compare these expected numbers with the observed frequencies and find a very close match. Statistical "closeness" can be confirmed by a chi-square goodness-of-fit test (Zar, 1984).

We now proceed to estimate N_i for each transect i, where n_i below refers to $n_{i.}$ in the table of data:

$$\hat{N}_i = \frac{A}{2L}\frac{1}{\hat{c}}n_i = \frac{A}{2L}\hat{\lambda}n_i = \frac{10^8}{2(10^3)}(0.01)(n_i) = 0.5\times 10^3(n_i)$$

$$\hat{N}_1 = 15{,}000 \;;\; \hat{N}_2 = 11{,}000 \;;\; \hat{N}_3 = 20{,}500 \;;\; \hat{N}_4 = 10{,}000$$

The mean abundance for the mean estimator, $\hat{\bar{N}}_m$, is

$$\hat{\bar{N}}_m = \frac{1}{k}\sum_1^4 \hat{N}_i = \frac{1}{4}[15{,}000 + 11{,}000 + 20{,}500 + 10{,}000]$$

$$= \frac{1}{4}[56{,}500] = 14{,}125 \text{ fishing boats}$$

with a variance of the estimate as follows:

$$\mathrm{var}\left(\hat{\bar{N}}\right) = \frac{1}{k(k-1)}\sum_1^4\left(\hat{N}_i - \hat{\bar{N}}_m\right)^2$$

which for these data is

$$\mathrm{var}\left(\hat{\bar{N}}\right) = \frac{1}{4\cdot 3}\left[(15{,}000 - 14{,}125)^2 + (11{,}000 - 14{,}125)^2 + (20{,}500 - 14{,}125)^2 + (10{,}000 - 14{,}125)^2\right] = 5{,}682{,}291.667$$

which yields a standard error of 2384 boats.

An alternative estimate is to pool the data as if they came from four consecutive transects so that

$$\hat{N}_p = \frac{A}{2L_p}\frac{1}{\hat{c}}n_p = \frac{A}{2L_p}\hat{\lambda}n_p$$

where λ is the same, $L_p = 4000$ m, $n_p = \sum_1^4 n_{i\cdot} = 113 = n_{\cdot\cdot}$. That is,

$$\hat{N}_p = \frac{10^8}{2(4\times 10^3)}(0.01)(113) = 14{,}125$$

Therefore, for these extremely well behaved data, $\hat{N}_p = \hat{\bar{N}}_m$. If one computed variances they would, in general, differ. The averaged estimate will be unbiased (more accurate) and will generally be more stable, especially if the patterns (histograms) of the data between transects are very different.

This last example serves as a brief entrance into transect sampling design, in particular, the placement of transects. There is no simple right answer on placement, only many placements which are clearly not appropriate. Let us sketch in Figure 4 an interpretation of the sample placements one might associate with the methods of estimation used to find $\hat{\bar{N}}_m$ and $\hat{N}_p$. The survey area A is schematically shown in two shapes; in Figure 4 the four transects with their 800 m wide strips are as shown, each 10^3m long. This design corresponds to $\bar{N}_m$. Transects are equally spaced over the width and staggered relative to starting point. The four transects **effectively** cover $\frac{(4\times 800\times 10^3)}{(10^4\times 10^4)} = 3.2\%$ of the universe to be sampled. Figure 4 also shows a rearranged sample area A and a schematic of how an $\hat{N}_p$ estimator might be imagined. The same fraction of the universe is sampled.

Regarding the choice of sighting models, in one sense, enough information about various sighting models has been presented to inform the interested reader, but in reality, only a few of the possible models have been presented. The real question one must first consider after determining that the appropriate assumptions are satisfied, is whether a parametric or non-parametric sighting model will be used.

At this time, non-parametric models (Burnham et al., 1980) and the parametric generalized exponential models (Pollock, 1978; Quinn and Gallucci, 1980) are commonly used. As a generalization, non-parametric models may provide better statistical fits but estimation problems may be more severe, more data is thus required, and efficiency may be lower. Parametric models of two parameters have a lot of

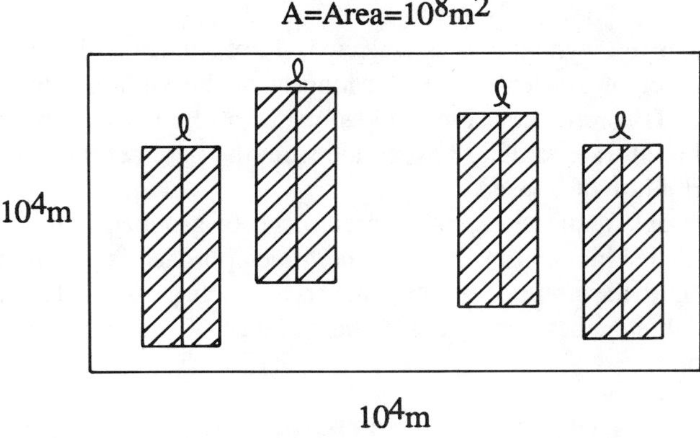

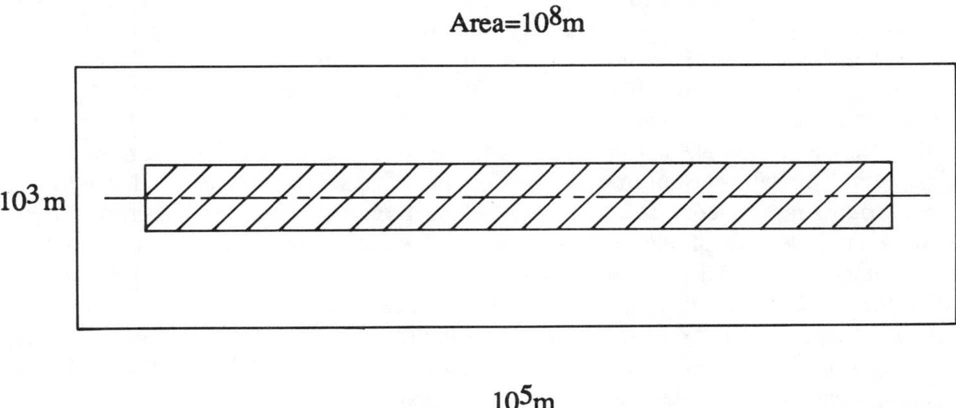

Figure 4. Two examples of placement of line transects.

flexibility and the advantage of being able to associate a particular parameter with some characteristic of the sighting model. But, ultimately, they are less flexible than non-parametric models. Another class of models using kernel functions has recently been described (Thompson, 1992) and shows promise for the future.

Regarding sampling designs, the simplest fact is that the best solution to questions about numbers of transects and transect length must be based upon a pilot study to provide initial estimates of errors (variances). Nevertheless, as can be seen from Figure 4, the area being sampled may provide guidance as to the number and placement of transects. For example, it is clear that intersecting transects merely reduce efficiency. Clearly, however, even in Figure 4 there is “overlap” between transects, but it occurs beyond the distances over which data is collected. Pilot studies will provide insight into the necessary number of transects from the longest of the sighting distances collected and the variances. Theoretical or empirical estimates of the half width $\hat{c}$ also allow an initial estimate of total transect length, separate from the parameter estimates that result from a given distribution of transects.

Line transects are frequently applied to clumped or grouped items, e.g., schooling mammals or grouped ships (Quinn, 1985). The basic approach is exactly the same as in the above except that the number of elements in each clump must be estimated. The final estimate of abundance must include the number of clumps and the number of elements per clump.

3. DATA COLLECTION OPTIONS AND UTILIZATION

3.1 GENERAL CONSIDERATIONS FOR ARTISANAL FISHERIES

Data management is an essential function in any natural resource management unit. It is a critical step, which sometimes can become a bottleneck in the orderly flow of information from the field, to the analysts, and thence to the decision-makers. Data management can also be expensive, sometimes rivaling the costs of the data collection and data analysis phases in terms of equipment and personnel resources.

The multiple scientific, bureaucratic, and political cultures in the management entities of an artisanal fishery create diverse data requirements. There is frequently a distinction between data that are conveniently available and data that are required for bioeconomic models relevant to stock assessment. Data sets can come in all shapes and sizes, originating in commercial records, formal market sampling programs, trawl surveys, tagging studies, laboratory culture experiments, and correlated environmental measurements. The basic data structures can vary from attributes of individual fish (such as species identification, morphological measures, and aging information or monthly length-frequency distributions of different species and stocks), to catch and effort measures on individual vessels, and include time series of fleet catch stratified by location, gear, and calendar month. Thus, databases are built from multiple sources and, in most cases when data are needed for analysis, the database system should present these data in an integrated, coherent form.

Sensitivity must be developed to the financial realities of management units supporting artisanal fisheries in developing nations since the resources for the collection and processing of fisheries data may be limited. For example, in the Gulf of Nicoya fishery, the management group is technically sophisticated but there exist finite limits on the human and computational resources available to address the diverse issues presented to it. Although a landing slip program is in place for monitoring the artisanal catch, data entry capacity is not sufficient to keep up with data flow. Further, it is not clear that the benefits of increased input capacity are commensurate with the costs, even if more funds were committed. Thus, technical data management issues cannot be isolated from the overall purposes and goals of the fisheries managers, from constraints imposed by the monitored bioeconomic system and from its environment and available resources.

The output characteristics of the data management function are also important. In addition to preserving functional capacity to meet the reporting needs of the organization, the modes of presentation should be designed to be maximally insightful and easy to interpret. An emerging requirement of data management systems is that they must have input and output channels to coordinate with other information technologies such as expert systems, decision support platforms, and statistics packages, without the need to transform the representation of the data structures.

The resource limitations and geographical isolation of the managers of many artisanal fisheries in developing countries sustain interest in microcomputer implementation of data management tools. Microcomputer environments induce both constraints and opportunities for data managers, although hardware and software functionality is rapidly increasing. A number of excellent microcomputer database systems are currently available, although not all may be appropriate for the mixture of data types and retrieval requests that naturally develop in a fisheries research and management unit.

The future of data management support for administrators of artisanal fisheries will undoubtedly see increasing functionality in computer hardware and software and the development of an extendible world-wide specialist network for communication of data, programs, and electronic mail.

3.2 CASE HISTORY: THE GULF OF NICOYA FISHERIES DATA

In this section we refer to a case history concerning the artisanal corvina fishery in the Gulf of Nicoya on the Pacific coast of Costa Rica. This fishery is primarily based on gillnets of three different mesh sizes: 3, 5, and 6 inch spread dimensions. There are primarily four corvinas caught ranging in adult size from 30 cm to 110 cm. Most frequently, the fishing boats drift in and out of the Gulf with their nets deployed, usually at night. A variety of sampling efforts has been in effect over various lengths of times. It is illustrative to consider these sampling programs since: (1) we used some of the data for analysis, (2) we shall comment on the value of some of the sampling designs, and (3) some of the problems encountered in a stock assessment can be alleviated at the sampling level. Thus, by our carefully specifying the problems we encountered in this example system, we may be able to minimize the prospect of our readers facing the same problem.

Data we denote as "catch-slip data" were collected by the Department of Fisheries in the Ministry of Agriculture (DOF/MAG). Fishermen were required to complete "catch slips" of paper as a necessary precursor for reduced fuel prices. A catch slip would contain information on:

1. vessel designation
2. amount of fuel used/vessel power
3. amount of time spent fishing
4. amount of ice loaded
5. types of gear used
6. number of crew
7. location of fishing activity
8. total landed weight of catch partitioned into catch categories.

The catch categories are:

1. primera grande
2. primera pequena
3. classificado
4. secunda
5. chatarra

Most of the primera category is composed of the larger corvina species and is the preferred catch of fishermen because of high market value, but the frequency of this category in contemporary catches is low.

Table 4: The corvina data presented in this chapter are from catch categories 1 and 2 only.

	Scientific name	Common name	Catch category
1.	*Cynoscion squamipinnis*	Aguada	Primera pequena
2.	*Cynoscion stolzmanni*	Coliamarilla	Primera grande and Primera pequena
3.	*Cynoscion phoxoocephalus*	Picuda	Primera pequena
4.	*Cynoscion albus*	Reina	Primera grande and Primera pequena

While some of the data was collected for purposes other than stock assessment, some of the information was adequate for statistical analyses. It is interesting to note that considerable difficulty was encountered in the interpretation of some of the data. This was especially true with respect to geographic locations which had colloquial or local names not well-known outside of the area, and with respect to the use of catch-categories. The geographic location problem included difficulties in both the landing sites and the locations of catch. The use of catch categories rather than species categories might make sense for certain economic analyses and for reasons of efficiency; it raises significant problems for stock assessment.

Transcription of the catch slip data to electronic form was a non-trivial operation requiring skill and knowledge of the fishery and of the software. The data management software DBASE 3 was used. The data selected for transcription from catch slips would ideally have formed a time series with representative catch/effort statistics from dry (December to April) and wet (May to November) seasons over as many years as possible. In fact, for reasons of needing a representative sample as soon as possible, the months that were transcribed are:

1984	January and October
1985	January and July
1987	January and July

Therefore, the catch slip data base is a total enumeration of the universe of catch slips; the data we worked with were sampled from the catch slip universe and put into electronic form.

A second type of data that was collected is denoted here as "size-selection data". This data set was based on sampling from the universe of arriving vessels ("landings data"). A sample of fish of each species was measured for length distribution and for total weight from selected arriving boats. The sampling program was carried out in the city of Puntarenas and at other landing sites in the Gulf of Nicoya. However, the fish selected for measurement were not randomly chosen.

Other important information was also collected, such as hours spent fishing (a measurement of effort), the types of gear employed (gillnet mesh size or hook and line), the fishing location in the Gulf, etc. At Puntarenas alone the following numbers of boats (Table 5) were sampled over 46 months, between 1980 and 1983 (except for January 1980 and August 1983),

Table 5. The number of boats sampled for catch at Puntarenas by year.

Year	1980	1981	1982	1983
No. sampled annually	475	357	336	519

Therefore, significant effort went into the collection of these data. Unfortunately, the arriving boats were not stratified into type, by size and gear, with random samples taken from each stratum. Further, the fraction of the total sampled from each arriving classification was not recorded. Such samples are called "judgment" samples.

Judgment samples occur when the samplers use their own judgment for selection of samples, rather than taking samples randomly. Therefore, the rules of ordinary statistical inference do not apply for the analysis of these data. Additionally, none of the months in this data set coincided with the 6 months of transcribed catch slip data. Hence it was not possible to estimate the sampling (fraction) intensity with respect to the entire fleet, since we did not know the total catch for any of the months from 1980–1983. Continued idiosyncrasies with species names forced us to make educated guesses in some instances. Some of the character names for the sample species were so long that standard statistical software tended to chop off the unique part of the name, causing further difficulties. Some form of subjective analysis of these data may eventually be possible; we did not attempt any analysis.

Besides the obvious point that sampling designs must be carefully made and carried out so as not to waste effort and money, another point should be noted. There is an intimate linkage between the sampling design and the data management operations. The linkage must be made so that the two components perform smoothly and efficiently. Ideally, this linkage should be specified prior to any major effort being exerted in either of the components.

Summary catch statistics from the data we did analyze are presented in Table 6. The remaining sections in this sampling component are devoted to different analyses of these data.

4. ANALYSIS OF CATCH SLIP DATA FOR THE GULF OF NICOYA

4.1 DESCRIPTION

The specific data set for the evaluations described in this section was the catch slip data for the months of January 1984, October 1984, January 1985, July 1985, January 1987, and July 1987. The elements were individual catch slips, each representing the landed commercial catch of one boat. The slips were gathered each month from at least a score of landing sites around the Gulf of Nicoya and brought together at a central location for processing by the department of Fisheries.

The descriptive statistics for mean catch are presented in Table 6 and contain several hints of distributional peculiarities. These include large differences between means and medians, right skewness and standard deviations that are much bigger than the means. (The skewness for any symmetric distribution, in particular the normal distribution, is zero.)

Table 6. Summary statistics for the primera catch category (kilograms per vessel landing).

Month	N	Mean	Median	Standard deviation	Skewness	Max value
Jan 84	2655	23.89	7.00	86.73	28.41	3600
Oct 84	4792	16.42	6.00	31.27	5.25	496
Jan 85	1862	18.97	6.50	38.21	6.09	572
Jul 85	4878	19.61	5.50	69.82	30.49	3605
Jan 87	7161	13.69	2.50	154.47	71.50	12322
Jul 87	5687	13.09	0.00	51.01	16.37	1896

4.2 ANALYSIS

4.2.1 Distributional Properties

It is quite possible for distributions of catch data to vary significantly from the Gaussian or normal distribution. Exploratory data analysis is necessary to get a better understanding of distributional properties. As a concrete example, we present results from 6 months of catch slip data from the Gulf of Nicoya.

Figures 5 and 6 depict histograms of the reported catch per boat for two of the months, January 1985 and July 1987. Inspection showed that the 6 months could be placed into two groups. January 1984, July 1985, January 1987, and July 1987 all required extremely large scales for displaying histograms of untransformed catches. Table 6 shows the high skewness values and large maximum values for those 4 months. The other 2 months (October 1984 and January 1985) were also skewed to the right but to a lesser degree; their maximum values were an order of magnitude smaller. In the interest of avoiding repetitive graphs, the histogram for July 1987 is shown representing the strongly right-skewed nature of the aforementioned 4 months. Similarly, the histogram for January 1985 also illustrates the general nature of the data for October 1984. The histograms on the left-hand side are those before data transformation. These empirical estimators of the probability density functions are nearly degenerate, with most of the samples lumped into a single bar corresponding to very small or zero catch, but with a non-trivial number of larger samples spread over several orders of magnitude. That is the reason for the extremely large scales.

A typical approach to analyzing skewed distributions is to look for a monotonic transformation that spreads the catch distribution more evenly through its range. Logarithmic transforms are frequently employed for random variables like catch that can only take on non-negative values (Mosteller and Tukey, 1977)

$$\text{modified } \log_e = \log_e\{1 + \text{catch}\}. \tag{5}$$

Note that zero catch maps to a value of zero through this modified logarithmic transform.

The insets in Figures 5 and 6 are histograms of the log-transformed variates. The distributions are much more symmetric, and the larger catch samples, which almost looked like outliers in the original plots, are now plausible members of the distribution. The histograms consist of two modes, one a dense mode containing about a third of the samples at the exact value of zero (corresponding to no catch), and a plausibly normal mode at higher logarithmic catch values.

Log-normal forms are not uncommon models of the catch distributions even in temperate industrial fisheries (Pennington, 1983; Smith, 1981). However, the probability mass at zero catch needs a more elegant expression than that achieved by the ad hoc exploratory procedure of adding a constant before logarithmic transformation.

Distributions with the configuration depicted in Figures 5 and 6 can be found in other areas of biostatistical application, and are especially well studied in the context of samples of plankton species with patchy spatial distribution (Pennington, 1983). A probability distribution known as the delta-distribution is often used as a model to describe statistical behavior in which a sample has a finite probability Δ (delta) of a null catch, and a log-normal distribution for the $p = (1 - \Delta)$ non-zero fraction of the catch.

To verify that the delta-distribution is a reasonable approximation of the Gulf of Nicoya catch data, it is necessary to check that (1) the data can be partitioned into finite proportions of presence and absence of catch, and (2) the log of the magnitude of the non-zero catch approximates a Gaussian (normal) distribution. Table 7 shows the delta-distribution summary statistics for the log-transformed, non-zero data. Note that

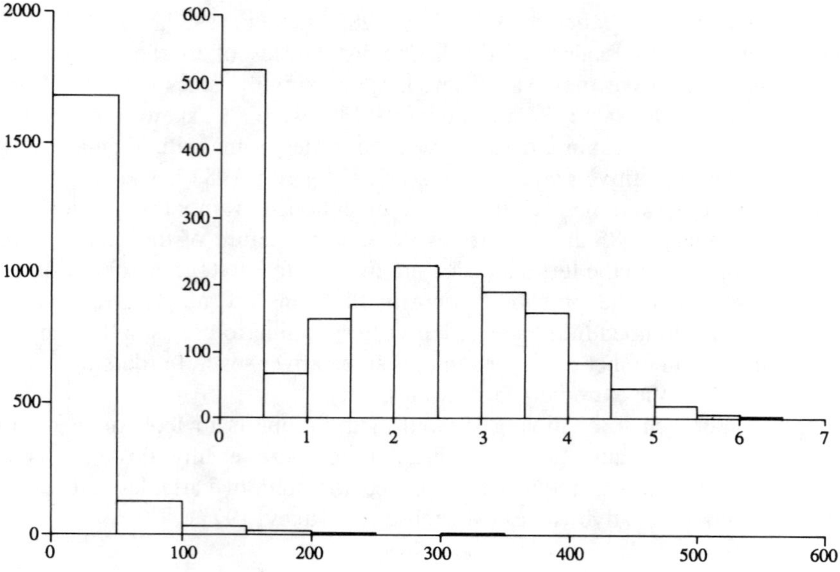

Figure 5. Histogram of the reported catch per boat for January 1985. The inset is the log-transformation of the catch per boat data.

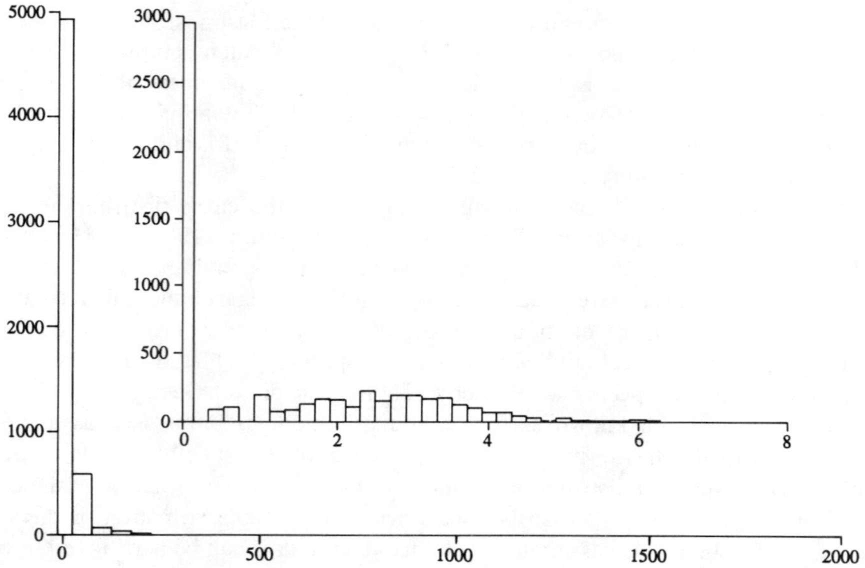

Figure 6. Histogram of the reported catch per boat for July 1987. The inset is the log-transformation of the catch per boat data.

primera catch was present in approximately one half to three quarters of the vessels in all months. The average proportion of non-zeroes was 0.66. The mean and standard deviation estimates in the log-transformed non-zero data display a similar consistency across the 6 months, with more similarity between the means and medians than for the untransformed data.

Table 7. Summary statistics for the log-transformed non-zero catches (p_{nz} is proportion non-zeroes, μ_{nz} is mean, σ_{nz} is standard deviation of log-transformed data).

Month	p_{nz}	μ_{nz}	σ_{nz}	Median
Jan 84	0.6904	2.71	1.29	2.77
Oct 84	0.7446	2.33	1.28	2.40
Jan 85	0.7315	2.49	1.27	2.56
Jul 85	0.7017	2.44	1.32	2.48
Jan 87	0.6361	1.93	1.32	1.95
Jul 87	0.4726	2.36	1.38	2.48

4.2.2 Estimators Based on the Delta-Distribution

Because of the distribution results detailed in the previous section, we considered using the delta-distribution as the underlying model for parameter estimates in designed sampling studies. By definition, a delta-distribution is a log-normal distribution with a spike at zero. It therefore differs from the log-normal distribution in the following manner. The ordinary log-normal (which does not allow zeroes) is a random variable X' whose natural logarithm has a Gaussian distribution with mean μ and variance σ^2. Then the expected value of X' is

$$E(X') = \exp\left\{\mu + \frac{1}{2}\sigma^2\right\}, \tag{6}$$

and the variance of X' is

$$\mathrm{Var}(X') = \left(\exp\{\sigma^2\} - 1\right)\exp\{2\mu + \sigma^2\}. \quad \text{(Lindgren, 1968)} \tag{7}$$

For the delta-distribution for a random variable denoted by X, let $p = \Pr\{X > 0\} = (1 - \Delta)$ be the proportion of non-zeroes. Then define X as follows:

$$X = \begin{cases} 0 & \text{with probability } \Delta \\ X' & \text{with probability } (1-\Delta) = P \end{cases}$$

Then the expected value of the untransformed log-normal with zeroes becomes

$$E(X) = p \times E(X') = p \times \exp\left\{\mu + \frac{1}{2}\sigma^2\right\}, \tag{8}$$

and the variance becomes

$$\begin{aligned} \mathrm{Var}(X) &= p(1-p)[E(X')]^2 + p\mathrm{Var}(X') \\ &= p(1-p)\exp\{2\mu+\sigma^2\} + p\left(\exp\{\sigma^2\}-1\right)\exp\{2\mu+\sigma^2\} \end{aligned} \tag{9}$$

(Aitchison and Brown, 1969).

What makes the mean of either the ordinary log-normal or the delta distribution difficult to estimate is that the mean is now a function of two unknown parameters, μ and σ^2, instead of one.

Pennington (1983, 1986) and Smith (1981, 1988) have investigated estimators of the mean and their efficiency for the delta distribution. In particular, Smith (1988) has studied the efficiency for small sample sizes and suggested modifications for the estimates of the variance of the mean. For a sample of n observations, let m denote the number of non-zero values and let $\bar{y}$ and s^2 denote the sample mean and sample variance, respectively, of the log-transformed non-zero values. Then unbiased estimates of the mean and variance of X (the delta-distributed variate) are

$$\hat{E}(X) = \frac{m}{n}\exp\{\bar{y}\}G_m\left(\frac{1}{2}s^2\right) \tag{10}$$

and

$$\hat{\mathrm{Var}}(X) = \frac{m}{n}\exp\{2\bar{y}\}\left[G_m\left(2s^2\right) - \frac{m-1}{n-1}G_m\left(\frac{m-2}{m-1}s^2\right)\right] \tag{11}$$

where

$$G_m(t) = 1 + \frac{m-1}{m}t + \sum_{j=2}^{\infty} \frac{(m-1)^{2j-1}}{m^j(m+1)(m+3)\ldots(m+2j-3)}\left(\frac{t^j}{j!}\right) \tag{12}$$

The large sample estimate of the variance of the estimated mean becomes (Pennington, 1983) the following:

$$\text{Estimated } \widehat{\text{Var}}[E(X)] = \frac{m}{n}\exp\{2\bar{y}\}\left[\frac{m}{n}G_m^2\left(\frac{s^2}{2}\right) - \frac{m-1}{n-1}G_m\left(\frac{m-2}{m-1}s^2\right)\right] \quad (13)$$

Table 8 shows results on catch slip data for the large sample Pennington estimates for the delta-distribution. In each case, the delta-distribution estimate is not too far from the ordinary sample average of the untransformed data, (repeated from Table 6 Section 4.1) but the delta-distribution estimator does better in terms of relative precision. The ordinary sample average for the untransformed data is aided here by the large sample sizes involved; Pennington (1983, 1986) has shown that for small sample sizes the performance of the ordinary sample mean estimator is much worse than other estimators.

Table 8 Pennington delta-distribution estimates for catch data.

Month	Estimated mean [a]	Estimated variance of [a]	Estimated standard deviation [b]	Estimated standard error [c]	Ordinary mean, std. deviation from Table 6
Jan 84	23.82	1.04	61.22	1.19	23.89, 86.73
Oct 84	17.49	.28	42.64	.62	16.42, 31.27
Jan 85	19.87	.92	48.03	1.11	18.97, 38.21
Jul 85	19.32	.39	51.98	.74	19.61, 69.82
Jan 87	10.39	.08	29.16	.34	13.69, 154.47
Jul 87	13.17	.07	47.25	.63	13.09, 51.01

[a] = estimate of $p \cdot \exp\left(\mu + \frac{1}{2}\sigma^2\right)$, the expected value of the delta-distribution (equation 10)

[b] = the square root of the estimated variance of the delta-distribution, (equation 11)

[c] = $\frac{[b]}{\sqrt{n}}$

4.2.3 Results of Subsampling Experiments

Subsampling procedures can be extremely useful in assessing required sample sizes for future sampling protocols, investigating the sampling distribution of an estimator, or comparing the sampling behavior of two or more estimators. Our sampling studies have all focused on estimates of mean catch per boat, since the estimate of a total catch is a linear combination of corresponding mean estimates. The data set used in this section are the 6 months of catch slip data for the Gulf of Nicoya.

Analytic predictions of the standard error of the sample mean for samples of any size n, assuming an infinite population, can be computed from the familiar equation

$$SE = \frac{\sigma}{\sqrt{n}} \tag{14}$$

This predicted standard error of the estimate of the mean is scaled in the units of measurement, (kg/boat for catch slip data in the Gulf of Nicoya). An expression of the relative standard error that is scale-invariant is the coefficient of variation (CV) of the estimate of the mean, which is the standard error expressed as a percentage of the mean value

$$CV = \frac{\frac{\sigma}{\sqrt{n}}}{\mu} \times 100 \tag{15}$$

Table 9 displays analytic predictions of the standard error and the coefficient of variation for each of the 6 months, assuming samples of size 30, 100, 300, and 1000. To guarantee one digit of accuracy corresponding to a coefficient of variation of no more than 10% would theoretically require an allocation of at least 300 samples, as this is the minimum sample size for which the CV approaches a value of 10%. Even so, this only happens for the 2 months with the lowest skewness, October 1984 and January 1985. For 3 of the remaining months (January 1984, July 1985 and July 1987), the CV approaches a value near 10% only for sample sizes of 1000; the CV for January 1987, the most highly skewed month, does not get near 10% at all. These results give some indication of what order of magnitude for the sample size would be required to reach certain desired levels of precision.

Systematic sampling has already been identified as procedurally convenient for the Gulf of Nicoya vessel catch slips. The reams of vessel landing reports can be thought of as a sampling list, and systematic samples can be drawn with minimal handling costs. In research conducted at the University of Costa Rica, a variety of systematic sampling protocols were simulated on frames of the available historical catch slip census. A representative analysis is presented in Table 10, which summarizes the estimates and variations experienced while sampling specified fractions of the data set.

Note that these sampling simulations differ from those presented in the previous section in that the summary statistic estimated here is the total catch over several catch categories, not just the primera catch category. Nevertheless, the variability of the estimates for systematic sampling at a given sampling fraction is quite comparable with the coefficients of variation noted earlier for simple random sampling. For this particular month, with systematic sampling, it would still require the collection of approximately a thousand samples to guarantee a CV of 10% ("one digit of accuracy") in our assessment of catch.

Table 9. Predicted standard error in catch/boat and coefficient of variation.

Sample size	Jan 84	Oct 84	Jan 85	Jul 85	Jan 87	Jul 87
30	15.83 (66.3%)	5.71 (34.8%)	6.98 (36.8%)	12.75 (65.0%)	28.20 (205.9%)	9.31 (71.1%)
100	8.67 (36.3%)	3.13 (19.0%)	3.82 (20.1%)	6.98 (35.6%)	15.45 (112.8%)	5.10 (39.0%)
300	5.01 (21.0%)	1.81 (11.0%)	2.21 (11.6%)	4.03 (20.6%)	8.92 (65.1%)	2.94 (22.5%)
1000	2.74 (11.5%)	0.99 (6.0%)	1.21 (6.4%)	2.21 (11.3%)	4.88 (35.7%)	1.61 (12.3%)

Table 10. Coefficients of variation observed in systematic samples of a subset of catch slip data for the Gulf of Nicoya artisanal fishery, January 1987. Based on all landing sites, all gear types, and all commercial catch categories. (Number of vessel landings = 4882).

Sampling fraction	Sample size	Coefficient of variation
10%	488	27.8%
20%	976	12.8%
33%	1627	6.1%

To evaluate the potential for variance inflation due to systematic patterns in the data, autocorrelation functions for various lags (see Section 2.3) of the available historic catch slip records are shown for the 6 months of catch slip data (Figure 7). In only one month (October 1984) is there a large correlation for short lags. Recall that if the autocorrelation for a particular lag is high, then the effect is to underestimate the variance for a population mean or total. If however, the autocorrelations are generally low or zero, then systematic sampling performs as well as simple random sampling. Thus, since there is no consistent pattern of high correlations, systematic sampling remains a viable alternative as a sampling protocol for this and similar data sets.

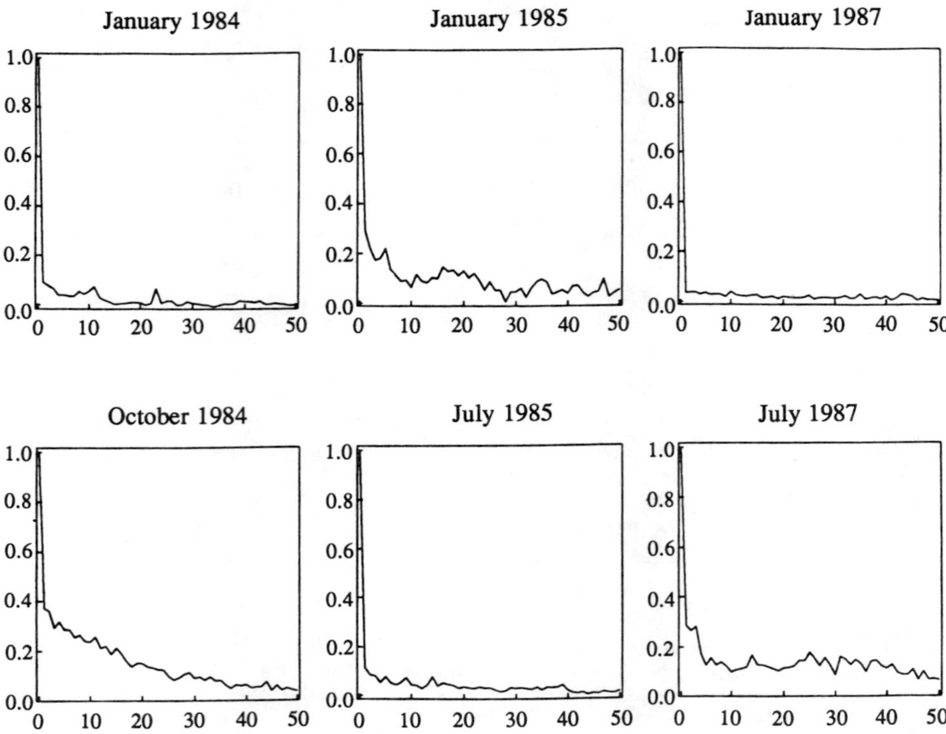

Figure 7. Autocorrelation functions for 6 months of data. Lag size is on the horizontal axis; autocorrelation value is on the vertical axis.

4.2.4 Comparison of Traditional Estimate with Delta-Distribution Estimate

We compared the results of (1) the stratified vs. the unstratified sampling approach, and (2) the ordinary sample estimate of the population mean fish catch vs. the delta-distribution-based estimate of the population mean. This section gives the results for the unstratified sampling approach comparing the ordinary sample average to a delta-distribution-based estimate. Section 4.2.5 below summarizes the results comparing the stratified sample average to a stratified delta-distribution estimate of the mean.

For each of the 6 months of catch slip data, we performed repeated random sampling (with replacement) for sample sizes of 30, 100, 300, and 1000, each sample size involving 1000 repetitions. For each sample, we calculated both the sample average (referred to as the "traditional estimate" of the mean) and the estimate suggested by Pennington (1983) based on the delta-distribution. Figures 8 and 9 show the frequency distributions of the estimates of mean catch for the two example months, for both approaches. Table 11 contains several statistics summarizing the results of the simulations for all of the months. For samples of size 30, the Pennington estimate yields a slightly lower observed standard error for 4 of the 6 months. When the traditional estimator yields a slightly lower observed standard error, we find that the original skewness is also the lowest (Table 6). The Pennington estimator seems to work better as the original skewness increases (for example, compare the results for all sample sizes for the sampling distributions of the two approaches for October 1984, January 1984, and January 1987, which have low, medium, and high skewness values, respectively). Similar remarks may be made concerning the ranges of the two estimators. For the larger sample sizes of 100, 300, and 1000, the sampling distribution of both estimators becomes more normal in appearance as would be expected from the Central Limit Theorem, but the distribution for the Pennington estimator approaches normality faster than that for the traditional estimator. Estimated mean-catch-per-trip values display narrower and more symmetric distributions based on the Pennington method compared with the traditional method. The relative efficiency of the Pennington estimate compared to the traditional estimate is sustained through all the sample sizes (dips slightly below unity for the 2 least skew months) and is quite substantial for months where the spread of catch-per-trip covers several orders of magnitude.

The 2 months with the lowest skewness, October 1984 and January 1985, do not improve very much using the Pennington estimate over the traditional estimate, but the other 4 months with higher original skewness do. This is evident by looking at Figures 8 and 9 and Table 11; the Pennington estimate at all sample sizes is less right skewed except for those 2 months that had the lowest skewness in their original distributions. In terms of the skewness attached to the sampling distributions for the two types of estimate, the Pennington estimate does at least as well as the traditional estimate; and where the skewness for the traditional estimate is quite large, the Pennington estimate tends to do much better (lower skewness for the sampling distribution across all sampling sizes). This is also borne out by the additional summary statistics (standard deviation, efficiency, range).

4.2.5 Comparison of Stratified Traditional Estimate with Stratified Delta-Distribution Estimate

This section gives sampling results comparing the stratified sample average to a stratified delta-distribution estimate of the mean. Table 12 shows the results of stratification by 9 landing sites in the Gulf of Nicoya for January 1987. The two estimates of mean catch differ by 4.5 units, but their estimated variances differ by an order of magnitude: the variance based on the ordinary stratified mean is 21.6 times the variance from the delta-distribution-based mean. For these data, an estimation approach using the delta-distribution apparently gives lower variance estimates than using the ordinary sample average, with or without stratification.

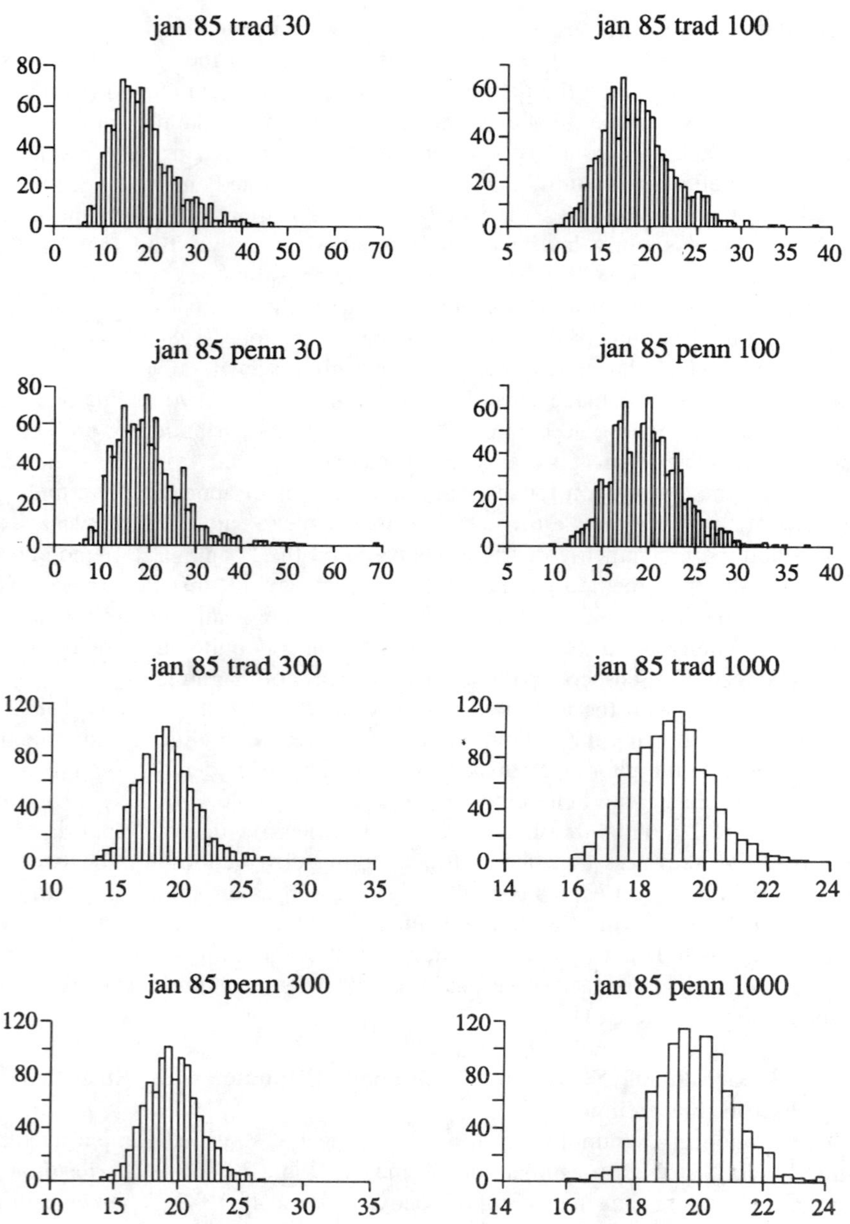

Figure 8. Frequency distribution for traditional and Pennington estimates of the mean catch per trip, based on sample sizes of 30, 100, 300, and 1000 for January 1985. Horizontal axis indicates values of estimates. Vertical axis indicates frequency.

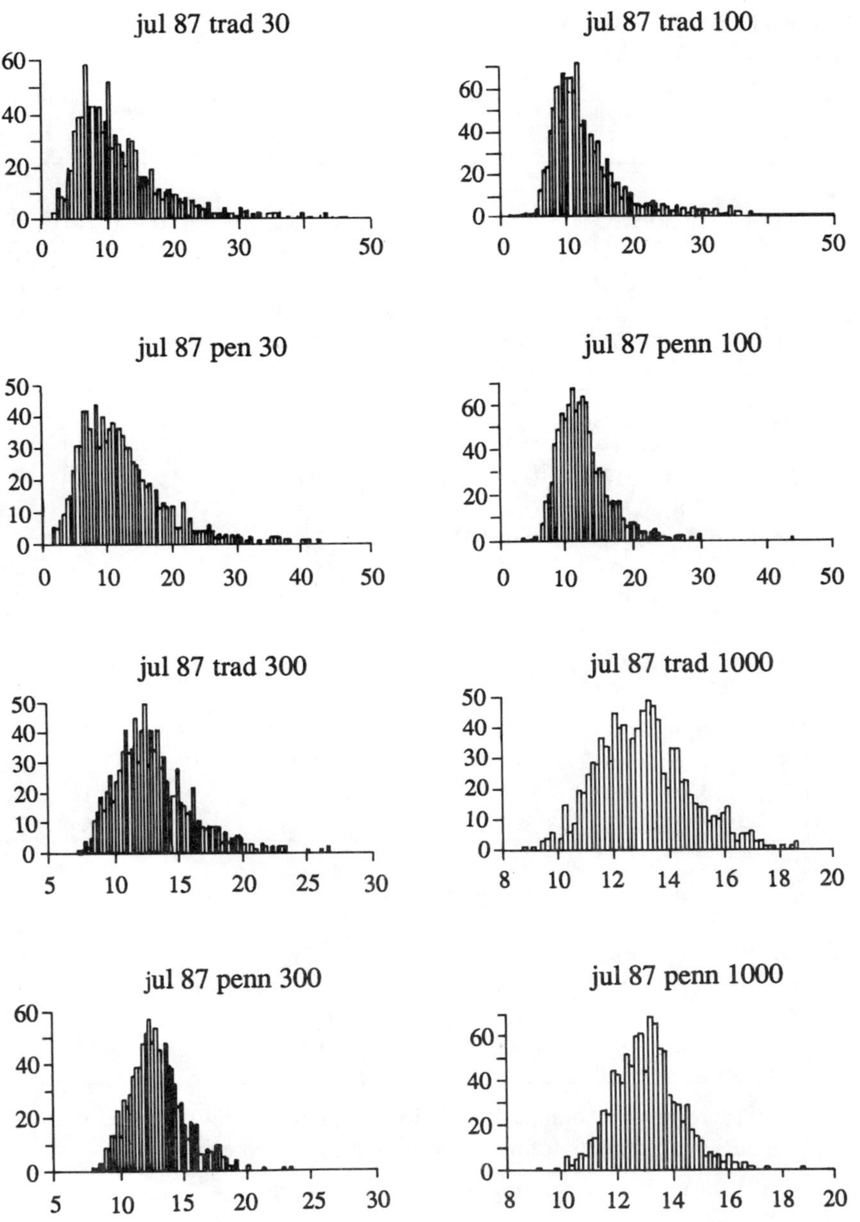

Figure 9. Frequency distribution for traditional and Pennington estimates of the mean catch per trip, based on sample sizes of 30, 100, 300, and 1000 for July 1987. Horizontal axis indicates values of estimates. Vertical axis indicates frequency.

Table 11. Comparisons of traditional and delta distribution estimates of the mean fish catch. Units = kg/boat.

Month	Std. error*	Skewness**	Efficiency***	Range (trad.)	Range (Penn.)
n=30					
[1]	15.30/9.65	4.91/1.35	2.52	6.0 – 152.3	6.2 – 7.4
[2]	5.46/5.91	0.83/0.78	0.86	4.8 – 37.3	5.2 – 40.6
[3]	6.90/7.00	1.28/1.24	0.97	5.4 – 69.7	5.3 – 68.9
[4]	14.24/7.50	5.96/1.55	3.63	5.0 – 155.7	4.9 – 80.5
[5]	35.90/6.28	10.60/3.50	32.70	2.3 – 449.2	2.3 – 78.5
[6]	8.60/6.72	3.13/1.87	1.64	2.2 – 78.6	2.3 – 67.6
n=100					
[1]	8.99/5.1	2.79/.068	3.07	11.7 – 92.6	12.9 – 41.8
[2]	3.06/3.27	0.68/.067	0.88	8.9 – 30.8	8.7 – 30.6
[3]	3.83/3.84	0.76/.062	0.99	10.1 – 38.3	9.9 – 37.3
[4]	7.24/4.20	2.92/.072	2.98	9.2 – 61.8	9.3 – 36.3
[5]	18.56/3.08	5.90/1.85	35.21	3.9 – 155.1	4.0 – 34.7
[6]	5.13/3.84	2.17/1.53	1.80	3.8 – 57.3	3.5 – 43.7
n=300					
[1]	5.09/2.99	1.44/0.40	2.88	15.0 – 47.4	15.9 – 34.8
[2]	1.76/1.90	0.36/0.37	.85	11.8 – 23.3	12.0 – 25.0
[3]	2.24/2.23	0.54/0.50	1.01	13.4 – 30.3	14.0 – 32.0
[4]	4.20/2.40	1.64/0.46	3.07	12.1 – 45.1	12.6 – 29.4
[5]	8.91/1.60	3.78/0.76	31.02	6.0 – 64.1	6.3 – 18.5
[6]	3.01/2.72	1.02/0.71	1.85	7.0 – 26.6	7.8 – 23.7
n=1000					
[1]	2.84/1.65	0.88/0.42	2.97	18.6 – 37.8	19.9 – 30.7
[2]	0.97/1.01	0.28/0.27	0.90	13.7 – 19.6	14.4 – 21.0
[3]	1.18/1.18	0.27/0.22	0.99	15.8 – 23.3	16.3 – 23.9
[4]	2.27/1.35	1.03/0.30	2.83	14.9 – 31.3	14.3 – 23.9
[5]	4.31/0.83	2.06/0.26	26.81	8.4 – 35.8	8.2 – 13.6
[6]	1.61/1.21	0.41/0.34	1.79	8.7 – 18.6	9.2 – 18.8

* Standard error = standard deviation of the sampling distribution of the statistic. Traditional estimate followed by delta-distribution estimate.

** Skewness of the sampling distribution. Traditional estimate followed by delta-distribution estimate.

*** Efficiency = variance $(\overline{x})$ / variance (delta-distribution estimate) from sampling distributions.

[1] = Jan.'84; [2] = Oct.'84; [3] = Jan.'85; [4] = Jul.'85; [5] = Jan.'87; [6] = Jul.'87

Table 12. Stratification of 4319 (out of total 7161) catch slips for January, 1987.

Landing Site	N_i	# of Zeroes	$\overline{X}_i$	s_i	y_i	S_y^2	$\hat{\mu}_\Delta$	$Var(\hat{\mu}_\Delta)$	$\hat{\sigma}_\Delta$
Chira	299	13	9.04	10.36	1.78	1.00	9.31	0.46	12.55
Chomes	284	12	5.49	9.10	1.30	0.74	5.04	0.10	5.51
Costa de Pajaros	732	29	16.22	24.1	2.28	1.12	16.40	0.68	24.17
Cuajiniquil	322	306	0.43	3.8	1.26	1.49	0.34	0.02	2.67
Manzanillo	716	41	10.40	11.4	2.01	0.81	10.60	0.20	12.46
Nispero	44	1	45.34	38.2	3.47	0.98	50.30	79.5	62.05
Puntarenas	1,102	445	42.5	390.2	2.53	2.46	25.50	5.74	108.30
San Pablo	334	53	8.73	10.36	1.78	1.00	5.44	0.31	11.64
Tarcole	486	202	5.09	10.79	1.29	1.87	5.36	0.40	16.63

Stratified mean using N_i: $\dfrac{\sum_{i=1}^{N} N_i \overline{X}_i}{\sum N_i} = 18.05$ kg/boat

Stratified Δ-based mean using N_i: $\dfrac{\sum N_i \hat{\mu}_\Delta}{\sum N_i} = 13.58$ kg/boat

Stratified variance using $\frac{1}{N^2}\sum N_i^2 Var(\overline{X}_i) = 9.03$; $\sqrt{\ } = 3.00$ kg/boat

Stratified variance using $\frac{1}{N^2}\sum N_i^2 Var(\hat{\mu}_\Delta) = 0.417$; $\sqrt{\ } = 0.645$ kg/boat

$$\frac{\text{Stratified var based on } \overline{X}_i}{\text{Stratified var based on } \hat{\mu}_\Delta} = \frac{9.03}{.417} = 21.65$$

Table 13 shows the proportion of a total sample allocated to each of the 9 sites using Neyman allocation, where the sample size per stratum is proportional to the standard deviation for that stratum. Two situations are explored, one using an ordinary estimate of the standard deviation, and a second using a delta-distribution based estimate. The delta-distribution estimate gives less "lopsided" results since the estimates for variability are not nearly as extreme as those using the ordinary standard deviation. For example, using ordinary standard deviation estimates, the port of Puntarenas would be allocated 91 % of the sampling effort. Using delta-distribution based estimates, 72 % of the sampling effort would be allocated to Puntarenas, with the remainder being spread out to other sites.

Table 13. Neyman allocation for 11 landing sites.

Landing site	Using s_i for $\hat{\sigma}_i$	Using $\sqrt{\sigma^2}$ via delta estimate
Chira	0.0065	0.0225
Chomes	0.0055	0.0094
Costa de Pajaros	0.0373	0.1060
Cuajinquil	0.0026	0.0051
Manzanillo	0.0172	0.0535
Nispero	0.0035	0.0164
Puntarenas	0.9089	0.7153
San Pablo	0.0073	0.0233
Tarcoles	0.0111	0.0484

4.3 Recommendations and Comments

The size of the collected catch slip information was the principal impediment to timely entry, analysis, and utilization of the data by fisheries managers in the Gulf of Nicoya. The volume of the data set can seriously tax, and often overwhelm, the human and computer resources of the data processing unit of the Department of Fisheries. In fact, much of the historical record of this invaluable data set has yet to be entered and analyzed, and current information can be months out of date by the time it is entered.

We considered how much data must be collected to accurately estimate the catch-per-vessel landing of the valuable primera catch category in the Gulf of Nicoya in Costa Rica. We were able to simulate and evaluate a number of sampling approaches to this problem on a 6-month data set that was essentially a complete census of the Gulf of Nicoya for those time periods. Classical sampling approaches were not very efficient, requiring as much as a thousand samples to achieve a 10% coefficient of variation. Stratified sampling techniques produced moderate gains over simple random sampling, but did not improve performance enough to justify as low as a 30-boat-per-month landing sampling plan. Using catch estimates based on the delta-distribution, whether stratified or not, seems to reduce variances considerably.

Systematic sampling protocols showed coefficients of variation comparable to random sampling designs. Systematic sampling may be an attractive alternative to simple random sampling when sequence effects do not predominate. Modest losses in estimation efficiency can be more than offset by reduced costs of handling in many fisheries data analysis scenarios. Systematic sampling can be coupled with stratification to capitalize on homogeneities in catch between, say, different landing sites. If clumps of landing site data are naturally aggregated into a sampling frame list, systematic sampling induces a natural *stratified sampling proportional-to-size* protocol, since samples are extracted from the frame in close approximation to the strata proportions.

The enormous variation and the large component of zero catches necessitated the use of nontraditional sampling methodology. The distribution of catch-per-trip was modeled using the delta-distribution investigated by Pennington (1983, 1986) and Smith (1981, 1988). We were able to use an estimate for the mean developed by Pennington (1983) to obtain reasonably accurate estimates of the mean catch-per-trip from these highly skewed data. Because our sample sizes were so large, the issue of variance estimator performance (as in Smith, 1988) did not enter into our analyses using the delta-distribution. From our results, we recommended sample sizes in the neighborhood of at least 300 catch slips per month and use of the Pennington (1983) method to estimate mean catch. We expect that this approach will be useful in other artisanal fisheries where the success of each fisherman in catching a given species or category is fairly low, but where the catches can be quite large if that species or category is encountered during fishing.

A final comment is in order regarding what is known as a design-based approach (not assuming any particular underlying probability distribution) versus a model-based approach. The former relies on the Central Limit Theorem in that parameter estimates are assumed to be normally distributed for a large enough sample size. The latter relies on the additional model-based information to provide better estimates, usually in terms of efficiency. Discussions of design-based and model-based approaches for sample surveys may be found in Hansen et al. (1983) and Little (1983). Myers and Pepin (1990) address the robustness of log-normal-based estimators of abundance. A comparison of both model- and design-based estimators for central tendency for the log-normal distribution with respect to trawl surveys may be found in McConnaughey and Conquest (1993).

For the Gulf of Nicoya data set, use of the delta-distribution was verified at the beginning and resulted in definite improvements in terms of efficiency. Our results showed that even for sample sizes as large as 1000, the estimator based on the delta-distribution exhibited more Gaussian-like behavior than that based on the sample average. Stratification further improved the efficiency somewhat, but the major gains occurred from using the model-based estimator. Exclusive use of any particular probability distribution with fish catch data is certainly not recommended without first verifying that the probability model assumptions are indeed met by the data.

ACKNOWLEDGMENTS

The authors are indebted to Dr. Terrance Quinn II, who reviewed an early version which resulted in significant improvements. Staff and students at the Centro des Investigaciones en Sciencias del Mar y Limnologica of the University of Costa Rica (CIMAR - UCR) processed landing slips from the artisanal fishery. Special thanks are due Jorge Campos, University of Costa Rica, for providing much of the needed data. Benyounes Amjoun and John Hedgepeth contributed via participation in many discussions. Katherine Peterson, Erin Moline, and Steven Anderson typed the many drafts. This publication was supported by the U.S. Agency for International Development under grant #DAN-4146-G-SS-5071-00, Fisheries Stock Assessment, Collaborative Research Support Program (CRSP). This publication is Contribution No. 897, University of Washington, School of Fisheries. Correspondence should be addressed to Loveday Conquest, Box No. 355230 University of Washington, Seattle, WA, 98195.

REFERENCES

Aitchison, J. and Brown, J.A.C. 1969. *The Log-normal Distribution.* Cambridge University Press, London.

Anderson, D.R. and Pospahala, R.S. 1970. Correction of bias in belt transect studies of immobile objects. Journal of Wildlife Management 34: 141-146.

Buckland, S.T. 1985. Perpendicular distance models for line transect sampling. Biometrics 41: 177-195.

Burnham K.P., Anderson, D.R. and Laake, J.L. 1980. Estimation of density from line transect sampling of biological populations. Wildlife Monographs 72: 1-202.

Caddy, J.F. and Bazigos, G.P. 1985. Guidelines for statistical monitoring: Practical guidelines for statistical monitoring of fisheries in manpower limited situations. FAO Fish. Tech. Paper No. 257.

Cochran, W.G. 1977. *Sampling Techniques*. 3rd edition. John Wiley and Sons, New York.

Eberhardt, L.L. 1968. A preliminary appraisal of line transects. Journal of Wildlife Management 32: 82-88.

Eklund, C. and Atwood, E.A. 1962. A population study of antarctic seals. Journal of Mammalogy 43: 317-328.

Gates, C.E. 1969. Simulation study of estimators for the line transect sampling method. Biometrics 25: 317-328.

Gates, C.E., Marshall, W.H. and Olson, D.P. 1968. Line transect method of estimating grouse population densities. Biometrics 24: 135-145.

Hansen, M.H., Hurwitz, W.N. and Madow, W.G. 1953. *Sampling Survey Methods and Theory*, Vols. I and II. John Wiley and Sons, New York.

Hansen, M.H., Madow, W.G. and Tepping, B.J. 1983. An evaluation of model-dependent and probability sampling inferences in sampling surveys. Journal of the American Statistical Association 78: 776-793.

Kish, L. 1965. *Survey Sampling*. John Wiley and Sons, New York.

Lindgren, B.W. 1968. *Statistical Theory*, 2nd Edition. The Macmillan Company, London.

Little, R.A. 1983. Comment on: An evaluation of model-dependent and probability sampling inferences in sample surveys. Journal of the American Statistical Association 78: 797-799.

McConnaughey, R.A. and Conquest, L.L. 1993. Trawl survey estimation using a comparative approach based on log-normal theory. Fishery Bulletin 91: 107-118.

Mosteller, F. and Tukey, J.W. 1977. *Data Analysis and Regression*. Addison-Wesley Publishing Company, Reading, MA.

Myers, R.A. and Pepin, P. 1990. The robustness of log-normal-based estimators of abundance. Biometrics 46: 1185-1192.

Pennington, M. 1983. Efficient estimators of abundance, for fish and plankton surveys. Biometrics 39: 281-286,

Pennington, M. 1986. Some statistical techniques for estimating abundance indices from trawl surveys. Fishery Bulletin 84: 519-525.

Pollock, K.H. 1978. A family of density estimators for line-transect sampling. Biometrics 34: 475-478.

Quinn, T., II. 1985. Line transect estimators for schooling populations. Fisheries Research 3: 183-199.

Quinn, T., II. and Gallucci, V. 1980. Parametric models for line-transect estimators of abundance. Ecology 61: 293-302.

Scheaffer, R.L., Mendenhall, W. and Ott, L. 1990. *Elementary Survey Sampling*, 4th edition. PWS-Kent, Boston, MA.

Seber, G.A.F. 1973. *The Estimation of Animal Abundance and Related Parameters*. Griffin, London, England.

Smith, S.J. 1981. A comparison of estimators of locations for skewed populations—with applications to groundfish trawl surveys, in Bottom Trawl Surveys. Can. Spec. Publ. Aquat. Sci. 58: 154-163.

Smith, S.J. 1988. Evaluating the efficiency of the delta-distribution mean estimator. Biometrics 44: 485-493.

Thompson, S.K. 1990. Adaptive cluster sampling. Journal of the American Statistical Association 85: 1050-1059.

Thompson, S.K. 1991a. Adaptive cluster sampling: Designs with primary and secondary units. Biometrics 47: 1103-1115.

Thompson, S.K. 1991b. Stratified adaptive cluster sampling. Biometrika 78: 389-397.

Thompson, S.K. 1992. *Sampling*. John Wiley and Sons, New York.

Zar, J.H. 1984. *Biostatistical Analysis*, 2nd edition. Prentice-Hall, Englewood Cliffs, NJ.

CHAPTER 5

CORAL REEF FISHERY SAMPLING METHODS

John W. McManus[1/2], Cleto L. Nañola[1], Annabelle G.C. del Norte[1], Rodolfo B. Reyes, Jr.[1], Joseph N.P. Pasamonte[1], Nygiel P. Armada[3], Edgardo D. Gomez[1], and Porfirio M. Aliño[1]

[1]Marine Science Institute, University of the Philippines, [2]University of Rhode Island, [3]College of Fisheries, University of the Philippines in the Visayas

1. BACKGROUND: THE CORAL REEF FISHERY SITUATION

Coral reef fisheries represent approximately 20 to 25% of the potential fish harvests available to developing countries (McManus and Arida, 1988). Recent work indicates that harvests of invertebrates and seaweeds can potentially exceed those of finfish. The total number of commercially valuable species involved can reach into the hundreds.

The fishery on a reef can be divided among several very different habitat types, including: sandy lagoons, lagoonal coral patches, seagrass beds, mangrove roots, backreef zones, forereef slopes, reef channels, and others. Each of these tends to support distinctive faunal assemblages, and each fauna has components which utilize other habitats at various life stages or times of migration or foraging (McManus, 1988).

Fringing reefs tend to attract large numbers of very poor coastal people, who take advantage of the low investment costs involved in gathering reef organisms. They often accumulate until competition drives profits to minimal levels and destructive harvest methods become prevalent. The total number of fishing gear types in use on a reef may exceed 30, of which a few important ones are often illegal and difficult to obtain accurate data on. All of these factors add complexity to the problems of analyzing reef

1-56670-151-1/96/$0.00+$.50

fisheries and to the development of optimal management recommendations (McManus et al., 1988; del Norte et al., 1989).

On the positive side, there are some features of a fringing reef which are advantageous from a management standpoint. A coral reef tends to be an integral system, which differs substantially from the soft bottom or open water environments which surround it. Most of the important reef habitats important in small-scale fisheries are found in waters above 20 m. Because these waters are often fairly clear, the boundaries of each important habitat type can often be detected from aerial or satellite imagery (Jupp et al., 1985; McManus, 1989). Coral reefs tend to be resilient to nondestructive overfishing. As high-priced, larger fish are overfished, there may be a steady influx of new recruits of juveniles or smaller species on a seasonal basis which maintains the fishery (del Norte et al., 1989; Campos et al., 1989; del Norte and Pauly, 1989). As coral cover becomes depleted by blast fishing or fish poisoning, fishermen, women, and children may fall back upon the resources of the seagrass beds which are not so easily destroyed by these methods. Thus, many coral reef fisheries appear to be buffered against sudden collapse and in some cases may permit limited recovery of resources following the institution of appropriate management measures.

Fisheries management on coral reefs tends to require a broader set of analytical tools than are generally employed in large-scale marine fisheries. Most common fishery analysis methods are useful on coral reef fish. For example, the current tropical fishery paradigm of length-based population analysis (Pauly, 1984; Pauly, 1982) can be used on some dominant species to determine levels of overfishing, provided the stocks are not depleted to levels wherein adequate sampling becomes difficult (del Norte et al., 1989; del Norte and Pauly, 1989). However, this is only one aspect of fishery management on coral reefs.

A major concern on many reefs is the use of destructive fishing methods such as blasting (Armada, 1989; McManus, 1988; McManus and Arida, 1988). The quality and possibly quantity of production of many reefs are directly (although not necessarily linearly) related to the cover of living coral. A proper assessment of the status of the fishery, therefore, would necessitate an evaluation of the amount of coral cover and an assessment of the major causes of its destruction.

In coral reefs, as in lakes and estuaries, a fishery manager must be as concerned about factors such as pollution as he or she is about fishing pressure. In some cases, the fishery biologist will be able to rely on others to assess the water quality. However, because most coral reefs are in developing countries with limited funds and manpower, he or she must always be prepared to assume the role of a general ecologist as well.

One fairly successful approach currently available for managing coral reef fisheries is that of fishery restriction by area. There is now strong evidence that maintaining a fishing reserve near a coral reef fishing ground can substantially improve both the quality and quantity of fish production (Russ and Alcala, 1989). The establishment of fishery restricted areas requires a knowledge of the distributions of reef-associated habitats and of the migration patterns of local species. Determining these factors will invariably lead the fishery biologist far astray of the chapters of standard fishery textbooks (e.g., Gulland, 1983; Pauly, 1984; Ricker, 1975). The biologist is led even further astray when tasked with weighing biologically based options against local sociological and economic limitations. These limitations vary radically from village to village along a fringing reef.

The current boundaries which separate fisheries and wildlife biologists from "mainstream" ecologists in the U.S. and some other countries are historically based, and are substantially attributable to personality conflicts and rivalries from the beginning of this century (McIntosh, 1985). This separation between biologists and ecologists has been maintained primarily through specific training programs and methodological paradigms. The differences can be such that a fishery biologist may tend to avoid pursuing an objective which leads him or her into unfamiliar ecological methodologies. The reverse is also true, as is evident in the fact that length-based population analyses have only recently seen wide use among those who study benthic invertebrates (e.g., Gallucci and Hylleberg, 1976; del Norte, 1986; Brey, 1986).

As mentioned before, for coral reef fisheries management, the analyst must assume the role of an applied ecologist. The research strategy must be based on a clear definition of objectives and a choice of methods appropriate to the objectives.

The success of a field project depends on a great many factors. Prominent among these are the adaptability and ingenuity of a research leader and his or her ability to tap available expertise. The literature on field techniques is vague in many areas. A fieldworker has to deal with a multitude of problems which are generally omitted from the "methods" sections of publications. It is only after considerable experience that a researcher is even aware of many of the potential biases involved in the use of a method and of the options available to minimize these biases.

The following chapter considers three approaches to gathering data on the ecology of multispecies fisheries on coral reefs and other coastal ecosystems. The techniques include remote sensing, fishery independent sampling, and fishery dependent sampling. These general approaches are highly complementary to each other. However, only a few specific methods are discussed in each section. The chapter has been written with the realization that each researcher should adopt whatever methods are best suited to a project's specific objectives and the circumstances under which the work will be undertaken. It is not fruitful to prepare a detailed "cookbook" of what to do in any given circumstance. It is more important to concentrate on presenting a variety of options for which some experience has been gained recently and to illustrate how the choice of methods should be molded to specific project objectives.

The objectives may be to answer questions of immediate value to a specific fishery management situation. Examples might include determining the range of species and biomass available for harvest, determining the fish community structure so that changes can be detected as the fishery progresses, determining the migration habits of dominant species, determining the distribution of subsurface habitats so that reserve areas may be selected, etc. Alternatively or additionally, the objectives may concern ecological questions of importance to future management of such fisheries. An example might be a study concerned with determining the patterns of resilience in the fish community to changes in fishing pressure. In either case, the clarity with which the objectives are defined can have much to do with the success of the study.

The material which follows is concerned primarily with sampling techniques employed during the Fisheries Stock Assessment Collaborative Research Support Program reef monitoring activities in the Philippines. Additionally, some alternative methods are briefly mentioned, such as the section on aerial photography from kites and balloons. These sections are included to emphasize the fact that alternative methods exist and to direct the reader to the appropriate literature. Many of the techniques described here are covered in varying degrees in the UNESCO *Coral Reefs: Research Methods* (Stoddart and Johannes, 1978). No attempt has been made to replicate that material. Therefore, reference to that text is highly recommended.

A strong emphasis has been placed on the need to balance statistical considerations against the practical limitations of the project and the complexity of the sampling environment. We assume that the reader will study this chapter in conjunction with the other chapters on fishery sampling. No attempt has been made to make the problems appear any less complicated than they really are. Rather, the approach is to outline the range of problems and potential biases one might encounter in the sampling, so that proper learning or consultative help can be obtained to ensure optimal success in a project. Upon reading the chapter, if a researcher finds that he or she is suddenly concerned about more problems than before, then perhaps the chapter has been successful. If those problems are resolved early, there will undoubtedly be fewer problems later.

2. DEFINING THE ENVIRONMENT OF THE FISHERY

2.1 INTRODUCTION

Before a fishery is to be investigated, it is generally advisable to gather as much information as possible about the environment of the fish being harvested. The intensity of this aspect of the investigation will often vary with the specific objectives of the research and the cost of gathering detailed environmental information. In most cases, a considerable amount of information can be obtained at very low cost from available coastal charts and topographic maps. Orthographic maps and sediment charts are available for some countries and areas. There often exist aerial photographs of a given coastline, usually taken sometime during or immediately after World War II, when the lack of such coverage for most of the world suddenly became a major military concern. Unfortunately, a surprising number of countries have severe restrictions on the use of such photographs, and obtaining legal access may be difficult. The U.S. military maintains files of photographs taken in countries it has operated in, but there are so many photographs that they might be accessible only if very specific information is known such as dates and mission designations.

The average coastline around the world probably has very little detail on coastal charts, and topographic maps will be considerably out of date in terms of defining coastal ecosystems such as reefs, mangroves, nipa swamps, and marshes. Two means of obtaining basic information on the coastal ecosystems include professional aerial mapping and satellite mapping. The former will yield more detailed information, but is usually prohibitively costly. The latter is much less expensive and may provide information of sufficient detail for many fishery investigations. The state-of-the-art is such that satellite mapping of coastlines still requires considerable qualitative field verification from aircraft, but this can be done with low cost techniques (McManus, 1989).

2.2 PROFESSIONAL AERIAL SURVEYS

Whenever possible, standard aerial mapping surveys should be undertaken. Such surveys would normally involve standard aircraft using vertical and low oblique cameras to obtain black-and-white photos for stereo interpretation and photomosaic mapping. A professional aerial survey company would generally be involved. In addition to standard photographs, the company might be requested to provide supplemental polarized color footage to permit greater interpretation ability. The company should be briefed to avoid times when reflective glare on the water surface would limit the usefulness of the photographs for interpreting subsurface features.

Properly obtained photographs can be interpreted with the aid of modern digitizing and image enhancement equipment, including that used for satellite image analysis. Stereo pairs can be used for topographic mapping on land. With appropriate corrections, subsurface bathymetry can be obtained for clear, shallow areas. However, quantitatively useful photographs must be obtained with standardized methods, and the involvement of a professional survey company will generally be mandated. A mapping survey of 50 miles of coastline may cost on the order of $50,000 to $200,000, even in developing countries with indigenous companies. This cost is much greater than will generally be feasible for a coastal fishery study. General information on the use and interpretation of standard aerial photographs may be found in Lillesand and Kiefer (1979) and Avery and Berlin (1985).

2.3 SATELLITE MAPPING

There are two broad classes of satellite imagery of interest to fisheries management, depending upon resolution. Low resolution data is based on a resolution on the order of 1–4 km, and includes most available sea surface temperature data and the now defunct ocean color scanner data. The most common form of sea surface temperature data currently available is Advanced High-Resolution Radar (AVHRR) data. In spite of the name, this data is of too low a resolution to be useful in studies of small-scale fisheries, except in defining oceanographic conditions on a scale of hundreds of km. AVHRR is the data of choice for many studies of commercial pelagic fisheries. However, it can also be useful for identifying coastal upwelling and potential larval entrainment features such as seasonal gyres (Amidei, 1985). These oceanographic features are strongly suspected of governing reproduction timing in coral reef and other tropical coastal fishes (Longhurst and Pauly, 1987). AVHRR receivers are very inexpensive, and the data is becoming available in many developing countries. Work is currently underway at the National Atmospheric and Space Administration (NASA) and elsewhere to develop microcomputer-based algorithms for interpreting the data.

For studies of shallow-water fisheries, high resolution Landsat-type data can be invaluable. In addition, there are four commonly used data sources (Butler, circa 1987).

Multi-Spectral Scanner data (MSS) has a resolution of approximately 80 m and has been available to many parts of the world since 1972. Thematic Mapper (TM) data has a resolution of 30 m for most bands and has been available since 1982 in selected areas. The French Systeme Probatoire, d'Observation de la Terre (SPOT) has a resolution of 20 m for color and 10 m for panchromatic data. It began surveys in 1985. The fourth type of data is from a series of satellites launched by the USSR and is not widely available outside the country.

A general problem with coastal ecosystems is that for any given habitat type, such as coral reef, swamp, etc., much or most of the total area along a coastline may be made up of small, isolated patches of growth. Thus, there may be substantial differences in the estimates of total area made from 80 m resolution versus 10 m resolution data. For small-scale studies, high-resolution data is preferred. However, the 10 m panchromatic data of SPOT carries far less information than the 20 m color data. The 30 m TM data penetrates water further and with more bands than the SPOT data, and many areas of the world have only MSS coverage available. Therefore, the choice of data will depend on several factors, and it may be useful to use more than one data type to make use of the different characteristics.

In each case, a single computer tape with geographically rectified data may cost a few thousand dollars, and it may cover 100–200 km on each side of a parallelogram. Until recently, the major cost factor was that of the analysis system, which was generally based on a mainframe computer. Recently, however, a number of microcomputer-based image analysis systems have become available. These range in cost from a few thousand to tens of thousands of dollars, depending upon options. Generally, a very flexible system including a reader for 9-inch (mainframe) computer tapes can be obtained for less than $40,000. This same unit can be used for any of the Landsat-type data, as well as for the analysis of aerial photographs, medical scanning data, microscopic analyses, morphometric studies, etc. It may thus be cost-shared with other funded projects in an institute. The files of analyzed satellite data are generally compatible with Geographic Information Systems (GIS), and the same hardware system can form the basis of a Coastal Management GIS/Resource Mapping System, which can ultimately be used in further fishery analyses.

Analyses of Landsat-type data generally consist of subsetting a scene of interest and then manipulating one or more bands of data to enhance features of interest. Common manipulations include: contrast stretching, smoothing, edge-enhancement, dehazing, ratioing bands, principal components data reduction, and supervised or unsupervised classification. The areal coverage of a feature or set of features can be easily calculated, and the enhanced images can be printed out as black and white or color maps with line or text annotation. In coral reef work, a technique of considerable importance is band slicing. In clear water, the shorter bandwidth of MSS data penetrates to approximately 15–20 m. The next longer band penetrates to approximately 4 to 5 m. The third band penetrates only very thin layers of water (< .5 m), and the fourth band does not penetrate seawater. This information can be used to isolate land, intertidal, shallow, moderate, and deep waters (Biña et al., 1978; Biña and Ormac, 1979; Jupp et al., 1985).

One common problem with shallow water data is the prevalence of spectral highlights in the data. The same information on the order of depth penetration can be used to identify errors from spectral highlights and other sources which do not represent bottom features. Basically, if a feature appears on a given band and not on one of deeper penetration, then the feature is obviously a surface effect or band error. This technique can be programmed to provide a rapid identification of errors (Roa, 1989).

Analyses of the Great Barrier Reef generally involve unsupervised classifications (Jupp et al., 1985). The technique is based on classifications of data points based on their variances in multidimensional space, where the dimensionality depends upon how many bands there are, and modified bands are analyzed concurrently. There is no particular reason to believe that discrimination on the basis of mathematical variance will always yield a classification of optimal usefulness in ecology. For example, an ecologist who is asked to make a single division in the set {sand, rock, and seagrass} will probably divide seagrass from sand and rock, because the division will fundamentally divide biotic from abiotic features. Statistical classification methods will generally divide out the sand first, because of the substantial difference in reflectance values. There is often a substantial difference between reflectance and ecological significance.

The Great Barrier Reef analyses have been successful partly because bottom flora and fauna on its reef flats tend to vary sharply with minor variations in depth. Philippine fringing reefs are usually heavily dominated by seagrass. The seagrass forms intricate patterns of density at many scales but the patterns do not correlate well with depth. This makes it very difficult to obtain ecologically useful classifications of such reefs by unsupervised classification (McManus, 1989).

In supervised classification the computer samples the data characteristics among the bands searching for a small area of known bottom type or vegetation. The computer calculates the multivariate mean and variance for that feature. Once a variety of known areas have been sampled, the computer can be tasked to divide an entire map into areas with similar reflectance spectra (Lillesand and Kiefer, 1979). With this technique, one or more small areas known to contain seagrass can be used to define a class which will be extrapolated to a mapping of all the seagrass areas on an image. The technique is error-prone because it is partly dependent upon the complete set of samples used to divide up an image. The error tends to be stepwise rather than continuous. This means that an analysis which is not correct will generally be very substantially incorrect. The method can therefore be used in conjunction with aerial verification to produce reliable maps (McManus, 1989).

Ultimately, it would be better to avoid both supervised and unsupervised classification in favor of algorithms specifically designed to identify particular coastal habitat types. For example, one should be able to combine a knowledge of seagrass reflectance values (with a model describing the variation of these values) with tide height to classify seagrass in images taken at known tide heights. Very little work has been done in this direction. However, there is a very compelling need for such research. Worldwide, there is very little information on tropical marine habitats such as coral reefs. It is difficult to assess their resource values or statuses because no reliable estimates of reef area or location charts exist. Remote sensing represents a potentially very valuable tool in this regard. However, efforts must be made to minimize the need for field verification before global studies of coral reefs can be conducted (McManus, 1989).

2.4 LOW-COST AERIAL SURVEYS

2.4.1 Conventional Light Aircraft

Studying a shallow-water ecosystem without aerial photographs would be like trying to study an elephant by touch while wearing a blindfold. Skin and SCUBA diving are important for investigating the structure of the subsurface ecosystem on 10-meter scales. However, it is rarely possible with diving surveys to obtain accurate information on larger features or on the geographical relationships between local habitat types. Without this information, it is difficult to confidently investigate such things as interhabitat migration or nutrient flow.

Formal aerial photographic surveys are extremely useful for quantitative analyses, but the cost is generally prohibitive unless the survey objectives are shared with well-funded coastal studies. In many cases, one can obtain qualitative or semi-quantitative data from light aircraft (Hopley, 1978). Mounted cameras on a board can be passed outside the aircraft door to achieve (approximately) vertical photographs. Any quantitative analyses to be conducted will require either a knowledge of aircraft altitude and lens focal length, or the inclusion of an object of known size in the photograph or photographic series. Generally, however, vertical photographs are extremely difficult to interpret usefully unless a large set has been taken for photomosaicking. For example, imagine having a 5 cm x 8 cm photograph of one small section of a classroom floor 0.5 m on each side. The photograph alone will show you very little about the classroom floor, unless it is accompanied by a large number of other photographs with specific information on which part of the photo each floor part is from. On the other hand, a photograph taken at about 45 degrees across the floor will give a great deal of qualitative information about the arrangement of cracks, stains, and wear spots on the floor. The information may be sufficient to lead to a conceptual model for construction methods and deterioration patterns of the floor. It is much easier to obtain quantitative information from a vertical photograph, but the quantities may be of little use without a general conceptual framework to which they may be applied. Therefore, the priority with informal aerial photography should generally be placed on obtaining oblique photographs.

For many purposes, conventional aerial photography is best done with high-resolution black-and-white film. No color films exist with the resolution of such specialized film, and inspection of the photographs under high magnification is often required for such purposes as mapping roads and trails. However, in the marine environment, it is often a major task to identify even the largest features in a photograph. For this reason, color film is generally preferred (Hopley, 1978). Unless one's purpose is specifically mapping an ecosystem from vertical photographs, it is generally more economical to use slide than color print film. For studies of coastal vegetation or aquatic vegetation in waters less than 3 meters deep, it is often useful to use color infrared film. However, this film requires the use of a yellow filter.

The camera to be used will generally be a 35 mm model with manually adjustable speed and aperture. The lens should be set (and sometimes taped) on an infinity setting. A major concern is vibration from the airplane. Shutter speed should be kept as high as possible. Depth of field is not a problem. However, some lenses show visible distortion at the widest aperture. It is generally safe to set the aperture on one or two stops less than the widest setting and the speed at 1/1000 or 1/2000 second for film of ASA 100. Films with higher ASA often yield high contrast pictures, and colors such as light blue often shift to white. These settings should be varied (mostly in terms of aperture) to account for available light conditions. However, there is usually little to be gained from taking photographs of shallow-water ecosystems except on clear, sunny days with very little wind to make waves.

It is sometimes helpful to have small shadows in the picture to aid in interpretation. However, surface glare is to be avoided as much as possible. One should generally try to fly between 9 a.m. and 3 p.m. to avoid glare. In some areas and seasons, a more restricted time bracket may be necessary. Some glare can be eliminated with the aid of a polarizing filter set on maximum effect. The lens should generally be wide-angle, such as a 35-mm lens. This permits more spatial information to be encompassed at low altitudes. High altitudes cause increases in atmospheric haze, which will diminish resolution in the photographs. In some cases, a wide-angle to short telephoto zoom will be useful for properly framing photographs of specific features. The lens should generally be held by the photographer in such a way as to allow the photographer's body to absorb the vibrations of the aircraft. At no time should the camera be rested against a window or other part of the aircraft.

Oblique photographs are generally interpreted by viewing while studying some form of map, chart, or satellite image of the same area. Vertical photographs can be projected onto paper so that significant features can be traced and measured. Alternatively, they may be projected onto a digitizing board for measurement with microcomputer software. Stereo pairs can also be projected, with the use of mirrors, to be visible with a stereoscope and traceable on the digitizing board. This procedure assists in the discrimination of objects separated vertically and permits the true measurement of heights and areas of obliquely oriented features.

2.4.2 Kites and Balloons

Photographs can be taken from kites or balloons under ideal circumstances. A kite requires a stiff, steady breeze to be deployable. An unmanned balloon is best handled with little or no breeze (Rutzler, 1978). However, a small blimp with a remotely controlled motor can extend the usefulness of the balloon approach (Preu et al., 1989). In either case, a camera can be mounted with a radio-controlled trigger.

The major disadvantages of these approaches are the restriction in altitude and restrictions in mobility. The higher up the device is deployed, the more difficult it is to target. This makes it difficult to take useful series of photographs. However, skilled operators have obtained useful photographs using these techniques in several studies. In cases where no other source of photographs is available, these approaches are certainly worthy of consideration.

Manned balloons offer considerable potential for high-quality, vibration-free photography. This approach is frequently used by researchers on the Cousteau expeditions. However, the cost in equipment and level of training are comparable to or exceed those required for ultralight aircraft, which offer much greater mobility. Gas balloons may cost on the order of a few thousand dollars per trip to operate. However, with hot air balloons, the chief operating cost is fuel for the heater, and this may amount to only a few tens of dollars per trip (Waligunda and Sheenan, 1981). The chief advantage of the manned balloon over the ultralight aircraft is probably the ease with which the former can be stored on a ship. Some of the maneuverability problem may be alleviated with the use of hot air blimps, which include propellers for horizontal motion. (Waligunda and Sheenan, 1981).

2.4.3 Remote-Controlled Aircraft

In the last decade, there has been a worldwide upsurge in the use of remotely operated aircraft. Such aircraft are used extensively by military units worldwide for aerial photography. Some private companies in the U.S. and Japan are offering low-cost aerial photographic coverage using this equipment. One recent device is an hourglass-shaped machine with a propeller whirling around its "waist" and a vertically mounted camera in the base, specifically designed for forestry surveys (Gwynne, 1987). Another aircraft has an aimable camera in its nose and a steerable parachute recovery system designed to permit landings on difficult targets such as ship decks.

A great variety of hobbyist aircraft can be equipped with cameras and radio-controlled triggers. However, it is generally difficult to know in advance what the camera target will be. For this reason, most commercial systems incorporate a "real-time" TV camera system to permit aimed photographs to be taken. This limits the use of hobbyist aircraft to large models with strong engines. It also usually requires possession of a television broadcasting license.

One major disadvantage with most remote-controlled (RC) aircraft is the skill required to operate them. The landing of such a machine is particularly difficult. All of the steps used in landing a standard aircraft must be undergone with a scaled-down version, but often at some distance from the operator. This difficulty with landing can be circumvented in a number of ways. Some military units use aircraft designed to be crashed and rebuilt after each use. Other aircraft are caught with cargo nets to minimize damage to important instruments. Parachutes are becoming increasingly popular. The parachutes add considerably to the payload and complexity of the aircraft, but are practical with well-tested systems.

Take-offs and actual flights are not particularly difficult for a novice operator. Some difficulties have been reported from operators of planes designed to be flown by video camera because of electronic delays in reception of images and transmission of control signals. However, these problems can be overcome with practice.

For operation in protected waters, such as coves or reef lagoons, it is feasible to use a float plane or flying boat. In general, this requires that the waves be no more than about 10 cm high and the winds be less than a few km/hr. Military RC aircraft are often launched from catapults. These generally consist of a track of about 3 m length and a spring-loaded propulsion unit.

A desirable characteristic of RC planes in coastal operations is the ability to take-off and land in water. It may be feasible to design an aircraft for aquatic use. This would consist of a vertically flying helicopter camera platform shaped like a lunar landing module with rubber car tires for landing pods. More research in this area is clearly needed.

2.4.4 Ultralight Aircraft

The term "ultralight" or "microlight" aircraft refers to a class of aircraft weighing less than a few hundred pounds. Ultralights include a broad range of aircraft from flimsy home-designed constructions to state-of-the-art aerodynamically designed commercial models. The broadness of the classification and the lack of license requirements in many countries has led to a general concern about the safety of flying an ultralight. In reality, there are safe ultralights and trained, skilled pilots, and there are also unsafe models and untrained, unskilled pilots. There are a variety of well-tested ultralights commercially built by companies with excellent safety records; however, we will not be concerned with other models here.

This section will review features of ultralights of particular concern to coral reef scientists. An ultralight has been previously used for reef surveys by Starck (1984). Information in this section comes from our use of the ultralight in the Fisheries Stock Assessment CRSP reef monitoring project in the Philippines (McManus, 1989). More general information can be found in recent books, including Coombs (1984) and Millspaugh (1987).

2.4.4.a *Safety Aspects*

A common group of ultralights are essentially small, distinctive standard aircraft, but with open or lightly covered cockpits and engine compartments. The construction is generally of aircraft-quality aluminum alloy frame, high-grade steel bolts and cables, and specially modified, extra-strong sailcloth covering. They are available with reliable engines of up to 65 hp. By comparison, the Piper Cub, the most popular aircraft in history, originally came with a 37-hp engine (Moll, 1987), and was constructed of aluminum poles covered with Irish linen and a doping compound.

A typical model will offer emergency backup control systems for turning and attitude control. The plane will have been designed to fall into a pattern of "natural stalls" if the operator has lost consciousness. This means that the plane will dip downwards, pick up speed, level out horizontally, and then dip again. The process is intended to bring the pilot gradually towards the earth. The safety factor involved in such stalls depends to a large degree on the stall speed of the aircraft. A standard private aircraft will often have a stall speed exceeding 60 mph. Most ultralights have stall speeds of 20 to 30 mph. Thus, the dipping process is much more restricted and safer in an ultralight. However, ultralights also offer the capability of being equipped with ballistically launched parachutes which can safely bring down the entire aircraft with the pull of a release grip. These devices have been so successful with ultralights, that they are now becoming available for larger aircraft.

Safety aspects of ultralight operation are quite similar to safety in SCUBA diving. A good training program for either activity is of approximately the same length and level of technical detail. Like a SCUBA diver, an ultralight pilot has complete access to all the equipment he will use. A complete, systematic check of all exterior nuts and bolts on the plane and engine takes about 15 minutes of preflight time. In this respect, ultralights differ from standard aircraft, for which the latter's engine is only inspected at most a few times a year by a licensed mechanic. As in SCUBA diving, safety is almost entirely under the control of the operator. It is notable that one very avid promoter of ultralight aircraft is retired Air Force General Chuck Yeager. In addition to having flown more different aircraft than any man in history, General Yeager was formerly in charge of aircraft safety for the entire U.S. Air Force (Yeager and Janos, 1986).

2.4.4.b *Ultralight Equipment for Shallow-Water Surveys*

Several models of ultralight have been designed specifically for amphibious operations. Other popular models can be equipped with floats for water or amphibious operation. The latter system is commonly used by fish spotters operating in the Georges Bank Fishing Grounds a few hundred miles off the New England coast. The aircraft are carried out aboard fishing craft and operated for up to 8 hours at a time using fuel tanks in the floats. When waves become excessive, the fishing craft push forward into the wind, creating a smooth landing area in their wake.

The major attractions of the ultralights are low cost, slow speed, and high visibility. The trade-offs are primarily the restricted load and the fact that most ultralights, especially those operating from water with floats, should not be operated in winds much beyond 10 miles per hour. An amphibious ultralight will cost less than \$20,000. A top-of-the-line ultralight equipped with floats will generally cost less than \$10,000. Annual upkeep will usually be less than \$2,000. Fuel costs are generally less than \$5 per hour in regular gasoline. By contrast, a helicopter can use more than \$100 per hour in specialized fuel. Because of this difference in hourly cost, the British Navy is currently investigating the use of ultralights for routine reconnaissance purposes in conjunction with an aircraft carrier.

For most purposes it will be useful to have an altimeter and air speed indicator. Very inexpensive models are available. If funds are available, a radio system can be installed or carried. This will permit coordinated studies with researchers in boats, such as the photography of die releases, or search operations. It is important to remember that there is less noise shielding in most ultralights than in standard aircraft. Therefore, a high-quality noise-cancelling microphone will be required, such as those designed for use on helicopters. A pair of radios and one good headset will total up to about \$1000–\$2000.

Oblique (and vertical) photographs are easily obtained with a helmet-mounted camera, inside a box mounted on a wing. The camera can be cushioned with foam rubber inside the box. The minimum requirements for the camera are that it: be 35 mm, have an automatic film advance, support a remote trigger of some kind, have a wide-angle lens with an infinity focus setting and have camera speeds of 1/125 second or faster. We have found suitable cameras for under $100. A bulb-type trigger release is satisfactory. The camera box can be designed to direct light from the camera flash toward the pilot, so as to confirm that a picture has been taken. Using mercury switches, it is possible to construct a device to indicate that the plane is level before vertical photographs are taken.

The costs of equipment are kept low in case the ultralight capsizes on the water. Such events are not uncommon. In particular, there is a point immediately following a landing when the plane has residual momentum in a single direction. If the wind suddenly shifts at this time, the plane will be caught between the forward momentum of the landing and the sideways pressure of the wind. This can lead to a capsize. This is a problem of concern for all floating aircraft. However, the ultralight is more likely to capsize at lower wind velocities because of its low weight. A capsize is not a major problem, because it is generally quite easy to disconnect the wings underwater so that the plane can be righted. The engine can be completely disassembled by a competent general mechanic and flushed with oil. The ultralight can be functional within 24–36 hours after capsizing.

2.4.4.c *Data Collection from Ultralights*

A variety of information can be gathered from the ultralight. Many mesoscale subsurface structures are visible on maps, charts, or satellite images which are difficult to interpret at the scale of an individual diver. A brief investigatory flight aboard a slow-flying ultralight can resolve many questions raised by these structures. The ultralight is an excellent platform for checking the accuracy of vegetation and substrate maps produced from satellite imagery. Many subsurface features which are not discernable on satellite imagery, such as patterns of ridges and rifts or topographic depressions, are visible from the air and can be useful in interpreting such things as fish migration patterns or biomass distributions. In many cases, it is easier to do interpretative work from the air rather than from static aerial photographs. For this reason, it is desirable that the pilot be a knowledgeable investigator. However, it may be possible to record the flight with videotape in order to preserve some dynamic aspects of feature investigations, such as preserving the record of viewing a topographic feature at different altitudes and from different directions.

A major problem is the limitation of a pilot's ability to record observations and quantitative information while at the same time operating the aircraft. In the U.S., an ultralight cannot include a passenger to record observations, and a two-seat ultralight is considered to be a conventional aircraft, requiring more stringent certification and a licensed pilot. Regular pilot training is certainly not undesirable for an ultralight operator, but in some circumstances it may be too expensive or time-consuming to achieve. This is not the case in Canada and some other countries, where two-seat ultralights are permitted. Fortunately, many ultralight training programs include the use of a two-seat aircraft for initial flights. This has kept a variety of such aircraft available on the market, including both flying boat and float-adaptable models. In the final stage, however, it may very well be restrictions on insurance availability or conditions imposed by funding agencies or institutional safety committees which prevent the inclusion of a passenger on a project-funded ultralight. An alternative approach to recording quantitative information during the flight (such as counts of fishing craft or altitude records during aerial photography) would be to use a radio continuously during the flight. The cost of an appropriate system might then be a problem.

2.4.4.d *Parts and Licenses*

The use of an ultralight in a country other than that of its manufacturer can entail problems in acquiring spare parts. Many countries will require a series of licensing procedures and clearances from military authorities. It may not even be possible at the outset to be certain that the ultralight operation will be permitted at all, until the aircraft has been presented and test-flown. Fortunately, many developing countries are encouraging the use of ultralight aircraft as economical aids in development. For example, a company in Thailand is reportedly building ultralights for use in activities such as crop spraying. However, whenever an ultralight is to be imported, it is strongly advisable to bring in two complete units. One can be cannibalized for parts while the new orders for parts undergo months of delays in ordering, shipping, and clearance in customs facilities. Fortunately, the overall low cost of the unit makes this a reasonable approach in many cases. It is also advisable to include an adequate and tested radio system along with the initial equipment purchase so as to ensure that optimal use may be made of the ultralight as a scientific tool.

3. FISHERY INDEPENDENT SAMPLING

Fishery independent sampling refers to any program of sampling the fish community (of a fishery) which does not involve catches taken by the fishermen in their routine activities. In this section, we will discuss selected aspects of interest in the monitoring of fish communities involved in coral reef and trawl fisheries.

3.1 MONITORING CORAL REEF FISH VIA VISUAL TRANSECTS

3.1.1 Introduction

This section covers various aspects of visual censusing of coral reef fish. The section proceeds from the general to the detailed. One must clarify the project objectives, the sampling approach, and the sampling design before one can make rational decisions on the choice of sample unit size, shape, and materials. The section begins with discussions of transects versus traps, and turns to considerations important in matching the sampling approach to the objectives. General aspects of the use of visual censusing in experiments and pattern analysis are followed by a consideration of problems of scale. Finally, there is a discussion of sample units and other details; these are generally not available in the methods sections of published papers. No attempt has been made to present an exhaustive review of the literature. Thus, the reader is referred to key articles such as Sale (1980) for transects and Bohnsack and Bannerot (1986) for stationary censusing. Other articles of interest concerning sample design and the choice of sample units in visual censusing include Bell et al. (1985), Bellwood and Alcala

(1988), Brock (1982), Fowler (1987), Kimmel (1985), Russell et al. (1978), Sale and Douglas (1981), Sale and Sharp (1983) and Thresher and Gunn (1986).

3.1.2 Visual Transects vs. Trap Sampling

The most widely used method for determining the abundances and distributions of fish species on coral reefs is the visual transect. Under some circumstances, an alternative method such as the use of standard traps is more practical. In particular, when the objective of the study is to determine factors affecting fish caught in trap fisheries, and traps are the overwhelmingly dominant gear type affecting the species of concern (as in many parts of the Carribean), then trap sampling may be preferred. However, in many areas, trap fishing is only one of a broad spectrum of fishing methods. Trap sampling only gives information on a very selective subset of the fish community in those cases. Visual transects also have biases, such as inadequate sampling of sharks and other large, highly mobile fish. They are also difficult to conduct at night. However, they have been shown to be adequate for many purposes.

3.1.3 Visual Transect Sampling Design

3.1.3.a *Matching the Method to the Objectives*

As with any sampling problem, the design of a visual sampling program must be based upon clearly defined objectives. If an objective is to gather information to conduct an inferential statistical test of differences, then it is generally possible to choose the specific test in advance, and then to determine the optimal sampling design for that test. For example, one might wish to show that a heavily fished reef has a lower abundance of target fish than a lightly fished reef. The test of choice might be a t-test. From advanced knowledge of the variances of abundances in the two areas, one can determine how many samples should be taken in each area to provide for a comparison. This comparison would account for the power of the chosen test to discriminate differences at a given probability level. With additional information on the cost of sampling, an optimal sampling intensity can be determined for situations slightly less clearly defined. A discussion of sampling optimization is presented elsewhere in this book, and further details can be found in most statistical textbooks.

The time factor can add complexity to the problem. In the example above, it might be necessary to account for seasonal interannual variability before one could safely make conclusions about differences between the two fish communities. A minimal sampling scheme might involve quarterly samples over a 2-year period. The operation should involve at least two divers plus a boat operator. A diver can often be expected to dive on no more than three sites a day because of safety and health considerations. This means that several man-days will be involved on each sampling trip. In this case, as in most operations involving diving, cost becomes a major factor. It may not always be practical for a project to spend several man-months over a 2-year period merely showing that a significant difference did or did not exist between two reefs. In most cases such as these, the study would be designed to have multiple objectives.

This complexity should never inhibit the activity of clearly defining the objectives of a study. If anything, the multipurpose nature of the study makes a clear definition of objectives more important. Neither does the complexity necessarily prevent the optimization of the sampling scheme with respect to the multiple objectives of the project. Unfortunately, very few fishery studies have incorporated formal attempts to optimize a multipurpose sampling scheme, and the methods for doing so are generally described in statistical texts too advanced for digestion by the average fishery biologist. This means that in the short term, there is a need to include statistical sampling experts in the planning of fishery studies, and in the long run, a need to simplify the steps in such optimization procedures for consumption by fishery scientists.

3.1.3.b *Visual Transect Sampling for Experiments*

Visual transect sampling can be used in conjunction with a formal experiment, i.e., an experiment in which one or more potentially significant variables have been manipulated by the researchers. In general, this type of experiment would be designed for analysis in a form of an analysis of variance test of significance. For this type of study, the determination of optimal sample size is relatively straightforward, unless complicated by multiple objectives.

More commonly, an opportunity arises to conduct an a natural experiment or a so called experiment of opportunity. One positive aspect of visual transect studies of coral reefs is that it is very likely that some major change in the reef will occur within a few years of the study, leading to an opportunity for resampling and substantially meaningful conclusions. One example of this is the series of visual transect studies of Sumilon Island in the Philippines (Alcala and Luchavez, 1981; Russ, 1985; Russ and Alcala, 1989). In those studies, a set of visual transects were conducted in a Philippine marine reserve and an adjacent reef fishing ground. At one point, a change in municipal government caused a loss of reserve status for the area, and destructive fishing pressure was abruptly applied. A second set of visual transects demonstrated a substantial loss in standing stocks both in the reserve and the adjacent fishing ground. The study was used to support a reinstatement of reserve status for the area. More importantly, this is one of the clearest examples available for demonstrating the value of marine reserves as "seed areas" for adjacent fishing grounds.

From the aspect of the natural experiment or experiment of opportunity, the initial sample design is generally based on objectives with more immediate significance. However, the resurvey can certainly be modified for optimality. For example, if a reef study initially incorporated twenty transects, only ten of these may need be resampled to show that a significant change has occurred. The criteria for optimizing the sample plan can probably be derived for the results and expenses of the prior survey. Of course, the results would have limited interpretability unless both the initial and later samplings were repeated in such a way as to account for seasonal and interannual variation. Further confidence in the results could be achieved if both sampling periods were long relative to the life spans of the reef fish.

It would be instructive to look at a hypothetical example of a study to show the effect of fishing pressure on visually estimable standing stock on otherwise pristine reefs. For example in this study, we have set up a monitoring scheme for a large number (20 or so) of fringing reefs in some area of low coastal population, such as the

Andaman Islands. The initial period of observation was set at 4 years (or the length of time the average fish in a the community take to undergo at least two generations). This period is expected to be longer in a pristine reef than in a heavily fished reef, because the average fish in a pristine reef is expected to live longer (perhaps 2 years) than that in the heavily fished reef (6 months to 1 year). A preliminary sampling trip was used to set an optimal number of random or stratified random sample units (transects) to be surveyed based on the variance between reefs. The criteria for optimizing this was the need to discriminate differences between reefs at a 95% level of confidence.

The sampling intensity over time would be more difficult to optimize initially, so an arbitrary frequency of every 2 months is selected to ensure adequate coverage during both phases of a highly variable monsoon system. This frequency may have conceivably been reduced upon analysis of the first 2 years' data, after showing that there was no significant difference across stretches of certain months.

After 4 years of monitoring, a solid concept and measure of variability over time will presumably have been achieved. At this point, a subset of the reefs (10 or so) is set aside to support a limited fishing industry. We shall assume for simplicity that the fishing pressure is instantaneously and constantly applied equally to all reefs. Monitoring on the entire set of reefs may be continued for another 4 years.

At this point, the importance of monitoring across generation times becomes clear. As an example, imagine a community of unfished fish which is suddenly subjected to fishing pressure. We expect that several large, dominant species to suffer immediate losses in population. For a few months, the total biomass of fish tends to be dramatically low. Within about 6 months, a variety of small species with 6-month reproductive cycles have become uncommonly abundant. Any of the former large dominants also begin to recover in numbers, but are still small in size and probably not yet approaching dominance. As these former large dominants grow, they are reinforced with recruits from annually reproducing species. However, the continual fishing pressure keeps all populations low. Conceivably, the populations will fluctuate until some form of constancy is achieved, and a community specifically adjusted to the constant fishing pressure is instated. Equally conceivable is that no particular state of constancy will ever be achieved. There is, in fact, very little evidence that coral reef fish communities in pristine areas have any substantial degree of constancy across generations. In either case, it may be expected that no true concept or measure of variability can be considered to be meaningful unless based on monitoring across generations.

The assumption of instantaneous fishing pressure with constancy in space and time is, of course, completely unrealistic. Fishing pressure nearly always responds to fish abundances and vice versa. In some cases, the response of one to the other might reduce fluctuations, and in others it will probably enhance it. The presence of such feedback loops in a complex system is generally characteristic of a chaotically behaving system (Prigogine and Stengers, 1984). Such a system is likely to exhibit sporadic patterns, or behavior which is "patchy" over time. A general principal in dealing with patchy systems is that a large number of small sample units widely dispersed tends to yield more representative samples than larger, less dispersed ones (e.g., Kinzie and Snyder,

1978). This logic should be extrapolated to the time dimension and may become a consideration in determining the sampling interval. In point of fact, it is very likely that the unfished community is also highly chaotic over time, indicating that sampling over time should perhaps be more intense than indicated in optimization analyses under the usual assumptions.

The above experiment has a single, clear objective which unfortunately would be very costly to achieve. Very few researchers would have the time or funding to undertake an eight year study for such a limited objective. However, the long-term objective can easily serve to provide a framework for a variety of other ecological objectives to be achieved concurrently. Whenever possible, a long-term study should be oriented similarly.

3.1.3.c *Sample Designs for Pattern Analysis*

Ecology is a process of successive approximation. At one end of the spectrum are a variety of studies where most of the groundwork has been already laid down, and a simple, final test is all that is required. At the other end is a whole universe of vital questions where the initial steps have yet to be initiated. Included in this universe are many questions of fundamental importance to tropical fisheries — such as those dealing with the relationship between fishing pressure and the predictability of fish community structure. Initial investigations in this field generally are of the correlative or exploratory type.

The term Exploratory Data Analysis (EDA) has been applied to studies of data for purposes of hypothesis generation in the absence of hypotheses. In visual transect studies, the researchers generally have a number of hypotheses in mind. However, the study is also usually designed to generate new hypotheses when preconceived hypotheses are found to be inadequate to explain the patterns observed. We shall use the term "Pattern Analysis" (*sensu* W.T. Williams, 1976) to refer to a study of patterns in space and/or time wherein the set of hypotheses under consideration is an open set. In this sense, it bridges the gap between EDA and inferential statistics.

The utility of a sample design for pattern analysis can be illustrated by example. Imagine a forereef slope 10 km long, 1 km wide, with a gradual slope from 3 to 20 m depth (Figure 1). On one side is a brackish water cove fed by a stream. Low mounds of coral are regularly dispersed across the hard substrate, but two large patches of branching coral are situated at 3 and 6 km from the cove. The hard substrate is interrupted by sand patches, irregularly distributed. The objective of the study is to determine the major factors which explain the distributions and abundances of fish species across this forereef slope.

Because interpretation of the distribution patterns will be greatly enhanced by reference to a map, the study of the gradients of fish composition is quite similar to the study of physical gradients such as depth. Generally, the most productive way to study depths is to produce contour lines on a map. This is also a reasonable approach for studying changes across an ecological community. This being a primary strategy, it is usually better to disperse the sample units in a regular, rather than a random pattern. One reason for this is that, on a contour map, the accuracy of interpreting contours

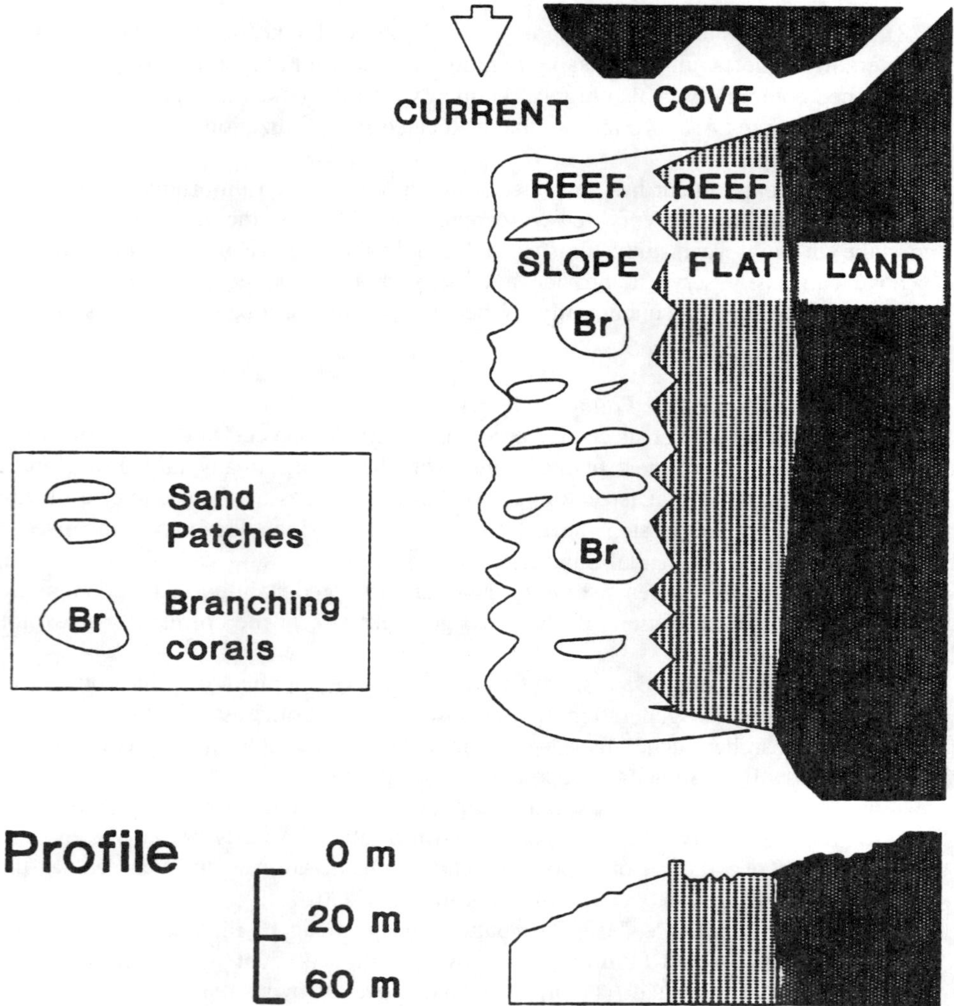

Figure 1. Hypothetical fringing coral reef.

depends upon the greatest intervals between adjacent points used to construct the contours. Another reason for using a regular pattern in such studies is the ease with which estimations of values and variances (associated with the values) can be made using two-dimensional moving-average methods such as kriging (Davis, 1973). The optimal packing pattern for points in two dimensions, a diamond pattern, is also best for contouring. However, there are cases when a square pattern is more feasible, and the loss of information is not substantial.

Depth is generally an important variable in marine community analyses. In many cases, it is desirable to insure that the sampling of depth strata is comprehensive. In the case of the example study, this means dispersing the sample units along lines perpendicular to shore, such that the sample units are more closely situated along the depth gradient than they are longitudinally along the shore. This means that the contours will be less accurate on the map. However, an analysis of the sample units on a graph, where depth forms a major axis, will be greatly facilitated. For some purposes, the sample units are best determined by depth. For example, McManus et al. (1981) marked fish counting quadrats at approximately 2-m depth intervals as measured by divers. Quadrat depths were later corrected for tide height, which caused slight shifting of the points up or down. However, the design permitted a broad dispersion of sample units for analysis with respect to the depth gradient. This approach is particularly useful in areas where the depth gradient (or any other gradient of primary interest) is steep or uneven with respect to geographic distances. A general rule is that one should be certain to have as wide as possible a dispersion of sample units with respect to the hypothesized gradients.

In our example study, one can on an a priori basis determine the types of patterns to be expected depending on which factors are dominant. We will assume that two multivariate methods will be used in the study: TWINSPAN (Hill, 1979, see also Pielou, 1984b), a divisive clustering approach which permits extensive analyses of the relationships between groups of species and groups of sites, and detrended canonical correspondence analysis or DCCA (ter Braak, 1988), an ordination technique which permits extensive analyses of the relationships between gradients of sites and species and gradients of environmental variables (e.g., Nañola et al., 1989; Licuanan and Gomez, 1988). Anticipated results are shown in Figure 2.

Upon analysis of data following the study, one might find that a pattern of community gradients appears which was not anticipated. This may indicate: (1) a general lack of stable patterns in species distributions and abundances, (2) a community structure which is fairly uniform across the gradients, (3) a synergistic effect of two or more strongly influential physical gradients, or (4) an effect of a gradient which was previously unhypothesized.

Of the four possibilities, the last is the one that is the most straightforward to investigate. In some cases, it may be possible to investigate the pattern with respect to auxiliary data gathered either by the investigator or by other researchers working in the area. For example, it is almost a ritual for field workers to measure bottom temperature in marine ecological studies, even when no particular hypothesis is expected to be tested. It is also a good practice to qualitatively describe each site and to at least provide semi-quantitative estimates of bottom cover types.

In our case, let us assume that no likely relationships can be developed from existing information (Figure 2). One can simply hypothesize that there is a physical difference across reef which explains why the central portions of the reef differ markedly in composition. On this basis, one can return to gather more data. We will assume that nothing obvious appears on an underwater assessment survey. However, an overflight of the reef reveals that there is a distinctly higher concentration of sand patches towards the central portions of the reef. Vertical aerial photographs are taken, and the images analyzed to determine the ratio of hard to soft substrate in the vicinity (some fixed distance) around each site. This ratio can be included in a reanalysis of the data by DCCA, and a final decision can be made as to its importance based on the correlation analysis which forms part of the program output. This relationship could be checked by either skewer analysis (Pielou, 1984a) or ordered similarity analysis (Pielou, 1979, 1983) or both. A consistent acceptance would indicate that the conclusion is robust with respect to analytical technique.

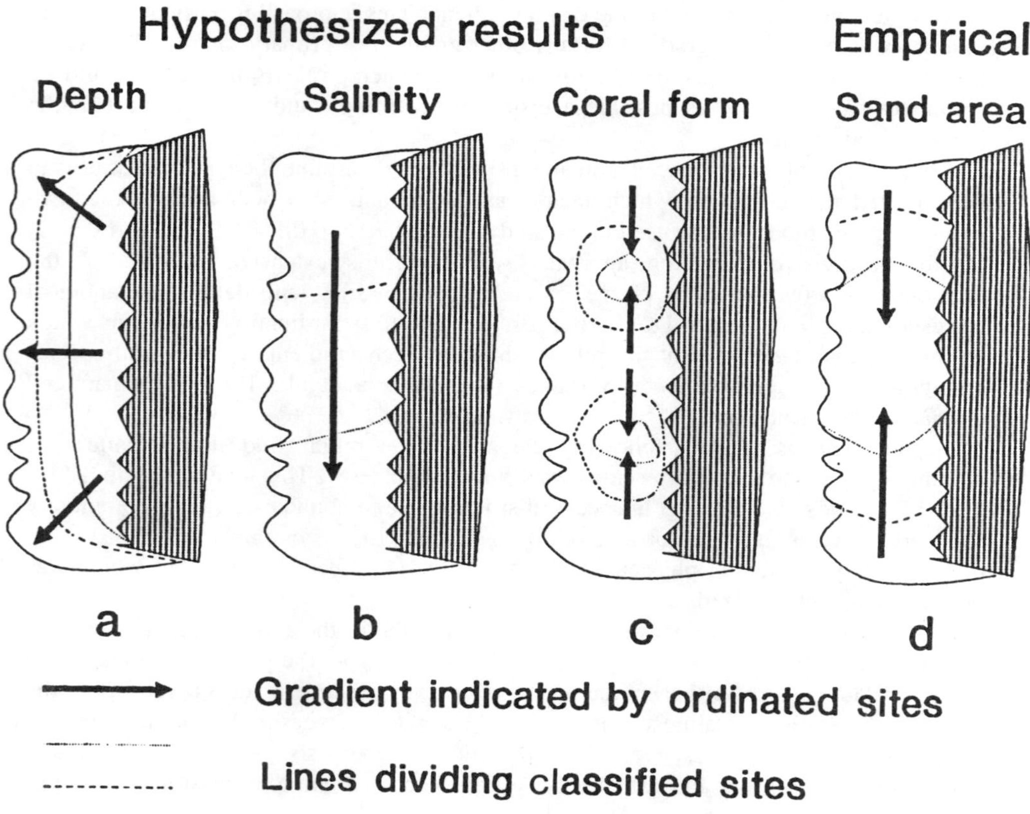

Figure 2. Hypothesized and empirical results obtained from analyzing patterns of species abundance and distributions on the hypothetical reef.

3.1.3.d *Problems of Scale*

One final consideration of importance in the design of the visual census study is that of scale. It has become increasingly clear that ecosystems in general, and coral reefs in particular, are subject to structuring forces which change in importance at different scales. One example of this may be found in the series of papers by Sale and his co-authors (e.g., Sale, 1972, 1977, 1978, 1980; Sale and Dybdahl, 1975) on the structuring of coral reef fish communities on small patches of coral. Contrary to the prevailing paradigm, Sale showed that the recruitment of coral reef fish species on such patches was highly unpredictable and subject to a variety of fluctuating physical and biological factors. It would be difficult to predict from one year to the next what species or groups of species will be found on a given patch of coral. However, on a much larger scale, there definitely is a useful degree of predictability. Deep water reefs tend to have deep water species and shallow reefs generally have shallow species. Across the entire Great Barrier Reef, there are assemblages of fish which are characteristically found near, moderately far, and far from the mainland (D. McB. Williams, 1982). A fundamental question arises as to how much predictability is possible at any given scale. Regardless of this, however, it is clear that the density of sites and their relative width of dispersion across gradients will have a potential effect on the conclusions one reaches concerning the importance of various structuring factors. Similarly, the exact nature of the sample unit will be of concern in the problem of scaling the study.

3.1.4 Sample Unit Selection

3.1.4.a *Quadrats*

The standard sampling unit in most community ecology studies is the quadrat. A quadrat is a geographical area within which intensive inventories will be made. A typical quadrat for small organisms is 2 m on each side and is often marked by a portable frame of wood or metal. In this case, the term "quadrat" is often applied both to the area sampled and the apparatus used to delimit the sample unit. Quadrats are often divided into smaller squares for subsampling, and some studies use intersections of an internal grid as points for point sampling (Mueller-Dombois and Ellenberg, 1974; McManus et al., 1981). However, for coral reef fish studies, it is generally difficult to use a portable frame without severely disrupting the distribution patterns of the fish. In these studies, the quadrat is generally marked with some combination of corner weights, small buoys, and ropes. This is usually done at least 30 minutes before censusing so that the fish have returned to normal activities. Most studies define quadrats in terms of a transect. However, Risk (1972) used scattered quadrats of 15 m on each side in order to compare fish and coral distributions.

3.1.4.b *Transects*

A transect is generally either: (1) a long, thin quadrat, or (2) a set of quadrats arranged in a straight line, or (3) a contiguous set of curved line segments. For benthic organisms of low mobility, the transect is often reduced to a single line marked with a rope or cable, and organisms are quantified in terms of intersections with the line (Randall, 1973; Loya, 1978), and with points along the line (Ross and Hodgson, 1981). The higher mobility of fish makes this approach difficult.

For one-time only fish surveys, the usual method is to lay two parallel lines of considerable length (often 50–200 m) at some fixed distance (often 1–5 m). The lines are usually nylon ropes with small lead weights wrapped around the lines at 1 meter distances. More recently, transect lines using negatively buoyant fiberglass measuring tape have become popular. These are available with reels containing a minimum of metal parts which could corrode. Some form of reel is essential for lines of any substantial length. Reels mounted on boats are generally difficult to use on reefs because of entanglement with corals. However, they are feasible when there is no intention of retrieving the line after a survey. Some groups use a double reel system mounted at the ends of a 2-m pipe, which lays parallel lines at the appropriate width. This system may be augmented with the use of a spreader bar consisting of a 2-m pole with nearly closed loops at either end to help guide the laying of the line. The spreader bar is used by a diver following the one carrying the reels.

When the reef exhibits extreme patchiness it is best to minimize sampling time at each site and to maximize the number of sites to be studied. For temporary transects, this often means limiting to a single line to reduce laying and recovering times. In this case, the laid transect line is taken to represent either one side or the center of a sampling swath of estimated width. This permits fewer divers to be involved in censusing longer transects of greater areal coverage on each dive. It has the disadvantage that subjective decisions must be made as to whether to include or exclude borderline fish. Variability from such decisions can be minimized through a comprehensive training program; all study divers strive for comparable, repeatable results on a study transect.

A single transect can be treated as a single quadrat, as a set of contiguous quadrats, or as a guideline for the selection of isolated quadrats. Generally, it is more useful to divide the transect into several contiguous quadrats than to simply lump all counts together for the entire quadrat. This permits analysis in terms of within-site variability and studies of structuring at multiple scales. It is possible to lump the data from subdivided transects after the survey if desireable. The reverse however, is not possible. Some concern arises over the ease with which a diver can record the counts on underwater slates or paper, when the transect has been subdivided. However, an experienced observer will generally not show a major increase in time between tallying a 100-m transect and twenty 5-m quadrats. Much of this depends on the way groups of fish and their positions along the transect are recorded. Some experimentation is advisable within a group of observers preparing to do surveys. McManus et al. (1981) used quadrats within transects marked with stakes at approximately two m depth intervals. The purpose of the study was to compare depth structuring of fish and coral assemblages. However, for most studies, the contiguous quadrat approach will be the most labor- and cost-effective.

3.1.4.c *Quadrat Size, Transect Length, and Cumulative Curves*

The use of cumulative curves in community sampling can be best illustrated by example. Suppose that the purpose of a study is to determine the differences in fish species richness between two reefs. Species richness will be defined here as the number of species found in a closely defined assemblage. An ecologist may go to each reef and perform species inventories in a large number of quadrats. When the cumulative number of species is graphed as a function of total sample size, the ecologist may find that the number climbs rapidly until an area of 10,000 m^2 is reached, at which point very few more species are found by taking larger samples. The two areas can now be compared in terms of richness, using a variance and improved estimate from a jackknife estimation routine (note that if the information used is from a single 10,000 m^2 sample, there will not be an estimate of variance between portions of the same forest).

When dealing with a new forest, if there is not time for preliminary sampling, the ecologist may have some idea of the sample size needed to make the estimate. Of course, numbers of species virtually always increase with sample size, therefore a jackknife or other estimation procedure is desirable. It would, in fact, be better to graph the **estimate** of species number against sample size, not the counts themselves.

This is the basic concept behind determining sample size from cumulative curves. There are at least two problems with this. One concerns the interpretability of the curve, and the other concerns its applicability to community structure analyses in general.

The curve of species richness, as defined above, from a very homogeneous assemblage would appear to curve off in most cases at a clearly defined region. However, if the data are reanalyzed with a longer x-axis and shorter y-axis, the point one chooses may change. This has been of some concern in vegetation ecology. Some suggest using an arbitrary cutoff value on which to base a decision, such as the point at which a 10% increase in area yields only a 5% increase in species (Mueller-Dombois and Ellenberg, 1974). However, not all properties of the community curve off so clearly. A species richness estimate may increase stepwise in a very patchy environment, especially when quadrats are arranged in a long quadrat. The curve may level off for a length and then shoot up rapidly again. The longer the transect, the more chances of intersecting a slightly different assemblage, and the more likely another set of steps will occur. Some indices, such as diversity based on information content, even decline occasionally with increased area.

The use of a species area curve to select samples for species richness determination can be otherwise quite helpful. However, there is not always a rational justification for using this type of curve to predict sample sizes for other purposes. If the objective is to classify an area based on a classification technique, then a reasonable way to determine if the sample size is adequate is to reclassify the same area. Reclassification is based on various sample sizes for numerical comparisons of classifications. It is theoretically possible to do this, but apparently rarely done — if at all. It is by no means certain that using a species area curve to define transect length will necessarily improve the robustness of the classifications. However, in the absence of other guidelines, such an analysis may aid in ensuring that reasonable samples have been obtained. In cases where one transect will be used in the classification to represent one important set of variables, it might be helpful to use this method to determine transect length.

As for the determination of quadrat size, the cumulative curve may or may not be relevant. A general guideline is simply to break the optimal sample size into a large number of sample units. These units cover as wide a range of conditions as possible, representative of the area or variables of interest. Pielou (1974) suggests only that the quadrat size be large enough to include more than one of the target organisms, but not so large that local density variations become blurred. Beyond this, she emphasizes the high degree of ambiguity involved in choosing a quadrat size or shape and its strong impact (in some cases) on conclusions concerning the patterns being investigated (Pielou, 1974).

For visual censusing, the quadrat will generally be at least as long as the width dictated by the transect width. Transect widths are generally set based on regional experience. The factors to consider include: the objectives of the study, the difficulties in laying parallel transects at various widths or estimating to various distances, water clarity, the density of the fish in the area as it affects the ability to record the numbers accurately, prior observer training, the usual sizes and habits of the target fish populations, and the trade-off between intensive local sampling and extensive broad area sampling.

Sale and Sharp (1983) have conducted comparative studies of various transect widths. They found that they were able to inventory fish more thoroughly at a width of one m than they could using wider transects. However, a smaller width generally would favor the inclusion of smaller species (especially cryptic fishes) and the undersampling of larger, more mobile fishes. To prevent this, the number of transects are increased to compensate for the decrease in total area sampled. The studies were conducted on the Great Barrier Reef in nearly pristine reef conditions. The fish in such areas are often so tightly packed that it is very difficult to identify and count them. On a fringing reef in the Philippines, however, fishing pressure has reduced populations to the point that it is often possible to swim for 5 minutes at a time without seeing any fish. The Association of South East Asian Nations (ASEAN)-Australian Cooperative Program on Marine Science (ACPMS) recommends that a diver record fish along a 100-m measuring tape, recording fish to a width of 5 m on either side (i.e., a 10-m swath). One important consideration in that program is the need to standardize so that researchers in different areas produce comparable data. This need outweighs local variations in an optimal procedure, because of the objectives of the study. In contrast, the more locally intensive FSA-CRSP monitoring program has used two divers on either side of the tape, each counting within a 5 m swath. This design accounts for small-scale variations and variations between observers. There are times when the visibility of waters over Philippine fringing reefs diminishes to about 5 m. For many studies, it may be desirable to conduct local analyses of optimal transect widths prior to field sampling, particularly when prior experience in sampling in the area has been limited.

3.1.4.d *Vertical vs. Horizontal Transects*

Depth is almost a universal concern in stratifying community variables. Therefore, there is often a major difference in sampling design between studies that use within-depth (horizontal) and across-depth (vertical) transects. The intent of a horizontal transect is generally to obtain data from within a particular depth stratum. This being an objective, it is often best to make the transect meander somewhat with topography to ensure depth constancy. There is a limit to this, however, and the meandering in and out of densely packed ridge and rift systems can occasionally be impractical. More commonly, the problem arises that a researcher laying a transect runs out of hard substrate, and he or she bends the transect up or down a slope to achieve the required length. This can be avoided in some cases by breaking the transect up into smaller, discontinuous segments at the correct depth. For temporary transects of 100 m length, it might be helpful to carry an extra 100 m fiberglass measuring tape reel in case a gap must be created in the primary transect.

An across-depth (vertical) transect is usually used either to insure that a single sample unit encompasses depth variations or to provide a guideline for depth stratified quadrats. The latter approach can be thought of as a form of regular sampling. However, it can also provide data for a variety of statistical analyses oriented to contiguous quadrat information (Pielou, 1974).

3.1.4.e *Temporary vs. Permanent Transects in Repetitive Surveys*

Repetitious surveys in an area can be conducted using an approximate location approach. To do this, one first determines the geographic error involved in relocating a site based on compass, LORAN, NAVSAT, GPS, or other location equipment. Then, an effort might be put into insuring that the transects in that area cover enough ground to be representative of the area for purposes of comparisons over time.

However, at one extreme, there are studies where the goal is general enough that one-for-one matching of transects over time is not necessary. For example, to show that a decline in fish abundance has occurred from one year to the next (disregarding differences in the decline over area), it is only necessary to ensure that a representative random or systematic sample has been taken each year, and positioning becomes irrelevant.

At the other extreme, there are studies of multiple purposes where one would like to correlate local changes in abundances or composition with local variables (such as abundances of particular species of corals or algae, or heterogeneity measurements). In these cases, permanently marked transects are desireable.

The transect itself may be made of rust-resistent cable, or cable which is so thick that it will not rust away during the study period. Usually, however, transects are marked with weighted nylon lines which are anchored to the bottom with concrete or stakes. Free-standing concrete blocks tend to shift during strong storms, even at depths of 100 feet. Stakes, however, are often difficult to drive into very hard substrates or pull out too easily from sand patches. Some workers actively pour concrete into holes drilled in the substrate, but this requires some prior training. Regardless of the anchoring method, it must be expected that some maintenance will be necessary. To facilitate this, the transects should be placed in a known cardinal direction from some reference point underwater, so that they can be realigned if necessary.

The major problem with permanently marked transects is that of losing the marking buoy. If problems are anticipated with losses of field markers, it is often better to design the study to avoid the need for permanent markers entirely. Otherwise, a week-long sampling trip can easily drag into 3 weeks for merely locating or reestablishing the field sites. However, some steps can be made to minimize loss of field sites. The anchoring method is critical. Some researchers drill large holes in the substrate and pour concrete to anchor a strong bolt. Many professional divers involved in construction projects are familiar with procedures for pouring concrete underwater.

For many purposes, the anchor can be a large block of concrete towed to the site and dropped. The block should be very heavy, such that the average dimension is at least 50 cm. However, in order to avoid the block being pushed around by large waves, it is best to keep the profile low relative to the other dimensions. The FSA-CRSP in the Philippines uses a block of 30 cm x 60 cm x 60 cm (Figure 3), which is transported on a small steel towed barge (less than 2 m long; commonly available in port areas). The barge is flat, which facilitates the process of pushing the block into the water. Some success has been reported elsewhere with placing concrete structures on bamboo rafts, which are split in the middle for easy off-loading on site. The block must have a built-in loop of some sort for attaching the mooring rope. A metal loop is more permanent, but a loop of rubber tire will minimize line chafing. It is best to construct the block with a variety of attachment options.

The rope for tethering the marking buoy is usually nylon. If a chain or cable is to be used, it must be balanced against a very buoyant buoy. The line should be long enough to accommodate extreme high tides, but too much slack can cause subsurface entanglement.

Buoys can range from commercial types to those made of natural materials. In the Philippine study, it was anticipated that expensive buoys might be stolen by fishermen. There was an additional concern that the usual buoy would entangle the propellers of fishing craft, and the long horizontal scare lines frequently being dragged through slightly subsurface waters. It was determined that a long bamboo pole attached several feet below surface would tip over if struck, avoiding entanglement problems. The poles were attached through two holes drilled through the shaft, and they were replaced at 6-month intervals as they lost buoyancy. The usefulness of the poles to fishermen still led to a theft problem, until the poles were distinctively painted fluorescent orange and marked with institutional labels.

The cost of losing a site is very high in terms of time, labor, and associated field expenses. Each site should have a subsurface backup buoy to facilitate relocation. A buoy suspended 10 m below the surface is easily spotted by a towed diver, and it eliminates the need for long bottom times at great depths. A separate line for this buoy might lead to friction between lines. However, the subsurface buoy can be attached to a loop in the primary tether. For programs with ready access to a geographic positioning system (GPS), it may be possible to record the locations of the buoys to within a few meters, so that the anchor blocks may be relocated if the buoys are lost. In other cases, it is advisable to use two complete buoy systems at each site, each with subsurface buoys. The subsurface buoys can be made from plastic bottles, styrofoam in plastic bags, or the wood and cork arrangements commonly used by New England lobster fishermen. Further insurance against site loss can sometimes take the form of various electronic underwater locating devices. However, these are generally much costlier than bamboo and concrete, are subject to battery failure, and are very tempting objects for intentional or unintentional theft by local divers.

In all cases, attention must be focused on the problem of modifying the behavior of fish and fishermen with the sampling system. Underwater, any object with more than average vertical height for the local habitat will attract fish. All anchoring devices along the transect lines must have low profiles. The marking buoy anchoring device should have a low profile as well. It is likely that fishermen will take advantage of the marking buoys by tying up to them, rather than risk losing their own anchors. Therefore, it is a good practice to offset the transect line some reasonable distance from the buoy anchor (Figure 3). The distance must be as far as possible without requiring substantial amounts of bottom time to proceed from the anchor to the transect. For some purposes, 20–50 m might be reasonable. A thin line can be attached to the anchor, leading to the start of the transect line. It should not, however, be attached to the transect line. Otherwise, the movement of the anchor in a storm may dislodge the transect line.

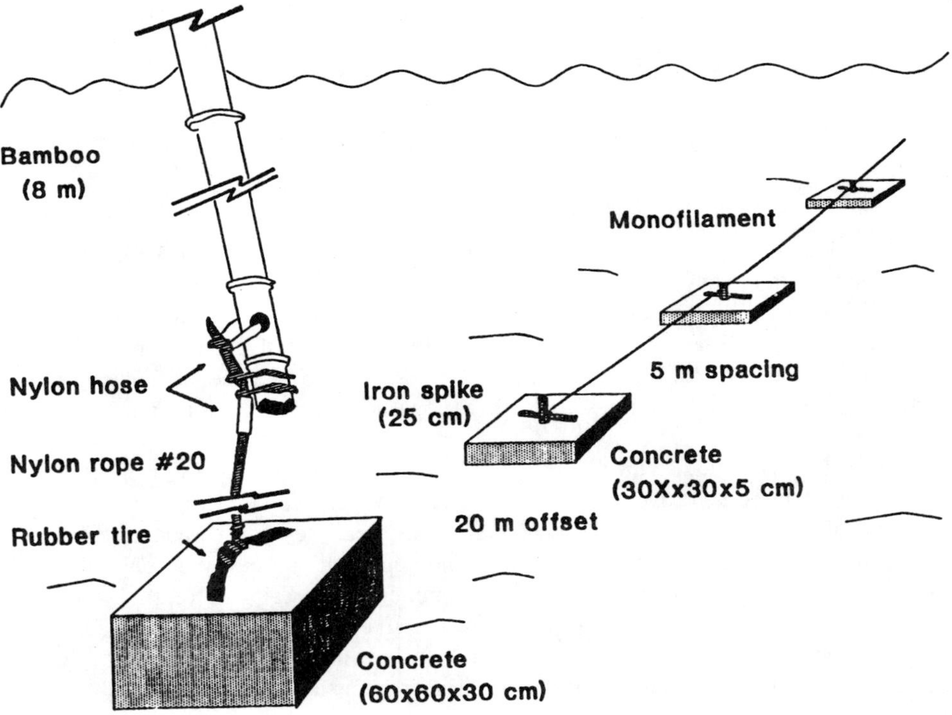

Figure 3. Transect marking devices used during the FSA-CRSP coral reef fishery monitoring program in the Philippines.

3.1.4.f *Unmarked Transects*

The extreme heterogeneity of a coral reef makes it important to obtain large numbers of samples whenever regional comparisons are to be made. In determining the patterns of fish distribution across the Great Barrier Reef, D. Williams (1982) and Russ (1984a,b) used a transect technique that required no markers. This method has evolved from earlier informal survey methods, such as the general assessment survey (Kenchington, 1978; McManus and Wenno, 1981).

Each transect consists of a swim of timed duration (e.g., 30 minutes) during which abundances of species are recorded on an abundance scale. The log 3 abundance scale used by Russ (1984a) is shown in Table 1.

Table 1. Fish abundance scale used by Russ (1984a).

Abundance Scale	Number of Fishes
1	1
2	2 – 3
3	4 – 9
4	10 – 27
5	28 – 81
6	82 – 243
7	244 – 729

Russ included all fish on either side of the diver to a distance of 5 m, based on a previous cumulative curve study of *Acanthuridsn.* A 30-minute swim generally covered from 300–400 m, yielding a coverage of 3,000–4,000 m^2. Data were cumulated per species, such that each transect represented a single quadrat. Transects on the reef flat, crest, and slope were parallel to shore, while those on the backreef meandered among coral aggregations or "bommies". Each zone was censused in four nonoverlapping transects.

This approach (i.e., with no markers) has the advantage of being rapid, allowing for very cost effective coverage of large areas. Whatever concerns arise about the intensity of sampling necessary to characterize an area, they can be addressed through preliminary trials and statistical analysis. The approach can easily be paired up with broad-area estimation techniques for coral community description (e.g., Done et al., 1981). It is also optimal for those who would like to characterize a house without worrying about the bricks used to make it. All results will be valid at a scale of hundreds of meters or greater. Studies of greater intensity will generally not cover as much area with the same funding and time limitations, and they will probably be less effective for large area comparisons than this "no marker" method.

3.1.4.g *Target Populations*

Generally, it is not feasible to assess all species of fish on a reef concurrently. Many fish are cryptic and are rarely seen except when flushed out of holes with rotenone or cyanide. Other fish, such as reef sharks, may rarely be encountered in transects. It is best to define the general group of fish to be studied based on the study objectives and practicality of sampling. For example, one may study all non-cryptic species large than 5 cm which appear over 5 m wide transects to a height of 5 m. As the assemblage may vary slightly depending on the transect width and time spent sampling it, consistency of technique becomes important. The established technique will further define the assemblage in ways that will be difficult to precisely define.

Many observers find that they can cover larger areas more rapidly if they limit the study to a set of, for example, 50 or 100 dominant species. This certainly facilitates data handling below and above surface. Care must be taken to insure that fish which are dominant at one period of time are not going to be replaced by other dominants later on. This approach would be of limited value in diversity studies.

A major consideration will be whether or not to include night fish assemblages. These are invariably much different than daytime assemblages. It is possible to a limited degree to census fish visually with flashlights at night. However, it will be difficult to compare these quantitatively with daytime assemblages, because of differences in the scale of coverage and technique used.

Fish can be sized to four or five categories after appropriate observer training. A decision must be made as to whether or not the sizing will vary between species. The latter approach may provide more useful data from a population analysis standpoint, but it requires more familiarity with each species.

3.1.4.h *Training*

Most of the concerns about the validity of a visual transect approach can be dealt with through proper training and field evaluation. A training course must involve considerable taxonomic work, generally using color slides as flash cards. A sample transect can be set up, and observers can be required to do censusing along it until consistency between paired observers is achieved. At various times during the survey, a similar approach can be used to assess the influence of a learning curve on the data obtained. Sizing of fish can be dealt with using training models of cardboard attached along a transect area. In all cases, the objective should be to minimize error to a reasonable level, and then to estimate the error so that the error in the final data set can be properly accounted for.

3.2 MEASURING THE HETEROGENEITY OF A CORAL REEF

One of the factors of greatest concern in analyzing the distributions and abundances of benthic fishes is the physical heterogeneity of their environments. It is well known that a structure of high relief tends to attract benthic fish. A structure with many small holes can support lots of small or eel-like fish. A structure with larger holes, such as a rubber tire construction, is likely to support larger fish such as groupers.

In order to quantify the relationship between heterogeneity and fish community structure, some measure of heterogeneity must be used. The most common technique involves the use of two measuring strings (Carpenter et al., 1981). A single line of known length is stretched out to define a distance parallel to the general trend of the bottom. A second string of equal length is draped along the length of the first, in such a way as to follow the topography.

This second string is then stuffed into holes to some set distance and laid up over small protrusions. If it falls short of the first, and the difference in distance covered or the ratio between distances covered by the two strings then becomes a measure of heterogeneity.

Bradbury and Reichelt (1983) were concerned with determining the fractal index of a coral reef. The fractal index is the scale-invariant component of heterogeneity. The approach came about because many uneven surfaces or boundaries, such as coastlines, have no true length in an absolute sense. The length that one measures for a coastline depends directly on properties of the measuring device used. If straight measuring sticks are used, then the smaller the ruler used, generally the longer the determination becomes. Logically, this paradox has been extended to conclude either that coastlines are infinitely long, or that the measurement of distance along a complex line is meaningless only when expressed in terms of a particular scale (Gleick, 1987). Many studies have found that there is a consistent relationship between the measurement and the scale of measurement along a complex line. This relationship can be expressed in terms of fractal dimensionality.

Bradbury and Reichelt measured the heterogeneity of the reef using calipers of different scales. A caliper can be used to measure distance as a chord (sensu geometry) through an uneven structure. Scale-specific heterogeneity can then be expressed in terms of the number of such chords which make up a distance parallel to the substrate. In this study, the researchers used calipers of 10 cm, 1 m, and 10 m underwater. Using aerial photographs, they obtained measurements at 100 m and 1000 m scales. The relationships between the scales and the measurements were plotted, and the slope of the line was used to estimate the fractal index. This index was surprisingly close to that of a human lung.

For purposes of studying fish habitats, it is generally the variability in heterogeneity at different scales which is of interest. Logically, one would expect small species to be favored in areas which are heterogeneous at the 1 cm or 10 cm scales, and larger fish to prefer areas which are heterogeneous at a scale of 1 m or more. A method for measuring this heterogeneity should be very simple, so that repetitive measurements at each visual transect site are possible in a short time.

McManus et al. (personal communication) settled on a method which combined features of the comparative string technique with the fractal measurement technique of Bradbury and Reichelt (1983). They (McManus et al.) assumed that a small brass chain with links of 1 cm each provides a valid measurement of heterogeneity from some conceptual aspect. The chain is laid alongside a 10-cm portion of a wooden meter stick weighted with a fishing weight in the center to keep it anchored. The problem then becomes to make chain-like readings at larger scales. This is solved at the 10-cm scale by using a 10-cm portion of a meter stick to simulate the links of a chain. It is rotated on end after each count and laid tangentially to the surface for the next count. There are cases in which a decision must be made as to how to span a chain across a rough object. For this reason, the actual counts are made with a pair of such sticks connected end to end with a loop of fishing line. Any time a decision must be made about how a chain would be distributed over an object, the sticks are used as a pair to assist in conceptualizing the answer. As with the small chain, the index is simply the ratio of the links counted against a straight line measure in tens of cm. A 1-m stick (stainless steel) is used for the straight measurement.

A straight measurement of 10 m can be made with a fiberglass measuring tape. Two divers hold the tape such that each diver is equally high above the substrate. Each sights toward the bottom perpendicularly to the line. After sending and receiving tugs on the line signifying preparedness, both divers release the line and mark the sighted spots with wire stakes. The tape is then stretched in as straight a line as possible between the stakes to serve as a guideline for the 1 m counts. Note that the tape must be then pulled out longer than 10 m to mark the rough bottom, so that a 20-m tape on a plastic hand-winding reel is recommended. One diver then counts links along the tape using a 1-m stick as a link. In practice, this is a convenient time for the other diver to make the smaller scale readings.

Note that it is very important that both ends of the straight line are always equally high above the substrate. Otherwise, it is easy to mistakenly measure less than 10 m, and a poor ratio be obtained.

McManus (personal communication) used ten measurements at each scale distributed parallel to a 100-m visual transect line. With considerable practice, it became possible to complete one set of such readings on a single 80 m^3 SCUBA tank at 80 ft (24 m). The readings were incorporated into a set of environmental parameters that were contrasted against species counts by detrended canonical correspondence analysis (DCCA). The first species-sites axis was correlated with depth and temperature. The second, third, and fourth axes were correlated with heterogeneity at the 10 cm, 1 cm, and 1 m scales, respectively (Nañola et al., 1989).

3.3 BLAST FISHING STUDIES

3.3.1 Significance

Blast fishing is a world-wide problem of growing importance along developing coastlines. As coastal populations rise, fishing pressure increases, and efforts are directed at finding ways to compensate for declining catches per fishing hour through nontraditional fishing methods. A rapidly changing local society tends to lose its concern for long-term goals, and quick-profit methods such as blast fishing gradually become socially acceptable. The illegality of the method is rarely a concern under such circumstances.

The fisheries manager must be able to assess the effects of blast fishing for several reasons. Length- or age-based population analysis methods are generally aimed at determining the levels of fishing pressure based on estimations of total mortality and natural mortality. The fishing pressure for any particular gear cannot be determined without a knowledge of mortality due to blasting. The catches of blast fishermen are often hard to survey because of the illegal nature of the gear, leaving studies of fish landings incomplete. Blast fishing on reefs is habitat destructive, and the progressive decline in suitable habitat area for important fish may be a primary concern of the fishery manager. The effects on the populations and habitats of blasting may be important in affecting the major goal of blast fishing management, that of making blast fishing socially unacceptable.

3.3.2 Effects of Blasting on Fish

The range of injury or mortality due to blasting will vary between species depending on the presence or absence of a swimbladder. Furthermore, fish with swimbladders open to the alimentary canal (physostomous) are expected to be less sensitive to blasting than those with a closed (physoclistous) bladder (Armada, 1989).

3.3.3 Types of Blast Materials

The nature and intensity of the blast will vary depending on the materials used to build the bomb. Explosives found in soft-bottom fisheries in the Lingayen Gulf of the Philippines include (Armada, 1989):

(1) dynamite sticks — these are often obtained from mining operations,

(2) lump type bombs — contents of dynamite sticks lumped and rewrapped,

(3) boogey type bombs — explosives from unexploded military bombs or high-caliber ammunition gathered from military bases, and

(4) oxidizing chemical bombs — often in bottles with commercial fuses obtained from mining operations. Chemicals include potassium chlorate, potassium nitrate, sodium nitrate, ammonium nitrate, ammonium phosphate, or ammonium sulfate.

Blasting operations at the nearby Bolinao Reef Complex often involve either pint- or liter-sized beer and gin bottles with commercial fuses. The explosive chemicals are often in the form of either fertilizer or boat-building foam (used for floatation in fiberglass boats). These are generally mixed with gasoline and gun powder to form a thick slurry.

3.3.4 Blast Effects in Different Habitats

The effects of a blast will vary depending on where and how it is used. Over soft bottoms, a bomb is often dropped directly on a sighted school of fish. Alternatively, bombs may be used to kill fish in encircling nets, or those residing in artificial reefs of natural materials or more permanent construction. On coral reefs, the target may be a school of fish or the fish congregating around a dense growth of coral. Surveys of the blast patterns in Bolinao confirm the frequently cited figures of 1 to 2 m diameter mortality for corals. The usage continues until only a few dense stands of corals exist.

The blast effect on the fish populations will depend upon the nature of fish dispersion in the blast area. This will tend to vary with the type of fish habitat. For example, blasting is not generally effective in seagrass beds where fish tend to be broadly dispersed. Coralline areas tend to have a high rate of fish recruitment, but it is the low rate of coral recruitment that poses the ultimate problem. Because most usage is based on visual targeting, the frequency of blasting tends to be higher on calm, wind-free days with high water clarity. These type of days probably also encourage use in deeper waters. In general, blasting usage decreases sharply beyond depths of 15–20 m because of difficulties in sighting targets and recovering catch. Even in shallow waters, the method is extremely wasteful because dead fish are often difficult to see and recover from the bottom.

The type of bottom can also be expected to strongly affect the nature of the blast. The blast energy may be more effectively absorbed by some bottom types than others. A flat bottom will have a more predictable blast pressure effect than an uneven one, and a sloped bottom may have a channeling effect on the blast.

3.3.5 Field Surveys of Blasting Effects

There is often much confusion in reef areas between natural craters and blast patterns. Coral reefs consist of limestone which has been exposed to erosive processes over thousands of years. Most reef substrates tend to be naturally pitted. On the other hand, limestone is very difficult to damage with explosives. The effects of the blast are generally channelled up from the bottom. Therefore, a crater found in the limestone substrate is usually not an effect of blasting.

Evidence of blasting is generally available only in areas of standing corals; they have recently been broken in a circular pattern (as might be expected) from a blast. Less than 30% of an average reef supports the type of coral growth which would yield such evidence. In other places, there will simply be a higher than normal amount of dead standing coral or a lower than normal cover level left. The latter occurs after corals have been broken by blasting and washed away by wave action during storms. Because the term "normal" is not well defined, blast damage estimation is generally a very crude art.

Lagoonal areas tend to have clumps of standing corals; these are relatively easy to assess for effects of blasting. In one study, we surveyed approximately 100 stands of corals in the Bolinao lagoon and found that 60% (+ or – 10%) of the cover of such stands was killed coral, but these were still recognizable as recently killed. There were signs of blasting in most of the stands, and the majority of the dead branching coral had been cracked into small sections. Thus, we concluded that more than half of the coral of the lagoon had been killed by blasting within the last 5 or so years.

Field studies of the effects of blasting on fish are complicated by the illegal nature of the method. It is difficult to obtain appropriate funding and difficult to gain the cooperation of fishermen familiar with the nature of the local use of the explosives. Fish may be surveyed before and after the blast. However, the extremely variable nature of fish distributions and blasting application mandates a very large number of samples to yield a reliable model of effects. Finally, there are difficulties in estimating such factors as unutilized fish, because these fish are often difficult to see as they lie motionless on the bottom or among the coral.

3.3.6 Experiments for Blast Fishing Models

The large number of factors to be accounted for in predicting the effects of blast fishing on fish communities leads one to favor an empirical, experimental approach over one which is purely theoretical. However, experimental design in this case should be based upon both theoretical considerations and experimental laboratory results. A general approach might include:

(1) determining the composition and distribution of the fish communities in the fishery; emphasizing the types of swimbladders among the species and its variability over time,

(2) determining the levels of damaging pressure changes for each group of species,

(3) developing an average effects model for the pressure changes occurring during each major category of blast (by material); these blast categories will be specific to area subunits corresponding to major fish communities,

(4) verifying mortality effects from field studies before and after blasting; as well as blasting effects on the fish habitat and factors such as recoverability of the fish by the fishermen,

(5) determining the rate of usage of each blast material in each subarea, and

(6) combining this information to estimate mortality by species group.

Step (2) may be best facilitated through laboratory studies. Information for step (3) may be obtained from field studies in various fish habitat types using arrays of pressure sensors. Armada (1989) used caged fish in a simple array for a combined approach to steps (2) and (3). Step (5) can be facilitated through the use of hydrographic equipment. Blasts can be recorded over time, and wave forms of blasts might be stored digitally to discriminate between types of blasts. Arrays of listening devices might be useful in mapping blasting by subregion.

The major difficulty in all of this may be the need to coordinate the blasting media with the study. The blasting materials used by fishermen are often dangerous to handle. Funding agencies do not generally wish to be involved with studies involving blasting, and there may be difficulties in obtaining permits. Finally, agencies involved in convincing people not to use blast fishing can quickly lose credibility if they are concurrently involved in using or even organizing blasting operations, regardless of the purpose. For these reasons, it may often be desirable to minimize empirical testing in favor of laboratory studies and theoretical considerations.

4. FISHERY DEPENDENT SAMPLING

Any form of sampling which depends upon fishing activities for data collection is fishery dependent sampling. Most coral reef fisheries surveys involve shore-landing surveys, and some approaches to this are discussed below. Other information on shoreline sampling may be found elsewhere in this volume.

4.1 STATISTICAL CONSIDERATIONS

Statistical optimization should be attempted whenever practical in this type of survey. If data are to be obtained to answer a specific question based on an inferential statistical test, then an effort should be made to ensure that the data collection over time and space will be intense enough to provide for a reasonable evaluation of the hypotheses. This usually requires *a priori* knowledge of variance, which is obtainable from preliminary sampling or reference to similar studies conducted earlier. In most cases, the survey will be designed for multiple purposes, and optimization could become more complex. However, it is often wise to make certain that the sampling will be sufficient for one or two specific statistical tests to be conducted, and then expand the sampling scheme to ensure that multiple concerns are also addressed.

In very complex small-scale fisheries, formal statistical optimization often becomes a secondary concern in the face of a larger problem in order to informally attain reasonable coverage in all areas of concern. For example, in the Bolinao monitoring program, an effort was made to be certain that replicate samples were obtained from each major fishing gear. There were eleven fish landing sites in the area, each the domain of fishermen from primarily one village. The question was, how many landing sites must be randomly selected in order to be sure that each of eight major gears used in each part of the reef was sampled more than once each time. The answer was eleven. It turned out that each village favored a particular gear over all others, and where gears were the same, fishing grounds differed. From the standpoint of landing site selection, censusing rather than sampling was necessary (del Norte et al., 1989). This case would presumably be avoidable in larger scale sampling efforts, wherein statistically optimal sampling would be more appropriate.

It is very important that sampling be designed to include both the complexities of the fishery and statistical considerations. In many cases, this will necessitate that a team of two or more people (of different specializations) work closely to design the study. The team should be prepared to continue working around unforseen problems as they arise. Hopefully, the same team will be involved in the data analysis at the end of the study. However, two scenarios are to be avoided as much as possible. The first is that of a researcher who does not involve a statistically knowledgeable person until it is obvious that the sampling design is deficient and the objectives cannot be met. The second is that of a consultant who is hired to advise on the design, but is not retained long enough to modify the study over time as changes or new information invalidate the initial assumptions. Many studies are carried out blindly after the initial design, only to yield highly questionable material at the end. Therefore, it is extremely important to ensure (as much as possible) that: (1) the study be designed with access to both field and statistical expertise, and (2) those involved with the design of the study be involved throughout to the final stages of data analysis.

4.1.1 Regulated Fish Landings

In some areas, fishermen are required to record all sales of fish for purposes such as taxation. These records may be used as the basis for a cost-effective sampling scheme for catch, effort, and even fishing ground determinations. It is important to determine how much of the fishery is actually represented by the bills of sale or other records. Illegal landings are a problem in most countries, and some effort must be made into estimating it. Other losses to be estimated include: personal consumption of fish caught and fish used to directly feed families, relatives, or other persons on a non-cash basis. The bills may not reliably break down fish to the species level, necessitating periodic auxiliary sampling by researchers. An example of the use of bills of sale in shore sampling is that of Campos and Stagg (1988).

4.1.2 Unregulated Fish Landings

In areas where no bills of sale are required, or where the requirement is not well-enforced, other means must be found to estimate the landings. Some studies use the approach of directly sampling from boats as the catches are off-loaded. This approach is sometimes important as an auxiliary sampling approach, even when other means are used for the primary method. Boat sampling should be based on a thorough knowledge of potential biases. For example, boats which arrive early in the morning generally carry different species of fish caught by different methods than those landing at midday or evening. Furthermore, the transition between gear sources may be abrupt and inconsistent from one day to the next. A sampling scheme involving a random hour each day will generally yield little data because the landings are often extremely patchy over time and processing often very rapid. A boat sampling scheme must account for all these factors.

For some areas, it is possible to base sampling on a survey of middlemen purchase records. Many middlemen keep such records, partly to facilitate who owes money to whom, or for a multitude of other reasons. Middlemen can be even more distrustful of outsiders than fishermen, and many months may be expended to gather their confidence. It may be wise to hire a few ambitious villagers to be the ones to copy the middlemen records. This becomes an excellent way to involve the villages in studies of their fisheries, and the talliers become effective conduits of important information of concern to the village. The Philippine monitoring program employs three research aides, hired from the villages, to gather the middlemen data. These research aides also perform a great many other tasks, such as operating boats, building field sampling devices, and key punching the data into microcomputers.

It is important to have overlapping estimates of the data of interest. This helps in the identification of biases from insufficient sampling or improper assumptions. For example, the Philippine monitoring program involves boat surveys on a random day each week. All fishing craft outside the reef flat are mapped by compass triangulation and classified by activity. This provides a check on effort determinations based on shore landings. The general distributions of fishing boats have also been cross-checked by overflights on an ultralight aircraft. Additional forms of cross-checking and data reinforcement include questionnaires answered by selected fishermen. Each sampling approach has its particular strengths and weaknesses. Once a study has been established with resident personnel, it is generally cost-effective to keep them active in a variety of sampling activities. The result will be information with known robustness and estimable error and bias.

4.1.3 Length-Frequency Sampling

In addition to information on catch composition, production, and effort, shore sampling in tropical countries generally involves length-frequency studies. Length-frequency data can be used for a wide variety of population analysis methods, and they are becoming increasingly important in community level studies. Various forms of modal progression analysis are currently being used to estimate population parameters of key species. The major objective is to estimate the degree of over- or underfishing that a population is subjected to.

Much research is needed on determining optimal sample sizes for length-based population methods. In the meantime, however, it is generally important to obtain at least 50 measurements for a species each month. The fishing gear must yield samples which are relatively unbiased with respect to size, or there must be a way to mathematically adjust for the bias. Specific guidelines for sampling are available in the literature associated with the microcomputer program to be used.

A common approach is to purchase the fish to be used to obtain the measurements. However, this is generally not necessary, unless the fish are to be used for other purposes such as gut content analysis, otolith studies, or food for the researchers. Most fishermen will require a minimal amount of incentive to allow their fish to be quickly measured as they are off-loaded. This helps to reduce costs and also prevents the regular purchase of fish by investigators to bias the fishery under study. The fish can be rapidly measured on a measuring board, and lengths can be recorded by a second researcher.

4.1.4 Data Handling

The problem of data handling is often overlooked in the design of a study. It is quite easy to become thoroughly familiar with the analysis of simplified data sets in papers and textbooks without ever fully appreciating the vast complexity involved in obtaining, maintaining, and simplifying the data. Complexity of the problem increases exponentially as one progresses from population to community studies and from temperate to tropical fisheries.

Realistically speaking, over the next 10 years, the average fishery scientist in a developing country will have limited access to a microcomputer and virtually no access to a mainframe computer. Most complex databases are kept on mainframe computers. However, there is no reason why large databases cannot be kept on microcomputers. It is, in fact, quite easy to program a 32K computer to ask a researcher to insert disk number 327 so that values in column 2568 and row 39638 can be added to a running sum. However, both programming and using such systems become extremely labor-intensive and error-prone. Even at the abysmally low pay scales of most tropical scientists, it becomes cost-effective to invest in a modern microcomputer with a large-capacity hard disk. It is quite reasonable to set aside $5,000 out of any $50,000 project to facilitate the analysis and thus effectively utilize the data to be gathered. Furthermore, although database management can be performed on a slow microcomputer, there are a growing number of analytical methods which can be performed in a few hours on a fast microcomputer or several days on a slow one. (A researcher in a blackout-prone area may spend several weeks obtaining an answer on a slow machine, even with the aid of battery backup systems.) Other analytical methods, such as recent approaches to eigenvalue extraction, are so difficult to program for small memory sizes that it is unlikely that anyone will bother doing so in the future.

Fisheries data sets can be extremely complex, particularly if they involve shoreline sampling of multigear, multispecies landings, and length-frequency data. It appears to be reasonable to have a custom database system developed for each study. This must be done with extreme caution. A custom database system is generally very dependent on prior decisions about the exact nature and the volume of data to be handled. However, fisheries studies must always be flexible enough to be modified in response to new information as it becomes available. Therefore, the most flexible and simplest possible database system is often the best. These are the characteristics of commercially available spreadsheet programs.

Spreadsheet programs are simple to use, and they are so universally important that most researchers involved in microcomputer work are already familiar with their use. A single spreadsheet can be used to represent a table of any type. The table could mimic the data sheet being inputted, or the table may be in more of a summary form, wherein a data sheet yields a single column or row of data. It is generally wise to store all data in as raw a form as possible, so that retrospective analyses at any level are facilitated. At most, a procedure may be developed to check for extraneous values after each keypunching session. The design of the table should be done with considerable thought, and some experimentation is advised. The table then becomes the basis for all future work.

All major spreadsheets can be converted for summary in major database systems. At this point, it may be worthwhile to involve a professional programmer. The programmer can develop programs to create the summary tables to be printed out in data reports or to be inputs into various analytical routines. The programmer can be contracted on an as-needed basis to facilitate specific activities. However, should he or she not be available at any point, nearly all necessary work can be done by the researchers themselves by manipulating the spreadsheets.

One important concern is the bulk of the data set. A complex fishery study can easily generate over 100 diskettes full of spreadsheets in a single year. All diskettes should have a minimum of two backup copies available at all times. This is because it is surprisingly easy for a power fluctuation to occur during a backup procedure, destroying both the original and the backup. Insurance against catastrophes such as fires can be best had by keeping yet another copy or two in a separate building. Diskettes for 2 years of data and backups can easily cost on the order of $1000. however, commercial or custom database systems can reduce this data volume, but at the cost of greater dependence on personnel more computer-qualified than the average fisheries scientist. It is true that a simple database can be maintained by a minimally trained person on a commercial database, but shoreline fisheries databases are rarely simple. The alternative is to use archiving programs to reduce data bulk. The Philippine monitoring program researchers have found that a set of over 90 diskettes can be reduced to approximately 20 through a popular archiving program. The data must be unarchived before use, so it is still wise to keep working disks in the bulkier form.

Recently, a number of computer "viruses" have become rampant. It is extremely important that the core data set never be exposed to a virus, as all data could conceivably be lost on all backup copies simultaneously. Some programs are available to screen diskettes for some known viruses. Procedures can be adopted which ensure that one or more sets of the data are never exposed to analytical programs which might be carrying a virus. Ultimately, the final insurance is the paper printout. With large databases, maintaining paper copies of everything may be impractical. However, it is important that summary tables of all data be printed out and maintained in useable form at an early stage in the data handling process. The final printouts can be arranged in the form of data reports which, upon appropriate distribution with accompanying diskettes, can play the essential role of providing fuel for the creative research of other scientists worldwide.

ACKNOWLEDGMENTS

This chapter summarizes a diverse body of information, suggestions, and experience which has been brought to bear on the coral reef fishery monitoring program in Bolinao, Pangasinan, Philippines. We are indebted to everyone who has contributed to the evolution of the project, especially: Porfirio M. Aliño, Nygiel B. Armada, Sonia S. Formacion, Miguel Fortes, Wilfredo Licuanan, Liana T. McManus and Daniel Pauly. We thank the support staff of the International Center for Living Aquatic Resources Management (ICLARM) and the International Center for Marine Resources Development (ICMRD) at the University of Rhode Island for keeping things moving on our behalf. Most of the faculty and staff of the Marine Science Institute of the University of the Philippines contributed ideas, identifications, labor, or moral support at various times. Most especially we would like to thank Wilfredo L. Campos for his major role in helping to design the field sampling program, train the fieldworkers, and supervise the first 2 years of field activities. This publication was supported by the U.S. Agency for International Development under grant #DAN-4146-G-SS-5071-00. Correspondence should be addressed to John W. McManus, ICLARM, MCPO Box 2631, Makati, 0718, Metro Manila, Philippines.

REFERENCES

Alcala, A.C. and Luchavez, T. 1981. Fish yield of the coral reef surrounding Apo Island, Negros Oriental, Central Visayas, Philippines. Proceedings of the Fourth International Coral Reef Symposium. Manila. I: 67-73.

Amidei, R. (Ed.) 1985. Applications of remote sensing to fisheries and coastal resources. Report no. T-CSGCP-012, California Sea Grant.

Armada, N.B. 1989 (ms). Analysis and evaluation of the use of explosives in illegal fishing in Lingayen Gulf, Philippines. College of Fisheries, University of the Philippines in the Visayas.

Avery, T.E. and Berlin, G.L. 1985. *Interpretation of Aerial Photographs*, 4th edition. Burgess Publishing Company, Minneapolis, MN.

Bell, J.D., Craik, G.J., Pollard, D.A. and Russell, B.C. 1985. Estimating length frequency distributions of large reef fish underwater. Coral Reefs 4: 41-44.

Bellwood, D.R. and Alcala, A.C. 1988. The effect of minimum length specification onvisual estimates of density and biomass of coral reef fishes. Coral Reefs 7: 23-27.

Biña, R., Carpenter, K.E., Zacher, W., Jara, R.S. and Lim, J.B. 1978. Coral reef mapping using Landsat data: follow up studies. Proceedings of the 12th International Symposium on Remote Sensing of the Environment, ERIM. Ann Arbor, MI. III: 2051-2070.

Biña, R. and Ormac, E. 1979. Effects of tidal fluctuations on the spectral patterns of Landsat coral reef imageries. Proceedings of the 13th International Symposium on Remote Sensing of the Environment, ERIM. Ann Arbor, MI. III: 1293-1308.

Bohnsack, J.A. and Bannerot, S.P. 1986. A stationary visual census technique for quantitatively assessing community structure of coral reef fishes. NOAA. Technical Report NMFS 41: 1-15.

Bradbury, R.H. and Reichelt, R.E. 1983. Fractal dimension of a coral reef at ecological scales. Marine Ecology Progress Series 10: 169-171.

Brey, T. 1986. Estimation of annual P/B ratio and production of marine benthic invertebrates from length-frequency data. Ophelia, Supplement 4: 45-54.

Brock, R.E. 1982. A critique of the visual census method for assessing coral reef fish populations. Bulletin of Marine Science 32: 269-276.

Butler, D.M. (Ed.) 1987. From pattern to process: the strategy of the Earth Observing System. Earth Observing System Science Steering Committee Report, III. National Aeronautics and Space Administration. Washington D.C.

Campos, J. and Stagg, C. 1988. A strategy for organizing and analyzing Golfo de Nicoya fishery data. Fisheries Stock Assessment CRSP Working Paper 20. Collaborative Research Support Program. University of Costa Rica/Chesapeake Biological Laboratory, University of Maryland.

Campos, W.L., Cabansag, J.B., del Norte, A.G., Nañola, C. and Reyes, R.B., Jr. 1989. Stock assessment of the Bolinao reef flat fishery: yield estimates and the use of dominant species in accessing coastal multispecies resources. Fisheries Stock Assessment CRSP Working Paper 49. Collaborative Research Support Program. University of the Philippines.

Carpenter, K.E., Miclat, R.I., Albaladejo, V.D. and Corpuz, V.T. 1981. The influence of substrate structure on the local abundance and diversity of Philippine reef fishes. Proceedings of the 4th International Coral Reef Symposium. Manila. II: 497-502.

Coombs, C. 1984. *Ultralights: the Flying Featherweights*. William Morrow and Company, New York.

Davis, J.C. 1973. *Statistics and Data Analysis in Geology*. John Wiley and Sons, New York.

del Norte, A.G. 1986. Some aspects of the growth, recruitment, mortality, and reproduction of the Asian moon scallop, *Amusium pleuronectis* in Lingayen Gulf, Philippines. University of the Philippines, Quezon City. M.S. thesis.

del Norte, A.G. and Pauly, D. 1989. Virtual population estimates of monthly recruitment and biomass of rabbitfish (*Siganus fuscescens*) off Bolinao, Northern Philippines. Fisheries Stock Assessment CRSP Working Paper no. 50. Collaborative Research Support Program, University of the Philippines.

del Norte, A.G., Nañola, C.L., McManus, J.W., Reyes, R.B., Campos, W.L. and Cabansag, J.B. 1989. Overfishing on a Philippine coral reef: a glimpse into the future. Coastal Zone '89: Proceedings of the 6th Symposium on Coastal and Ocean Management, ASCE. V: 4847-4861.

Done, T.J., Kenchington, R.A. and Zell, L.D. 1981. Rapid, large area, reef resource surveys using a manta board. Proceedings of the 4th International Coral Reef Symposium. Manila. I: 299-308.

Fowler, A.J. 1987. The development of sampling strategies for population studies of coral reef fishes: a case study. Coral Reefs 6: 49-58.

Gallucci, V.F. and Hylleberg, J. 1976. Quantification of some aspects of growth in bottom-feeding bivalve *Macoma nasuta*. Veliger 19: 59-67.

Gleick, J. 1987. *Chaos: Making a New Science*. Viking Penguin, New York.

Gulland, J.A. 1983. *Fish Stock Assessment: A Manual of Basic Methods*. John Wiley and Sons, New York.

Gwynne, P. 1987. Remotely piloted vehicles join the service. High Technology 7: 38-43.

Hill, M.O. 1979. TWINSPAN—A FORTRAN program for arranging multivariate data in an ordered two-way table by classification of the individuals and attributes. Section of Ecology and Systematics, Cornell University, Ithaca, NY.

Hopley, D. 1978. Aerial photography and other remote sensing techniques. In *Coral Reefs: Research Methods*. (Stoddart D.R. and R.E. Johannes, R.E., Eds.) UNESCO, Paris: 23-44.

Jupp, D.L., Mayo, K.K., Kuchler, D.A., Heggen, S.J., Kendall, S.W., Radke, B.M. and Ayling, T. 1985. Landsat based interpretation of the Cairns Section of the Great Barrier Reef Marine Park. Natural Resource Series, no. 4. CSIRO Division of Water and Land Resources, Canberra.

Kechington, R.A. 1978. Visual surveys of large areas of coral reefs. In *Coral Reefs: Research Methods* (Stoddart D.R. and R.E. Johannes, R.E., Eds.) UNESCO, Paris: 149-161.

Kimmel, J.J. 1985. A new species-time method for visual assessment of fishes and its comparison with established methods. Environmental Biology of Fishes 12: 23-32.

Kinzie III, R.A. and Snider, R.H. 1978. A simulation study of coral reef survey methods. In *Coral Reefs: Research Methods*. (Stoddart D.R. and R.E. Johannes, R.E., Eds.) UNESCO, Paris: 231-250.

Licuanan, W.Y. and Gomez, E.D. 1988. Coral reefs of the northwestern Philippines: a physiognomic-structural approach. Proceeding of the 6th International Coral Reef Symposium. Townesville, Australia. III: 275-280.

Lillesand, T.M. and Kiefer, R.W. 1979. *Remote Sensing and Image Interpretation*. John Wiley and Sons, NY.

Longhurst, A.R. and Pauly, D. 1987. *Ecology of Tropical Oceans*. Academic Press, San Diego, CA.

Loya, Y. 1978. Plotless and transect methods. In *Coral Reefs: Research Methods*. (Stoddart D.R. and R.E. Johannes, R.E., Eds.) UNESCO, Paris: 197-217.

McIntosh, R.P. 1985. *The Background of Ecology: Concept and Theory*. Cambridge University Press, London.

McManus, J.W. 1988. Coral reefs of the ASEAN region: status and management. Ambio 17: 189-193.

McManus, J.W. 1989. Earth observing system and coral reef fisheries. Coastal Zone '89: Proceedings of the 6th Symposium on Coastal and Ocean Management, ASCE. Charlottesville, NC. V: 4936-4949.

McManus, J.W. and Arida, C.C. 1988. Philippine coral reef fisheries management. Fisheries Stock Assessment CRSP Working Paper no. 39. Collaborative Research Support Program. University of Rhode Island/University of the Philippines.

McManus, J.W., Ferrer, E.M. and Campos, W.L. 1988. A village-level approach to coastal adaptive management and resource assessment (CAMRA). Proceedings of the 6th International Coral Reef Symposium, Townsville, Australia. II: 381-386.

McManus, J.W., Miclat, R.I. and Palaganas, V.P. 1981. Coral and fish community structure of Sombrero Island, Batangas, Philippines. Proceedings of the 4th International Coral Reef Symposium. Manila. II: 271-280.

McManus, J.W. and Wenno, J.J. 1981. Coral communities of outer Ambon Bay: a general assessment survey. Bulletin of Marine Science 31: 574-580.

Millspaugh, B. 1987. *Ultralight Airman's Manual*. Tab Books, Blue Ridge Summit, PA.

Moll, N. 1987. Seeing the light: the airplanes that proved light weight is the right weight. Flying 114: 46-50.

Mueller-Dombois, D. and Ellenberg, H. 1974. *Aims and Methods of Vegetation Ecology*. John Wiley and Sons, New York.

Nañola, C., McManus, J.W., Campos, W.L. del Norte, A.G., Reyes Jr., R.B. and Cabansag, J.B. 1989. Spatio-temporal variations in community structure in a heavily fished forereef slope in Bolinao, Philippines. Fisheries Stock Assessment CRSP Working Paper no. 59. Collaborative Research Support Program. University of Rhode Island/University of the Philippines.

Pauly, D. 1982. Studying single-species dynamics in a tropical multispecies context. In *Theory and Management of Tropical Fisheries*. (Pauly, D. and Murphy, G.I., Eds.) International Center for Living Aquatic Resources Management, Manila: 33-70.

Pauly, D. 1984. Fish population dynamics in tropical waters: a manual for use with programmable calculators. ICLARM Studies and Reviews 8. International Center for Living Aquatic Resources Management, Manila.

Pielou, E.C. 1974. *Population and Community Ecology: Principles and Methods*. 3rd printing with corrections. Gordon and Breach, New York.

Pielou, E.C. 1979. Interpretation of paleoecological similarity matrices. Paleobiology 5: 435-443.

Pielou, E.C. 1983. Spatial and temporal change in biogeography: gradual or abrupt? In *Evolution, Time, and Space*: *The Emergence of the Biosphere*. (Sims, R.W., Price, J.H. and Whalley, P.E.S., Eds.) Academic Press, New York: 29-56.

Pielou, E.C. 1984a. Probing multivariate data with random skewers: a preliminary to direct gradient analysis. Oikos 42: 161-165.

Pielou, E.C. 1984b. *The Interpretation of Ecological Data: A Primer on Classification*. John Wiley and Sons, New York.

Preu, C., Sterr, H. and Zumach, W.-D., 1989. Monitoring of the coastal environments by means of a remote controlled balloon-borne camera (LAP technique). Coastal Zone '89: Proceedings of the 6th Symposium on Coastal and Ocean Management, ASCE. V: 4847-4861.

Prigogine, I. and Stengers, I. 1984. *Order Out of Chaos: Man's New Dialogue with Nature*. Bantam Books, New York.

Randall, R.H. 1973. Distribution of corals after *Acanthaster planci* (L.) infestation at Tanguisson Point, Guam. Micronesica 9: 213-222.

Ricker, W.E. 1975. Computation and interpretation of biological statistics of fish populations. Bulletin of the Fisheries Research Board of Canada 191.

Risk, M.J. 1972. Fish diversity on a coral reef in the Virgin Islands. Atoll Research Bulletin 153: 1-6.

Roa, P.A. 1989. A study of band noise associated with LANDSAT bathymetric images. Fisheries Stock Assessment CRSP Working Paper No.54. Collaborative Research Support Program. University of Rhode Island/University of the Philippines.

Ross, M. A. and Hodgson, G. 1981. A quantitative study of hermatypic coral diversity and zonation at Apo Reef, Mindoro, Philippines. Proceedings of the 4th International Coral Reef Symposium. Manila. II: 281-291.

Russ, G. 1984a. Distribution and abundance of herbivorous grazing fishes in the Central Great Barrier Reef I: levels of variability across the entire continental shelf. Marine Ecology Progress Series 20: 25-34.

Russ, G. 1984b. Distribution and abundance of herbivorous grazing fishes in the Central Great Barrier Reef II: patterns of zonation of mid-shelf and outer shelf reefs. Marine Ecology Progress Series 20: 35-44.

Russ, G. 1985. Effects of protective management on coral reef fishes in the Central Philippines. Proceedings of the 5th International Coral Reef Congress. Tahiti. IV: 219-224.

Russ, G.A. and Alcala, A. 1989. Effects of intense fishing pressure on an assemblage of coral reef fishes. Marine Ecology Progress Series 56: 13-27.

Russell, B.C., Talbot, F.H., Anderson, G.R.V. and Goldman, B. 1978. Collection and sampling of reef fishes. In *Coral Reefs*: *Research Methods*. (Stoddart D.R. and R.E. Johannes, R.E., Eds.) UNESCO, Paris: 329-345.

Rutzler, K. 1978. Photogrammetry of reef environments by helium balloon. In *Coral Reefs: Research Methods*. (Stoddart D.R. and R.E. Johannes, R.E., Eds.) UNESCO, Paris: 45-52.

Sale, P.F. 1972. Influence of corals in the dispersion of the pomacentrid fish, *Dascyllus aruanus*. Ecology 53: 741-744.

Sale, P.F. 1977. Maintenance of high diversity in coral reef fish communities. Am. Nat. 111: 337-359.

Sale, P.F. 1978. Coexistence of coral reef fishes—a lottery for living space. Environmental Biology of Fishes 3: 85-102.

Sale, P.F. 1980. The ecology of fishes on coral reefs. Oceanographic Marine Biology Annual Review 18: 367-421.

Sale, P.F. and Dybdahl, R. 1975. Determinants of community structure for coral reef fishes in an experimental habitat. Ecology 56: 1343-1355.

Sale, P.F. and Douglas, W.A. 1981. Precision and accuracy of visual census technique for fish assemblages on coral patch reefs. Environmental Biology of Fishes 6: 333-339.

Sale, P.F. and Sharp, B.J. 1983. Correction for bias in visual transect censuses of coral reef fishes. Coral Reefs 3: 37-42.

Starck W.A., III. 1984. Aussie ecstasy. Ultralight Aircraft 4: 44-57.

Stoddart, D.R. and Johannes, R.E., (Eds.) 1978. *Coral Reefs*: *Research Methods*. UNESCO, Paris.

ter Braak, C.J. 1988. CANOCO—A FORTRAN program for canonical community ordination by [partial] [detrended] [canonical] correspondence analysis, principal component analysis and redundancy analysis (Vers. 2.1). Agricultural Mathematics Group. Wageningen, The Netherlands.

Thresher, R.E. and Gunn, J.S. 1986. Comparative analysis of visual census techniques for highly mobile, reef-associated piscivores (*Carangidae*). Environmental Biology of Fishes 17: 93-116.

Waligunda, B. and Sheenan, L. 1981. *The Great American Balloon Book: An Introduction to Hot Air Ballooning*. Prentice Hall, Englewood Cliffs, NJ.

Williams, D. McB. 1982. Patterns in the distribution of fish communities across the central Great Barrier Reef. Coral Reefs 1: 35-43.

Williams, W.T. 1976. *Pattern Analysis in Agricultural Science*. Elsevier Science, Amsterdam.

Yeager, C. and Janos, L. 1986. *Yeager*. Bantam Books, New York.

CHAPTER 6

THE APPLICATION OF SOME ACOUSTIC METHODS FOR STOCK ASSESSMENT FOR SMALL-SCALE FISHERIES

John B. Hedgepeth[1], Vincent F. Gallucci[1], Richard E. Thorne[2], and Jorge Campos[3]

School of Fisheries, University of Washington[1], Biosonics Corporation[2], Research Center for Marine Sciences and Limnology (CIMAR), University of Costa Rica[3]

1. INTRODUCTION

Fisheries acoustics is the application of sound to study the dynamics of fishes. A major focus of fisheries acoustics is the estimation of fish abundance both for immediate use and for use as auxiliary information in larger stock assessment analyses.

Optimally, a fisheries acoustician collects data from monospecific aggregations of similar size fish. However, one often encounters assemblages comprised of many sizes and species. Acoustic methods efficiently detect and quantify fish but provide little fish size and species information. Samples for size and species composition can be taken by independent techniques such as trawls or video.

The field of fisheries acoustics is broad and is not limited to stock assessment. For example, by analyzing Doppler shifts, one can infer the direction of fish motion or tail beat frequency, which may indicate size. Bioacoustics, ultrasonic fish tracking, and medical ultrasound for investigation of structures and responses of fish are other applications of sound to fisheries biology.

1-56670-151-1/96/$0.00+$.50

This chapter will explain the use of low-cost and relatively simple methods for using acoustics in stock assessment. Applications are drawn from developing country small-scale fisheries and smaller, temperate water fisheries. This work is not encyclopedic since good "manuals" on acoustics do exist, and are listed at the end of the introduction. We do propose to present enough technical background so that the methods can be understood by biologists and can be put into perspective with other available methods. Our target audience extends from students interested in a scientific understanding of how technology is applied to stock assessment to professionals who make decisions about sampling gear and about how to utilize acoustic results with results from other methods of estimating abundance.

Section 2 is an overview linking acoustic survey estimates to fisheries stock assessment models. Some essential properties of underwater sound propagation are presented in Section 3, and Sections 4 and 5 lay a foundation for the analysis and use of single-beam methods, including deconvolution. Two multibeam systems are discussed for comparison purposes. Subsequent sections present a case study in the Gulf of Nicoya, Costa Rica, where a single-beam system was used. The final section discusses trends in technology and acoustic equipment.

Acoustics is often referred to as "sonar" (for sound navigation and ranging). When sonar is directed downward it is also termed "echosounding". Sonar is an "active" method in that it uses the reflection of a projected sound wave. A "passive" method would simply listen for sounds.

In principle, a serious acoustic investigation might incorporate expertise in the following areas: (1) sampling design, (2) electrical engineering, (3) computer programming and data processing, (4) fisheries ecology and (5) population dynamics. However, practitioners of acoustics find that they develop a sufficient understanding of electronic technology, computer methodology, and familiarity with signal processing equipment and data analysis. Indeed, commercially available hardware and software automate the process so that a biologist (with some additional training) can treat the equipment and computer-based data analysis as a "black box" and usually be correct.

The dangers of "black box" analyses can be minimized by a sufficient recognition of what can be accomplished with the available resources and when outside expertise is required. Since sonar-specific data analysis packages can be expensive, an alternative is to develop a system of echo analysis particularly adapted to the local application. This is practical because contemporary sonar methods such as echo integration and echo counting (references at end of Introduction) are documented and computer programs are available in the literature. A collection of some of the programs we have used appears in Hedgepeth (1993, CRSP Working Paper No. 105). An annotated listing of these programs is in Appendix III. A complete listing of all CRSP Working Papers is in Annex A of the book.

Population dynamics computer programs which accept acoustics data as auxiliary information are also available (for example, Deriso et al., 1985, 1990; Megrey, 1989; Sullivan, 1988; Sullivan et al., 1990; Kimura, 1989, 1990). Sullivan et al. (1989 CRSP Working Paper No. 66), and Amjoun et al. (1990) developed the CASA program and user guide, respectively, as part of a focus on size based analyses.

1.1 FIRST CONCERNS, LIMITATIONS, EQUIPMENT AND APPLICABILITY

One of the concerns of persons contemplating acoustic research is equipment. The limitations of gear, such as those imposed by boundary conditions (e.g., the air-water interface) should be considered when purchasing equipment. Fish too close to the surface or to the bottom may be impossible to detect. As examples of the reality of the use of acoustics, note that the reflection of the sound waves from a fish are more easily detected in species with a gas bladder. Fish behavior will constrain the survey's configuration and timing. The presence of multiple and diverse species and sizes will complicate interpretation. Patchiness will constrain the sampling design and the choice of equipment. Although there are cases where acoustics is inappropriate, in most situations acoustics will provide complementary information.

Table 1 presents three commonly used types of fisheries sonar or echosounding equipment. The main categories are single-beam and multiple-beam systems. Multiple-beam systems are of two types: dual-beam and split-beam systems.

1.2 SINGLE-BEAM DECISION

A single-beam system was selected for the case study in the Gulf of Nicoya, Costa Rica. The Gulf is a relatively shallow, tidally influenced estuary that supports a substantial artisanal fishery. It is the largest Pacific Ocean gulf of Costa Rica, covering about 1500 square kilometers. Data was collected in the inner or northern half of the Gulf which is more shallow than the outer parts, with typical depths of 4 to 20 meters. The case study shows both the capabilities and the limitations of single-beam echosounders for stock assessment purposes in a tropical environment.

We developed two low-cost processing systems to estimate fish abundances. One system used a digitizing tablet on echograms to look at diurnal and tidal variations in the number of fish detections, the behavior of fishes around an artificial reef, and the estimation of the velocity and direction of fish movement (Thorne et al., 1990, CRSP Working Paper No. 10; Thorne et al., 1989, CRSP Working Paper No. 14). The other system carried out the echo integration and target strength analyses. We also used commercially available products for signal processing.

The future will bring new technology for fisheries acoustics such as imaging arrays and remotely operated vehicles (Thorne, 1988). There has been recent success using pattern recognition to identify fish to the species level. Rose and Leggett (1988) suggest that single-beam equipment will classify fish schools to species, so their work is reviewed in a subsequent chapter. As computer systems become both more sophisticated and more "user friendly", the feasibility of expert systems will be felt in acoustics, where the user will be led though a series of decision support questions and answers.

The limitation to single-beam sonar places constraints on what can be done. The ping-to-ping variation in estimated target strength for a tracked fish cannot be as easily measured as it can with multiple beam systems. Although individual targets can be tracked with a single beam (see, for example, Furusawa and Miyonohana, 1988; Thorne et al., 1990; CRSP Working Paper No. 10), exact location is often difficult to estimate. Target strengths can be indirectly estimated from single-beam data but the estimates may be biased. On the other hand, multibeam systems can also include biases, so one must weigh the costs and benefits of any technique.

Table 1. Decision table for system selection (systems and terminology are explained in Section 3).

System	Cost	Complexity	Target strength
Single Beam	Low	Low	Indirect (requires many targets)
Dual Beam	Medium	Medium	Direct
Split Beam	High	High	Direct

System	Single fish detection features
Single Beam	Range (distance from the transducer)
Dual Beam	Range, off-beam-axis angle
Split Beam	Actual location

System	Capabilities with tracking
Single Beam	Echo trace statistics, duration-in-beam
Dual Beam	Ping to ping variations in target strength
Split Beam	Ping to ping variations in target strength Location as passes through beam

System	Impact of measurement error and noise
Single Beam	Impacts echo peak statistics
Dual Beam	Theoretically, more noise impact than split beam
Split Beam	Theoretically, less noise impact than dual beam

In the next section we discuss ways to use acoustic survey data for fisheries stock assessment. The ability to identify or classify the fish species is key. Methods for determining fish species include the use of nets such as trawls, purse seines, gillnets, and traps, some with video or scuba observation. The ability to identify fish species and to take representative samples from the population are two major factors which must be considered when surveying with acoustics.

1.3 ADDITIONAL READINGS NOT FOUND IN REFERENCES

Forbes, S.T. and Nakken, O. 1972. Manual of Methods for Fisheries Resource Appraisal. Part 2. The Use of Acoustic Instruments for Fish Detection and Abundance Estimation. FAO Manuals in Fisheries Science No. 5.

Margetts, A.R. (Ed.) 1977. Hydro-acoustics in Fisheries Research. Rapp. P.-V. Reun. Cons. Int. Explor. Mer. 170.

Mitson, R.B. 1983. *Fisheries Sonar*. Fishing News Books Ltd. Farham, England.

Mitson, R.B. and Hood, C.R. 1989. *Progress in Fisheries Acoustics. Proceedings of the Institute of Acoustics*. Vol. 11. Pt. 3. Edinburgh.

Nakken, O. and Venema, S.C. (Eds.) 1983. Symposium on Fisheries Acoustics. FAO Fisheries Report No. 300. Food and Agriculture Organization of the United Nations, Rome.

Thorne, R.E. 1983. Chapter 12: Hydroacoustics. In *Fisheries Techniques*. (Nielsen L.A. and Johnson D.L., Eds.) American Fisheries Society, Bethesda, MD.

2. FISHERIES ACOUSTICS AS AUXILIARY DATA INCLUSION IN CASA (CATCH-AT-SIZE-ANALYSIS)

The statement, "Today, quantitative assessment of fish population size is necessary worldwide in support of fisheries management" (Sissenwine et al., 1983), will be true for some time to come. Population assessment for fisheries management often relies on a combination of sampling of catch and an independent method of abundance estimation. Fisheries acoustics is one of several ways used to determine the abundance of fish. Other methods include trawl and seine surveys and ichthyoplankton surveys.

Virtual population analysis, cohort analysis, catch-at-age analysis and catch-at-length are members of a group of methodologies which model abundance using exponential survival. This group of models uses a time series of catch data (distributed with respect to age or size) for input and may include added input such as fishing effort, recruitment to the fish population, and population surveys. Such added types of data are termed auxiliary data, and the process is commonly called "tuning" the model. Catch-at-age and catch-at-length models have become common tools for fisheries workers to examine alternative scenarios, each of which offers plausible explanations for the observed time series of catch, in conjunction with survey, effort or other observations. An added advantage of catch-at-age and catch-at-length models is that they usually include estimates of the variability of the observations in parameters of the function used in "fitting" these models (Deriso et al., 1990; Kimura, 1989; Kimura, 1990; Sullivan, 1988; Megrey, 1988; Methot, 1990).

Megrey (1988) performed one of the first large-scale analyses using techniques in Deriso et al. (1990) in which acoustics was used as auxiliary data. This catch-at-age analysis used auxiliary data from both trawl survey and acoustic assessments of the north Pacific Ocean fish, *Theragra chalcogramma*, or walleye pollock.

Sullivan et al. (1990) presented a catch-at-length method of analysis which modeled growth stochastically. This model does not require that fish be categorized by age, but rather used length observations as input. In order to apply the model, the software

(Amjoun et al., 1990) known as CASA (catch-at-size analysis) was written and distributed.

The degree of importance of survey observations to the CASA model depends on the precision of the catch observations. The effects of bias and variation can be easily addressed by computer simulation. Figure 1 shows the error and coefficient of variation (CV) for one estimate, the initial population size for a simulated gillnet fishery, using the CASA program. The important feature in this figure is that the inclusion of survey data minimizes the error and variability of estimates for moderately to highly variable catch observations.

If the acoustic sampling is satisfactory, some of the parameters of population models can be estimated directly with acoustic data. For example, acoustic and net surveys have also been incorporated into north Pacific whiting management. Acoustic estimates are used to estimate total mortality for the whiting, by following the decay of abundance of cohorts using data from triennial acoustic/trawl surveys (Megrey et al., 1989).

A major concern in the use of acoustic survey data is the reliability of the estimated fractions associated with length or age groups, whether applied to a stock assessment model such as CASA or when estimating natural mortality. Multispecies situations, such as can be found in tropical environments, present special problems. At a minimum, the acoustician needs to perform surveys at times and locations when fish species can be most easily discerned and the fish can be most easily detected by the acoustic gear.

When acoustic survey data is used as auxiliary information in the CASA program, the user must input the observations as a function of time and length group. The program uses an objective function to produce the best fit between observations and the program generated estimates. That is, the best fit results from minimizing the least squares difference between the observed catch, denoted as C, and its estimate from the algorithm, $\hat{C}$, and simultaneously minimizing the difference between the survey observations, N, and their estimates, $\hat{N}$. An example is seen in (1) below.

Another version of the CASA program allows the user to fit the objective function using the least squares difference between the natural logs of the catch and survey observations and their estimates. This is (2) below. The two objective functions are:

$$f(\theta) = \sum \left(C - \hat{C}\right)^2 + \phi \sum \left(N - \hat{N}\right)^2 \tag{1}$$

$$g(\theta) = \sum \left(\ln C - \ln \hat{C}\right)^2 + \phi \sum \left(\ln N - \ln \hat{N}\right)^2 \tag{2}$$

where all of the summations are over length and time. Both objective functions are used to estimate a set of parameters, i.e., θ. C, $\hat{C}$, N and $\hat{N}$, all of which are vectors which are functions of time and the elements are length categories. The constant figure ϕ is a weighting term whose purpose is to optimize convergence to a minimum. Survey data need not be grouped into the same length categories as catch observations, nor is it necessary to have all time periods represented by auxiliary information. For example,

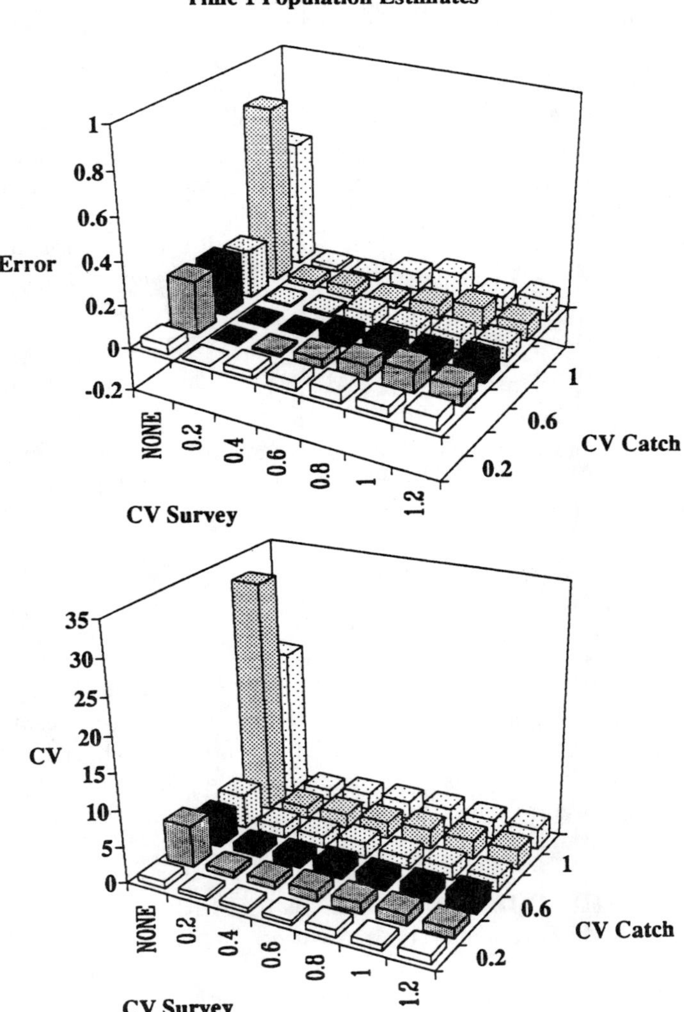

Figure 1. Error and CV for the initial population estimate as a function of variable catch and survey observations. This figure is the result of 4200 simulated datasets and estimations. Estimates were made by the nonlinear least squares program CASA described in the text. Error was computed as the relative difference of the mean estimate and the population parameter which generated the synthetic datasets. CV of the time 1 population estimate was the standard deviation of the estimates divided by the generating parameter. Notice that for moderately variable catch observations, survey data inclusion gives estimates with less variation and error.

Figure 1 results came from a simulation which pooled survey data as an observation over all lengths; i.e., one observation for each time period. These results were applied in the CASA model which was run through 4200 simulation runs to generate the results shown in this figure.

Parameter estimation, using the two types of objective functions, is usually referred to as nonlinear least squares (Kimura, 1990). ϕ is usually thought of as a variance ratio (Deriso et al., 1990). This is similar to weighted regression using the reciprocal of the variances of the observations as weights. In the CASA model, the observation vector consists of catch and survey observations. The first objective function (1) assumes that the observations of catch and survey include normally distributed error, and that the covariances are negligible. It can be then shown that the best value of ϕ is

$$\phi = \frac{VAR(C)}{VAR(N)}$$

where C and N are the observed data. The second function (2) assumes that the log of the catch and survey observations are normally distributed with negligible covariance, and the best estimate of ϕ is

$$\phi = \frac{VAR(\ln C)}{VAR(\ln N)}$$

The point is, surveys should be planned so that the observation errors can be estimated. That is, observations should be taken as part of an experimental design so that the appropriate variances can be estimated for use in such stock assessment analyses. Furthermore, auxiliary data are desirable for a wide range of variation in catch observations, because the precision and accuracy of the stock assessment estimates are increased.

2.1 ADDITIONAL READINGS NOT FOUND IN REFERENCES

Anderson, T.W. 1984. *An Introduction to Multivariate Statistical Analysis*. John Wiley and Sons. New York.

Fournier, D. and Archibald, C. P. 1982. A general theory for analyzing catch at age data. Can. J. Fish. Aquat. Sci. 39: 1195-1207.

3. ACOUSTIC THEORY

3.1 BACKGROUND

This section presents some of the theory underlying fisheries acoustics. This presentation is will serve as an introduction to the methods of simple single-beam systems. What is known as “sonar equations” will be explained, and their use will be illustrated by a beam detection zone example at the end of the section. Parameters which are included in a sonar equation, such as target strength or the "acoustic size" of a fish, will also be briefly explained.

3.2 SOUND PRODUCTION AND PROPAGATION

A fisheries research acoustic system typically consists of transmitter, receiver and signal processor. Electrical energy from the transmitter is converted into acoustical energy by an underwater transducer. This energy is projected into the water, and reflected by fish, plankton or other "targets" such as the bottom. The returning reflected energy is converted back into electrical energy by a transducer (usually the one used for projection), and output in a form which the signal processor can use. Fisheries research acoustic systems are available in either modules consisting of a transmitter/receiver, a signal processor, and an interpreter, or in a combined system. The trend today is toward real-time data processing, although output signals from the receiver are often saved on computer storage media or tape recorded so that they can be processed after the acoustic survey is completed.

3.3 BACKGROUND FOR THE "SONAR EQUATIONS"

Of the many formulae which relate the acoustical properties of water, the transducer, and the organisms in the water, it is the sonar equations which are standardly used to relate the "acoustic size" of a fish to the acoustic energy in an echo return (Urick, 1983). Other equations which are important in fisheries acoustic work, such as the echo integration equation and the equations for echo counting, will be explained after the sonar equations are introduced.

A brief explanation of the variables and parameters found in the sonar equations follows: (1) (SL) Source level is the energy projected by the transducer. (2) (αR) Alpha times range or distance is an energy loss to the water, called absorption loss, caused by exchange of energy. (3) (TL) Transmission loss is an attenuation of the projected and reflected energy due to wave spreading. (4) (TS) Target strength is the acoustic size or reflectivity of the "target", which in fisheries acoustics is usually an individual fish. (5) (NL)(NF) Noise level and noise figure are noise in the water and the electrical equipment. (6) (E) Efficiency of the transducer converting electrical to acoustic energy and vice versa is not 100 percent. (7) (RL) Reverberation level is an additional reflection of sound due to the presence of other biological or nonbiological "scatterers". (8) (G) Gain and (TVG) time-varied-gain are measures of amplification or attenuation of the receiver. (9) (DI) Directivity index and ($B(\theta)$) beam pattern factor are measures of how the transducer distributes and receives sound. (10) (V) Voltage level is the (EL) echo level corrected by (RS) receiving sensitivity, and is the voltage that is used by the signal processor. (11) (DT) Detection or voltage threshold is the level of voltage above which fish targets can be seen by the acoustic system. In addition, there are other physical parameters that will be considered below, such as transducer cavitation which constrains the system's design.

The "sonar equations" commonly relate the acoustic variables in a logarithmic form. Logarithmic units termed "decibels" (dB) are used because of the wide range of acoustic intensities encountered. A decibel quantity is formed by taking the log base 10 of a quantity (usually a ratio: the denominator is a reference value) and then multiplying by 10. The sonar equations may also be used in non-transformed units, such as MKS.

Sound propagates better than any other form of wave radiation in water, moving almost five times faster in water than in air, from 1400 to 1500 m/s.

When sound is propagated in a straight line, the distance (D) it travels is a function of time (t) and speed (c).

$$D = c\,t$$

The speed of sound propagation (c) is a function of the compressibility of the medium. The medium and speed of sound are related by

$$c = \sqrt{(p' / \delta')} = \sqrt{(B / \delta)}$$

where p' = acoustic pressure dp
δ' = acoustic component of density $d\delta$
B = bulk modulus
δ = density

The speed of sound can also be estimated from empirical equations using temperature, salinity, depth, and so forth. The literature contains several of these equations. Kinsler et al. (1982) present the following approximation.

$$c = 1449.05 + 4.57T - 0.0521T^2 + 0.00023T^3 + (1.33 - 0.0126T + 0.00009T^2)(S - 35) + 16.3L + 0.18L^2 \quad (3)$$

where T = temperature in degrees Celsius
S = salinity in ppt
L = $D(1 - .0026\cos(2\phi))/1000$
D = depth in meters
ϕ = latitude in degrees
c = sound propagation speed in meters per second

Urick (1983) suggests several other empirical estimates for the speed of sound which are less complicated than equation (3). The speed of sound for depths less than 1000 meters, temperatures less than 35 degrees and salinity less than 45 ppt is

$$c = 1449.2 + 4.6T - 0.055T^2 + 0.00029T^3 + (1.34 - 0.01T)(S - 35) + 0.016D$$

"Active" as opposed to "passive" acoustic systems depend on the distance, and therefore time, to and back from an object. Thus, the equation for active echosounding which relates time, distance and speed is

$$D = .5\,c\,t$$

Therefore, the practitioner of acoustics uses sound speed and time to estimate the range or the distance to the fish target.

Fisheries acoustic sonars or echosounders commonly use pulsed transmission, for which pulse durations range roughly from 0.2 ms to 2.0 ms. The acoustic energy is transmitted by vibrational waves in the pulse measured in cycles per second or Hertz (Hz). Sound waves from about 20 Hz to 20 kHz are audible, those above 20 kHz are ultrasonic and those below 20 Hz are infrasonic. Fisheries acoustic systems usually broadcast ultrasonic waves. The frequency used is important because some frequencies are attenuated more than others by the water and because fish reflectivity is increased at certain frequencies.

Frequency is measured in Hz (cycles per second) or in angular frequency (radians per second) (there are 2π radians in one cycle). The relations between frequency (f), angular frequency (ω), wavelength (τ), period (T), acoustic wave transmission speed (c) in meters per second, and wavenumber (k) are:

$$f = \frac{\omega}{2\pi} \text{ cycles per second or Hz}$$

$$T = \frac{1}{f} \text{ second}$$

$$\tau = \frac{c}{f} = cT \text{ meters}$$

$$k = \frac{\omega}{c} = \frac{2\pi}{\tau} = \frac{2\pi f}{c} = \frac{2\pi}{cT} \text{ radians per meter}$$

Decibel levels (dB) are formed by $10\times\log$ of the ratio of acoustic intensity levels (I) to a particular reference intensity. The reference (I_{ref}) is the intensity that corresponds to 1 μPascal (μPa) for underwater applications. Intensity in dB is related to intenstiy in μPa by

$$I_{dB} = 10\times\log I / I_{ref}$$

$$1 \text{ Pascal} = 1 \text{ Newton} / m^2$$

Acoustic intensity is defined as the average rate of flow of energy (power) per unit area normal to the propagation direction, with units of watts/m^2. Acoustic or effective pressure is the "excess pressure" (relative to equilibrium pressure) at any point in the water. For a spherical wave, effective pressure (pressure amplitude P_e) is related to acoustic intensity by density and sound speed (Kinsler et al., 1982)

$$I = P^2/(2\delta c) .$$

Because acoustic intensity is proportional to the effective pressure squared, one can express acoustic pressure in dB as

$$P_{dB} = 20\log P_e / P_{ref}$$

which is equivalent to the equation for acoustic intensity. The preferred reference intensity in underwater acoustics is that of a 1 μPa plane wave (about 6.76 x 10^{-19} W/m^2). The following table shows the relation between dB, intensity and pressure for the range of decibel values from 10 to 20.

Decibel values (dB) are given in reference to a fixed constant (I_{ref} or P_{ref}) for both intensity or pressure values (intensity = pressure squared). dB = 10 log(intensity), where the reference is a 1 μPa plane wave.

dB	I/I_{ref}	P_e/P_{ref}
−10	0.1	0.316
−6	0.25	0.5
−3	0.5	0.707
0	1.0	1.0
3	2	1.414
6	4	2.0
10	10	3.16
20	100	10.0

3.3.1 The Energy Projected by the Transducer

In dB notation the energy projected by the transducer is called source level (SL). These levels are extrapolated to one meter from the source; in non-dB notation the source level is the average transmitted power (P_T) or pressure (P_e):

$$SL = 20\log(P_e / P_{ref})$$

$P_T = 4\pi r^2 I$ for a nondirectional transducer at range r.

One could express power in dB for the sonar equation at 1 meter as

$$SL = 10\log(P_T / (4\pi 6.76\,E - 19)) = -170.7 + 10\log P_T$$

Given the power of the echosounder in watts, the efficiency of conversion (E), and a measure of the concentration due to the shape of the transducer's beam (DI), a revised source level SL′ can be expressed as

$$SL' = SL + E + DI$$

Both E, efficiency of conversion, and DI, the directivity index, are discussed in 3.3.5 and 3.3.10.

3.3.2 Losses to the Water, or Absorption Losses, Caused by Exchange of Energy

Urick (1983) says that in seawater, absorption is caused by three agents: shear viscosity, volume viscosity and chemical relaxation. The first two agents are related to the physical properties of water and the absorption caused by relaxation is due primarily due to $MgSO_4$ in saltwater. Absorption (α) is usually expressed in terms of dB/m, so that in the sonar equation the one way loss is $-\alpha r$. Urick cites a study that empirically relates absorption to salinity and frequencies up to 25 kHz:

$$\alpha\,(\text{dB/m}) = \{ASf_t f^2 / (f_t^2 + f^2)\} + \{Bf^2 / f_t\}$$

where S is salinity in ppt
f is frequency in kHz
f_t is relaxation frequency $= 2.19E(6 - 1520 / (T - 273))$kHz
T is temp Celsius
$A = 1.70\,E - 5$
$B = 2.45\,E - 5$

3.3.3 Losses of the Projected and Reflected Energy due to Wave Spreading (including 3.3.2)

The amplitude in pressure of a spherical wave decreases with distance. If r equals range, the pressure amplitude (P_r) at range r is

$$P_r = (1/r)$$

where P_1 is the pressure at 1 meter.

Transmission loss (TL) in dB can be defined as $20 \log(P_1/P_r)$.

$$TL = 20 \log r$$

Absorption also is included in the transmission loss:

$$TL = 20 \log r + \alpha r$$

The ideal part of the transmission loss could also be written (in terms of acoustic intensity) as

$$TL = 10 \log(I(1)/I) = 10 \log r^2 = 20 \log r$$

Some acoustic references explain this part of the transmission loss by conservation of energy. Power crossing concentric spherical shells, the source at the center, would then be equal since power is flow of energy, and flow rate is assumed not to vary. Because power equals acoustic intensity times area,

$$\text{constant power} = 4\pi r_1{}^2 I_1{}^2 = 4\pi r_2{}^2 I_2{}^2$$

Such a loss in intensity with range is referred to as spherical or inverse-square spreading. In dB terms it is a one-way spreading loss.

However, many of the "basic" acoustics texts discuss spreading loss by a solution to a partial differential equation, including complex variables. A discussion of this acoustic wave equation is found in Appendix I.

3.3.4 Target Strength: The Acoustic Size or Reflectivity of the "Target"

Target strength (TS) is the term in the sonar equation which describes the acoustic size of a fish or other target. It characterizes the fraction of the incident signal (I_i) reflected back (I_r) toward the transducer and measured 1 meter from the target:

$$TS = 10 \log(I_r / I_i)$$

Target strength has been called the key quantity in the acoustic assessment of fishes (Foote, 1987). A commonly used method to estimate the abundance of fish, known as echo integration, requires a measure of target strength or its equivalent, in linear terms, backscattering cross section. On the other hand, some researchers (e.g., Thorne, 1988) use methods which, while ultimately depending on the acoustic size of the fish, do not involve a computation of target strength.

Target strength is analogous to the non-dB quantities, acoustic cross section (σ) with area units or backscattering cross section (σ_{bs}) in relative units. σ_{bs} is relative because it is divided by the surface area of a sphere having a 1 m radius:

$$\sigma_{bs} = \sigma/(4\pi) = (I_r / I_i)$$

$$TS = 10 \log \sigma_{bs},$$

where σ_{bs} is defined as the ratio of σ to the area of a sphere with 1 meter radius, which implies that an ideal sphere with radius of 2 meters has acoustic cross section of $\pi r^2 = 4\pi$ or a TS equal to 0 dB. This ideal model may apply to large (relative to the wavelength), perfectly reflective spheres. Other shapes, with different compressibility, size and density will behave differently. For example, Anderson (1950), Machlup (1951) and Johnson (1977) developed models to estimate σ_{bs} for small spheres with acoustic properties near those of the medium. Such models were used by these and subsequent researchers to model zooplankton. Stanton (1989) reviews some of these models and presents some simple approximations for estimating σ_{bs} based on simple shapes. These simple relations may be valid for sizes and acoustic frequencies which are used in fisheries applications.

A fish swimbladder may account for 90–95% of echo energy returned by a fish with a swimbladder (Foote, 1985). Fish (and most zooplankters), unlike spheres, are not omni-directional backscatterers. That is, σ is not constant, and depends on the angle that the fish is insonified, the interaction of the various scattering elements in the fish, the activity of the fish, the presence and size of a swimbladder, the acoustic frequency, and so on. Modelers have considered the fish as a flexible line array of small scatterers (Huang and Clay, 1980) or a line array with point scatterers and concentrated scatterers (Clay and Heist, 1984). Foote (1985) used a Kirkoff approximation to model a fish with a swimbladder, from digitized swimbladder measurements. He found that a fish can be modeled solely by its swimbladder when using acoustic frequencies with wavelengths that range from 1/8 to 1/36 of the length of a fish. Both Stanton (1989) and Furusawa (1988) model a fish swimbladder as prolate spheroids.

Although measurements of target strength of fishes have been made on several species of caged and tethered fishes, at various angles (Haslett, 1964; Love, 1971a; Love, 1971b; Love, 1977; Naaken and Olsen, 1977; Goddard and Welsbey, 1977; Huang and Clay, 1980) there is a trend toward using target strengths measured in situ and in concert with physical models of target strength. Regardless of the way target strength is measured, the target strength fish length relationship is often estimated in the form

$$TS = a \log(\text{length}) + b$$

A frequency term can be added to the above equation (Love, 1971a and b). For the dorsal aspect TS, Love (1971a) presented the relation

$$TS = 20.1\log(\text{length, cm}) - .9\log(\text{frequency, kHz}) - 62$$

In addition, the relationship for target strength may be estimated from kilograms of fish biomass and used when catch estimates are better made by weight than by numbers.

Some models which relate fish length (or swimbladder length) to σ_{bs} use a length squared term which is expressed in dB notation as

$$TS = 20 \log \text{length} + b_{20}$$

Foote (1985) presented for pollack

$$TS = 20 \log(\text{ fish length in cm}) + b_{20}$$

where a typical b_{20} for 0 tilt angle and dorsal insonification was between –64 and –65 dB, and smaller for larger tilt angles. Foote (1985) states that his model results were "indistinguishable from those based on conventional constrained-single-fish measurements". Foote and Traynor (1988) report a b_{20} of –66.0 dB for walleye pollock based on dual-beam and split-beam in-situ measurements and swimbladder morphology. These measurements are from a frequency of 38 kHz. If a different frequency were used, one might expect a different b_{20}. Foote (1987) presents the relations

$$\text{physoclists} \quad TS = 20\log(\text{length}) - 67.4$$

$$\text{clupeoids} \quad TS = 20\log(\text{length}) - 71.9$$

Stanton (1989) and Furusawa (1988) model the swimbladder as a rigid prolate spheroid. Furusawa showed measurements for several fish species which clustered around the model prediction.

The frequency used in an acoustic survey should be chosen to maximize the target strength of the organisms being surveyed. Some compromise may have to be made, trading attenuation (3.3.2) for gain in target strength. Frequency projected by the echosounder can have a dramatic effect on measured target strength, especially when the swimbladder, or acoustic size is around the same size as the acoustic wavelength (Greenlaw, 1979; Penrose and Kaye, 1979).

Frequency can be used to provide additional information about the fish in the echosounder beam. For example, multifrequency systems, combined with minimization techniques and acoustic models which include the target strength, can estimate the abundance and size frequencies of fish and zooplankton (Greenlaw, 1979; Greenlaw and Johnson, 1983; Holliday et al., 1989; Kalish et al., 1986).

3.3.4.a In Situ *Target Strength Estimation*

One of the most common ways to estimate fish target strength is from field data under natural conditions, that is, *in situ*. Two classes of approaches have been used: indirect and direct. Indirect methods require a large number of measurements from individual fish to construct a frequency distribution of raw voltage peaks. Theoretically, these raw voltage peaks are the convolution (also called folding) of the beam pattern distribution of the transducer and the acoustic cross section distribution of the fish. The beam pattern distribution is then removed by an inverse process called deconvolution. The result is an estimate of the voltage distribution, proportional to the square root of the acoustic cross section distribution. Some researchers fit a parametric distribution (usually Rayleigh or Rician) after deconvolution. Other methods have solved the convolution integral equation by assuming a distribution for the acoustic cross section *a priori*, modifying the convolution integral with this distribution in mind, and numerically finding a solution (Ehrenberg et al., 1981). There is a rich history of the implementation of indirect methods (Craig and Forbes, 1969; Ehrenberg, 1972; Petersen et al., 1976; Robinson, 1983; Clay, 1983; Lindem, 1983; Clay and Heist, 1984; Degbol et al., 1985; Stanton and Clay, 1986; Rudstam et al., 1987; Foote, 1987; Rudstam et al., 1988; Hedgepeth and Thorne, 1989 and others).

Direct methods are often the choice for fisheries acoustics analyses. They remove the beam pattern directivity using multiple beams to estimate location in the beam. The two currently used methods in fisheries acoustics are split-beam and dual-beam. The advantage of direct methods is that an estimate of backscattering cross section can be made on individual targets. These methods use the voltage amplitudes (dual-beam) or phases (split-beam) corresponding to the individual beams in a relationship which determines the angular location of the target (Ehrenberg et al., 1981; Traynor, 1985).

Other innovative approaches have been used to estimate target strength. Furusawa and Miyanohana (1988) used echo trace or shape analysis to estimate target strength and fish swimming velocity. Others have combined the number of fish counts, the estimated volume, and the echo integration value to estimate target strength.

3.3.5 Noise in the Water and the Electrical Equipment

Noise and interference are unwanted signals from a number of sources that are added to the acoustic signals of interest. The noise level controls the ultimate or full-range performance of a sonar system and is extremely important in target strength measurements. The cause, effects and measurement of noise have been reviewed by a number of authors (Kinsler et al., 1982; Urick, 1983; Johannesson and Mitson, 1983; Ross, 1976). Noise enters the sonar equation as

$$\mathrm{NL} = \mathrm{NSL} + 10\log w - \mathrm{DI}$$

where NSL = the noise spectrum level
w = bandwidth (the frequencies that are detected)
DI = the directivity index defined below.

3.3.5.a *Bandwidth, Pulse Duration and Fish Resolution*

Bandwidth is the range of frequencies that the echosounder uses. Bandwidth and pulse duration are approximately inversely related because the smaller the pulse duration, the wider the bandwidth must be. Some echosounders have receivers which have selectable bandwidths and pulse durations, therefore it is important to set the bandwidth large enough for a chosen pulse duration, but small enough to avoid added noise input.

Both pulse duration and bandwidth affect fish resolution, or detectability of single targets. When the pulse duration is longer, the chance of getting several targets in the beam increases. When the bandwidth is wider, more noise can be collected, and the chance of observing small signals is smaller.

Resolution (range for separating targets in range) depends on pulse duration and sound speed:

$$\text{resolution distance} = c\,t\,/\,2$$

If the desired resolution is one meter, the maximum pulse duration is 1.33 ms using

$$1 \text{ meter} = c\,t/2,$$

where c is about 1500 m/s and is the pulse duration. This resolution is in terms of range or "vertical" separation, while inter-animal spacing will affect "horizontal" resolution, which can only be controlled by choice of the directivity of the transducer.

Although, one never transmits a sinc wave form (sin(x)/x), a sinc function approximation for the shape of a transmitted pulse would give a required minimum bandwidth equal to the inverse of the pulse duration. The 1.33 ms pulse corresponds to a minimum 750 Hz bandwidth (base band). If a rectangular shape were used instead, the effective minimum bandwidth would be 3/(2t), which would be 1125 Hz bandwidth for 1 meter vertical resolution. Other pulse envelope shapes give different bandwidths. Clay and Medwin used a Gaussian approximation to the echo envelope, which will give a bandwidth of $4/(t\pi)$ or 955 Hz for a meter resolution. As pulse duration lessens, the distance at which two targets can be distinguished (in range) decreases but bandwidth must increase. The increased bandwidth allows more noise into the system. The bandwidth is also a measure of information content. A smaller bandwidth contains less information per unit time. If all one were interested in were resolution between targets, then one would tend to a function producing wider bandwidth. However, we are also interested in measuring amplitudes of signals with the least amount of noise. The "optimum bandwidth" is a compromise between recognition of a signal above noise and resolution of single targets from multiple ones.

3.3.6 Efficiency of the Transducer Converting Electrical to Acoustic Energy and Vice Versa

Transducers are far from completely efficient in converting electrical power (P_e) to acoustic power (P) and vice versa. The electroacoustic efficiency can be measured over a range of frequencies and one finds that there will be a point where the efficiency will be optimum. Efficiencies of transmission range from 0.2 to 0.7 (Urick, 1983). Most fisheries acoustic systems have efficiencies in the lower part of that range, from 0.3 to 0.4. If

$$E = P/P_e$$

or in dB terms,

$$E_{dB} = 10 \log E$$

3.3.7 Presence of Other Biological or Nonbiological "Scatterers"

The presence of targets in which we are not immediately interested causes a type of noise called reverberation level. We can estimate the reverberation level in the same way in which we employ density estimation by the echo integration technique. In dB terms this reverberation level is denoted by RL.

Often, the density of fish is so great that the resolution of single fish is impeded. This is not as much a range resolution problem as it is a horizontal resolution problem. That is, a number of fish are occurring at or near the same depth or distance from the transducer. The technique of echo integration is applied to estimate the number of the fish of interest.

3.3.8 Gain (Amplification or Attenuation) of the Receiver

There are three measures with which one can express the effect of the echosounder's receiver. The first is the controllable or fixed gain, G in dB, because the receiver is usually part of the echosounder and there is a control on this fixed gain. On some receivers this control switches in dB steps while on others the gain control is continuous over some interval. The second type of gain measures the receiver's overall performance in converting an echo level returning at the transducer to the production of voltage which we monitor. This is the echosounder's receiving sensitivity, RS in dB. The receiving sensitivity is a function of the efficiency of the transducer converting sound energy into electrical energy, including related losses or gains but excluding the controllable gain, in the receiver. The third type of gain is time-varied gain, TVG in dB. The time varied gain is designed to correct for spreading losses (TL) and absorption losses. We saw that TL was range dependent, which is why this is called time-varied gain. As the range (or, equivalently, time) increases to the target the amount of gain needed to correct for transmission loss increases. A TVG called 40 logR time varied gain would correct for the two-way spreading loss (that is, to the target and back; it also often includes absorption losses, ($TVG = 40 \log r + 2\alpha r$). A 20 logR time-varied gain would correct for one way spreading loss ($TVG = 20 \log r + \alpha r$).

3.3.9 Transducer Cavitation

When too much power is applied to a transducer, negative pressures generated by the transducer cause bubbles to form in the water. The point at which too much power is applied is called the cavitation threshold (Urick, 1983). The importance of cavitation threshold is that it constrains the power that can be sent to a transducer of certain size. The relation between electrical power in watts, pressure (p) in atmospheres and transducer diameter is

$$10 \log P = 10 \log(p^2(.34)((D/2)(100))^2 \pi)$$

Although cavitation threshold is important for system designers, the acoustic practitioner should be aware of the problem so that simulations can be realistically enacted.

3.3.10 Beam Directivity

The beam directivity factor is defined as "the ratio of the total power actually radiated over all directions to that which would be radiated if the transducer had a uniform response in all directions, equal to that at the maximum of its main lobe". The directivity index, DI_T, is

$$DI_T = 10 \log(I / I_{non\text{-}directional}) = 10 \log D$$

where I is the intensity measured on the acoustic axis and $I_{non\text{-}directional}$ is the intensity that would be measured from an equivalent spherical source of the same power. D is the ratio and is called directivity.

The directivity, D, of a theoretical circular transducer can be modeled as (Kinsler et al., 1982)

$$D = \frac{I_{directional}}{I_{non-directional}} = \frac{4\pi}{\int_0^{\pi/2} \left| \frac{2J_1(ka \sin\theta)}{ka \sin\theta} \right|^2 2\pi \sin\theta d\theta} = \frac{(ka)^2}{1 - J_1(2ka)/ka},$$

where ka = k times the effective transducer radius, a.

Recall k is the angular frequency divided by sound speed. J_1 is a Bessel function of the first kind order. For ka >> 1, D = $(ka)^2$. This model of a circular transducer using the Bessel function will be explained below, in the section on transducer design.

3.4 THE SONAR EQUATIONS

In Section 3.1 we noted that the objective of this chapter is the development of the sonar equations. The detectable signal is related to noise by an equation constructed as follows. Assuming no noise, if a target is located at angle θ in a symmetric acoustic radiation (as from a circular transducer), the echo level in the receiver is measured as

$$V_1 = 20\log \text{volt}$$

$$V_1 = SL + TS + RS - 2TL - 2\alpha R + 2B(\theta) + G + TVG$$

If targets were absent, only thermal and receiver noise would be measured, called V_2,

$$V_2 = NL + NF + BW + RS - DI + G + TVG$$

Detection of the fish can take place when the signal from the fish is greater than the noise by a certain factor. The minimal level for signal detection is an additive factor in the sonar equation, the detection threshold, DT.

$$V_1 = V_2 + DT$$

$$DT = SL + TS - 2TL + 2B(\theta) - NL - NF - BW + DI$$

A second condition accompanies the detection threshold relation. That is, the echoes must exceed a certain level, a threshold (V_T), to be registered by the echosounder

$$V_T \leq SL + TS + RS - 2TL - 2\alpha R + TVG + 2B(\theta) + G$$

3.4.1 Visualization of the Sonar Equation

A particular application of the sonar equations is seen with the help of the computer program, CRTBEAM.EXE, which simulates the shape of the volume in which an acoustic system can detect and resolve a fish of a certain target strength. This simulation lets the novice user of acoustic equipment manipulate the variables of the sonar equation and learn some of the consequences of different configurations of input variables. The computer simulation in CRTBEAM.EXE relies on the two sonar equations. One is an expression for the detection of a fish echo above the level of noise. The other is the detection of the fish echo above a voltage threshold of the echosounder or signal processor (which can be set during analysis as well).

The computer program is written with the purpose of determining and then displaying on the computer the limits of range for the detection of an object with fixed TS. This requires input values for several variables, including the power of the sounder and beam angle corresponding to the half transmitted power point of the beam directivity.

The program was compiled in Microsoft QuickC, version 2.5. This language was chosen because it is relatively simple to produce graphics and the mathematics library contains Bessel functions so that the theoretical beam pattern for a circular transducer would be easily computed. Graphics consist of a cross section through the detection volume, the transducer at the top of the graph. Main lobe and sidelobes, and three levels of target strength are shown. Source code for the program appears in the CRSP users guide by Hedgepeth (1993, CRSP Working Paper No. 105). The following chart shows the logical flow behind CRTBEAM.EXE.

PROGRAM CRTBEAM

Initial Fixed Values:

NF = 4 dB, RS = –170 dB, E = 10log(.15)
Pulse Duration = .4 ms

Input:

Sound Frequency (kHz)
Half-Power Full Beamwidth (degrees)
Sounder Power (Watts)
Temperature (degrees Celsius)
Salinity (parts per thousand)
Fish Length (cm)

Compute:

$ka = 59\pi/\text{Beamwidth}$
$\alpha = .0000017\ \text{Salinity}\ AB/(A+B^2) + .00000245A/B$
where $A = \text{Frequency}^2$
$B = 21.9\ 10^{(6-1520/(T+273))}$
$BW = 10\log(1/\text{Pulse Duration})$
$DI = 10\log(ka^2)$
$TS = 20\log L - 70$
$SL = 170.7 + 10\log(\text{Sounder Power}) + E + DI$
if Frequency > 70 $NL = -15+20\log(\text{Frequency})$
else $NL = 55.11 - 18\log(\text{Frequency})$

Find Detection Ranges:

$\theta = 0$
Loop1: $\theta = \theta + \delta$
GOTO FINISH if $\theta = \pi/2$
$B = 10 \log b(\theta)$ Use a routine to compute Bessel function: J_1
$b(\theta) = [2 J_1(ka \sin\theta)/ka \sin\theta]^2$
$R = 1$
Loop2: $R = R + \varepsilon$
$TL = 20 \log R - \alpha R$
$TVG = 2TL$
if $(2TL > 2B(\theta) - NL + DI - BW - NF + TS + SL - DT)$
OUTPUT R, θ
else GOTO Loop2

FINISH

To include voltage threshold in CRTBEAM:

$$TS + SL + RS + 2B(\theta) + TVG - 2TL > V$$

CRTBEAM Program Variables

α = absorption coefficient
δ = increment for θ
ε = increment for R
θ = angle off transducer axis
ka = wavenumber times effective transducer radius
$B(\theta)$ = beam pattern direction factor in dB
BW = bandwidth in dB
DI = directivity index in dB

DT = detection threshold in dB
E = transducer efficiency in dB
L = fish length in cm
NF = noise figure in dB
NL = noise level in dB
R = range in meters
RS = receiving sensitivity in dB
TL = transmission loss in dB
TS = target strength of fish in dB
TVG = time-varied gain in dB (set equal to 2 TL)

There are two levels of input in the detection zone program. The first level asks for frequency in kHz, half power full angle of the beam in degrees, power of the sounder in Watts, fish size, water temperature, and salinity. The second level allows change of the sonar variables: SL, TS, DT, DI, NL, NF, BW, RS, α and V_T.

There is also file output, containing records for three levels of target strengths TS which is calculated from total fish length, TS-6, and TS-12: theta, range, y-coordinate, x-coordinate followed by level 1 and level 2 input data.

Screen output is graphics of the detection ranges, and level 1 and level 2 information. The screen output was captured to produce the next several figures. Figures 2 to 4 show a beam from a 70 kHz echosounder, with a circular transducer having half power half angle of 11 degrees or a full "beamwidth" of 22 degrees. Notice that as the threshold is lowered, proceeding to Figure 4, two series of sidelobes appear. The inner lines in the figures are due to the smaller target strengths. We expect to see the first sidelobe appearing in the detection volume at –35.14 dB below the maximum on axis return (VMAX), the second at –47.62 dB and so on. We can predict these levels using the beam direction factor. This relation was presented in the section on circular transducers.

Figure 5 shows a detection zone for a transducer with half the beamwidth of Figure 4, but with the same threshold. Notice that the beam is directed deeper and narrower.

Figure 6 shows the opposite with increasing beamwidth. In addition, there are more sidelobes on the narrower beam, fewer on the wider beam.

Figure 2 shows the commonly portrayed conical beam. Smaller voltage returns do not appear in Figure 2, but do in Figures 3 and 4. One must keep in mind that the beam is not actually a cone, but that the smaller voltages will be influential as the density of fishes increases. We will discuss this effect under the topic of echo integration. In addition, the expected signal from a fish is not a fixed quantity, but may vary as the fish undulates or changes aspect to the beam. Thus the simulated beam detection volume produced by CRTBEAM.EXE is a first approximation and useful mainly for visualizing the relationships of the sonar equation.

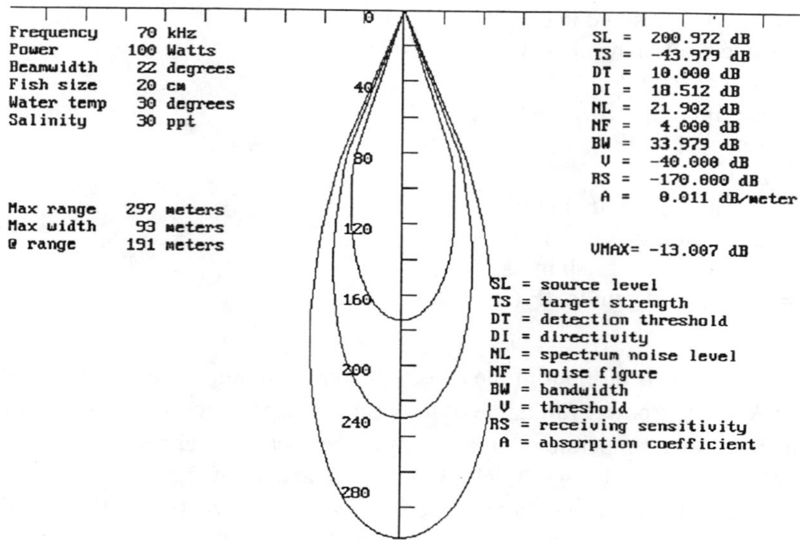

Figure 2. Single beam transducer detection zone, for an idealized 20-cm fish. Voltage threshold is well above sidelobe level, creating a conical detection zone in the upper 80 meters of depth. Output is from program CRTBEAM.EXE. Inner two series of lines represent smaller fish, with target strength decrementing 10 dB.

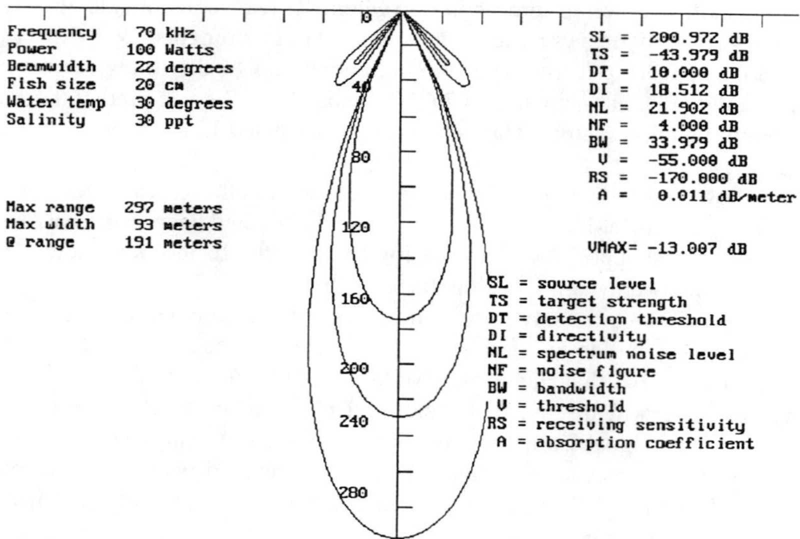

Figure 3. Single beam transducer detection zone, for an idealized 20-cm fish. Voltage threshold is below the first sidelobe level. Output is from program CRTBEAM.EXE. Inner two series of lines represent smaller fish, with target strength decrementing 10 dB.

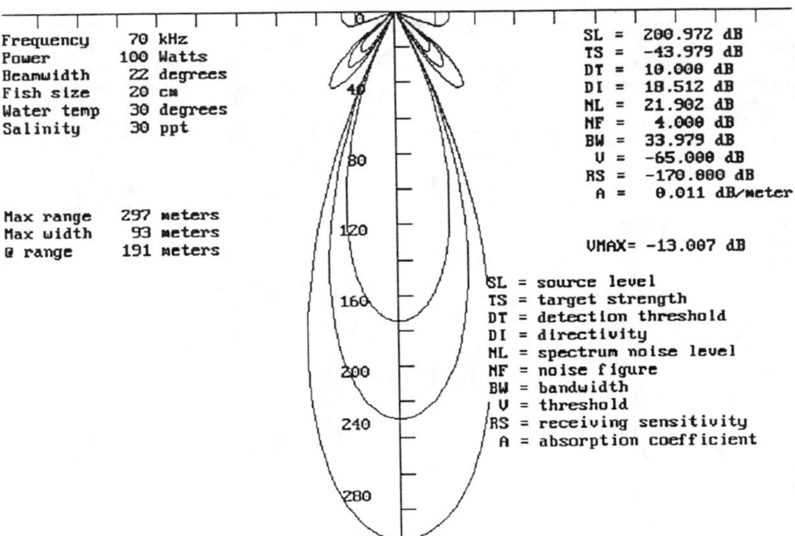

Figure 4. Single beam transducer detection zone, for an idealized 20-cm fish. Voltage threshold is barely below the second sidelobe level. Output is from program CRTBEAM.EXE. Inner two series of lines represent smaller fish, with target strength decrementing 10 dB.

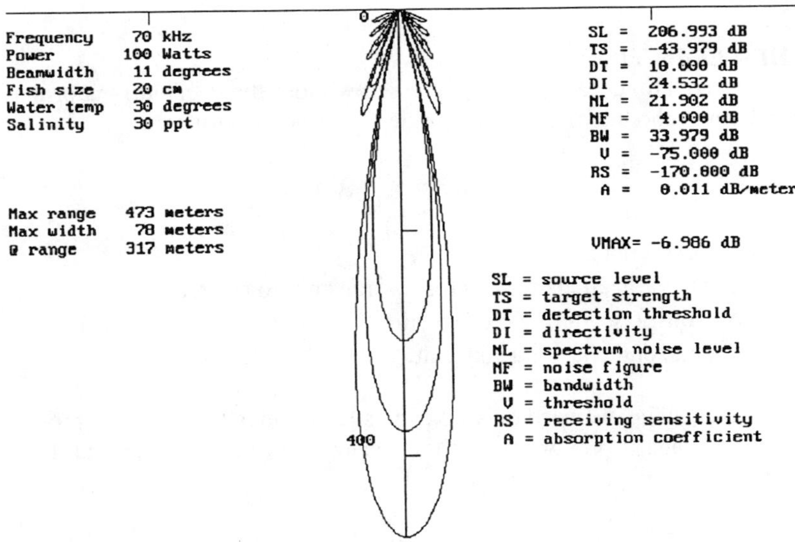

Figure 5. Single beam transducer detection zone, for an idealized 20-cm fish. Full beamwidth is 11 degrees. Output is from program CRTBEAM.EXE. Inner two series of lines represent smaller fish, with target strength decrementing 10 dB.

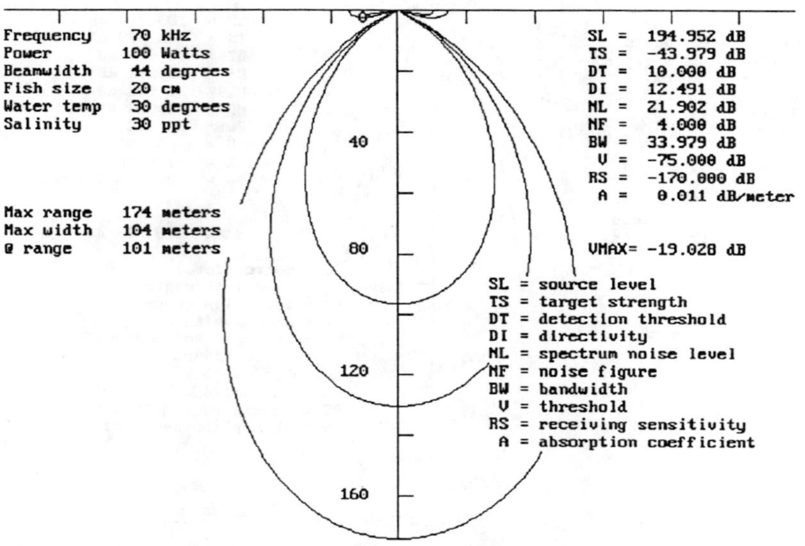

Figure 6. Single beam transducer detection zone, for an idealized 20-cm fish. Full beamwidth is 44 degrees. Output is from program CRTBEAM.EXE. Inner two series of lines represent smaller fish, with target strength decrementing 10 dB.

3.5 OTHER EQUATIONS

The sonar equations can also be expressed without the decibel transformation. An equation relating echo intensity from a single target can be written as

$$i = kb_t(\theta,\phi)b_r(\theta,\phi)\sigma$$

where i = acoustic intensity

$b_{t,r}(\theta,\phi)$ = beam pattern factor for transmission or reception

σ = acoustic scattering cross section

k = gain, source level, a constant

In terms of voltage, we need only take the square root since power is proportional to intensity, and voltage is proportional to the square root of power. Setting the reception and transmission beam pattern factors equal,

$$v = k' b(\theta,\phi)\sigma^{1/2}$$

Or one can express the root mean squared voltage squared return with k expanded, and use the backscattering cross section which when transformed is target strength, as

$$v^2 = p^2b^2(\theta)\sigma_{bs}rs^2g^2tvg^2 10^{-2\alpha R/10}R^{-4}$$

where p = pressure projected by transducer
rs = receiving sensitivity
tvg = time-varied-gain
g = fixed receiver gain
R = range

This last equation assumes that the directivity factor is symmetric, and that the reception and insonification patterns are the same

4. CALIBRATION AND TRANSDUCER CONFIGURATIONS

4.1 CALIBRATION, APPLICATIONS AND UNCALIBRATED ANALYSES

Echosounders, transducers and receivers must be calibrated (Urick, 1983). Calibration mainly means to measure the source level and the receiving sensitivity. In addition, the beam directivity pattern may be plotted or measured. The fixed gains and the time-varied gain on the receiver can be checked as well. The two major methods of calibration are via hydrophone (Johanneson and Mitson, 1983, for example) and by standard targets (MacLennan, 1981; MacLennan, 1982; MacLennan and Dunn, 1984). Using standard targets is difficult for single beam units in the field. Generally, a single-beam transducer will only allow an estimate of an on-axis measurement, based on the largest returns from a standard target. The best way to perform target calibration for a single-beam system is in a control tank. In the field it is easier to look at standard targets with dual-beam or split-beam echosounders. This is because the beam pattern effect can be removed since angular coordinates are estimated by amplitude or phase comparisons. The most common forms of targets are tungsten carbide and stainless steel spheres. Although not as highly recommended, standard ping-pong balls have often been used. Analyses which do not require complete calibration require system stability and accurate time-varied gain. A beam pattern description is also assumed. In general, any system should be checked for some minimal suite of measurements, which may require the assistance of an electronics technician.

One means of calibrating echo integration values for abundance estimates (see section on echo integration below) is by analysis of sections of data where densities are low enough to allow the counting of individual fish targets. It is necessary to have some means of computing the volume sampled during counting of the individual fish targets, such as duration-in-beam (see section below on duration-in-beam). This method essentially finds the average number of times a fish target is "seen" by the echosounder and estimates an "average" volume. Alternatively, one could use the method of deconvolution to arrive at a calibration measure (Hedgepeth et al., 1989, for example). This procedure is discussed more fully below. Thorne (1983) presents several references to studies using calibration of echo integration by fish counts.

Uncalibrated analyses include presence/absence studies, relative echo integration (that is, without target strength calibration factor), and relative echo counts (Thorne et al., 1990, for example). Measurements of Doppler shift use frequency shifts and do not require calibration of the echosounder and transducer. Although these types of analyses can be useful for fisheries management purposes, quantitative echo surveys require some type of calibration, e.g., by echo counting techniques, by in field or experimental target strength measurements, and by transducer directivity pattern measures.

4.2 TRANSDUCERS: GENERAL DESIGN AND APPROACH

The transducers used for fisheries research are commonly constructed from piezoelectric ceramic materials, with frequencies from 38 kHz to 1 MHz, and may be powered up to 4 kW. Single-beam, dual-beam and split-beam transducers are available with either circular or elliptical patterns. Dual-beam transducers transmit on the narrow beam element and receive on both narrow and wide-beam elements. Although elliptical shaped beams can provide broad coverage in one plane in shallow water or over confined sites, circular beams are probably the most common designs. The discussion that follows focuses on theoretical circular single-beam transducers. Both dual-beam elements can be modeled as two different single beams, and the split-beam can also be modeled as a single-beam, but with phase information entering by splitting the transducer into four sections or quadrants.

4.2.1 Single-Beam Systems

Real transducers perform more or less closely to the following theoretical model. Calibration of the beam pattern can tell the user if it is necessary to substitute empirical expressions for the theoretical ones when calculations require a description of the transducer beam pattern.

A circular transducer has a directivity factor which relates the angle off-axis, θ , to sound intensity or pressure squared (voltage squared). The direction factor, $h(\theta)$, can also be used to predict the peak pressure at θ in the sound field. This can be expressed as a function of a Bessel function of the first kind, order 1, denoted by $J_1(\ )$. When we use the same transducer to receive echoes, the direction factors for transmission and reception are combined by assuming they are identical and therefore squaring $h(\theta)$. In terms of power, the directivity is denoted by $b(\theta)$. In terms of voltage, the squared direction factor can also be denoted by $b(\theta)$. In particular, the direction factor is

$$h(\theta) = \left|\frac{2J_1(u)}{u}\right|$$

where $u = ka \sin(\theta)$ and the directivity is

$$b(\theta) = h^2(\theta) = (2J_1(u)/u)^2$$

The derivation for the direction factor models a piston type transducer with radius a, and wavenumber k (angular frequency divided by sound speed). The function, $h(\theta)$, with the Bessel function, follows from the phase interaction over the distance of the transducer face (Kinsler et al., 1982). The two-way direction factor $b(\theta)$ has nulls where $b(\theta)$ is equal to 0. This implies that there are local maxima of $b(\theta)$, called sidelobes.

Sidelobes can be seen in Figures 7a and 7b. Figure 7a shows a planar slice of the directivity of a typical transducer from three-dimensional space, superimposed over the output of CRTBEAM.EXE, for a transducer with ka about 8.28. Figure 7b shows the relation between an angle in radians and the two-way direction factor for a circular transducer with a ka of 8.28.

The first few values of u for nulls and sidelobes are:

u (nulls)	u (sidelobes)	b(u) at sidelobes	(and in dB)
3.8317	5.1356	0.0174981	–35.14 dB
7.0156	8.4172	0.00415814	–47.62 dB
10.1735	11.6198	0.00160066	–55.91 dB
13.3237	14.7960	0.000779466	–62.16 dB

The nulls and sidelobes correspond to zeroes of the Bessel functions (Abramowitz and Stegun, 1965). The sidelobes occur where the derivatives of $h(\theta)$ or $b(\theta)$ are 0, i.e.,

$$\frac{d\frac{J_1(u)}{u}}{du} = -\frac{J_1(u)}{u} = 0$$

In other words, the zeroes of the first order Bessel function give the nulls, and zeroes of the second order Bessel function give the sidelobes.

Before using a theoretical model, actual transducer directivities should be compared. There are several ways of comparing the actual transducer function to the theoretical prediction. Most rely on measurements taken when the transducer is activated. One procedure is to take data from one or several cross section "cuts" by rotating the transducer and using a second hydrophone to transmit to the transducer.

The transmit pattern is then assumed to be similar to the receive pattern. The direction factor (specifically ka as a single parameter) can then be fit, e.g., using a least squares criterion. Both the reception and transmission patterns can be measured and fit using a model. In addition, ka can be estimated from the half power level, using the θ that corresponds to –3dB on a one-way beam pattern,

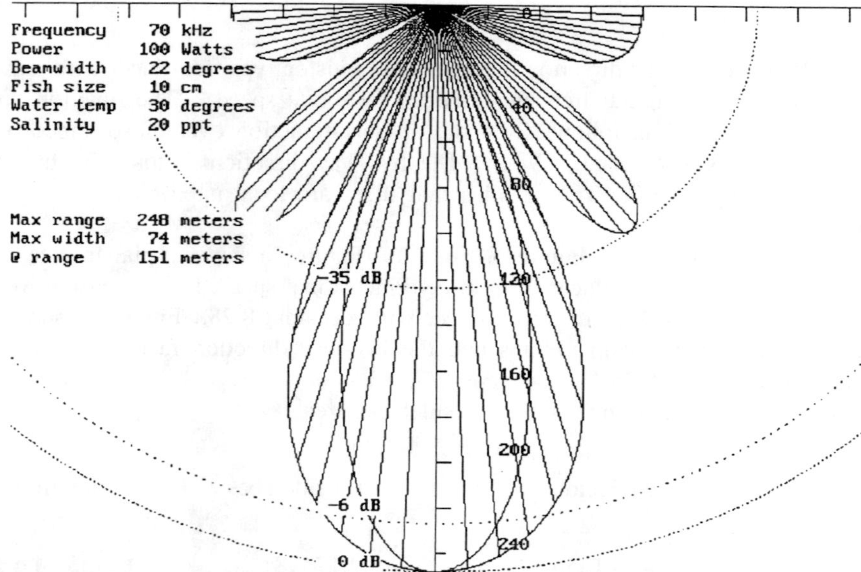

Figure 7a. Directivity pattern of the two-way beam factor, b, is superimposed over the detection volume. The directivity pattern is presented in polar coordinates, with lines at angles every 3 degrees. ka, the wave number times the effective radius of the transducer, is 8.28 for this example. The scale of the beam pattern is dB, while the detection volume scale is meters.

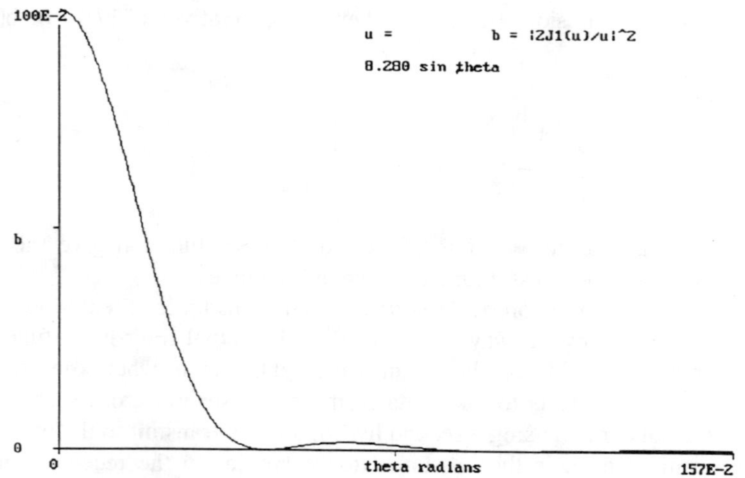

Figure 7b. Directivity pattern factor, b, represents the two-way voltage amplitude reduction, assuming the transmission and reception patterns are alike. ka, the wave number times the effective radius of the transducer, is 8.28 for this example. The first and second sidelobes are barely evident.

$$0.5 = b(\theta) = \left|\frac{2J_1(ka \sin\theta)}{ka \sin\theta}\right|^2$$

The likelihood or probability density function (PDF) of the beam pattern can be based on the above theoretical function. The PDF is used in the deconvolution technique to estimate the fish backscattering cross section distribution and fish abundance. A derivation of the PDF is in the deconvolution section. The fish are assumed to be distributed randomly in the beam, and the likelihood expression for a value of $b(\theta)$ is computed as a function of θ. Computer iteration is used to map $b(\theta)$ to θ. Peterson et al. (1976) plotted the likelihood function, $f_B(b)$, versus b for a transducer with ka of 2π. Clay and Medwin (1977) (in their Appendix 7) outline the steps to numerically estimate this likelihood, using Newton's method to estimate θ from $b(\theta)$.

An iterative technique was used to solve the likelihood of a circular transducer b(u) (Hedgepeth et al., 1994). It was implemented in a C-language program, PDFBPLOT.C (Hedgepeth, 1993; CRSP Working Paper No.105) with sample output shown in Figure 8.

The program includes provision for the first sidelobe to –47.6 dB and could be extended for several more sidelobes, but this usually is not necessary because signals collected below the first sidelobe are usually close to the noise level. The method used in PDFBPLOT starts on the acoustic axis and moves toward the sidelobes. If sidelobes are present, the probabilities for all b are summed. The probability of the beam pattern factor, as a function of u, i.e., of ka $\sin(\theta)$, is derived in the section below on deconvolution. The PDF is

$$f_B(u) = \frac{u^3}{8(ka)^2 J_1(u) J_2(u)\sqrt{\left(1-(u/ka)^2\right)}}$$

The computer program PDFBPLOT.C needs an input of ka, with output to the screen (Figure 8) and file. Notice that the PDF peaks locally at the first sidelobe, b equal to 0.0175, and that the probabilities are small for b greater than the sidelobe. One can see that the probability of b over the main lobe is not quite uniform, but that it increases gradually going from a b of 1 toward smaller b's. Echoes found in the sidelobes are harder to detect because they return small signals, which can be lost in the presence of noise. Generally, the probability up to the level of the first sidelobe (Figure 9) is what is used.

Empirical functions can be fit to the beam pattern, especially when working above the level of the first sidelobe. One simple nonlinear equation which can be fit for two parameters, a and b, is

$$b(\theta) = \exp(a\theta^b)$$

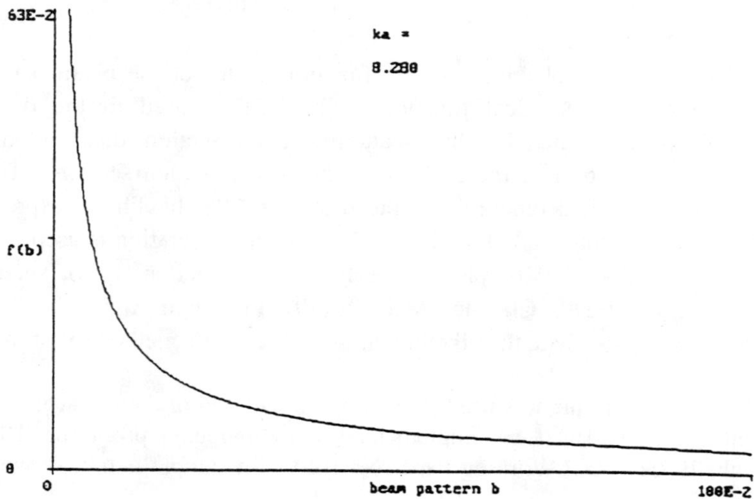

Figure 8. The probability distribution of the beam pattern factor is plotted to slightly above the level of the first sidelobe.

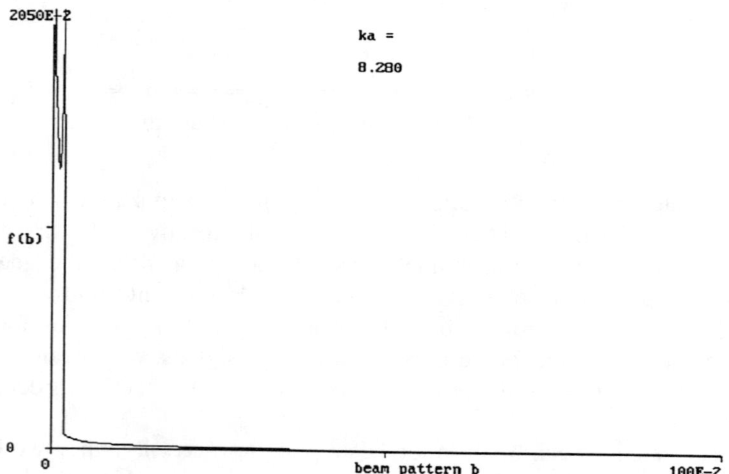

Figure 9. The probability distribution of the beam pattern factor is plotted to slightly above the level of the second sidelobe.

This function may, in fact, fit a directivity pattern better than the theoretical one presented above because real transducers often depart from ideal behavior. Both empirical and theoretical relationships can be used in deconvolution and in other "inverse" procedures which attempt to estimate target strength and fish abundance. These methods are explained in a following section. Although relatively recent in use, deconvolution procedures predate the use of multibeam systems for the estimation of the acoustic size of fish targets. Some of the technical advantages of multibeam systems will be considered in the next section.

4.3 MONOPULSE OR MULTIBEAM SYSTEMS: DUAL-BEAM AND SPLIT-BEAM

Another way to find the acoustic size of fish is by the use of multiple beams, as mentioned above. Like single-beam, multibeam estimation requires knowledge of the strength of the beam pattern. The system hardware is more complex because the angular coordinates of the fish target are estimated. The following presents two multi-beam systems for comparison to the single-beam process.

The simplest multi-beam system uses two beams called dual-beam. A more complex system is a four-lobed system called split-beam. Both the dual-beam and split-beam systems were first introduced by Ehrenberg (1979b). Both systems would be considered in radar terminology to be monopulse systems.

Monopulse systems can (1) estimate the fish's position in the water, and (2) using the estimate of fish position, estimate the acoustic cross section of an individual fish. The term "monopulse" means using simultaneous beams to determine the angular direction of a target. Estimates are usually made one echosounder pulse at a time, thus the term "monopulse".

The accepted IEEE definitions (Sherman, 1984) are:

4.3.1 Amplitude-Comparison Monopulse

This is a form of monopulse employing receiving beams with different amplitude-versus-angle patterns. If the beams have a common phase center, the monopulse is pure amplitude-comparison; otherwise, it is a combination of amplitude-comparison and phase-comparison.

4.3.2 Phase-Comparison Monopulse.

This is a form of monopulse employing receiving beams with different phase centers, as obtained, for example, from side-by-side antennas or separate portions of an array. If the amplitude-versus-angle patterns of the beams are identical, the monopulse is pure phase-comparison; otherwise it is a combination of phase-comparison and amplitude comparison.

The definitions of monopulse imply that multiple beams are used only on receiving the reflected sound. The transducer transmits as if it were a single beam. Dual-beam echosounders utilize the amplitude-comparison monopulse approach. Dual-beam transducers use two receiving beams with different directivity patterns. Normally these are called the wide and narrow beams. The ratio of the received voltages determines off-axis angle to the fish target. Voltage amplitude from the narrow beam is used to estimate the acoustic size of the fish. Split-beam sounders use a four-lobe (or four quadrant) approach. Most split-beam sounders implement their angle estimation procedure using a phase-comparison. Amplitude information from the composite beam is used to measure biomass or backscattering intensity.

4.3.3 Dual-Beam, Amplitude Approach

The dual-beam method (Ehrenberg, 1978a) is a widely used method for processing echoes from individual fish. The method uses two receiving transducers of different sizes. Because the receiving transducers are of different sizes, the beam pattern directivities are different. The two transducers can be either side-by-side, or a single circular array of elements can be configured so that there is an inner, circular array which makes the receiving wide beam, and the full circular array makes the narrow beam. The full array transmits a narrow beam, and the receiver collects the returns on both wide and narrow-beam.

Dual-beam theory is relatively simple. Section 3 explained that the detected voltage is related to the position of the target and its acoustic size:

$$v^2 = P_t b^2(\theta) G \sigma$$

The narrow beam transmits, but both the wide and the narrow beams receive. Then,

$$v_n^2 = P_t b_n^2(\theta) G_n \sigma$$

$$v_w^2 = P_t b_n(\theta) b_w(\theta) G_w \sigma$$

where v_n = narrow-beam receiver voltage
v_w = wide-beam receiver voltage
P_t = transmitted power
$b_n(\theta)$ = beam pattern factor, narrow beam
$b_w(\theta)$ = beam pattern factor, wide beam
G_n = receiver gain, narrow-beam channel
G_w = receiver gain, wide-beam channel

The narrow and wide receiver gains, G_n and G_w, are measured during system calibration. We can solve for the beam pattern factor, $b_n(\theta)$, after eliminating the common factors from the two equations: i.e., fish target size, σ, and the transmitted power, P_t.

The relationship between the wide and narrow beam pattern factors is first estimated using the directivity patterns of the wide and narrow beams, taking values every 0.5 degrees to about –10 dB (Ehrenberg, personal communication), and using the following empirical relationship:

$$b_n(\theta) = \left| \frac{b_n(\theta)}{b_w(\theta)} \right|^k$$

or, in dB:

$$B_n(\theta) = k(B_n(\theta) - B_w(\theta))$$

where k is fit using linear regression through the origin. The equation is expressed in dB and the data is from the beam pattern. The ratio of the squared voltages from the wide-beam and narrow-beam receivers and the ratio of their gains can then be used to estimate $b_n(\theta)$. That is,

$$\frac{v_n^2}{v_w^2} = \frac{P_t b_n^2(\theta) G_n \sigma}{P_t b_n(\theta) b_w(\theta) G_w \sigma} \Rightarrow \frac{b_n(\theta)}{b_w(\theta)} = \frac{G_n v_n^2}{G_w v_w^2} \Rightarrow \left| \frac{b_n(\theta)}{b_w(\theta)} \right|^k = \hat{b}_n(\theta)$$

Once $b_n(\theta)$ is estimated the acoustic cross section, σ, or the target strength can be estimated using the sonar equation.

4.3.4 Split-Beam, Phase Approach

The split-beam method is another widely used method to extract acoustic size information directly from single targets by estimating the angles to the fish, using the single beam directivity. The transducer array is usually divided into four quadrants each of which transmits a composite narrow beam. Received echoes from quadrant pairs are processed separately in order to estimate the direction to the target. The sum from all quadrants is used for echo amplitude. By processing the phase difference between halves of the transducer (comparing two sets of adjacent quadrant pairs), the arrival angles α (left or right of the main axis) and ß (above or below the main axis) can be estimated. Acoustic waves, returning from targets, cause the phase angle of the electrical output signal to differ between the four quadrants depending on the target's location. Once the arrival angles α and ß are estimated, one can use the radar equation to estimate the acoustic size of the fish:

$$\sigma = P_t b^2(\alpha,\beta) G / V^2$$

4.3.5 Comparison: Dual-Beam, Split-Beam and Single-Beam

Although dual-beam is simpler and less costly, the split-beam is theoretically superior for estimating target strength. Either system should be equally effective using echo integration as that method uses the composite return. Ehrenberg (1978b) modeled noise effects on a dual-beam sounder and compared split-beam and dual-beam processors using simulation and analytical models. Although the separate split-beam angle estimates were shown to be unbiased with a Gaussian distribution by both

Ehrenberg and by Sherman (1984), noise will cause bias in the ultimate target strength estimate. Molloy (1977) is an example of a particular result.

Today, differences in typical dual-beam directivities are smaller than those used in the original simulations. BioSonics, Inc. commonly manufactures 6 degree x 15 degree and 3.5 degree x 10 degrees dual-beam transducers. These configurations may be less biased than suggested by the simulations. Field applications also show that split-beam and dual-beam results compare closely. Both systems can produce erroneous results if beams are misaligned or eccentric and if phase or amplitude are not detected properly.

Ehrenberg's (1981) results suggest that the data from targets near the acoustic axis, with either dual or split-beam technology, are superior compared to targets farther from the acoustic axis. For example, if noise is –70 dB, target strength below –50 dB may be severely biased, especially when beam pattern threshold in a dual-beam system is below –6 dB.

Both split-beam and dual-beam transducers function as a single-beam when collecting echo integration data. The major difference between single and multiple beams is the approach to single target statistics such as target strength. Multibeam methods are able to estimate the acoustic size of fish targets with far fewer samples than single-beam methods require.

Both multibeam and single-beam methods assume that voltage data is taken from a single fish for target strength estimation. In the previous section on general acoustic theory, vertical and horizontal resolutions were shown to be dependent on fish position. As fish become denser and more closely packed, their signals begin to overlap. Differentiation of individual fish targets from multiple fish targets and noise is performed by the signal processing software. In multiple target environments, we generally employ echo integration, which is considered in the next section.

5. METHODOLOGIES FOR DATA PROCESSING

5.1 ECHO INTEGRATION FOR BIOMASS DETERMINATION

Echo integration is a way of using the mean of the squared voltages as a measure of the density of fishes. Both single-beam sounder and multiple-beam echosounders can provide the raw data for echo integration.

The assumptions for echo integration are (Urick, 1983)

(1) Straight line propagation paths

(2) Random, homogeneous distribution of scatterers throughout the volume at any insonification

(3) Large numbers of scatterers in the volume

(4) Relatively short pulse length

(5) No multiple scattering

When the fish (and individual fish structures) are located randomly throughout the volume of water, their intensities add. Intensity is proportional to volts squared. The average volts squared is proportional to the average intensity from fish in a volume of water. With knowledge of the mean backscattering cross section, one can then estimate the average number of fishes in the volume of water.

In Section 3 the sonar equation was developed from which the voltage of a single target can be expressed as

$$v^2 = P_t b^2(\theta) G \sigma_{bs}$$

When this expression is expanded to portray all gains and losses which are included in the gain term, G, we can write it as

$$v^2 = p^2 b^2(\theta) \sigma_{bs} rs^2 g^2 tvg^2 10^{-2\alpha R/10} R^{-4}$$

where p = rms pressure
g = fixed gain
tvg = time varied gain
R = range
α = absorption coefficient
rs = receiving sensitivity
v = rms voltage (peak voltage is $v/.707$)

To convert this expression to the echo integration relationship, see Ehrenberg (1979a), "The average squared voltage at a time corresponding to a range R is equal to the average squared voltage for a single fish times the product of the sampling volume and the fish (or biomass) density. The sampling volume in this case is a hemispherical shell of thickness $c\tau/2$, where c is the velocity of sound and τ is the length [duration] of the acoustic pulse. Therefore, the average squared voltage for range R is"

$$\overline{v^2} = \delta \cdot \text{Vol} \cdot p^2 \overline{b^2}(\theta) \overline{\sigma}_{bs} rs^2 g^2 tvg^2 10^{-2\alpha R/10} R^{-4}$$

where δ = fish density (fish/m^3)
Vol = volume of a hemispherical shell = $2\pi R^2 c\tau/2$

The thickness of the volume, $c\tau/2$, is explained by the assumption that the front of the echosounders pulse will arrive from the back of the volume insonified at the same time as the back of the pulse arrives from the front of the volume.

The average beam pattern factor can be computed by three methods. The first method uses numerical integration (Ehrenberg, 1979a) to approximate the following integral and obtain a mean value:

$$\bar{b}_2(\theta) = \int_0^{\pi/2} b^2(\theta)\sin(\theta)d\theta$$

For example, trapezoidal numerical integration was done for a mesh size of $\pi/2000$ and ka = 8.28; the average beam pattern factor is 0.0137. A C language program that estimates the average beam pattern factor is in the collection of acoustic-related programs (Hedgepeth, 1993, CRSP Working Paper No. 105). Urick (1983) presents two other methods for estimating the beam pattern factor. Rephrasing these in non-dB terms

$$\bar{b}^2(\theta) = \frac{5.8884(1/ka)^2}{2\pi}$$

which gives 0.0137 for our example of ka = 8.28. The third formula is

$$\bar{b}^2(\theta) = \frac{2.27115\, y^2}{2\pi}$$

where y is the half power beam angle in radians. If the measured –3dB half angle was 11.25 degrees on the one way directivity diagram, the beam factor is 0.0139 with this equation.

In a following section on applications we give an example based on echo integration. The average backscattering cross section may be estimated by many of the ways discussed in the target strength section of this document. However, the "echo integration" equation must be rearranged so that the object to be estimated is density, δ. The time-varied gain, tvg, is intended to correct for spreading losses. However, this correction frequently is not ideal. In those cases, one tries to measure the tvg using the calibration oscillator since it provides a constant level into the receiver. Then a correction factor can be applied. Ehrenberg (1979a) explains this method as follows. One first assumes that the tvg (and thus voltage received using the calibration oscillator) is correct at some range. Then one computes the ideal voltages based on the result of that particular range, for all ranges. The voltages received using the calibration oscillator are then compared to the ideal voltages and a correction factor can be included in the echo integration equation. For ranges R_1 to R_2, the estimate of density (δ) is

$$\hat{\delta} = \frac{\bar{v}^2}{p^2\bar{b}^2(\theta)\bar{\sigma}_{bs}\, rs^2 g^2 \dfrac{\int_{R_1}^{R_2} Vol\ tvg^2\ 10^{-2\alpha R/10}\ R^{-4}\ dr}{R_2 - R_1}}$$

This expression simply says that each measure of v^2 can be used to compute a density estimate, and that the best estimate is the average of many per ping estimates. If the tvg is ideal, the expression simplifies to

$$\hat{\delta} = \frac{\overline{v}^2}{p^2\overline{b}^2(\theta)\overline{\sigma}_{bs} rs^2 g^2 c\tau\pi \dfrac{\int_{R_1}^{R_2} R^2 dR}{R_2 - R_1}}$$

The expression would be much simpler if the tvg that was applied correctied for the spreading loss in only one direction. This type of tvg is called 20logR tvg (as opposed to the 40logR tvg used in echo counting).

$$\hat{\delta} = \frac{\overline{v}^2}{p^2\overline{b}^2(\theta)\overline{\sigma}_{bs} rs^2 g^2 c\tau\pi}$$

Depending on how the signal processing is done, tvg can be applied using analog devices, combination analog-digital devices, or by software. Therefore one particular "echo integration" equation is not unique. For example, if one estimated the back-scattering cross section in terms of average estimated peak on-axis voltage squared, s^2, the expression could be further simplified.

$$\hat{\delta} = \frac{\overline{v}^2}{\overline{b}^2(\theta) s^2 g^2 c\tau\pi}$$

Urick (1983) presents echo integration theory by explaining "volume reverberation". Using yet another different approach, he integrates the expression for directivity ($b^2(\theta)$) over the volume being considered, instead of estimating the expected directivity. This is essentially the same equation because $2\pi \sin\theta \, d\theta$ is the equivalent increment of the solid angle in his formulation. Burczynski (1982) also presents an excellent description of echo integration in his fisheries acoustics manual.

5.2 ECHO COUNTING (DECONVOLUTION AND DURATION-IN-BEAM)

5.2.1 Deconvolution

The application of deconvolution to acoustic echo signals has a rich history. The procedure attempts to solve the "inverse problem". The "inverse problem" is to use the beam pattern PDF as a "kernel" to unfold the distribution of on-axis voltages from the raw peak voltages. The beam pattern PDF is the likelihood of values of the beam pattern in the volume occupied by fish. Ehrenberg (1982) showed that the Craig and

Forbes technique (deconvolution using 10 steps per decade of voltage) requires sample sizes greater than about 800 echo voltage peaks to avoid biases greater than 20%. Examples of early work using deconvolution appear in Ehrenberg (1972), Petersen et al. (1976), Robinson (1979), Huang and Clay (1980) and Ehrenberg (1983). Some of these researchers tried to reduce bias by assuming a PDF which would describe the on-axis voltage distribution. Clay (1983) and Stanton and Clay (1986) showed that their deconvolution method should be able to recover any PDF of on-axis voltages, without the need to assume one beforehand. In our case study (Hedgepeth and Thorne, 1989; CRSP Working Paper No. 65), we apply Stanton and Clay's reasoning.

Ehrenberg (1972) used a polynomial of degree n to approximate the distribution of on-axis voltages. A minimization technique fit the parameters of this polynomial. Robinson (1978) used an alternative, but similar, technique where the unknown distribution was subdivided into a number of intervals.

Rudstam et al. (1988) found remarkable agreement between length-target strength relations from both the modified Craig and Forbes technique and from Rician PDFs fit after deconvolution. Some unconstrained methods, however, such as the modified Craig and Forbes technique (Lindem, 1983), set negative estimates to 0 in an ad hoc fashion. In addition, the user of any inverse method, like deconvolution, should be aware of any propagation of errors effect which will produce artificial modes in the resulting on-axis voltage (or target strength) distribution.

In addition to target strength estimation, deconvolution can give estimates of abundance, especially where good quality data are available (Stanton and Clay, 1986; Hedgepeth and Thorne, 1989). This is because the beam pattern PDF is used to infer the total numbers.

Key to the deconvolution analysis is the beam pattern PDF. In order to construct a beam pattern PDF, we first assumed that the fish were uniformly distributed in the insonified volume of water into which the transducer transmitted sound. Then the PDF of the angle off-axis, θ, could be described by $\sin(\theta)$. This relationship can be understood by assuming that the density of θ is proportional to the circumference of a circle, because the scatterers are uniformly distributed. The circumference of a circle is 2π multiplied by the radius, and the radius equals the range multiplied by $\sin(\theta)$. θ lies between 0 and $\pi/2$; hence, the PDF of θ is $\sin(\theta)$. The beam pattern direction factor then can be thought of as a random variable which is the function of the random variable θ. We could then construct a PDF of the beam pattern directionality by a transformation technique.

To find the two-way beam pattern PDF, $f_B(b)$, one begins with the two-way direction factor, b, as a function of θ,

$$b = g(\theta)$$

The inverse function is:

$$g^{-1}(b) = h(\theta)$$

The beam pattern PDF can be solved analytically, using methods in Mood et al. (1974), which leads to

$$f_B(\theta) = \left|\frac{dg^{-1}(b)}{db}\right| \sin(\theta)$$

which expresses the beam pattern PDF as a function of θ, and not of b. One could also estimate values for the derivative from a beam pattern directivity pattern.

5.2.1.a *Theoretical Circular Transducer PDF*

A theoretical circular transducer has a two-way directivity of

$$b = \left|\frac{2J_1(ka \cdot \sin\theta)}{ka \cdot \sin\theta}\right|^2,$$

with a corresponding PDF of

$$f_B(\theta) = \frac{ka \cdot \sin^3(\theta)}{8J_1(ka \cdot \sin\theta)J_2(ka \cdot \sin\theta)\cos\theta}$$

5.2.1.b *Empirical Circular Transducer PDF*

Other PDFs can be based on empirical relationships for the beam pattern factor, b. Empirical relationships can be fit to the measured transducer directivity pattern. These work adequately for non-ideal circular transducers (Ehrenberg, 1981). A simple expression for the beam directivity of a non-ideal transducer is an exponential function. This can provide good results above the level of the first sidelobe. (See Section 4 for a description of transducer beam sidelobes). For example, the empirical function might be

$$b = \exp(\alpha\theta^{\beta})$$

where α and β are fit using a least squares criterion and the exponential, empirical function. The derivative of the inverse of the exponential function is

$$\frac{dg^{-1}(b)}{db} = \frac{1}{\alpha\beta b}\left[\frac{\ln(b)}{\alpha}\right]^{1/\beta-1}.$$

The PDF computed with this inverse (Mood et al., 1974) is

$$f_B(b) = \frac{1}{\alpha\beta b}\left[\frac{\ln(b)}{\alpha}\right]^{1/\beta-1} \sin\left[\frac{\ln(b)}{\alpha}\right]^{1/\beta}$$

5.2.1.c *Deconvolution Procedure*

The deconvolution procedure is one of the inverse methods used to estimate target strength and fish density. The basic data are fish echo voltage peaks collected at the echosounder's receiver. These voltage peaks are then placed in voltage bins or intervals, so that the voltage distribution can be measured. Deconvolution is an inverse process which uses this peak voltage collection and the beam pattern PDF to estimate the on-axis voltage distributions. As an example of what is meant by voltage bins, our earlier methods placed the echo peak voltages into bins of .01 or .02 volts. Cubic spline interpolation was used to estimate the distribution between these relatively wide intervals. We use the natural 4096 bins of the 12-bit AD converter, where 4096 represents 9.998 volts and 0 is 0 volts, which has much finer resolution.

The deconvolution method operates as follows. The PDF of the raw echo voltage peaks can be expressed as (Clay and Medwin, 1977)

$$f_V(v) = \int_0^1 (1/b) f_B(b) f_s(v/b)\,db$$

or, with a change of variables,

$$f_V(y) = \int_0^1 f_B(x) f_s(y-x)\,dx$$

This is the form of a convolution integral (Mood et al., 1974) which, when discretized is,

$$f_V(n) = a\sum_{m=0}^{\infty} f_B(m) f_s(n-m)$$

where b = two-way beam pattern factor = $\exp(-am)$
v = raw echo peak voltage
s = on-axis echo peak voltage, $e_0\exp(-a(n-m))$
v = echo peak voltage, $e_0\exp(-an)$
e_0 = maximum voltage
n, m = integers 0, 1, 2, ..., maxm (= maxn)
a = $\ln(10)/N$
N = number of steps per decade

This convolution integral can be solved in several ways. One, transform the relationship using Z-transforms and polynomial division and then, backtransform (Clay, 1983; Hedgepeth and Thorne, 1989). Equivalently, a system of equations can be solved. Perhaps the system of equations is easier to see when the system is taken from the discretized convolution integral:

$$f_V(0) = a[f_B(0)f_S(0)]$$

$$f_V(1) = a[f_B(0)f_S(1) + f_B(1)f_S(0)]$$

$$f_V(2) = a[f_B(0)f_S(2) + f_B(1)f_S(1) + f_B(2)f_S(0)]$$

$$f_V(3) = a[f_B(0)f_S(3) + f_B(1)f_S(2) + f_B(2)f_S(1) + f_B(3)f_S(0)]$$

$$f_V(4) = \ldots$$

$$\ldots$$

Gaussian elimination is used. Beginning with the larger beam factors and voltages work through the matrix in a top-down procedure. The method is simple and equivalent to Z-transform technique. It is possible for the solution, f_S, to be negative, using this method. In addition, errors propagate through the system of equations.

Methods which include constraints can insure a positive solution. Matrix notation will be used to pose the problem to include a constraint such as non-negativity. Let A be the square matrix

$$A = \begin{matrix} af_B(0) & 0 & 0 & 0 & 0 \\ af_B(1) & af_B(0) & 0 & 0 & 0 \\ af_B(2) & af_B(1) & af_B(0) & 0 & \cdots \; 0 \\ \cdots & & & & \\ af_B(m\ max) & af_B(m\ max-1) & & & \cdots \; af_B(0) \end{matrix}$$

let x be the observation vector

$$x = [af_S(0)af_S(1)af_S(2) \;\ldots\; af_S(nmax)]^T \;,$$

and let the on-axis voltage peak PDF be the vector c

$$c = [af_V(0)\; af_V(2)\; af_V(3) \ldots\; af_V(nmax)]^T \;.$$

The discretized convolution written in matrix notation is

$$c = Ax$$

The problem is to minimize $\|A \cdot x - c\|$ subject to the constraint $x \geq 0$. Our experience with the recursive, unconstrained deconvolution has shown that it can produce some negative values for estimates of on-axis voltage peak probabilities. We developed a computer program to perform non-negative, non-linear least squares deconvolution estimation, where the estimates are constrained to be positive by using Box's complex method (Jacoby et al., 1972; Bunday, 1984). This is a direct search method in which the basic algorithm computes a number of convolutions, taking the convolution with the smallest squared deviations between the convolution estimates and the observed data. One of the starting x vectors is the one solved by Gaussian elimination.

A new method used for estimation of the unobserved voltage distributions uses the EMS (Expectation Maximization and Smoothing) algorithm (Hedgepeth, 1994). The EMS algorithm has been used for stereology and positron emission tomography (Silverman et al., 1990). We applied this technique and measured its performance by comparing results to the Gaussian Elimination and Smoothing (GES) (Craig and Forbes type analysis) using a sonar simulation to create the observed voltages. The sonar simulator (Hedgepeth, 1993; CRSP Working Paper No. 105) is a C language program which uses a length distribution of fishes to generate observed voltages. The distribution of fish lengths originates either from a population simulation or can be input from sample information. Fish length is transformed to a mean backscattering cross section, which is a parameter of a sampling distribution to generate voltages for a particular fish. The sonar itself is presumed to be a from-surface downlooking unit. All parameters are variable but, for the present simulation, a fixed group of parameters was used.

The EMS algorithm performs better than a smoothed Gaussian elimination routine, when used to estimate the distribution of "on-axis voltages" using "observation voltages" obtained from a fisheries sonar simulator (Figure 10). Sensitivity to parameter ka had been previously investigated by Hedgepeth and Thorne (1989) and the importance of accurately estimating the kernel function remains true, especially when using deconvolution to estimate numbers of fish. Stanton and Clay (1986) suggest using exponentially wide bins in order to avoid undersampling higher voltage bins. This suggestion appears to be useful when performing deconvolution by traditional Gaussian elimination with smoothing (GES). On the other hand, the EMS algorithm performs better than the GES with linear bins. In addition, the error in estimating the number of fish was less with the linear binned EMS than with the exponentially wide binned EMS. Perhaps the main contribution of the EMS inverse solution is that it appears to perform better with fewer samples than the Craig-Forbes technique. In addition, it is not strictly a "deconvolution" procedure except when bins are chosen with exponential widths.

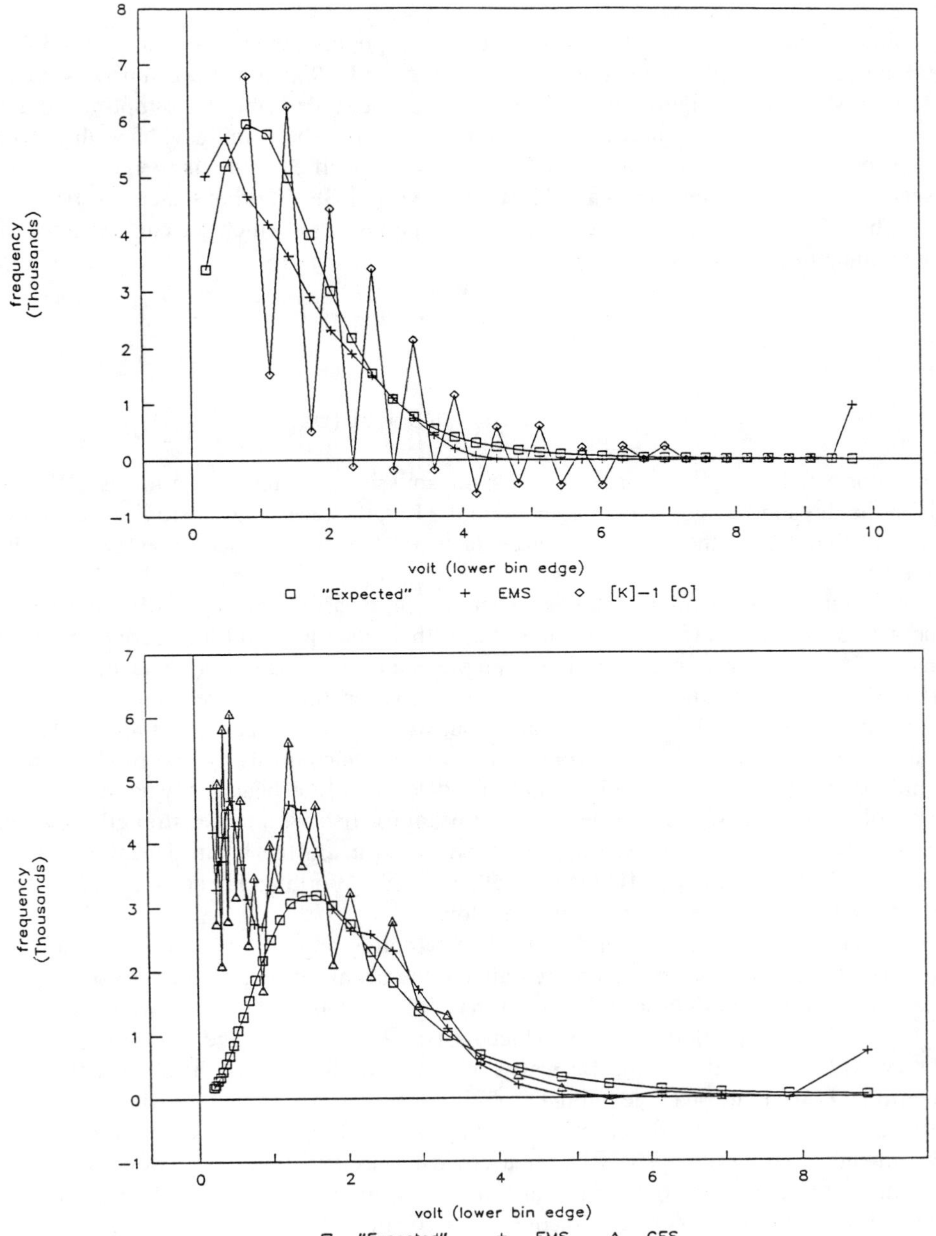

Figure 10. EMS (expectation maximization and smoothing) estimates of fish voltage distribution compared to traditional deconvolution results.

5.2.2 Duration-in-Beam

One of the problems in the estimation of fish population size is that the sampling volume of the acoustic beam needs to be determined. Though the beam possesses a complex sidelobe configuration, a cone may adequately describe the sampling volume when densities are small and inter-fish separation large, by using a voltage threshold. The acoustic beam, depicted as a simple cone with a specified angle, is seen in the above (Section 3) output of the program CRTBEAM.EXE (CRSP Working Paper No. 105).

The effective angle is a function of several parameters, which we can see from the sonar equation

$$EL_{max} = TS + SL + RS$$

or

$$EL_{min} = TS + SL + RS + 2B(\theta_{max}) ,$$

when corrected using time varied gain for transmission and absorption losses. We see, given the threshold (EL_{min}), the target strength (TS), the receiving sensitivity (RS) and a function that relates the beam pattern factor to θ, that we can determine the effective sampling angle, θ.

The manufacturer often specifies a beam angle, measured at the –3 dB point on the one-way directivity pattern. As we have seen, this value is useful to describe the beam pattern for a circular transducer. For example, we can estimate ka (wavenumber times effective transducer radius) using this value. However, the effective sampling angle is often larger than the –3 dB angle. In addition, we must recognize the possibility that, at shallow depths, the beam's horizontal cross section cannot always be modeled by a simple circle because of the likelihood of sidelobes of the beam. By increasing the threshold, however, we can create a conical beam for fish with target strengths above a certain critical value. This effect can be examined using the beam detection volume program CRTBEAM.EXE (Hedgepeth, 1993; CRSP Working Paper No. 105). This effect can be useful where fish densities are low.

An empirical way to obtain the effective sampling angle is to measure the longest duration that a particular fish target remains in the beam, correcting for ship speed. An alternative empirical approach is to estimate the effective sampling volume with a technique called duration-in-beam, which is based on the average duration a fish is observed (Thorne et al., 1988). Basically, the effective diameter of the acoustic beam is determined from the average number of times successive echoes are received from individual fish as the transducer moves at known speed over a fish, assuming that fish movement is negligible. The technique could be used in reverse as well, fixing the transducer to the bottom looking upward, in a moving body of water. Crittenden (1989) provides an extensive review and theoretical derivation of the method, including a look at the bias in the estimate of sampling volume caused by fish movement.

In general, assume that the transducer is stationary while the fish moves at known speed along the diameter of a continuously sampling acoustic beam. The diameter of the beam at that depth could be determined by the relationship

$$d = s\,t$$

where s is the swimming speed of the fish and time t is the time to swim through the beam. Time t could be approximated by

$$t = e / p$$

where e is the number of successive echoes on the fish as it crosses the beam and p is the echosounders pulse repetition rate. This gives the equation that would be used in the longest duration technique

$$d = s\,e / p$$

where s is now the ship's speed.

The duration-in-beam technique, based on average number of echoes, estimates the diameter d from an estimate of the average path through a circular cross section of the beam. If L equals the average path length, then Ld is the area of a rectangle with area equivalent to the circular cross section. Then equating the two expressions for area

$$Ld = \pi(d/2)^2 \quad => \quad d = 4L / \pi$$

which gives the duration-in-beam estimate

$$L = s\bar{e} / p \quad => \quad d = 4\bar{s}e / (p\pi)$$

The estimate diameter, d, is then used to compute the sampling volume. There are also a number of alternate ways to arrive at the duration-in-beam equation. Under the assumption that the fish pass through the acoustic beam uniformly, the PDF for the location along the radius a fish passes, r, is

$$f_R(r) = 1/R_{max} \qquad 0 \le r \le R_{max}$$

$$= 0 \qquad \text{elsewhere}$$

Each path length L' can be related to r as

$$L' / 2 = \sqrt{\left(R_{max}{}^2 - r^2\right)}$$

The average path length can be estimated by

$$E(L') = 2\int_0^{R_{max}} \sqrt{\left(R_{max}^2 - r^2\right)} f_R(r) dr$$

$$= 2\int_0^{R_{max}} \sqrt{\left(R_{max}^2 - r^2\right)}\left(1/R_{max}\right)dr$$

$$= R_{max}{}^{\pi/2}$$

$$= b\pi/2$$

This same technique could be used to model other average paths, for example, through an elliptical beam. This could be useful when using elliptical transducers. The equation of an ellipse can be used to map the cross-sectional area of an elliptical transducer:

$$x^2/a^2 + y^2/b^2 = 1$$

where a and b are parameters which describe the ellipse. The expected chord length parallel to one of the axes of the ellipse is:

$$E(L') = 2\int_0^a y\, f_R(x)\, dx$$

$$= 2\int_0^a b\sqrt{\left[1-\left(x^2/a^2\right)\right]}\left(1/a\right)dx$$

Elliptical transducers are less frequently used for research purposes than are circular transducers but they are found on many fishing vessels and are used for research where it is necessary to have a fan-shaped beam.

Besides serving as an estimate of the effective sampling volume due to thresholding, the duration-in-beam method could produce an estimate of target strength. Together with range, the diameter d can produce an estimate for θ since

$$\sin\theta = d / (2 \text{ x range})$$

If target strength were a constant, and from above,

$$EL_{min} = SL + TS + RS + 2B(\theta_{min}) \Rightarrow \hat{T}S = EL_{min} - SL - RS - 2B(\theta).$$

As outlined above, duration-in-beam methods are empirical methods. They have been used successfully in many surveys (see Thorne, 1988 for a review). However, the target strength distributions should be included for unbiased counting methods due to the thresholding effect on the beam angle (Keiser and Ehrenberg, 1987). Crittenden et al. (1988) have shown that random target strength variation should not bias the results otherwise. Therefore, the duration-in-beam method should work well in low noise, low density situations.

Weighted regression of durations on range is probably the best way to analytically apply the duration-in-beam method (Crittenden, 1989). Researchers often perform the duration-in-beam analysis over relatively small range intervals, assuming that a single species or single size class of fish occupies that range so that the detection volume can be estimated.

6. SAMPLING THEORY AND SURVEY DESIGN

Even though acoustic sampling is rapid and requires relatively little effort, a complete survey of large areas is impractical. A sampling design is then made to sample the fish population.

We can identify six valuable principles that one should remember in making a acoustic survey design (borrowed from ten principles of Green, 1979).

1. "Take replicate samples within each combination of time, location, and any other controlled variable."
2. "Carry out preliminary sampling to provide a basis for the evaluation of the sampling design and of the statistical analysis options. Those who skip this step because they do not have enough time usually end up losing time."
3. "Verify that the sampling device or method is sampling the population you think you are sampling, and with equal and adequate efficiency over the entire range of sampling conditions to be encountered. Variation in efficiency of sampling from area to area biases among-area comparisons."
4. "If the area to be sampled has a large-scale environmental pattern, break the area up into relatively homogeneous subareas and allocate samples to each in proportion to the size of the subarea. If an estimate of total abundance over the entire area is desired, make the allocation proportional to the number of organisms in the subarea."
5. "Verify that the sample unit size is appropriate to the sizes, densities, and spatial distributions of the organisms being sampled. Then estimate the number of replicate samples required to obtain the precision wanted."
6. "Test the data to determine whether the error variation is homogeneous, normally distributed, and independent of the mean. "

Most of these points refer to the estimation of the variance of the estimate of abundance. We mentioned in Section 2 that estimation in the population dynamics model CASA can be in terms of the raw observations or of their log transform. Estimates of variability can help define both the form of the model (log transform or not) and the size of the weighting factors which are a function of the variability of the observations. The point of this section on sampling is to estimate both the bias and the variability in our acoustic observations of fish population size. The proper sampling design contributes to minimizing variance and the cost of sampling.

We emphasize random sampling in the following section because our desire is to estimate the variation of the population estimate so that we can use these estimates in our population dynamics model CASA. However, a recent manuscript has pointed out that fixed grid sampling is used by many researchers (Simmonds et al., 1991). This is because their need is to simply "find" the fish stock. It is during the design process that the investigator must assess at what stage random sampling should play a role.

6.1 SAMPLING DESIGN STRATEGIES

Cochran's (1977) book on sampling theory is a recommended source for obtaining estimators of the population size and its variability for various sampling designs. The main categories for sampling are considered in this section.

The main categories for sampling are simple random, stratified random and systematic sampling. Stratification usually produces estimates with smaller variance than simple random sampling. Systematic sampling locates the acoustic samples evenly over the region the population occupies. However the behavior of the estimated precision of a systematic sampling estimate is difficult to predict without a prior knowledge of the population structure (Cochran, 1977).

Other sampling strategies can be used within these broad major categories. For example, post-stratification is useful when the acoustic survey suggests that an alternative grouping will be more appropriate than the original design. Ratio estimation is useful when two variables are being measured and one variable is correlated with another. We use it below with acoustic density estimates as an alternative to single variable stratified random sampling. Two-stage sampling designs often result in more uniform coverage over the sampling region while stratifying the samples. Systematic samples may also be taken in a stratified design (Cochran, 1977; Scherba and Gallucci, 1976).

6.1.1 Simple Random Sampling

Figure 11 shows a hypothetical area in which acoustic sampling is desired. For the three main sampling design categories, we will assign acoustic transects to the sampling area. A transect is the course the vessel moves along while the echosounder is being operated and the raw data for estimating fish population is being collected. For the purposes of this explanation, it will also be considered to be a sample unit.

Simple random sampling considers the surveyed region to contain a homogeneous population. There is no attempt to break the area into subareas or to consider sub-populations. Plot A in Figure 11 shows the assignment of 10 parallel transects to this area. We assign the transects randomly using ether a uniform random number table or a uniform random number generator in a computer program. Therefore, each transect is a sample unit so the sample size is $n = 10$.

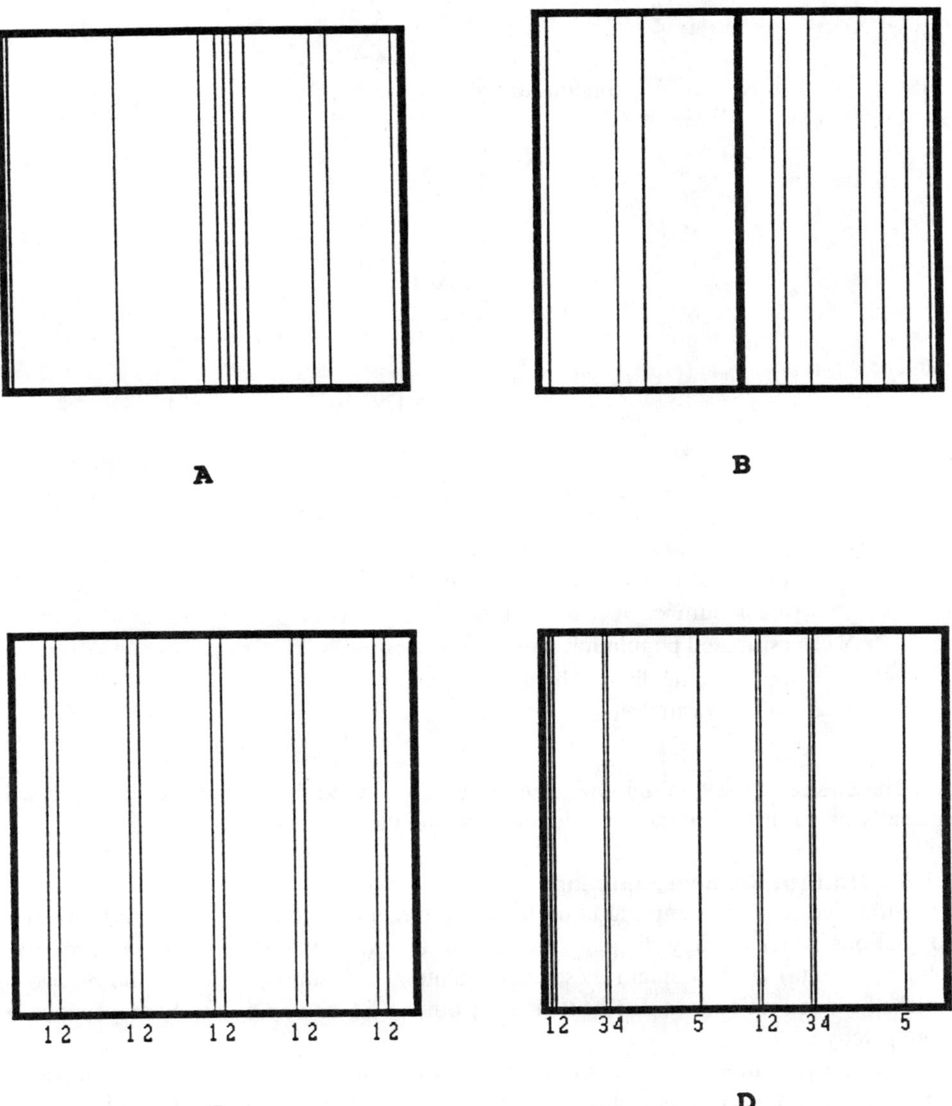

Figure 11. A hypothetical sampling area, with three sampling designs: (A) simple random sampling allocation of 10 transects; (B) stratified random sampling where the transects were allocated by the Neyman method (see text); (C) systematic sampling allocation of 10 transects using five sample units in two clusters (numbered); (D) systematic sampling allocation using two sample units in five clusters.

$$\text{Systematic } \bar{y} = \frac{\sum_{i=1}^{n} y_i}{n}$$

$$\hat{Y} = N\bar{y}$$

$$s^2 = \frac{\sum_{i=1}^{n} (y_i - \bar{y})^2}{n-1}$$

$$VAR[\bar{Y}] = \frac{s^2}{n} \frac{(N-n)}{N}$$

$$VAR[\hat{Y}] = N^2 VAR[\bar{y}]$$

where y_i = number of fish in transect i
$\bar{y}$ = mean number of fish per transect
$\hat{Y}$ = estimated population total
N = total volume divided by transect volume
n = number of transects

The choice of the total number of transects to be used in any particular survey will normally be a function of the precision desired and the resources available.

6.1.2 Stratified Random Sampling

Stratified random sampling is useful when the population can be divided into subpopulations which occupy distinct areas or strata. We divide our sampling area into subareas (strata) which contain these subpopulations. A subpopulation could be one in which the distribution was different than in others, for example, less homogeneous or more patchy.

In order to compute estimates based on a stratification, we must allocate samples to the different strata. The allocation scheme is affected by three variables: (1) number of sampling units in the stratum, (2) variability of the observations in each stratum, and (3) cost of obtaining the stratum samples (Scheaffer et al., 1986). In an optimal allocation scheme Cochran (1977) says to take more samples in a stratum if (1) the sample is larger, (2) the stratum is more heterogeneous and (3) if sampling is cheaper. Neyman allocation assumes that the cost of obtaining a sample (performing a acoustic transect) is the same among all strata. The number of transects to be allocated to a stratum could be estimated from a prior estimate of within-stratum variance (see Green, 1979) principle 5 above). The Neyman allocation is:

$$n_i = n\frac{N_i S_i}{\sum N_i S_i}$$

where N_i = number of sample units in stratum i
S_i = sample standard deviation
n = total number of transects
n_i = number of transects allocated to stratum i

As an illustration, suppose fish aggregate in a very patchy fashion in exactly half of our area, but that the numbers are the same in both halves. A hypothetical preliminary survey gave us a variance estimate in the very patchy area four times that of the other half of the area (two times the standard deviation). Using the formula for Neyman allocation we would allocate 2/3 or seven of the samples to the patchy stratum. Plot B in Figure 11 shows the sample allocation for the two hypothetical strata.

The population and variance estimates based on stratified sampling are:

$$\bar{y}_i = \frac{\sum_{i=1}^{n_i} y_{ij}}{n_i}$$

$$\hat{Y} = \sum N_i \bar{y}_i$$

$$S_i^2 = \frac{\sum_{j=1}^{n_i} \left(y_{ij} - \bar{y}\right)^2}{n_i - 1}$$

$$VAR\left[\bar{y}_{st}\right] = \frac{1}{N^2} \sum_{i=1}^{L} N_i^2 \frac{S_i^2}{n_i} \frac{\left(N_i - n_i\right)}{N_i}$$

$$VAR\left[\hat{Y}\right] = N^2 VAR\left[\bar{y}_{st}\right]$$

where y_{ij} = number of fish in transect j of stratum i
$\bar{y}_i$ = stratum i mean per transect
$\bar{y}_{st}$ = mean number of fish over all strata

$\hat{Y}$ = estimated population total
N = total volume divided by transect volume
n_i = number of transects in stratum i
L = total number of transects

6.1.3 Systematic Sampling

Systematic sampling is an alternative to simple and stratified random sampling in which it is often easier to allocate the samples and perform the sampling, but there is an element of unpredictability associated with the estimates of variance and bias. It has been noted that the variance estimate can be either more or less precise than simple random sampling with the same sample size, depending on the sample size and the distribution of animals. In fact, Cochran (1977) notes that variance can even increase with a larger sample size. Systematic sampling is a sampling strategy that is not as widely recommended as stratified random sampling for acoustic surveys (Jolly and Hampton, 1990) and is included here for comparison.

In a systematic sampling scheme of size n the first sample unit or transect is randomly placed and n–1 successive transects are run, each equally spaced from the preceding one. This would correspond to one cluster in a cluster sampling scheme (Cochran, 1977). The sampling variance VAR is the sum of the between sample variance VAR_b, the result of sampling more than one cluster, and the within sample variance VAR_w from all clusters. That is,

$$VAR = VAR_b + VAR_w$$

It can be shown that the sampling variance of the sample mean in systematic sampling VAR_b is reduced if the within sample variance VAR_w is increased, i.e., when sampling of a cluster is over a heterogeneous population. In theory, VAR_b could be zero if the samples between clusters were uncorrelated but, beforehand, one can never know that.

Figure 11 C shows 10 transects allocated as follows. Two clusters are sampled each with n = 5 sample units or transects. Figure 11 D shows an alternative allocation of the 10 transects using n = 2 sample units and five clusters. And this is the point. If one knows the population being sampled and the statistical relationship between and within cluster variances, an efficient estimator using systematic sampling can be found. Otherwise, alternative sampling designs are recommended.

Alternatives to systematic sampling include a two-stage randomization procedure to provide uniform coverage without invalidating variance estimates (Armstrong et al., 1988; Jolly and Smith, 1989). In the first stage, each stratum is divided into a number of strips of equal width. Strips are selected at random and then a transect is chosen at random within each selected strip. Figure 12 shows the transects and stratified sample design which resulted from a two-stage survey design off the coast off Africa (Jolly and Smith, 1989).

6.2 SURVEY DESIGN DECISIONS

The above examples gave a simple introduction to the many existing sampling designs for stock assessment estimation. There is a rich literature on sampling designs, that include the simple sampling designs we have discussed. For example, adaptive sampling and spatial statistics are among the newer additional topics. Simmonds et al. (1991) is a recent source of review of survey design and analysis.

In addition to the recommendations made by Jolly and his colleagues (Jolly and Hampton et al., 1990; Armstrong et al., 1988; Jolly and Smith, 1989), Brandt (1987) suggests that stratification should be made using factors which affect fish populations. For example one would not simply use bathymetry or geographical coordinates as stratum boundaries. He distinguishes between habitat-derived patchiness (difference between homogeneous environments — oxygen, light, species associations, salinity, etc.) and biological patchiness (diel and dispersion as examples). Essentially, he suggests that by using the habitat preferences of a target species, one can reduce the effort to estimate its abundance with a certain confidence.

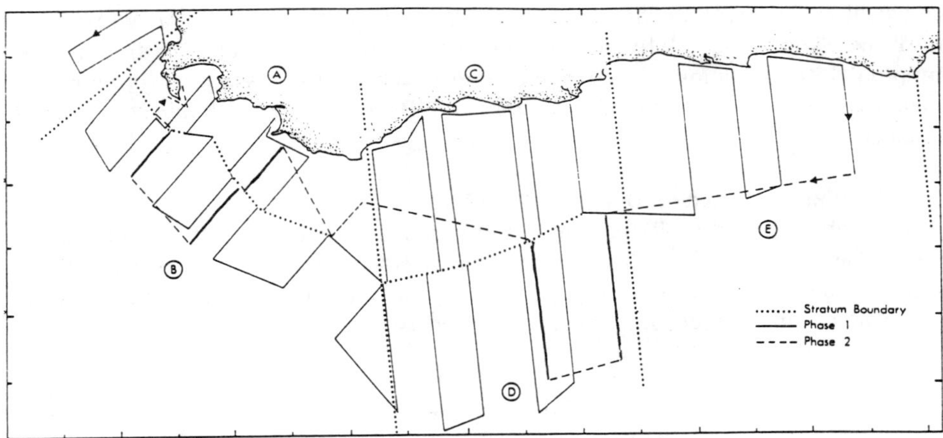

Figure 12. A sampling grid of parallel, stratified random transects. Strata are labeled A to E. (from Jolly and Hampton. 1990a. Can. J. Fish. Aquat. Sci. 47(7). With permission of the Minister of Supplies and Services Canada.)

Jolly and Hampton (1990b) present historical misconceptions about sampling and statistical analyses found in the acoustics literature. They suggest that the natural design for acoustic transects is a stratified random sample of parallel transects. Zigzag transects have the following disadvantages: poorer distribution of sampling effort, the lack of the advantage of parallelism to remove variation from density gradients in the direction of the transect, and complex boundary problems. Using the variation between random transects reduces the need to adjust for serial correlations. Other researchers have used an adjustment for serial correlations (Williamson, 1982; Burczynski and Johnson, 1986; MacLennan and MacKenzie, 1988).

Other sampling design considerations mentioned (by Jolly and Hampton, 1990b) are:

(a) "The observed values should be approximately normally distributed for unbiased means and variances and should be transformed if they are not"

(b) Criteria for "rejection/replacement of extreme values" should be established

(c) Criteria for "exclusion of zeros from data set" should be considered. Chapter 4 of this volume on sampling designs contains a thorough discussion of the delta distribution which is commonly used for this purpose.

The point is that acoustic survey design and analyses are amenable to the accepted statistical practices such as been practiced in other animal abundance estimations (statistical procedures: Cochran, 1977; Suhkatme and Suhkatme, 1970) (examples: Siniff and Skoog, 1964; Jolly and Watson, 1979; Armstrong et al., 1988). Jolly and Hampton (1990a) caution that other literature which discusses sampling designs (e.g., Johannesson and Mitson, 1983; Shotton and Bazigos, 1984) should be critically evaluated.

6.2.1 Estimation Based on Density: Ratio Estimation

Jolly and Smith (1989) present a ratio estimator (Cochran, 1977) for acoustic samples based on "density". They define density where the numerator is a function of biomass and the denominator is length of the transect. The mean biomass per unit surface area from transect j stratum i is expressed as:

$$\hat{d}_{ij} = \frac{b_{ij}}{L_{ij}}$$

where b_{ij} = biomass (fish / m^3)
L_{ij} = transect j length in stratum i
d_{ij} = mean weight per surface area of transect j in stratum i

Acoustic estimates are often presented in the form of a ratio or density (e.g., number of fish per unit volume or per unit area). The duration-in-beam procedure estimates the volume of a frustum of a cone, and then extrapolates that volume either to the volume swept (an approximation) or an estimate of volume sampled. The methods of echo integration and the deconvolution procedures use voltage data from many pings. The volume sampled by the echo integration and deconvolution methods can be pictured as a hemispherical shell volume. Therefore, all three methods use a measured volume, either assumed or estimated, and a natural estimate that is often output is the mean density per unit of transect length.

An estimate of mean stratum density and variance of the estimated mean density can be constructed using a ratio estimator (Sukhatme and Sukhatme, 1970; Cochran, 1977; Schaeffer et al., 1986) from which population estimates may be derived. Jolly and Smith (1990) use a ratio estimator with data from acoustic surveys. Refer to Appendix II for another we have used.

6.3 OTHER SURVEY CONSIDERATIONS

Simmonds et al. (1991) provide a historical perspective which includes the sampling approaches we have discussed. Many other topics are included: contouring, bootstrapping, transform methods, geostatistics, adaptive surveys, species identification using direct and indirect methods, and more. They state that most practitioners use systematic sampling, because little information is gained from having transects close together as often occurs in simple random sampling. We quote from their conclusions:

"From the wide range of stocks and surveys that we have reviewed it is clear that random spatial distribution of pelagic stocks on the scale of transect spacing is a commonly acceptable assumption and thus a systematic parallel survey grid is preferred. The major exception to this is for narrow shelf or fjord areas where the parallel grid is logistically wasteful and may be replaced by a carefully designed zigzag grid. In cases where the stock distribution cannot be assumed to be randomly distributed, local random positioning of parallel transects is required. For highly contagious distributions adaptive sampling may be preferred in the absence of significant stock migration during the survey."

These conclusions do not differ radically from any suggestions we have made. As an acoustic survey matures either over the course of a season, or from the point in time where the species begins to be assessed, we need to be flexible in choice of survey design. At first it may be wise to use a fixed grid that allocates equal effort over the area where we think the fish stock lives. When we obtain more knowledge about the distribution, we may be able to allocate random stratified or systematic samples.

We assume that nautical charts or maps are available. Areas which cannot be practically accessed should be omitted from the design and another form of survey can be used, e.g., nets in extremely shallow water. Navigation is based on the charts and is used to locate transects. Navigation can be performed as simply as using a hand bearing compass and known points of land. After positioning the transects on a chart use the bearings to known points of land to arrive at the transect start. Use a compass to steer by, remembering that the true course in degrees must be converted to the magnetic course using the magnetic variation for that locality. Principles of navigation can be found in texts such as Bowditch (1984) and Chapman (1977). Take and record bearings so that the actual transect position may be estimated using triangulation at the beginning and end of the transect, and perhaps at some points in between. This record is important so that the transect position can be confirmed and stratum assigned.

Previous knowledge of the fishing grounds can be used to design a stratified survey. Often bathymetry may give clues for stratification although caution should be exercised.

Ideally, presurvey data would be used. Three items need to be decided: the transect length and direction, the initial points, and the timing of the transect. Practically, we are usually constrained by the time allocated for completing a survey. This may be due to funds for personnel, equipment, or compatibility with fishing in the area. Refer to the methods for desired precision and optimal allocation (Cochran, 1977; Scheaffer et al., 1986) for transect selection. Additional trawl or net sampling can take considerable time and effort. If a 3 nautical mile transect takes half an hour to complete the acoustic sampling, a half hour trawl through the same area might take 2 hours or more to sort and record.

Surveys should be timed so that the data can be interfaced with other management type data and stock assessment models. This may require surveys to be made daily, weekly, monthly, yearly, or even at longer intervals.

In-field decisions can be made to alter survey design as the survey progresses, called adaptive surveys. Adaptive surveys require effort to be reallocated, so that sample design can be more appropriate to the species to be evaluated. Data processing can be made either on the working platform or at a land based facility, in real-time or in post-processing. In our Gulf of Nicoya survey, post-processing data took about twice as long as it took to perform the acoustic survey.

To use the acoustic survey effectively for the stock assessment of a target group of species, one should have a good idea of the areas occupied by that species. Ideally, there should be some way of separating those species from the others. Natural separation could be by depth or areal zonation, and acoustic separation could be by widely modal length distributions, which would then affect the amplitude estimation procedures (see for example, Clay (1983). Acoustic data taken using multiple frequencies can be input in a model to separate length classes (for example, Kalish et al., 1986). If it is not possible to isolate the targets of interest, probably the best one can do is to fractionate a total enumeration by using a secondary means of observation, for example, trawl sampling. Some attempt could be made at school/species classification. Examples of this kind of classification are methods of Rose and Leggett (1988), who found that target strength was not as important as distance measures. We will discuss methods for obtaining species information in the next section. It is an important feature of sampling and survey design because it addresses the bias of the estimates.

7. SPECIES IDENTIFICATION

We usually need to confirm or partition acoustics assessment, in order to allocate estimates to fish species or length groups. Probably the most common way of obtaining information about fish species and size is by the use of nets to sample and capture some of the fish under study. Other methods include video, scuba, poisons, explosions, pumps, stomach contents of predators, aerial observations, and so on. Often the choice for obtaining fish identification is limited. In turbid areas, visual observation may be impossible or environmental impacts may preclude the use of poisons and explosive devices.

For example, our case study location, the Gulf of Nicoya, Costa Rica, is a shallow and turbid estuary, making video observations of fish out of the question. We were restricted to net sampling for species identification. To partition the estimates from a single-beam echo sounder into various species and size groups we relied on midwater trawl captures and gillnet catches. We found that our trawl design was selective for fishes under 20 cm. We therefore relied on a combination of trawl and gillnet catches. The sampling design required the use of nets at least once during a 3-nautical mile acoustic transect. This was logistically pragmatic and not adaptive. An adaptive approach might be to sample each time the acoustic returns "looked" significantly different.

Murphy and Clutter (1972) showed that a small purse seine worked more effectively at capturing anchovy in Hawaii than did a 1-m standard towed net. Clarke (1983) used a 61 m long, 12.2 m deep purse seine with 3.2 square mesh to study anchovy in Hawaii. He says, "Like the PPS [plankton purse seine], the nehu purse seine (NPS) can be handled from a ~5 m skiff by two or three people. With some practice it can be set in 1–2 minutes, and with a battery-powered capstan can be pursed about 5 min after the set is completed." His results suggested that the net was capable of sampling from all of the larger fish in Kaneohe Bay.

MacLennan and Forbes (1987) discussed some of the practical limitations of species identification. The identification problem is difficult. However, the problem is best avoided by conducting the survey when the species of interest is the dominant component of the pelagic biomass. Limitations include the necessity of sampling the echo traces to establish the size composition as well as identifying the fish.

The echo traces may be sampled by trawling or other fishing methods. However, the catching efficiency of the trawl is not the same for all sizes and species (Wardle, 1983). Further, two vessels fishing in the same area can produce significantly different results. Thus the composition of the trawl catches may give only a rough indication of what has been detected by the echosounder.

Fortunately, there are other sources of information to supplement the trawl samples. Unwanted targets may be excluded by careful selection of the depth intervals for the integration, and it may be possible to differentiate between shoals of different species from characteristic features of the echo traces, particularly their height, horizontal extent and apparent density.

MacLennan and Forbes (1987) show figures of echograms of the different species, which convincingly separate the fish by their school features. Rose and Leggett (1988) have successfully separated schooling species by their features. They also include a review of some of the successful methods of fish school classification. Their techniques are particularly amenable when a digitizing system such as the one discussed in the signal processing section is available. They used a Techmar AD card and sampled v^2 signals from schools of cod, mackerel and capelin (20log R time-varied gain). They also estimated target strengths using a BioSonics dual-beam processor, using the wide and narrow beam voltages and 40log R time varied gain. They used the following characteristics:

40log R data

(1) target strength
(2) school depth [1]
(3) off bottom distance [1]

20log R data

(For each school, two sequences of 4–5 pings were analyzed. Schools signal boundaries were defined by a low voltage lasting 2.0 ms.)

(1) off bottom distance [1]
(2) depth of school [1]
(3) mean squared voltage
(4) SD of mean squared voltage — transformed to inverse CV
(5) maximum squared voltage
(6) mean distance in meters between school peaks [1]
(7) mean peak to trough squared voltage standardized to the mean squared voltage

[1] Variables were transformed using ln(x) + 1 due to non-normality.

Quadratic, stepwise discriminant functions were computed using the above variables and a SAS/STAT, a statistical package for the IBM-compatible microcomputer. Rose and Leggett found that 93% of the cod and capelin schools insonified during 1985 were correctly classified using 1986 data for the discriminant functions. One interesting finding is that target strength differences were not as useful as 20log R factors 7, 6, 2, 4, and 1, in that order. They suggested two reasons why target strength was not as important as the other factors: that large mackerel had target strengths similar to the smaller cod, and that selection of single fish echoes may be biased. They felt that nearest neighbor measures such as factors 7 and 6 reflected internal school structure best and thus gave the most meaningful identification of species. They caution that such a method be used on a case by case basis.

Simmonds et al. (1991) list several other references besides Rose and Leggett (1988) which classify fish echoes to fish species using discriminant analysis and features of the echoes from fish species. Some of these are: Simmonds and Armstrong (1987), Le Bourges (1990), Vray et al. (1987), Gerlotto and Marchal (1987), Souid (1989), Bjornoe and Kjaergard (1986).

Species discrimination methods may not work in tropical environments due to the huge species diversity. Nevertheless, although the Gulf of Nicoya in Costa Rica has a highly diverse species composition, we observed that species do aggregate, as was illustrated by artisanal fishing catches and by our own sampling nets. Assessment of unaggregated species will be an understanding of their life history so that an acoustic survey can be performed where and when the species are most isolated.

8. CASE STUDY: GULF OF NICOYA, COSTA RICA

We designed and undertook acoustic surveys in the inner or northern part of the Gulf. This area is relatively shallow with depths between 4 and 20 meters typically. Artisanal fishermen use a variety of gear to capture fish but the gillnet is the most widely used.

We performed acoustic experiments: (1) with fixed location transducers, and (2) with mobile surveys, using a small outboard powered vessel. In both cases we sampled with nets for species identification. We used: (1) a low-cost echogram digitizer for echo counting, (2) a low-cost analog to digital interface system for echo integration and echo counting, and (3) a commercial signal processor for echo integration and echo counting.

8.1 SIGNAL PROCESSING SYSTEMS

The following section deals with signal processing, making use of a digitizing tablet, an analog to digital system we developed, and a commercial product, the BioSonics ESP (Echo Signal Processor).

Signal processing is the recovery of information while dealing with problems caused by attenuation, distortion, interference and noise. Signal processing is to the acoustics practitioner what processing catch observations is to the fisheries analyst. Carlson et al. (1981) outlined some of the functions a signal processor might perform:

(1) Amplification to compensate for attenuation or reject common-mode interference.

(2) Filtering to reduce interference and noise, or to extract selected aspects of information.

(3) Equalization to correct certain types of distortion.

(4) Frequency translation or sampling to obtain a signal that better suits the characteristics of the system.

(5) Multiplexing to accommodate two or more signals in one system. (This could also could be dealt with by parallel or simultaneous processing.)

The sonar or other equations deal with some of the problems of signal processing mathematically, while others are dealt with using electronic hardware. Although the present trend is to solve many problems digitally in software with the aid of fast microprocessors, analog and digital hardware solutions can be cost effective. Some fisheries acoustic methodology, for example duration-in-beam, can rely entirely on extremely low-cost signal processing, such as a digitizing tablet.

8.1.1 Echograms and Digitizing Tablet

One of the simplest and least expensive methods of analyzing acoustic data on echograms is by use of a digitizing tablet connected to a microcomputer. A CRSP working paper by Thorne et al. (1990) explains a particular application of the digitizing tablet in the Gulf of Nicoya, Costa Rica. Beam patterns can also be digitized (Hedgepeth and Thorne, 1989). We have digitized maps and acoustic transects with the same assembly language interface to a Summagraphics Summasketch tablet. Crittenden et al. (1988) used a similar digitizing tablet to estimate lake fish population density using duration-in-beam methodology. In our case, as well as that of Crittenden et al., the lengths of echo traces on echo chart recordings were digitized as well as their range from the transducer.

The interface that we developed processes echogram data using the Summagraphics Summasketch digitizing tablet with an IBM-PC-type microcomputer. The analysis is relatively simple; however, digitization of many points on an echogram is time consuming. In addition, voltage amplitude data cannot be recorded on echograms. We developed a low-cost signal processing system to digitize voltage amplitudes.

8.1.2 Electrical Signal Processing with a Low-Cost Computer Interface

The next step up in complexity from digitizing tablets is the analog to digital (A/D) converter which digitizes the voltage signals from the receiver of an echosounder. This type of signal processing is more complex than processing echograms because greater electronics ability is needed.

Although we used a data collector, the BioSonics ESP echo signal processor, we also developed a system that was (1) low in cost, (2) easily built and (3) adaptable to user requirements (Hedgepeth and Meyhoefer, 1989). Part of the system was a Metrabyte A/D board that occupies a "half-slot" in an IBM-PC-type microcomputer. The computer programs were written by one of us (JH) and the prototype interface was built at the Department of Zoology, University of Washington by E. Meyhoefer. We estimated the costs for the interface at \$300, and when other components are used it can be considerably cheaper. The Metrabyte A/D board which is inserted into a microcomputer sells for \$400 and the modifications to make the sample faster cost about \$80. (Contec also makes a similar AD board with the faster converter, which sells for \$345). Total system costs were about \$800. All estimates of cost are circa 1990.

The data collection system we developed was limited to use on a single-beam sonar system. We assumed that the primary data could be stored on a tape medium, although many of the system functions can be operated in real time. Most other processors, such as the BioSonics ESP, operate in real time or can be used to post-process data. We chose to use at minimum an IBM-compatible PC. The data collection system consists of a signal interface, a microcomputer and an analog-digital acquisition and control board (Figure 13). Input is amplitude demodulated using a low level analog circuit. A digital circuit that interfaces with the computer allows for the correct timing of the sampling interval and provides an output for recognition of that interval. The design of the analog circuit included the ability to resolve low level signals so that we could later apply a deconvolution technique to resolved echoes. DC-offset and noise was limited by highpass filtering. The signal was then precision rectified and subsequently filtered with a 7th order 0.1 dB Chebyshev lowpass filter. A squared version of the output is available for echo integration. DC-errors were minimized to 100μV, and the precision of the circuit is better than 12-bit resolution.

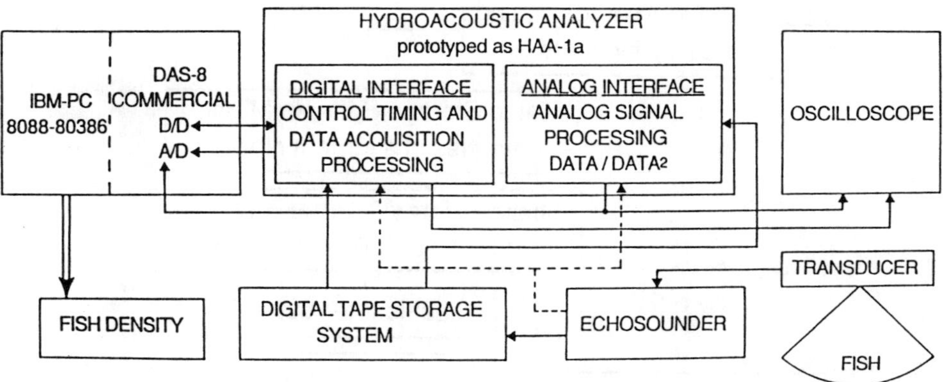

Figure 13. A low-cost data acquisition and processing system developed for analyzing fisheries hydroacoustic data.

We applied the data collection system to fisheries sonar data from the Gulf of Nicoya. Our system produced results similar to the commercially available echo signal processor, with the constraint that data for individual peak detection purposes has to be post-processed, although this could be done in the field. This constraint does have the benefit that very fast pingrates can be handled in order to avoid data loss. Therefore this system is particularly amenable to deconvolution techniques. Such a low-cost system could allow fisheries scientists with constrained budgets to complete analysis after buying an echosounder and transducer. The system is readily adaptable to various software schemes, and could be modified to process multibeam data. However, we recommend that interested persons first develop single-beam systems.

We exceeded our goal of 20 kHz digitization by writing the microcomputer program in a combination of assembly and high level language, using Microsoft Quick-C (Figure 14).

8.1.3 Use of Commercial Processor

We designed a two-stage sampling project for the inner Gulf of Nicoya during August and September 1988 (Figure 15). The region was divided into a number of strips in the first stage. Two randomly placed transects were allocated to each strip in the second stage. Each transect was approximately 3 nautical miles in length. We used a commercial signal processor to analyze tape recorded echo data. Details about the signal processor can be found by contacting BioSonics, Inc., Seattle, Washington, U.S.

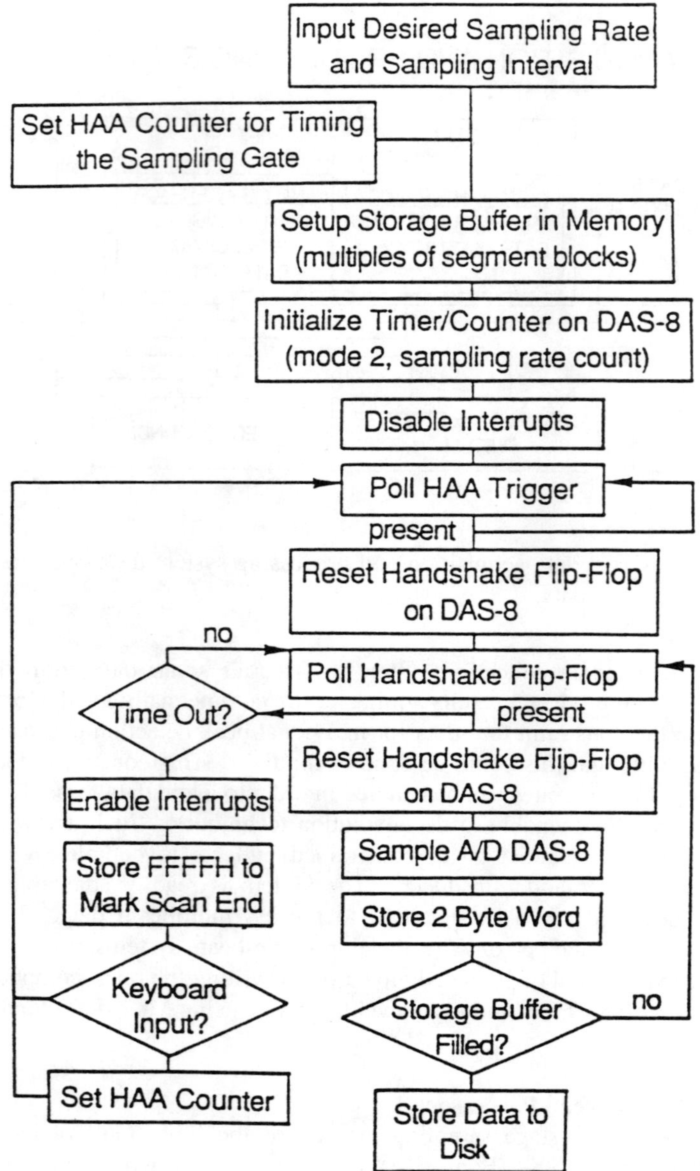

Figure 14. An assembly language program is used by the microcomputer to control data acquisition using the Metrabyte DAS-8. This accomplishes the data digitization in a low-cost data acquisition and processing system developed for analyzing fisheries hydroacoustic data.

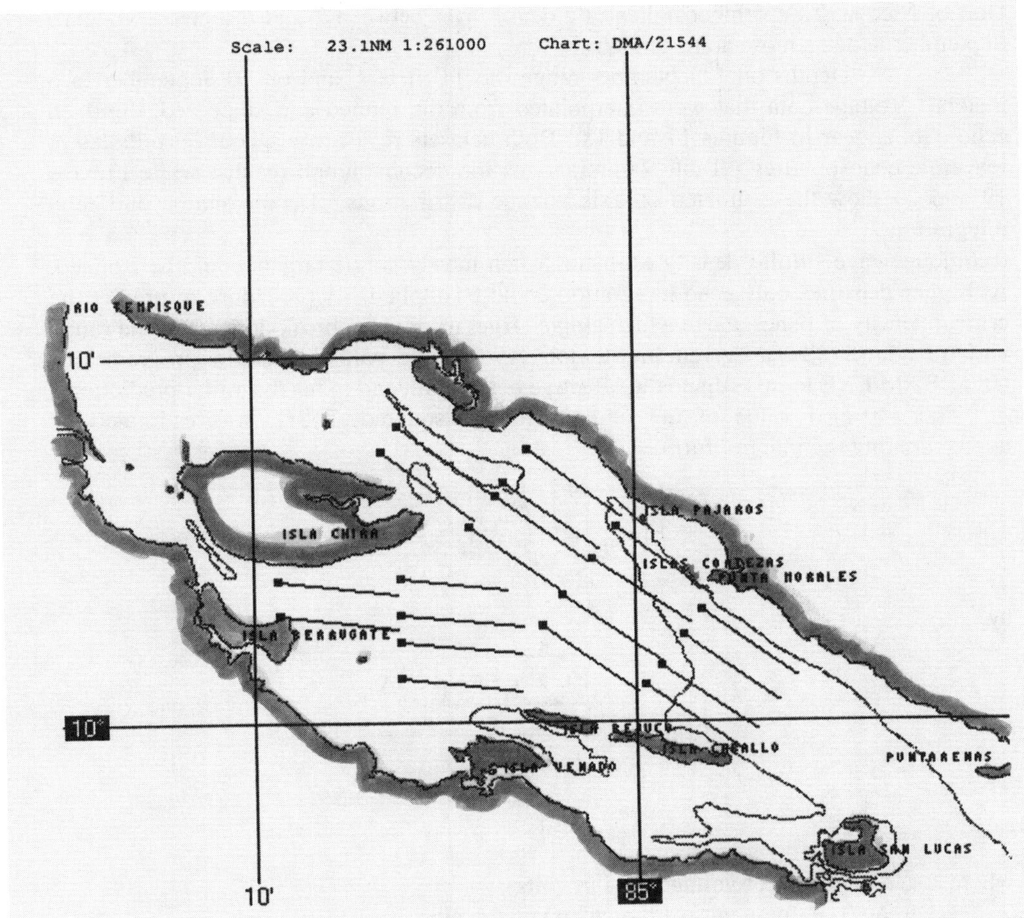

Figure 15. Survey design for collecting hydroacoustic samples in the inner Gulf of Nicoya, Costa Rica.

(Note: our use of this company's equipment does not imply endorsement of a particular product).

8.1.3.a *Signal Deconvolution and Echo Integration*

Data from the commercial signal processor were analyzed using the deconvolution filter. The signal processor also performed echo integration.

Figure 16 shows echograms from two typical transects taken in 1988 in the inner Gulf of Nicoya; 7 September indicates a dense layer between 2 and 6 meters, while 10 September appears more homogeneous.

Depths on 7 September ranged to 11 meters, and on 10 September to 6 meters. Voltage data that were interpolated from the binned and smoothed, digitized echo data appear in Figures 17 and 18. Both datasets result from about 30 minutes of real-time echosounding. Table 2 summarizes the deconvolution results, while Figures 19 and 20 show the estimated on-axis voltage distributions. Deconvolution and echo integration
techniques gave similar density estimates when mostly single targets could be isolated. At higher densities only echo integration could be applied. A procedure to predict the critical density of fishes above which single target techniques break down uses the range and transducer size as given in Stanton (1985). We can also use a program like CRTBEAM.EXE to investigate the effective beamwidth and to make similar predictions.

The expected value of the on-axis voltage squared, $E[S^2]$, was estimated by approximating the integral form

$$E\left[s^2\right] = \int_0^{\infty} s^2 f_s(s)ds$$

by

$$E\left[s^2\right] \approx \frac{\sum_{i=0}^{n} s_i^{\,2} \text{freq}(s_i)\Delta s_i}{\sum_{i=0}^{n} \text{freq}(s_i)\Delta s_i}$$

where Δs_i = width of voltage bin i in volts
freq(s_i) = frequency of targets in voltage bin i

The estimated expected on-axis voltage was then used in an echo integration equation to estimate density. We used data collected with a 40logR + $2\alpha R$ time varied gain, so the range bins (R_1-R_2) over which data were integrated were small, about 0.2 meters. Density estimates were made every 2 minutes, for each depth bin using

$$\text{Number/m}^3 = \sum V^2 \Big/ \left(pd(R_1/2+R_2/2)^2 Tc\pi E[S^2]E[b^2(\theta)]\right)$$

where T = pulse duration
c = sound speed
p = number of pulses from the echosounder
d = number digitized samples in a pulse from R_1 to R_2

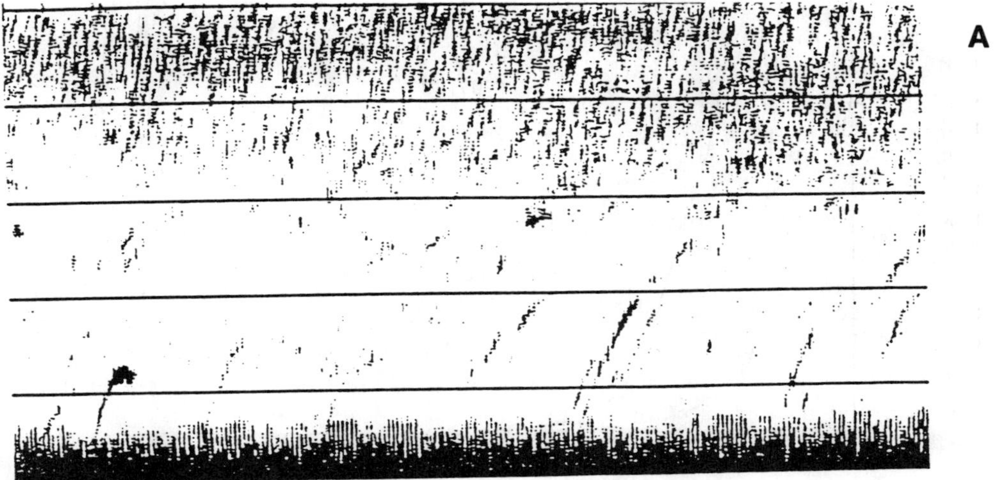

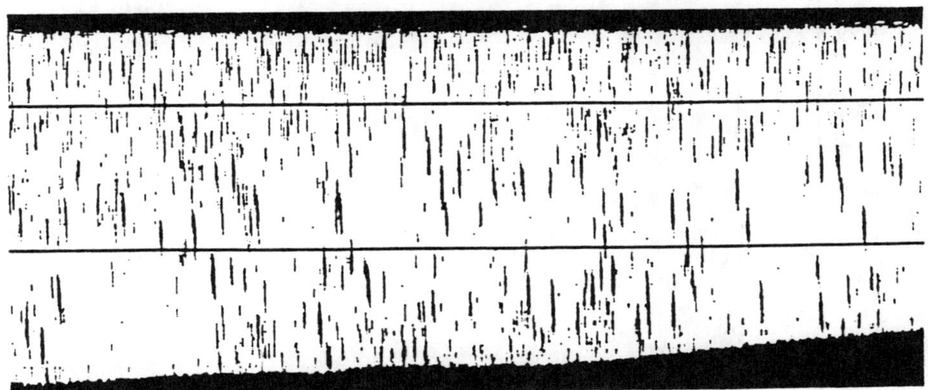

Figure 16. Nine-minute echogram samples from two transects in the Gulf of Nicoya, Costa Rica. Samples A and B were made on 7 September 1988 and 10 September 1988, respectively. Horizontal lines appear every 2 m of depth.

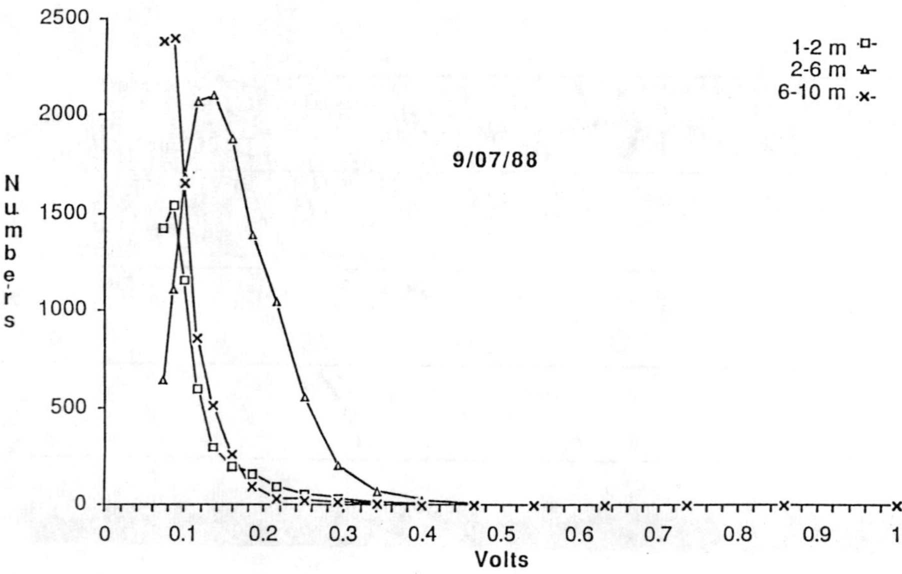

Figure 17. Interpolated raw data from one transect on 7 September prior to deconvolution.

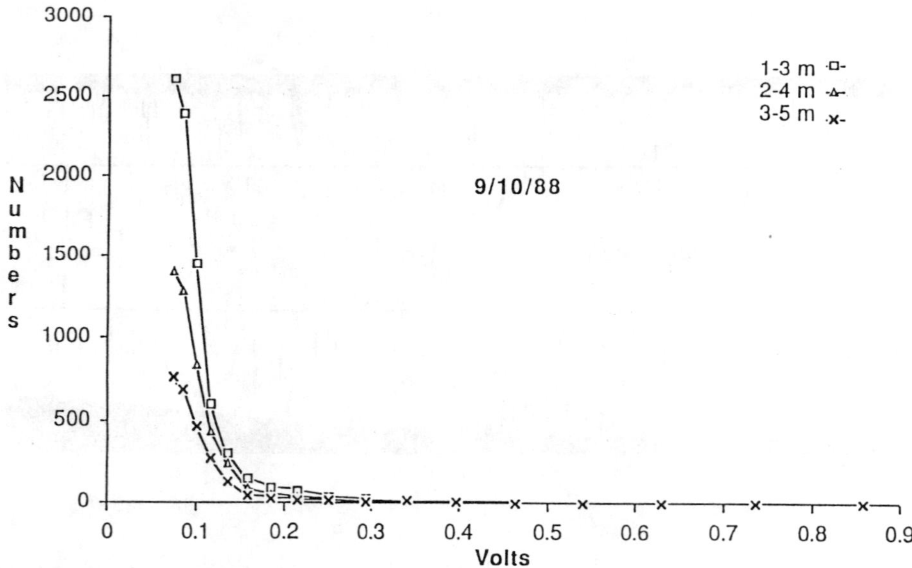

Figure 18. Interpolated raw data from one transect on 10 September prior to deconvolution.

Table 2. Summary of deconvolution results.

Transect Date	Depth m	Deconv. fish/m^3	Deconv. E[S^2]	Critical fish/m^3	Integration fish/m^3
7 Sept	1–2	1.93	0.032	1.85	1.76
	2–6	0.175[1]	0.063[1]	0.26	1.27
	6–10	0.035	0.020	0.065	0.048
10 Sept	1–3	0.79	0.027	1.04	0.847
	2–4	0.24	0.029	0.46	0.199
	3–5	0.13	0.030	0.25	0.112

[1]These estimates are from multiple targets based on both the appearance of the echogram and on the collection of fish target peak voltages. Critical fish density (Stanton, 1985) is below echo integration results which used 0.025 V^2 for E[S^2], the expected value of the on-axis voltage squared.

These densities were then summed over the time and depth of interest and the average reported.

We now collect the fish peak voltage data by the raw digitized steps, from 0 to 4095. This feature is useful when applying the EMS algorithm as well. The analog/digital converter is 12 bits and uses 2.44 mV quantization steps to 10 V. Therefore, 4095 equals 10 V and 0 equals 0 V, and so on. A sample deconvolution program, DECON.C, appears in the CRSP working paper by Hedgepeth (1993; CRSP Working Paper No. 105). Some of the advantages of binning by digitizing steps are (1) medians can be found easily, (2) rebinning is easily and rapidly accomplished and (3) the effects of bin size, voltage threshold, voltage maximum can be rapidly investigated. In addition, bootstrapped analyses can take advantage of data in this form.

8.2 SURVEYING THE INNER GULF OF NICOYA

Our efforts in the inner Gulf of Nicoya included both fixed location and mobile surveys. Fixed location surveys used transducers positioned on the bottom and "looking" upward, toward the surface as well as those "looking" downward. We analyzed the fish echo counts with a digitizer to see if there was a correlation between time of day and the number of fish detections, or between the tidal currents and fish detections. In addition to portraying diel and tidal variations in fish abundance, we discovered that fixed location gear can be hazardous in an area fished by drift gillnets. We learned the lesson that survey design is not always based solely upon analytical formulations, when a gillnet "captured" our anchored acoustic research boat.

Surveys based on mobile transects placed parallel to tidal currents are easy to execute because small boat operators will not have to fight against the current to maintain the course. We recorded boat speed with a knotmeter attached to a towed fin,

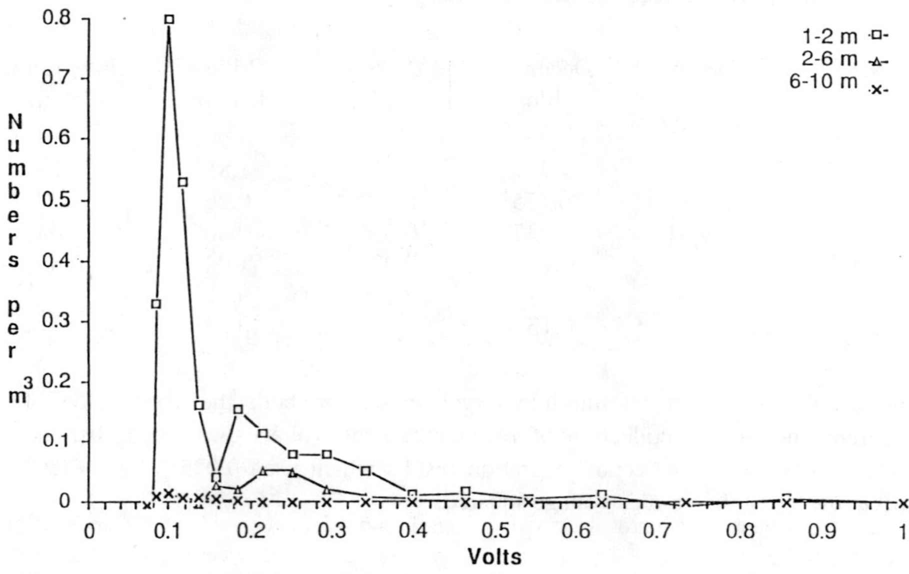

Figure 19. Estimated on-axis voltage distributions from one transect on 7 September after deconvolution.

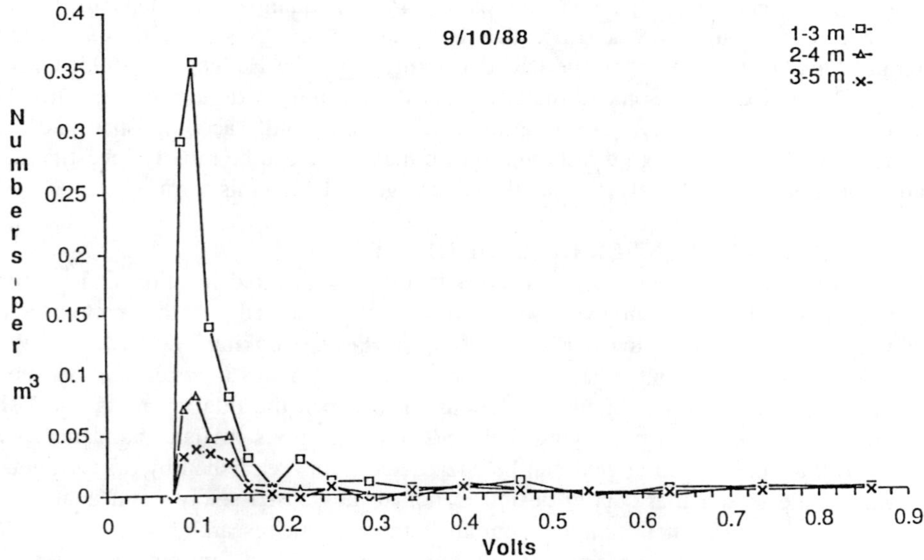

Figure 20. Estimated on-axis voltage distributions from one transect on 10 September after deconvolution.

which let us use the device on whichever boat was available. Recall that boat speed relative to fish speed is essential in the duration-in-beam method. Knowledge of boat speed also allows navigation with more certainty and provides sampling consistency.

8.2.1 Species Identification Using Nets

We initially felt that a midwater trawl might be the only way to sample to a depth of 20 m. In retrospect, the purse seine design of Clarke (1983) would have been more useful for most of the shallow waters of the inner Gulf of Nicoya, Costa Rica. Our trawl net had a mouth opening of 4 m across and 3 m deep. The pulling arrangement was to tow the net with a single 7-m boat powered by twin 75 HP outboard motors. The net mouth was spread open by horizontal and vertical bars. The net depth was adjusted by attaching variable length lines with "Polyform" plastic floats to the ends. The estimated towing speed using this configuration was about 2 knots. At the end of a tow the net was hauled up to the side of the vessel and the cod-end emptied onto the deck of an auxiliary vessel where the catch was enumerated and recorded. Subsampling was done when catches were large. When time was short, samples were placed into plastic bags and counted later.

To use acoustic estimates as auxiliary data in a model such as CASA (see Section 2) they need to be partitioned into appropriate species and length groups. When a multiple species assemblage is sampled this can be difficult. If the secondary allocation estimate is inadequate the acoustic data will be poorly used.

Our trawl sampled clupeids and engraulids quite well. The larger examples of our target species, corvinas, were caught mainly in gillnets.

Our primary navigation tool was the hand-bearing compasses. This type of compass can serve for both directing the course of the vessel along a predetermined transect and for triangulation of position using known points of land. We computed about three magnetic direction vectors to visible points of land, prior to surveying a transect. At night we used the lights in villages or industrial facilities around the inner Gulf of Nicoya. We confirmed our field positions using the depth to the bottom. The portable oscilloscope normally used to observe the echo signals made an effective fathometer. One simply multiplies the number of milliseconds to the oscilloscope bottom trace by 0.75 to read depth in meters since the speed of sound is about 750 m/sec for a two-way trip.

9. TRENDS IN TECHNOLOGY

The trend in hydroacoustics is toward rapid and real-time data analysis so that timely stock assessment estimates can be made. Acoustic data processing equipment coming to the present market is able to link satellite navigation, water quality sensors and other data inputs to present a complete data analysis.

Computer intensive statistical algorithms such as kriging and maximum likelihood estimations are now feasible for the practicing fisheries acoustician due to the availability of low-cost, fast and accessible microcomputers. The same computer that is able to "control" a digital scientific echosounder's ping rate, gain, source level and so on will be able to perform data analysis and stock assessment calculations. Smaller packages will mean greater portability and eventually reduced costs.

Although a number of manufacturers produce scientific echosounders, we will discuss only two systems, one built by Simrad, Horton, Norway and the other built by BioSonics, Seattle, Washington. They illustrate different approaches to the problem of acoustic assessment of aquatic organisms. Our review here does not in any way constitute an endorsement of either product. The Simrad system is an example of a split-beam processor, and the BioSonics equipment supports dual-beam processing. These two systems represent the "state of the art" for fisheries research echosounders. Although both systems are capable of multibeam target strength analysis, both can also be configured as single-beam echosounders. Both companies offer several hardware configurations, and the software for data processing is available.

Simrad's EK500 is a single unit which combines the echosounder, echo integrator and target strength analyzer. Color monitor, color printers, control cursor, interfaces to computers, and accessory software are optional. A number of frequency options are available and the unit can support up to three split beam modules. The existing software uses a networked (ethernet interface) Sun workstation operating under the Unix operating system with X-Windows (graphics interface) and Ingress (Relational Technologies database). Data storage can be on various media, e.g., optical disk or Super 8 video. The Simrad EK500 has color echogram displays on CRT and on two simultaneous color printers.

BioSonics provides the components to build an acoustic system. For example, there is a chart recorder which produces a three-shade echogram on paper. Color echograms show target strength on the monitor and can be output to a color printer. BioSonics' sounders provide access to the voltage signals for signal processing work or tape recording of the IF signal. Thus, a BioSonics sounder can be used with various signal processors.

Simrad's EK500 outputs raw power and phase information in a "superlayer" at the sampling rate. This is a digitized form of the data so the costs associated with the optional work station type computer are avoided. School volume and density, area density and biomass can be computed on the day of the survey.

A complete EK500 system might cost $200,000 and a minimal system about $80,000. A Biosonics dual-beam echosounder, transducer and signal processor might cost $70,000. Portable sounders and transducers can be purchased from either BioSonics or Simrad for approximately $20,000. Both of these sounders are interfaced to laptop computers and include all the signal processing software needed to perform acoustic surveys for fish abundance estimation using the single-beam methods discussed in the earlier sections of this chapter. The Simrad EY500 is based upon the EK500, whereas the Biosonics DT4000 has no predecessor. It utilizes an all-digital receiver housed in the transducer case. Both the BioSonics and Simrad portable sounders are light weight and easily operated from small boats with battery or generator power.

The University of Washington, School of Fisheries and the Applied Physics Laboratory developed a low-cost system that will run unattended for long periods (Robert Miyamoto, personal communication). The system uses a portable PC (like the

Biosonics and Simrad sounders). The detected fish signals come from a commercial echo sounder costing less than $1000. Total system costs are estimated at about $5000, plus the electronics expertise to integrate the pieces.

Thus, fisheries acoustics systems could cost almost $200,000 or as little as $5000. The high-priced system would include split-beam or dual-beam technology and a microcomputer workstation. The low cost system would use single beam technology and a portable personal computer. High cost systems provide both more acoustic power and more options for data processing.

ACKNOWLEDGMENTS

The authors express appreciation to the field personnel from the Punta Morales, University of Costa Rica marine field station for the collection of the hydroacoustic data from the Gulf of Nicoya. Edgar Meyhoefer assisted in the electronics design and Biosonics, Inc. supplied partial support for the development of the acoustic beam detection zone program. Drs. Robert Keiser, Jim Traynor and Edward Belcher gave us critical reviews. Dr. Keiser patiently read two drafts. Katherine Peterson, Erin Moline and Steven Anderson typed the many drafts. This publication was supported by the U.S. Agency for International Development under grant #DAN-4146-G-SS-5071-00, Fisheries Stock Assessment, Collaborative Research Support Program (CRSP). This publication is Contribution No. 899, University of Washington, School of Fisheries. Correspondence should be addressed to John Hedgepeth, Biosonics, Inc., 3670 Stoneway N., Seattle, WA, 98103.

REFERENCES

Abramowitz M. and Stegun, I.A. (Eds.) 1965. *Handbook of Mathematical Functions*. Dover Publications, Inc New York.

Amjoun, B., Lai, H.-L. and Gallucci, V.F. 1990. User's guide to program CASA, a catch-at-size analysis. Release 1.0 Management Assistance for Artisanal Fisheries, Center for Quantitative Science, School of Fisheries, University of Washington, Seattle.

Anderson, V.C. 1950. Sound scattering from a fluid sphere. J. Acoust. Soc. Am. 22(4): 426-431.

Armstrong, M., Shelton, P., Hampton, I., Jolly, G. and Melo, Y. 1988. Egg production estimates of anchovy biomass in the southern Benguela system. Rep.-CCOFI 29: 137-157.

Bjornoe, L. and Kjaergaard, N. 1986. Broadband acoustical scattering by individual fish. Associated Symp. on Underwater Acoustics, 12th Int. Congress of Acoust. Halifax, Canada.

Bowditch, N. 1984. American Practical Navigator. Defense Mapping Agency. Washington, D.C.

Brandt, S.B. 1987. Patchiness in fish distributions: a help or hindrance to acoustic stock assessment? Paper, International Symposium on Fisheries Acoustics. June 22-26, 1987, Seattle, Washington.

Bunday, B.D. 1984. *Basic Optimization Methods*. Edward Arnold, Baltimore.

Burczynski, J. 1982. Introduction to the use of sonar systems for estimating fish biomass. FAO Fisheries Technical Paper No. 191, Revision 1. Food and Agriculture Organization of the United Nations, Rome.

Burczynski, J.J. and Johnson, R.L. 1986. Application of dual-beam acoustic survey techniques to limnetic populations of juvenile sockeye salmon (*Oncorhynchus merka*). Can. J. Fish. Aquat. Sci. 43: 1776-1788.

Carlson, A.B., Gisser D.G. and Manasse, F.K. 1981. *Electrical Engineering: Concepts and Applications*. Addison-Wesley Publishing Company. Boston, MA.

Chapman, C.F. 1977. *Piloting, Seamanship and Small Boat Handling*. Motor Boating and Sailing, New York.

Clarke, T.A. 1983. Comparison of abundance estimates of small fishes by three towed nets and preliminary results of the use of small purse seines as sampling devices. Biological Oceanography. 2(2-3-4): 311-340.

Clay, C.S. 1983. Deconvolution of the fish scattering PDF from the echo PDF for a single transducer sonar. J. Acoust. Soc. Am. 73(6): 1989-1994.

Clay, C.S. and Heist, B.G. 1984. Acoustic scattering by fish: Acoustic models and a two-parameter fit. J. Acoust. Soc. Am. 75(4): 1077-1083.

Clay, C.S. and Medwin, H.. 1977. *Acoustical Oceanography: Principles and Applications*. John Wiley and Sons, New York.

Cochran, W.G. 1977. *Sampling Techniques*. John Wiley and Sons, New York.

Craig, R.E. and Forbes, S.T. 1969. Design of a sonar for fish counting. Fiskeri. Dir. Skr. Ser. Hav. Unders. 15: 210-219.

Crittenden. R.N. 1989. Abundance estimation based on echo counts. Ph.D. dissertation. University of Washington, Seattle.

Crittenden, R.N., Thomas, G.L., Marino, D.A. and Thorne, R.E. 1988. A weighted duration-in-beam estimator for the volume sampled by a quantitative echo sounder. Can. J. Fish. Aquat. Sci. 45: 1249-1256.

Degbol, R.B., Lassen, H. and Staehr, K.J. 1985. In situ determination of herring and sprat at 38 and 120 KHz. Dana. 5. 45-54.

Deriso, R.B., Quinn, T.J., II and Neal, P.R. 1985. Catch-age analysis with auxiliary information. Can. J. Fish. Aquat. Sci. 39: 1195-1207.

Deriso, R.B., Neal, P.R. and Quinn, T.J., II. 1990. Further aspects of catch-age analysis with auxiliary information. In: *Effects of Ocean Variability on Recruitment and an Evaluation of Parameters Used in Stock Assessment*. (Beamish R.J. and McFarlane G.A., Eds.) Can. Spec. Publ. Fish. Aquat. Sci. 127-135.

Ehrenberg, J.E. 1972. A method for extracting fish target distribution from acoustic echoes. Proc. 1972 IEEE Conf. Eng. Ocean Environ., 1: 61-64.

Ehrenberg, J.E. 1978a. The dual-beam system — A technique for making in situ measurements of the target strength of fish. In: Proceedings of the Conference, Acoustics in Fisheries. Held at: Faculty of Maritime and Engineering Studies, Hull College of Higher Education, Hull, England. Institute of Acoustics Underwater Acoustics Group. Paper 1.3.

Ehrenberg, J.E. 1978b. Effects of noise on in situ fish target strength measurements obtained with a dual-beam transducer system. Applied Physics Laboratory. University of Washington. APL-UW 7810.

Ehrenberg, J. E. 1979a. Calculation of the constants needed to scale the output of an echo integrator. In *Biomass Handbook*, No. 7.

Ehrenberg, J. E. 1979b. A comparative analysis of in situ methods for directly measuring the acoustic target strength of individual fish. IEEE Journal of Oceanic Engineering. 4(4): 141-152.

Ehrenberg, J.E. 1981. Analysis of split beam backscattering cross section estimation and single echo isolation techniques. Applied Physics Laboratory, University of Washington, APL-UW 8108.

Ehrenberg, J.E. 1982. New methods for indirectly measuring the mean acoustic cross section of fish. FAO Fisheries Report 300: 91-97.

Ehrenberg, J. E. 1983. A review of in situ target strength estimation techniques. FAO Fish. Rep. 300: 85-90.

Ehrenberg, J. E., Carlson., T. J., Traynor, J. J. and Williamson, N. J. 1981. Indirect measurement of the mean acoustic backscattering cross section of fish. J. Acoust. Soc. Am. 69(4): 955-962.

Foote, K.G. 1980. Importance of the swimbladder in acoustic scattering by fish: A comparison of gadoid and mackerel target strengths. J. Acoust. Soc. Am. 67(6): 2084-2089.

Foote, K.G. 1983. Linearity of fisheries acoustics, with addition theorems. J. Acoust. Soc. Am. 73(6): 1932-1940.

Foote, K.G. 1985. Rather-high-frequency sound scattering by swimbladdered fish. J. Acoust. Soc. Am. 78(2): 688-700.

Foote, K.G. 1987. Fish target strengths for use in echo integrator surveys. J. Acoust. Soc. Am. 82(3): 981-987.

Foote, K.G. and MacLennan, D.N. 1984. Comparison of copper and tungsten carbide calibration spheres. J. Acoust. Soc. Am. 75(2): 612-616.

Foote, K.G. and Traynor, J.J. 1988. Comparison of walleye pollock target strength estimates determined from in situ measurements and calculations based on swimbladder form. J. Acoust. Soc. Am. 83(1): 9-17.

Foote, K. G., Aglen, A. and Nakken, O. 1988. Measurement of fish target strength with a split-beam echo sounder. J. Acoust. Soc. Am. 80(2): 612-621.

Furusawa, M. 1988. Prolate spheroidal models for predicting general trends of fish target strength. J. Acoust. Soc. Jpn. (3) 9, 1: 13-24.

Furusawa, M. and Miyanohana, Y. 1988. Application of echo-trace analysis to estimation of behavior and target strength of fish. J. Acoust. Soc. Jpn. (E)9, 4 : 169-180.

Gerlotto, F. and Marchal, E. 1987. The concept of acoustic populations: its use for analyzing the results of acoustic cruises. Paper, International Symposium on Fisheries Acoustics. June 22-26, 1987, Seattle, WA.

Goddard, G.C. and Welsbey, V.G. 1977. Statistical measurements of the acoustic target strength of live fish. Rapp. P. - V. Reun. Cons. int. Explor. Mer. 170: 70-73.

Green, R.G. 1979. *Sampling Design and Statistical Methods for Environmental Biologists*. John Wiley and Sons, New York.

Greenlaw, C. F. 1979. Acoustical estimation of zooplankton populations. Limnol. Ocean. 24(2): 226-242.

Greenlaw, C.F. and Johnson, R.K. 1983. Multiple-frequency acoustical estimation. Biological Oceanography. 2(2-3-4): 227-252.

Hampton, I., Armstrong, M.J., Jolly, G.M. and Shelton, P.A. 1990. Assessment of anchovy spawner biomass off South Africa through combined acoustic and egg-production surveys. In: Int. Symp. on Fisheries Acoustics, Seattle, WA (USA), June 22-26, 1987, Developments in Fisheries Acoustics: A Symposium Held In Seattle. (Karp, W.A., Ed.) Rapp P. - V. Reun. Cons. Int. Explor. Mer., 189: 18-32

Haslett, R. W. G. 1964. Physics applied to echo sounding for fish. Ultrasonics. 11-22.

Hedgepeth, J.B. 1993. Programs for the Simulation and Estimation of Fisheries Acoustics: Parameters for Size Based Analysis. CRSP Working Paper No. 105.

Hedgepeth, J.B. 1994. Stock Assessment with Hydroacoustic Estimates of Abundance via Tuning and Smoothed EM Estimation. PhD. dissertation, University of Washington, Seattle, WA.

Hedgepeth, J.B. and Meyhoefer, E. 1989. A low cost, high resolution instrumentation system for analyzing fisheries sonar data. From Oceans '89 Proceedings. IEEE Publications. 1: 311.

Hedgepeth, J.B. and Thorne, R.E. 1989. Hydroacoustic assessment of fish stocks in the Gulf of Nicoya, Costa Rica. From Oceans '89 Proceedings. IEEE Publications. 4: 1039-1945.

Holliday, D.V., Pieper, R.E. and Kleppel, G.S. 1989. Determination of zooplankton size and distribution with multifrequency acoustic technology. J. Cons. Int. Explor. Mer. 46: 52-61.

Huang, K. and Clay, C.S. 1980. Backscattering cross sections of live fish: PDF and aspect. J. Acoust. Soc. Am. 67(3): 795-802.

Jacoby, S.L.S., Kowalik, J. S. and Pizzo, J.T. 1972. *Iterative Methods for Nonlinear Optimization Problems*. Prentice-Hall, Englewood Cliffs, N.J.

Johannesson, K.A. and Mitson, R.B. 1983. Fisheries acoustics: a practical manual for aquatic biomass estimation. FAO Fisheries Technical Paper No. 240. Food and Agriculture Organization of the United Nations, Rome.

Johnson, R.K. 1977. Sound scattering from a fluid sphere revisited. J. Acoust. Soc. Am. 61: 375-377.

Jolly, G.M. and Hampton, I. 1990a. A stratified random transect design for acoustic surveys of fish stocks. Can. J. Fish. Aquat. Sci. 47(7): 1282-1291

Jolly, G.M. and Hampton, I. 1990b. Some problems in the statistical design and analysis of acoustic surveys to assess fish biomass. In Int. Symp. on Fisheries Acoustics, Seattle, WA (USA), June 22-26, 1987, Developments in Fisheries Acoustics: A Symposium Held In Seattle, (Karp, W.A., Ed.) Rapp P.-V. Reun. Cons. Int. Explor. Mer., 189: 415-420

Jolly , G.M. and Smith, S.J. 1989. A note on the analysis of marine data. In Progress in Fisheries Acoustics, Proceedings of the Institute of Acoustics. Lowestoff, England. 11(3): 196-201.

Jolly, G.M. and Watson, R.M. 1979. Aerial sample survey methods in the quantitative assessment of ecological resources. *Sampling Biological Populations*. Int. Co-op. Pub. House. Fairland, MD.

Kalish, J.M., Greenlaw, C.F., Pearcy, W.G. and Holliday, D.V. 1986. The biological and acoustical structure of sound scattering layers. Deep-Sea Research. 33(5): 631-653.

Kieser, R. and Ehrenberg, J.E. 1987. An unbiased, stochastic echo counting model. Paper, International Symposium on Fisheries Acoustics. Seattle, WA.

Kimura, D.K. 1989. Variability, tuning, and simulation for the Doubleday-Deriso catch-at-age model. Can. J. Fish. Aquat. Sci. 46: 941-949.

Kimura, D.K. 1990. Approaches to age-structured separable sequential population analysis. Can. J. Fish. Aquat. Sci. 47: 2364-2374.

Kinsler, L.E., Frey, A.R., Coppens, A.B. and Sanders, J.V. 1982. *Fundamentals of Acoustics*. John Wiley and Sons, NY.

Le Bourges, A. 1990. Species recognition based on the spectral signature of individual targets. ICES C.M. 1990/B: 9.

Lindem, T. 1983. Successes with conventional in situ determinations of fish target strength. FAO Fisheries Rep. 300: 104-111.

Love, R.H. 1971a. Dorsal-aspect target strength of an individual fish. J. Acoust. Soc. Am. 49: 816-823.

Love, R.H. 1971b. Measurements of fish target strengths: A review. Fishery Bulletin. 69: 703-715.

Love, R.H. 1977. Target strength of an individual fish at any aspect. J. Acoust. Soc. Am. 62: 1397-1403.

Machlup, S. 1951. A theoretical model for sound scattering by marine crustaceans. J. Acoust. Soc. Am. 24(3): 290-293.

MacLennan, D.N. 1981. The theory of solid spheres as sonar calibration targets. Scottish Fisheries Research. Report No. 22.

MacLennan, D.N. 1982. Target strength measurements on metal spheres. Scottish Fisheries Research Report. 25: 1-11.

MacLennan, D.N. and Dunn, J.R. 1984. Estimation of sound velocities from resonance measurements on tungsten carbide calibration spheres. J. Sound and Vibration. 97(2): 321-331.

MacLennan, D.N. and Forbes, S.T. 1987. Acoustic methods of fish stock estimation. In: *Developments in Fisheries Research in Scotland*. (Bailey R.S. and Parrish B.B., Eds.) Fishing News Books, Farnham, England. 40-55.

MacLennan, D.N. and MacKenzie, I.G. 1988. Precision of acoustic fish stock estimates. Can. J. Fish. Aquat. Sci. 45: 605-616.

Megrey, B.A. 1988. Population dynamics of walleye pollock (*Theragra chalcogramma*) in the Gulf of Alaska. Ph.D. dissertation. University of Washington, Seattle, WA.

Megrey, B.A. 1989. Application of three catch-at-age models to stocks of walleye pollock (*Theragra chalcogramma*) in the western Gulf of Alaska. Paper for Gadoid Conference.

Megrey, B. A., Hollowed, A. B. and Methot, R. D. 1989. Integrated analysis of Gulf of Alaska walleye pollock catch-at-age and research survey data using two different stock assessment procedures. U.S. Department of Commerce, NOAA. Seattle, WA.

Methot, R.D. 1990. Synthesis model: An adaptable framework for the analysis of diverse stock assessment data. INPFC Bull. 50: 259-277.

Mood, A., Graybill, F.A. and Boes, D.C. 1974. *Introduction to the Theory of Statistics*. McGraw-Hill, NY.

Molloy, N.D. 1977. Effects of noise on an in situ technique for estimating the acoustic cross section of a fish population. M.S. thesis. University of Washington, Seattle, WA.

Murphy, G.I. and Clutter, R.I. 1972. Sampling anchovy larvae with a plankton purse seine. Fish. Bull. U.S. 70: 789-798.

Nakken, O. and Olsen. K. 1977. Target strength measurements of fish. Rapp. P.-V. Reun. Cons. Int. Explor. Mer. 170: 52-69.

Penrose, J.D. and Kaye, G.T. 1979. Acoustic target strengths of marine organisms. J. Acoust. Soc. Am. 65: 374-380.

Peterson, M. L., Clay, C. S. and Brandt, S. B. 1976. Acoustic estimates of fish density and scattering function. J. Acoust. Soc. Am. 60(3): 618-622.

Robinson, B.J. 1979. In situ measurements of fish target strength. In: Proceedings of the Conference, Acoustics in Fisheries. Held at: Faculty of Maritime and Engineering Studies, Hull College of Higher Education, Hull, England. Institute of Acoustics Underwater Acoustics Group. Paper 1.8.

Robinson, B.J. 1983. In situ measurements of the target strengths of pelagic fishes. FAO Fisheries Report No. 300: 99-103.

Robinson, E.A. 1983. *Multichannel Time Series Analysis with Digital Computer Programs*. Goose Pond Press, Houston, Texas.

Rose, G.A. and Leggett, W.C. 1988. Hydroacoustic signal classification of fish schools by species. Can. J. Fish. Aquat. Sci. 45: 597-604.

Ross, D. 1976. *Mechanics of Underwater Noise*. Pergamon Press, New York.

Rudstam, L.G., Clay, C.S. and Magnuson, J.J. 1987. Density and size estimates of Cisco (*Coregonus artedii*) using analysis of echo peak PDF from a single-transducer sonar. Can. J. Fish. Aquat. Sci. 44: 811-821.

Rudstam, L.G., Lindem, T. and Hansson, S. 1988. Density and in situ target strength of herring and sprat: a comparison between two methods of analyzing single-beam sonar data. Fisheries Research. 6: 305-315.

Scheaffer, R.L., Mendenhall, W. and Ott, L. 1986. *Elementary Survey Sampling*. 3rd Edition. Duxbury Press, Boston.

Scherba, S., Jr. and Gallucci, V.F. 1976. The application of systematic sampling to a study of infauna variation in a soft substrate environment. Fishery Bulletin 74(4): 937-948.

Sherman, S.M. 1984. *Monopulse Principles and Techniques*. Artech House, MA.

Shotton, R. and Bazigos, G.P. 1984. Techniques and considerations in the design of acoustic surveys. Rapp. P. - V. Reun. Cons. Int. Explor. Mer. 184: 34-57.

Silverman, B.W., Jones, M.C. and Wilson, J.D. 1990. A smoothed EM approach to indirect estimation problems, with particular reference to stereology and emission tomography. J. Roy. Stat. Soc., Ser. B. 52(2): 271-324.

Simmonds, E.J. and Armstrong, F. 1987. A wide band echosounder: measurement on cod, saithe, herring and mackerel from 27 to 54 kHz. Contrib. Int. Symp. Fish. Acoustics, June 22-26, 1987, Seattle, WA.

Simmonds, E.J., Williamson, N.J., Gerlotto, F. and Aglen, A. 1991. Survey design and analysis procedures: a comprehensive review of good practice. ICES. CM 1991/B: 54. Fish Capture Committee Theme Session U Application and Analyses of Acoustic Methods.

Siniff, D. B. and Skoog, R. O. 1964. Aerial censusing of caribou using stratified random sampling. J. Wildl. Manage. 28(2): 391-401.

Sissenwine, M.P., Azarovitz, T.R. Suomala, J.B. 1983. Determining the abundance of fish. In: *Experimental Biology at Sea*, (MacDonald, A.G. and Priede, I.G., Eds.) Academic Press, London: 51-101.

Souid, P. 1989. Automatisation de la description et de la classification des detections acoustiques de bancs de poisoons pelagiques pour leur identification. These Doct. Univ. Aix-Mairselle II, Dec. 1988.

Stanton, T.K. 1985. Volume scattering: echo peak PDF. J. Acoust. Soc. Am. 77(4): 1358-1366.

Stanton, T.K. 1989. Simple approximate formulas for backscattering of sound by spherical and elongated objects. J. Acoust. Soc. Am. 86(4): 1499-1510.

Stanton, T.K. and Clay, C.S. 1986. Sonar echo statistics as a remote-sensing tool: volume and seafloor. IEEE Journal of Oceanic Engineering. OE-11(1): 79-96.

Suhkatme, P.V. and Suhkatme, B.V. 1970. *Sampling Theory of Surveys with Applications*. Iowa State University Press, Ames, IA.

Sullivan, P.J. 1988. A Kalman filter approach to catch-at-length analysis. Ph.D. dissertation. University of Washington, Seattle, WA.

Sullivan, P.J., Lai, H.-L. and Gallucci, V.F. 1989. A catch-at-length analysis that incorporates a stochastic model of growth. CRSP Working Paper No. 66.

Sullivan, P.J., Lai, H.-L. and Gallucci, V.F. 1990. A catch-at-length analysis that incorporates a stochastic model of growth. Can. J. Fish. Aquat. Sci. 47: 184-198.

Thorne, R.E. 1983. Chapter 12: Hydroacoustics. In *Fisheries Techniques*. American Fisheries Society. Bethesda, MD.

Thorne, R.E. 1988. An empirical evaluation of the duration-in-beam technique for hydroacoustic estimation. Can. J. Fish. Aquat. Sci. 45: 1244-1248.

Thorne, R. E., Hedgepeth, J. and Campos, J. 1989. Hydroacoustic observations of fish abundance and behavior around an artificial reef in Costa Rica. Bulletin of Marine Science 44(2): 1058-1064.

Thorne, R. E., Hedgepeth, J. and Campos, J. 1990. The use of stationary hydroacoustic transducers to study diel and tidal influences on fish behavior. Rapp P. - V. Reun. Cons. Int. Explor. Mer, 189: 167-175.

Traynor, J.J. 1985. Dual beam measurement of fish target strength and results of an echo integration survey of the eastern Bering Sea walleye pollock (*Theragra chalcogramma*). Doctoral dissertation. University of Washington, Seattle,WA.

Urick, R.J. 1983. *Principles of Underwater Sound.* McGraw-Hill, New York.

Vray D., Gimenez G. and Person, R. 1987. Attempt of classification of echo-sounder signals based on the linear discriminant function of Fisher. Contrib. Int. Symp. Fish. Acoustics. June 22-26, 1987, Seattle, Washington, (USA).

Wardle, C.S., 1983. Fish reactions to towed fishing gears. In *Experimental Biology at Sea*, (MacDonald A. and Priede I.G., Eds.) Academic Press, London. 167-195

Williamson, N.J. 1982. Cluster sampling estimation of the variance of abundance estimates derived from quantitative echo sounder surveys. Can J. Fish. Aquat. Sci. 39: 229-231.

APPENDIX I. Explanation of the decrease in acoustic pressure using complex notation.

Because many of the "basic" acoustic texts treat the equations which we commonly use in fisheries acoustics as coming from solutions to partial differential equation known as wave equations, we include a short discussion. In this discussion, complex quantities will be denoted by capitals. For example, when the source of sound is treated as a small pulsating spherical surface, the wave equation (Helmholtz equation) in spherical coordinates is

$$\frac{d^2p}{dr^2}+k^2p = \frac{1}{r}\frac{d^2(rp)}{dr^2}+k^2p = 0 \qquad \text{where } k = \frac{\omega}{c},$$

$$= \frac{d^2(rp)}{dr^2}+k^2rp = 0 \qquad \text{since r is a constant}$$

The solution to this equation can be written as

$$p = A/re^{i(\omega t-kr)}$$

where $i = \sqrt{-1}$
- A = complex amplitude
- w = angular frequency in radians
- k = wave number
- r = range
- t = time

This immediately shows that acoustic pressure decreases inversely with distance, an observation we have called the spherical spreading loss. We can also include absorption losses by explaining that wave or sound speed c is complex with a dissipative out-of-phase component $\tau \ll 1$.

$$C = c(1 + i\tau)$$

This then causes the wave number k to become complex, and the spherical wave solution is

$$p = A/re^{i(\omega t-Kr)} = A/re^{k\tau r}e^{i(\omega t-kr)}$$

$$p(1)/p(r) = \left|Ae^{k\tau}e^{i(\omega t-k)}\right| / \left|A/re^{k\tau r}e^{i(\omega t-kr)}\right|$$

$$= re^{k\tau r}/e^{k\tau}$$

The transmission loss then becomes

$$\begin{aligned} TL &= 20\log p(1)/p(r) \\ &= 20\log r + 20\log e^{k\tau r} - 20\log e^{k\tau} \\ &= 20\log r + \alpha r - \alpha \qquad \text{letting } \alpha = 8.696\, k\tau \text{, since } k\tau \ll 1 \\ &= 20\log r + \alpha r \end{aligned}$$

APPENDIX II. Derivation of a ratio estimator based on density (numbers per unit volume); a useful reference for ratio estimation is Cochran (1977).

An estimate of density d_i is a ratio of two variates:

$$\hat{d}_i = \frac{N_i}{S_i}$$

$$\mathrm{Var}(\hat{d}_i) = \frac{L_i \sum \{S_{i,t}(N_i/S_i) - N_{i,t}\}^2}{(L_i - 1)S_i^2}$$

where $S_{i,t}$ = volume sampled in stratum i during transect t
S_i = volume sampled in stratum i
$N_{i,t}$ = estimated number in volume $S_{i,t}$
N_i = estimated number in volume S_i
L_i = number of transects in stratum i

One may wish to use population estimates via a ratio estimator if the variance of the estimated number of fish is proportional to the volume. The ratio estimator is unbiased as long as the covariance between d_i and S_i is 0. Sukhatme and Sukhatme (1970) say, "The bias to a first approximation vanishes when the regression of y on x [in our case N_i on S_i] is a straight line through the origin." The stratum population and variance can be estimated using the stratum volume.

$$P_i = \hat{d}_i V_i$$

$$\mathrm{Var}(\hat{P}_i) = V_i^2 \mathrm{Var}(\hat{d}_i)$$

where V_i = volume of stratum i
P_i = estimated population in stratum i

The stratified random sample estimate for the total population, P, would be:

$$\hat{P} = \frac{\sum \hat{P}_i V_i}{\sum_{1} V_i}$$

$$\mathrm{Var}(\hat{P}) = \sum_{1} \frac{V_i^2 \sum \{S_{i,t}(N_i/S_i) - N_{i,t}\}^2}{L_i(L_i - 1)\sum_{1} V_i^2}$$

This design allows post stratification using volumes sampled in each ping ($S_{i,t}$ = sum over all the pinged volumes in the transect). The actual weights were computed using known water column volumes in each stratum (V_i).

APPENDIX III. The CRSP Working Paper #105 by Hedgepeth (1993) contains a number of programs for use in the application of acoustics. The program names and a short annotation are listed below.

BINEW.C	A bisection and Newton algorithm for estimating population dynamics parameters F, Given C, N and M. Useful when applying some stock assessment packages, e.g., Deriso et al. (1985).
CRTBEAM.C	This program allows the user to construct a beam detection zone given some echosounder parameters and information about the size of fish.
DECON.C	A deconvolution program which uses both Gaussian elimination and least squares fitting to estimate the on-axis voltage distributions from fish echoes. See EMS.C.
EMSSVD.C	An inverse estimation algorithm using Expectation Maximization and Smoothing to estimate the on-axis voltage distributions from fish echoes. See DECON.C.
EBSQUARE.C	Numerical integration program for the expected value of the beam pattern factor squared.
PDFBPLOT.C	Program to plot the probability of the beam pattern factor.
SIMLEN.FOR	This FORTRAN program generates length distributions of fishery catch and fish populations and uses population parameters as input.
SONSIM.C	A single beam sonar simulator. Allows the user to simulate the field data collection of fisheries acoustic data. Input includes the distribution and abundance of fishes in the water column. Output includes the simulated voltage collections from a signal processor as well as oscilloscope display.

CHAPTER 7

THE APPLICATION OF TIME SERIES ANALYSIS TO FISHERIES POPULATION ASSESSMENT AND MODELING

Brian J. Rothschild[1], Steven G. Smith[1], and Helen Li[1]

[1]Chesapeake Biological Laboratory, Center for Estuarine and Environmental Studies, University of Maryland

1. INTRODUCTION

A primary concern of fishery science has been dynamic systems modeling, specifically the construction of mathematical models that determine values of fishing yield (harvest) through time. Consider the most common fishery dynamic system, that of fishing effort and catch diagrammed in Figure 1. In its most elemental form, this system is composed of an input, fishing effort, an output, catch, and a mathematical model which translates values of fishing effort to catch. The mathematical model of Figure 1 can be constructed from two very different approaches. The first, and most common in fishery science, is to derive the model from fundamental "laws of nature" or "principles of population dynamics". An example of this approach is the well-known equilibrium yield (or surplus production) model which relates fishing effort to catch (Schaefer, 1957; Ricker, 1975):

$$Y = FB \tag{1}$$

1-56670-151-1/96/$0.00+$.50

Thus, the contents of the "box" in Figure 1 are described completely by this equation. This is sometimes referred to as the mechanistic or **deterministic** approach to dynamic systems modeling in that the system is analogous to a machine which produces outputs in response to specified inputs. The second approach, referred to as the probabilistic or **stochastic** approach, takes the opposite point of view: the contents of the box are completely unknown, and no preconceived notion of the mathematical model exists. The model of Figure 1 is thus a "black box" which incorporates all of the factors acting in concert that account for the relationship between fishing effort and catch, including biological processes, physical environment variables, market demand, and angler behavior. From a mathematics point of view, a probabilistic model is used to describe the processes of the box. The fundamental difference between these two modeling approaches is that the deterministic approach describes a model in advance which is subsequently used to fit collected data, while the stochastic approach deduces a model solely from the actual data.

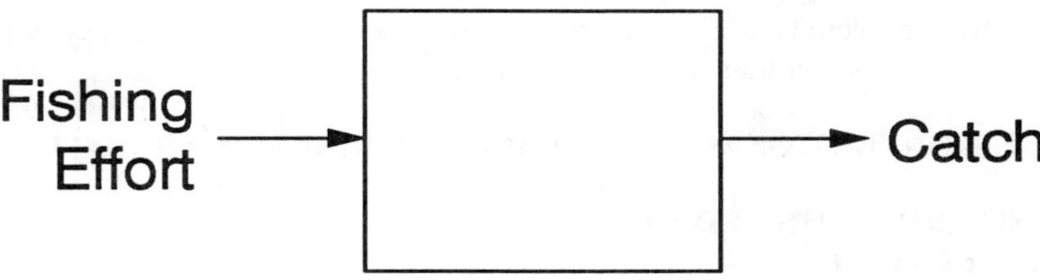

Figure 1. Black box diagram of fishery effort-catch input-output system.

Traditionally, the fishing effort and catch data used in dynamic systems modeling have been collected and compiled on an annual basis. A more recent trend is to sample and analyze catch and effort data over shorter time scales, such as months or weeks. As fisheries data sampling and information handling occur more frequently, data streams of catch and effort may be correlated in time. This poses a potential problem since virtually all of the statistical procedures employed in analyzing catch and effort data, as well as the estimation procedures for deterministic equilibrium yield models, assume that the collected data are independent and identically distributed. Time-correlated data thus break the important assumption of independence.

Time series analysis, the subject of this chapter, constructs stochastic models based upon time correlations of collected data. In the fisheries context, three important objectives of time series analysis are:

(1) to describe the underlying stochastic structure of a dynamic system (e.g. Figure 1),

(2) to simulate the output series of a dynamic system for different sets of input data, and

(3) to provide short-term forecasts of the output variable of a dynamic system.

Of these three objectives the third, forecasting, has been the primary focus of time series analysis applications to fisheries data, and is most commonly employed to predict catch or catch-per-unit-of-effort (CPUE) values for the next fishing season (Saila et al., 1980; Mendelssohn, 1981; Roff, 1983; Jensen, 1985; Stergiou, 1989). However, the first two objectives provide more insight into the fishery system's dynamics. Deducing the stochastic process that generated the data (objective 1) and simulating this process under a variety of input scenarios (objective 2) may eventually lead a fishery scientist to an understanding of the biological, physical, and economic processes which drive the dynamic system.

The intent of this chapter is threefold: (1) to present the fundamental concepts, models, and application procedures of conventional time series analysis; (2) to enhance the conventional procedures through the incorporation of system identification theory, procedures for handling real data, and spectral analysis; and (3) to illustrate the power of time series modeling to fisheries population assessment and management, with emphasis on input-output transfer functions as a generalized model of fishing mortality.

2. CONVENTIONAL TIME-DOMAIN TIME SERIES ANALYSIS

2.1 BASIC CONCEPTS AND MODELS

2.1.1 Basic Concepts

A **time series** is any sequence of data observations indexed with respect to time. A more formal definition of a time series begins by denoting z_t as an observation at time t. The set of consecutive observations from time 1 to n, $z_1, z_2, \ldots, z_n$, forms the time series $\{z_t\}$. In this paper, any general time series of data observations is represented by $\{z_t\}$, while $\{x_t\}$ and $\{y_t\}$ are reserved to represent an input and output time series, respectively. Table 1 lists the values of an example time series, commercial landings of blue crabs (*Callinectes sapidus* Rathbun) in the Maryland portion of Chesapeake Bay for the months of April to November, 1963 to 1987. The length of this time series is 200 observations, resulting from catch values obtained for an 8-month crabbing season over the course of 25 years. Since this dynamic system consists of only one variable, blue crab catch, the time series is termed **univariate**. The other type of dynamic system considered in this chapter, the **input-output** system of Figure 1, consists of two variables and is thus termed **bivariate**. Table 2 presents the values of an example bivariate time series, monthly commercial fishing effort and yellowfin tuna (*Thunnus albacares*) catch from a sampling region of the eastern tropical Pacific Ocean for January, 1969, to December, 1986. This example time series is actually two concurrent univariate time series, both of which have 216 observations.

The blue crab and yellowfin tuna data sets will be used as examples throughout this chapter for the following two reasons: (1) both are representative of the most common fisheries time series, commercial landings or effort-landings data; and (2) both are of sufficient length. It should be noted that the generally accepted minimum number of observations required to accurately fit a time series model is 50. Thus, fisheries information that is collected and compiled on an annual basis does not, in most cases, lend itself to this type of analysis (annual time series of 50 or more years being quite rare in fisheries). Time series analysis is, on the other hand, well suited to fisheries data sets collected more frequently, e.g. a time scale of months or weeks.

Table 1. Blue crab (*Callinectes sapidus*) monthly commercial landings (lbs.) in the Maryland portion of Chesapeaks Bay for April to November crabbing seasons, 1963–1987. (Source: Maryland Department of Natural Resources)

Year	April	May	June	July	August	September	October	November
1963	8581	541930	1656388	4084592	4888548	4639541	2721349	468696
1964	186027	1567915	3540648	3105096	6179783	7797087	2291430	388084
1965	220905	2136035	5417666	6678148	9121149	6223687	3555299	1213223
1966	60270	1880178	4642929	8583969	5981047	4235279	4569269	2200679
1967	383238	2470224	4917895	6872367	6772746	2482924	2498534	364631
1968	12637	344773	2081335	2945883	2628004	1572698	528350	200100
1969	195929	780365	979906	5017408	6832294	4578914	4095483	1902183
1970	259270	3560694	3654467	6043009	5079737	4254116	2734616	805220
1971	445493	1586607	4635382	6051777	5911680	4670174	2834042	1269497
1972	545878	1950601	3120308	5399573	7707344	3304002	1984358	856170
1973	539856	1182168	3147801	4412613	4987717	3179491	2491633	550844
1974	1934804	1568021	3727683	6170614	5321017	4142064	2835707	553374
1975	868264	1808333	3722854	6468861	5399323	3630721	3535267	449875
1976	397881	526577	3382479	5452798	4776210	3921409	2245028	200333
1977	16816	270695	2768912	5143092	5703569	3953872	2302034	229067
1978	64462	325537	1251523	3357138	4161662	4620955	3217492	410049
1979	677657	1032179	3192192	5234211	6058942	4769857	3545250	1153065
1980	857672	1122488	4101921	5639982	4967142	4079051	4618754	1054765
1981	188182	6488333	5373226	13803030	17458300	9394090	5364815	1624825
1982	190578	2761661	6928345	11343160	11922040	6866939	2756313	893090
1983	235478	1449493	8287215	11149660	12501260	9401170	7506991	1939056
1984	635322	1316413	7156308	9235485	12383110	9257446	6639539	2147144
1985	829103	3168820	8652327	11862140	13711690	9931947	6965815	3307550
1986	613526	3635434	8675078	12276110	10100410	8505802	5402885	1002864
1987	279906	1950342	7576751	9617362	10865500	7831503	6570675	1275298

Table 2. Yellowfin tuna (*Thunnus albacares*) (a) monthly commercial fishing effort (vessel days) and (b) monthly commercial catch (1000 metric tons) from a sampling region of the eastern tropical Pacific Ocean, 1969–1986. (Source: Inter-American Tropical Tuna Commission)

(a)

Year	Jan.	Feb.	Mar.	Apr.	May	June	July	Aug.	Sept.	Oct.	Nov.	Dec.
1969	74.0	113.0	39.5	40.0	35.5	31.0	30.5	9.0	4.0	2.0	3.0	22.0
1970	43.0	82.0	103.5	97.0	66.0	11.0	13.9	10.4	6.0	6.5	36.0	27.0
1971	16.0	61.0	46.5	69.5	427.5	306.0	34.0	22.5	25.0	10.0	0.0	6.0
1972	148.5	64.0	88.5	148.0	40.5	6.0	16.5	8.7	1.0	17.0	7.0	3.5
1973	33.0	93.5	80.0	105.5	49.0	13.5	15.5	32.0	42.5	22.5	67.5	21.0
1974	256.0	372.5	138.0	597.5	521.5	82.0	15.0	30.0	91.0	207.0	97.5	13.0
1975	35.0	44.0	95.0	147.0	186.0	48.0	10.5	5.5	86.0	255.0	136.5	7.0
1976	42.5	150.5	145.0	307.0	236.0	116.5	48.5	63.0	50.5	85.5	56.0	18.5
1977	44.0	101.5	43.0	44.5	59.5	112.5	118.0	177.0	149.0	79.0	50.5	18.0
1978	82.0	68.0	156.0	258.5	273.0	120.5	59.0	51.5	33.0	34.5	57.5	9.0
1979	218.5	320.0	124.5	103.0	68.0	26.0	44.0	50.0	309.5	90.0	26.0	22.0
1980	49.0	78.0	98.5	262.5	89.0	29.5	26.5	23.5	30.5	8.0	37.5	15.0
1981	33.0	9.5	44.0	28.0	30.5	70.0	69.5	30.0	31.5	14.5	13.0	57.0
1982	94.0	123.5	92.5	146.5	76.0	21.0	12.5	9.0	8.5	27.5	23.5	39.5
1983	24.0	30.0	97.0	17.0	5.0	6.0	7.0	5.0	12.0	2.0	3.0	2.0
1984	19.0	15.5	16.0	20.0	57.0	26.0	41.0	32.5	25.0	6.5	7.0	48.0
1985	77.5	26.0	56.5	44.0	25.0	32.0	16.5	12.0	21.5	13.0	22.0	55.0
1986	61.5	60.0	66.0	35.0	40.0	44.0	50.0	16.0	24.0	173.0	65.5	20.0

(b)

Year	Jan.	Feb.	Mar.	Apr.	May	June	July	Aug.	Sept.	Oct.	Nov.	Dec.
1969	0.976	1.810	0.466	0.802	0.652	0.213	0.055	0.000	0.001	0.000	0.000	0.050
1970	0.182	1.436	2.067	1.178	0.852	0.077	0.084	0.059	0.023	0.000	0.017	0.048
1971	0.237	1.094	0.791	1.125	2.741	0.557	0.043	0.042	0.018	0.000	0.000	0.000
1972	2.925	1.565	1.755	1.362	0.696	0.000	0.017	0.000	0.000	0.192	0.000	0.000
1973	0.033	0.466	0.624	1.280	0.839	0.000	0.028	0.207	0.412	0.005	0.000	0.039
1974	3.756	2.699	0.834	4.532	6.228	0.308	0.051	0.028	0.288	0.314	0.243	0.037
1975	0.230	0.199	0.591	0.485	0.603	0.120	0.016	0.011	0.124	0.328	0.157	0.010
1976	0.238	0.794	1.541	1.971	1.200	0.778	0.074	0.015	0.134	0.020	0.000	0.000
1977	0.045	0.383	0.057	0.721	0.302	1.291	0.640	1.042	1.404	0.795	0.007	0.000
1978	0.265	0.174	1.031	1.945	2.970	0.348	0.187	1.030	0.218	0.148	0.069	0.011
1979	1.137	1.015	1.095	0.161	0.543	0.240	0.342	0.122	1.757	0.440	0.133	0.069
1980	0.195	0.136	0.497	0.635	0.498	0.038	0.077	0.051	0.118	0.058	0.030	0.019
1981	0.225	0.022	0.281	0.119	0.007	0.605	0.141	0.125	0.089	0.000	0.029	0.712
1982	0.545	0.870	0.481	0.259	0.325	0.038	0.007	0.051	0.118	0.058	0.003	0.180
1983	0.009	0.113	0.336	0.064	0.040	0.119	0.199	0.018	0.037	0.000	0.000	0.000
1984	0.234	0.387	0.438	0.200	0.919	0.307	0.361	0.127	0.075	0.000	0.000	0.791
1985	1.095	0.143	0.529	0.075	0.369	0.564	0.089	0.156	0.122	0.052	0.312	0.768
1986	0.991	1.510	0.995	0.205	0.495	0.529	0.418	0.0	0.201	3.245	0.207	0.526

A **time series plot** of the blue crab data is presented in Figure 2. From the plot it can be observed that the monthly blue crab catch follows an annual periodic pattern of rise and fall that continues for the length of the series. This illustrates the basic precept of time series analysis: Historical data values may exhibit a stable pattern of time-dependency which can subsequently be used to predict future data values. Figure 3 is a plot of the blue crab catch for the 1965 and 1966 crabbing seasons. This small portion of the complete data series illustrates the two types of time-dependency: (1) low order and (2) seasonal. **Low order** time-dependency refers to time correlations among observations closely spaced in time. For example, the blue crab catch in May 1965 gives a good indication of what the blue crab catch will be in June of the same year, which in turn gives a good indication of the catch in July. Likewise, the blue crab catch in June 1965 provides a good indication of the catch in June 1966, indicating the existence of **seasonal** time-dependency in the blue crab series. In contrast, the blue crab catch in May 1965 does not provide a good measure of the catch in September 1965 or, for that matter, September 1966. Figure 3 thus illustrates that the relationship is stronger, or "weighted" more heavily, among observations closer together or related seasonally in time than among those farther away or not related seasonally. The low order and seasonal time dependencies of the blue crab series are further demonstrated by the scatterplots presented in Figure 4 in which the observations at times t+1 and t+8 are plotted against those at time t. In the language of time series analysis, time-dependencies are referred to as **lags**. Thus, Figures 4a, and 4b, respectively illustrate the lag 1 and lag 8 relationships of the blue crab data set.

Time series analysis falls under the general discipline of statistics; as a result, its discussion is immersed in mathematical notation. Much of the notation we use to describe time series models in this chapter follows that of Ljung (1987) rather than the more conventional notation of Box and Jenkins (1976). We use Ljung's notation for two reasons: (1) it provides for a clear and concise description of time series models, and (2) it facilitates the evaluation of Box-Jenkins models as a subgroup of the entire class of general linear system models.

2.1.2 Basic Models

In time series analysis, the "black box" of Figure 1 is represented by a stochastic mathematical model. Three basic models are commonly employed to describe univariate systems: (1) autoregressive, (2) moving average, and (3) mixed autoregressive-moving average. A fourth model, the Box-Jenkins transfer function, is typically used to describe bivariate systems. These models are elaborated in the following sections.

2.1.2.a *Autoregressive Model*

The **autoregressive** (AR) model assumes that the current value of an output series can be described by weighted values of past observations and a random error component. The general AR model is

$$y_t = a_1 y_{t-1} + a_2 y_{t-2} + \dots + a_{n_a} y_{t-n_a} + e_t \tag{2}$$

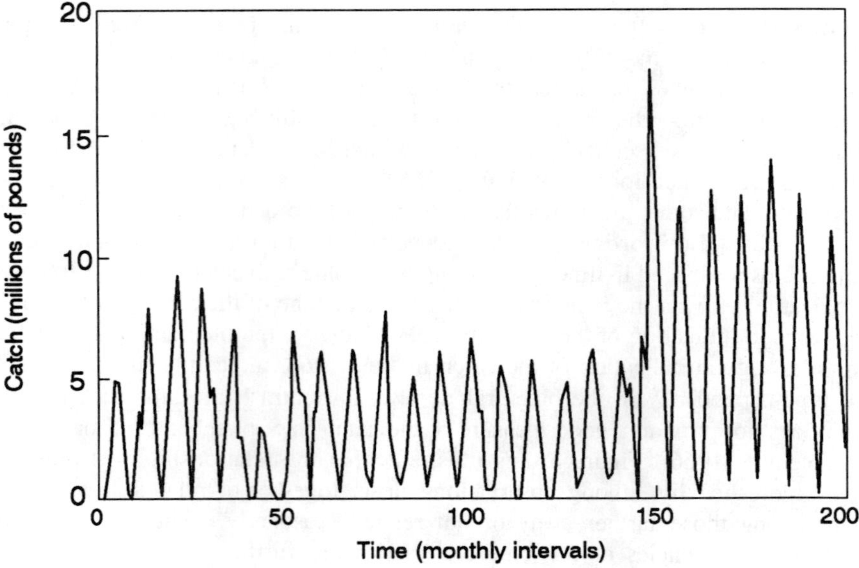

Figure 2. Time series plot of blue crab catch data from Table 1. Time values (x-axis) are consecutively numbered months for the April to November 8-month crabbing season for years 1963 to 1987 (i.e., month 1 = April 1963, month 9 = April 1964, etc.).

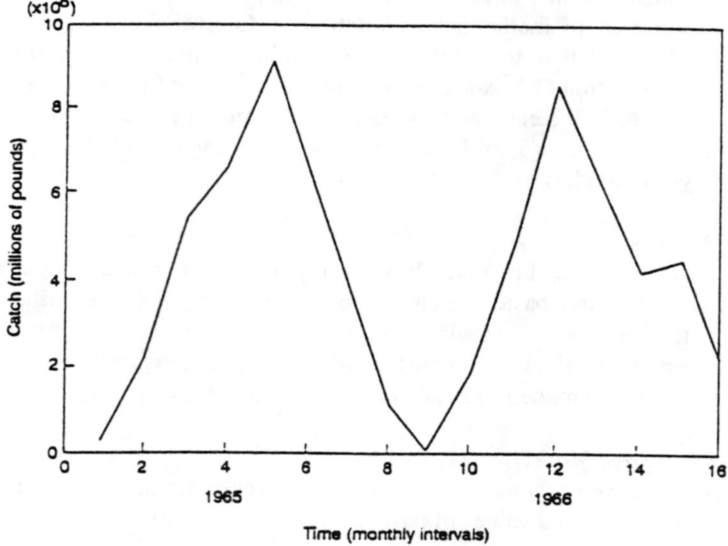

Figure 3. Time series plot of blue crab catch data (Table 1) for the 1965 and 1966 crabbing seasons (month 1 = April 1965, month 8 = November 1965, month 9 = April 1966, month 16 = November 1966).

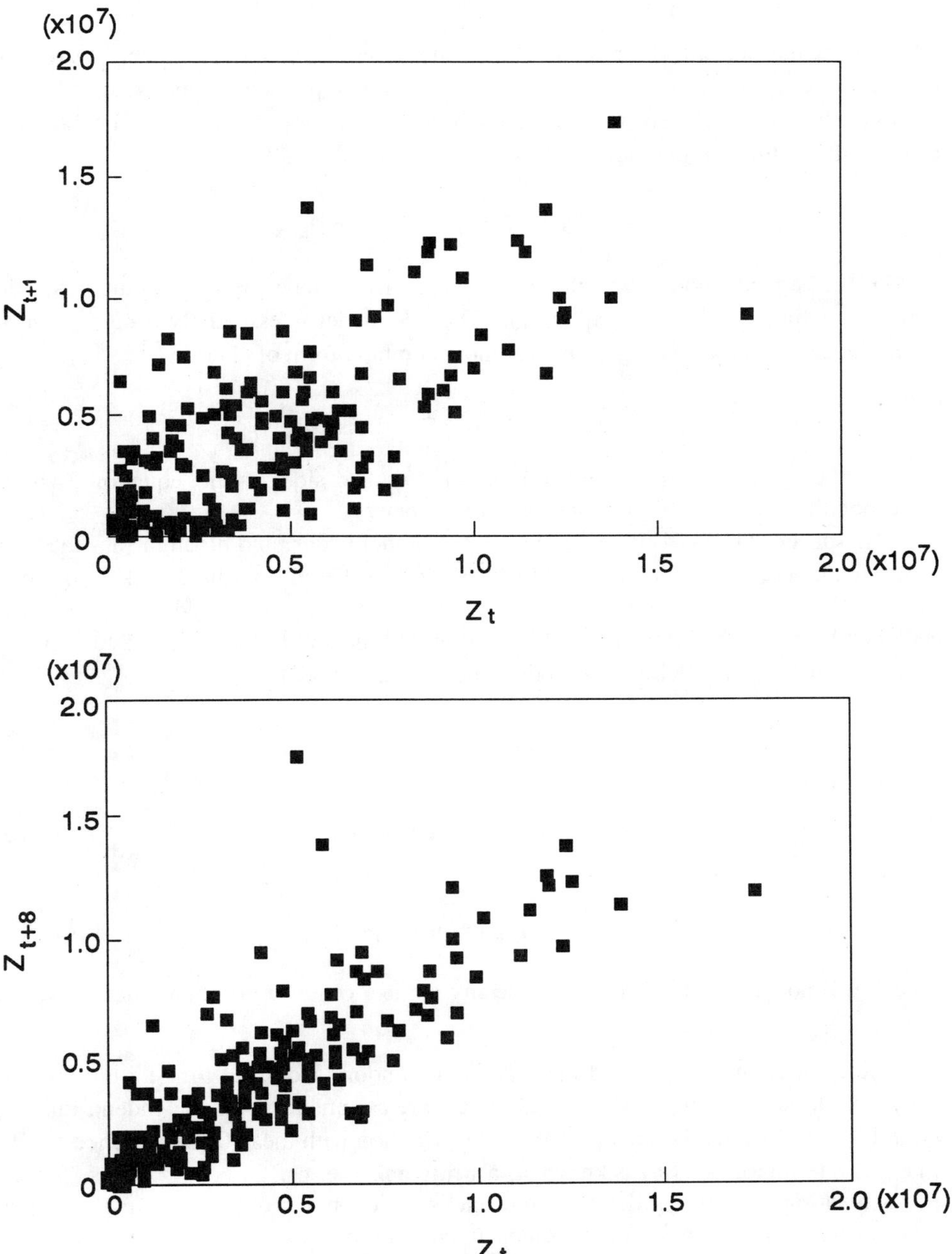

Figure 4. Catch values at (a) time t + 1 and (b) time t + 8 plotted against values at time t for the blue crab data series (Table 1), illustrating the (a) lag 1 and (b) seasonal lag 8 relationships within the data series

where y_t is the current observation of the output series; y_{t-1} , y_{t-2} ,..., y_{t-n_a} are past observations; a_1, a_2 ,..., a_{n_a} are the associated weights of past observations; and e_t is the random error term. Notice the similarity between equation (2) and the general first-order multiple linear regression model (Draper and Smith, 1981):

$$y = \beta_0 + \beta_1 x_1 + \beta_2 x_2 + ... + \beta_k x_k + \varepsilon \tag{3}$$

In (2), the lagged output variables y_{t-1}, y_{t-2} ,..., y_{t-n_a} have been substituted for the predictor variables x_1 , x_2 ,..., x_k of (3). The AR model is essentially a regression on itself; hence the name *auto*regressive. A more common form of (2) is

$$y_t + a_1 y_{t-1} + a_2 y_{t-2} + ... + a_{n_a} y_{t-n_a} = e_t \tag{4}$$

in which the output terms are collected on the left side of the equation (with a corresponding change in sign of the weight parameters).

The subscript n_a in (4) is referred to as the **model order**, and distinguishes the term in the model corresponding to the observation which lies the farthest back in the past from the current observation. AR models are classified by the model order n_a. The general AR model of (4) is thus referred to as an AR model of order n_a , or as an AR(n_a) model. Examples of different AR models are an AR(1) model,

$$y_t + a_1 y_{t-1} = e_t \tag{5}$$

an AR(2) model,

$$y_t + a_1 y_{t-1} + a_2 y_{t-2} = e_t \tag{6}$$

and an AR(12) model,

$$y_t + a_1 y_{t-1} + a_{12} y_{t-12} = e_t \tag{7}$$

Note that a model of order n_a may include any number of lower order parameters a_1, a_2 , ..., a_{n_a-1}.

The error term e_t in (2) and (4) – (7) is a "random shock" at time t from the un-observable time series $\{e_t\}$. The values of $\{e_t\}$ are assumed to be independent, random variables drawn from a Normal probability distribution with mean 0 and variance σ^2. In time series terminology, $\{e_t\}$ is known as a **white noise** series.

A notation-simplifying device is the **backshift operator** q^{-k} which, in its general form, performs the following operation on an observation z_t:

$$q^{-k} z_t = z_{t-k} \tag{8}$$

where k = 1, 2, ..., n. equation (4), in terms of backshift notation, is then

$$y_t + a_1 q^{-1} y_t + a_2 q^{-2} y_t + \ldots + a_{n_a} q^{-n_a} y_t = e_t \tag{9}$$

Let

$$A(q) = 1 + a_1 q^{-1} + a_2 q^{-2} + \ldots + a_{n_a} q^{-n_a} \tag{10}$$

be defined as the **autoregressive operator** or filter, which is a polynomial in q^{-k}, the backshift operator. The general AR(n_a) model can then be simply described in terms of the autoregressive operator, either by

$$A(q) y_t = e_t \tag{11}$$

or, equivalently, by

$$y_t = \frac{1}{A(q)} e_t \tag{12}$$

The black-box diagram of the AR model is presented in Figure 5a, illustrating that the output series is a result of two processes: (1) a white noise input series of random, independent shocks, and (2) an autoregressive filter operating on the output series.

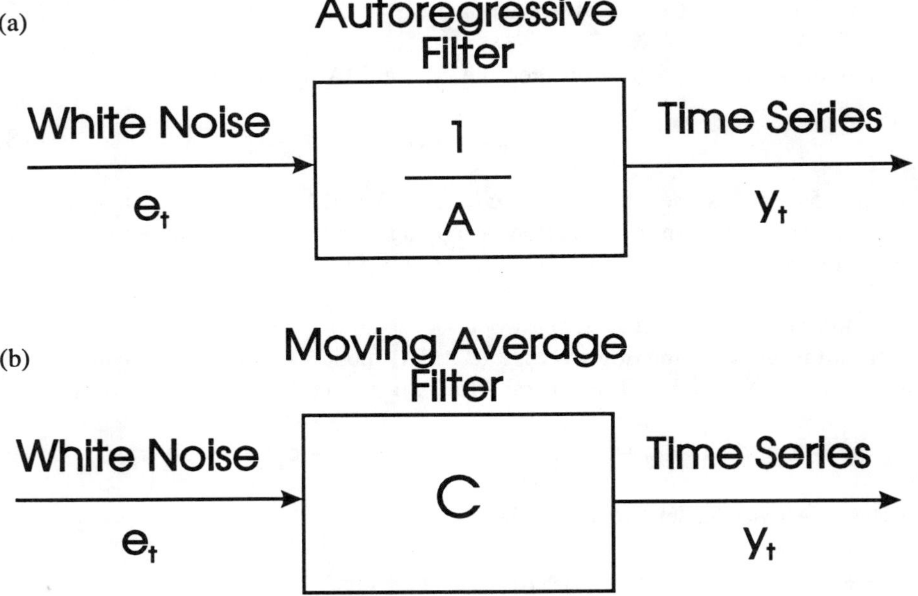

Figure 5. Black-box diagram of (a) autoregressive and (b) moving average time series models.

2.1.2.b *Moving Average Model*

The **moving average** (MA) model assumes that an output series can be modeled solely as a function of a white noise process. The general MA model of order n_c is

$$y_t = e_t + c_1 e_{t-1} + c_2 e_{t-2} + \ldots + c_{n_c} e_{t-n_c} \tag{13}$$

The name "moving average" has no particular meaning but, in the words of Box and Jenkins (1976), "this nomenclature is in common use, and therefore we employ it." Examples of MA models are an MA(1) model,

$$y_t = e_t + c_1 e_{t-1} \tag{14}$$

and an MA(12) model,

$$y_t = e_t + c_1 e_{t-1} + c_6 e_{t-6} + c_{12} e_{t-12} \tag{15}$$

In terms of the backshift operator, the MA(n_c) model becomes

$$y_t = e_t + c_1 q^{-1} e_t + c_2 q^{-2} e_t + \ldots + c_{n_c} q^{-n_c} e_t \tag{16}$$

Defining

$$C(q) = 1 + c_1 q^{-1} + c_2 q^{-2} + \ldots + c_{n_c} q^{-n_c} \tag{17}$$

as the **moving average operator** or filter, the general MA(n_c) model is

$$y_t = C(q) e_t \tag{18}$$

Figure 5b depicts the black-box diagram of the MA(n_c) model. The series of observations $\{y_t\}$ is shown to result from a moving average filter operating on a white noise input series.

2.1.2.c *Autogressive-Moving Average Model*

The **autoregressive-moving average** (ARMA) model is essentially a combination of an AR and an MA model. The general ARMA(n_a,n_c) model in explicit form is

$$y_t + a_1 y_{t-1} + a_2 y_{t-2} + \ldots + a_{n_a} y_{t-n_a} = e_t + c_1 e_{t-1} + c_2 e_{t-2} + \ldots + c_{n_c} e_{t-n_c} \tag{19}$$

Examples of ARMA models are an ARMA(1,1) model,

$$y_t + a_1 y_{t-1} = e_t + c_1 e_{t-1} \tag{20}$$

and an ARMA(2,12) model,

$$y_t + a_1 y_{t-1} + a_2 y_{t-2} = e_t + c_1 e_{t-1} + c_2 e_{t-2} + c_3 e_{t-3} + c_{12} e_{t-12} \tag{21}$$

In terms of the AR and MA operators of (10) and (17), respectively, the ARMA(n_a,n_c) model becomes

$$A(q)y_t = C(q)e_t \tag{22}$$

The black-box diagram of the ARMA(n_a,n_c) model (Figure 6a) demonstrates the model's flexibility in describing univariate systems in that the white noise input series and the output series are each filtered by their respective operators: the MA operator for the random noise series, and the AR operator for the data observation series.

(a)

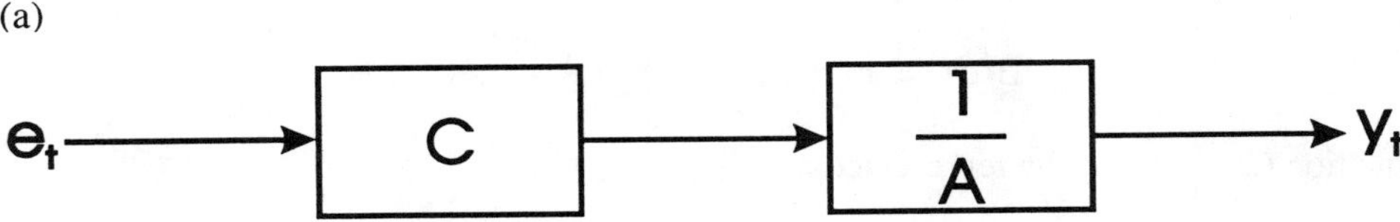

(b)

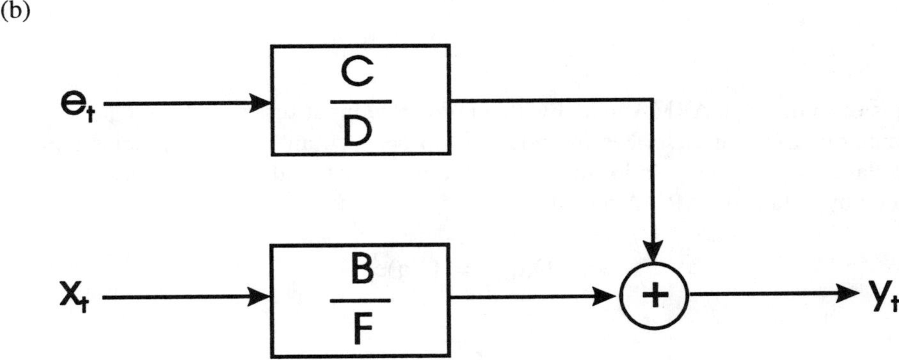

Figure 6. Black box diagrams of (a) the autoregressive-moving average model for univariate time series and (b) the Box-Jenkins transfer function model for bivariate input-output time series.

2.1.2.d *Box-Jenkins Transfer Function Model*

In the single-input single-output bivariate system (e.g., Figure 1), there are two concurrent univariate time series, each measured at equispaced times. Box and Jenkins (1976) developed a model in which an input-output relationship between $\{x_t\}$ and $\{y_t\}$ can be expressed in terms of present and past values of the input and output series:

$$y_t + f_1 y_{t-1} + f_2 y_{t-2} + \dots + f_{n_f} y_{t-n_f} =$$

$$b_0 x_{t-n_k} + b_1 x_{t-n_k-1} + \dots + b_{n_b} x_{t-n_k-n_b} \tag{23}$$

The term n_k is the **delay operator** which denotes the number of time lag delays that occur from input to output. In the effort-catch system there are no delays between an input and its corresponding output: the effort is applied and the fish are caught all in the same time step. The number of delays n_k is thus set to zero in the following discussion. Let us define

$$F(q) = 1 + f_1 q^{-1} + f_2 q^{-2} + \dots + f_{n_f} q^{-n_f} \tag{24}$$

and

$$B(q) = b_0 + b_1 q^{-1} + b_2 q^{-2} + \dots + b_{n_b} q^{-n_f} \tag{25}$$

equation (23) can thus be represented as

$$F(q) y_t = B(q) x_t \tag{26}$$

or

$$y_t = \frac{B(q)}{F(q)} x_t \tag{27}$$

which is essentially an ARMA model with an *observed* input series $\{x_t\}$.

equation (27) thus describes the relationship between an input data series and an output data series. The relationship of the output series and a white noise series is described by a standard ARMA model

$$D(q) y_t = C(q) e_t \tag{28}$$

or

$$y_t = \frac{C(q)}{D(q)} e_t \tag{29}$$

where

$$D(q) = 1 + d_1 q^{-1} + d_2 q^{-2} + \dots + d_{n_d} q^{-n_d} \tag{30}$$

The **Box-Jenkins transfer function** model incorporates both (27) and (29), and assumes that the observed input series $\{x_t\}$ is *independent* from the white noise series $\{e_t\}$:

$$y_t = \frac{B(q)}{F(q)} x_t + \frac{C(q)}{D(q)} e_t \tag{31}$$

The black-box diagram of Figure 6b illustrates more clearly the independence of $\{e_t\}$ and $\{x_t\}$. The Box-Jenkins transfer function model is also referred to as a BCDF model of order (n_b, n_c, n_d, n_f).

2.1.3 Stationarity

A critical assumption of the models presented above is that a constant mean and variance exist for the entire data record of a time series. This assumption is termed **stationarity** in time series nomenclature. Two types of stationarity are at issue: (1) mean stationarity and (2) variance stationarity. A classic example of a nonstationary time series in both the mean and variance is the airline passenger data set from Box and Jenkins (1976), shown in Figure 7. It is readily observed from the time series plot in Figure 7a that an increasing trend occurs over the course of the data set, indicating that the mean is not constant. It is also observed that the number of airline passengers tends to vary more and more as the series moves through time; thus, the variance is nonstationary also.

The airline passenger data must be made stationary before it can be described by a time series model. The variance nonstationarity can be corrected by applying an appropriate transformation from the family of power transformations introduced by Box and Cox (1964). The transformation is defined by

$$h(z) = \begin{cases} z^{\lambda}, & \lambda \neq 0 \\ \log(z), & \lambda = 0 \end{cases} \tag{32}$$

for a given value of λ. For example, $\lambda = -1$ performs a reciprocal transformation, while $\lambda = ½$ yields the square root transformation. A natural logarithm transformation was applied to the airline passenger data. The plot of Figure 7b demonstrates that the variance is now stationary for this series.

Mean nonstationarity is best corrected by applying the **difference operator** ∇_s^d, where d is the level of differencing (i.e., first difference, second difference) and s is the time period to which d is applied. Thus,

$$\nabla_1^1 \cdot z_t = z_t - z_{t-1} \tag{33}$$

takes the first difference of each observation,

$$\nabla_1^2 \cdot z_t = z_t - z_{t-2} \tag{34}$$

takes the second difference of each observation, and

$$\nabla_{12}^1 \cdot z_t = z_t - z_{t-12} \tag{35}$$

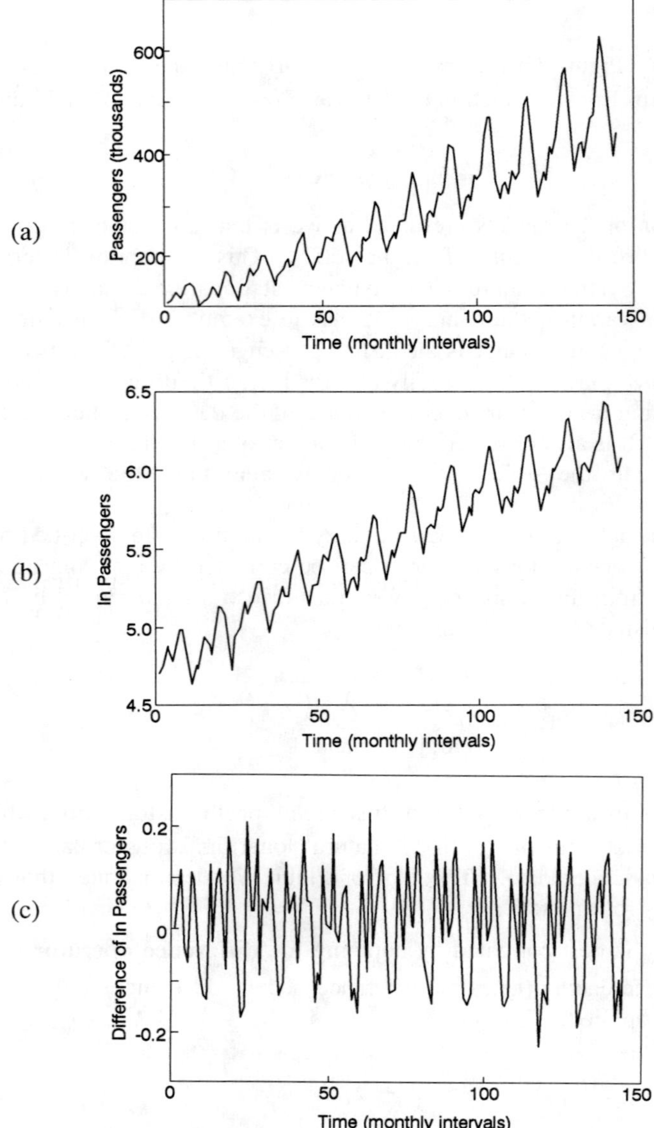

Figure 7. Time series plots illustrating nonstationarity corrections using the airline passenger data from Box and Jenkins (1976). (a) Original data series showing both a general upward trend (nonstationary mean) and greater fluctuations (nonstationary variance) in monthly airline passengers through time. (b) Log-transformed passenger values illustrating correction for variance nonstationarity. (c) First difference (equation 33) applied to log passenger values illustrating additional correction for mean nonstationarity.

takes the first difference of every 12th observation. The first difference of the airline passenger data removes the upward trend, solving the problem of a nonstationary mean (Figure 7c). A useful heuristic for differencing is the following (Schumway, 1988): a linear trend is eliminated by a first difference, and a quadratic trend is eliminated by a second difference. equation (35) is an example of a seasonal difference, which can be used to remove seasonal nonstationarity in the mean. In practice, d is usually no higher than 2.

2.2 BOX-JENKINS MODELING APPROACH

The procedures used to fit time series models to actual data owe largely to the work of G.E.P. Box and G.M. Jenkins (1976). They outlined the following model-building strategy:

1. Model identification
2. Parameter estimation
3. Model validation

This process is repeated until an "adequate", valid time series model is specified. The primary heuristic in specifying a time series model is the **principle of parsimony**. Simply stated, this principle guides a time series analyst in choosing the simplest model which adequately represents the dynamic system in question.

2.2.1 Model Identification

Model identification involves two issues: (1) the model structure (e.g., AR model, ARMA model) and (2) its corresponding model order. In univariate systems, the flexible ARMA(n_a , n_c) model (22) is usually the model structure of choice. Specifying the model order (n_a , n_c) is facilitated by the following "tools": the autocorrelation function, and the partial autocorrelation function. The **autocorrelation function** (ACF) describes the relationship of two values of a time series at two different times, t and t – k, where k is the time lag. It is defined as

$$\rho_k = \operatorname{Corr}(z_t, z_{t-k}) = \frac{\operatorname{Cov}(z_t, z_{t-k})}{\sqrt{\operatorname{Var}(z_t)\operatorname{Var}(z_{t-k})}} \tag{36}$$

The **partial autocorrelation function** (PACF) describes the relationship between z_t and z_{t-k} after removing the effect of the intervening variables, z_{t-1}, z_{t-2} ,..., z_{t-k+1}. It is defined as

$$\phi_{kk} = \operatorname{Corr}(z_t, z_{t-k} \mid z_{t-1}, z_{t-2}, \ldots, z_{t-k+1}) \tag{37}$$

(see Cryer, 1986; for more details).

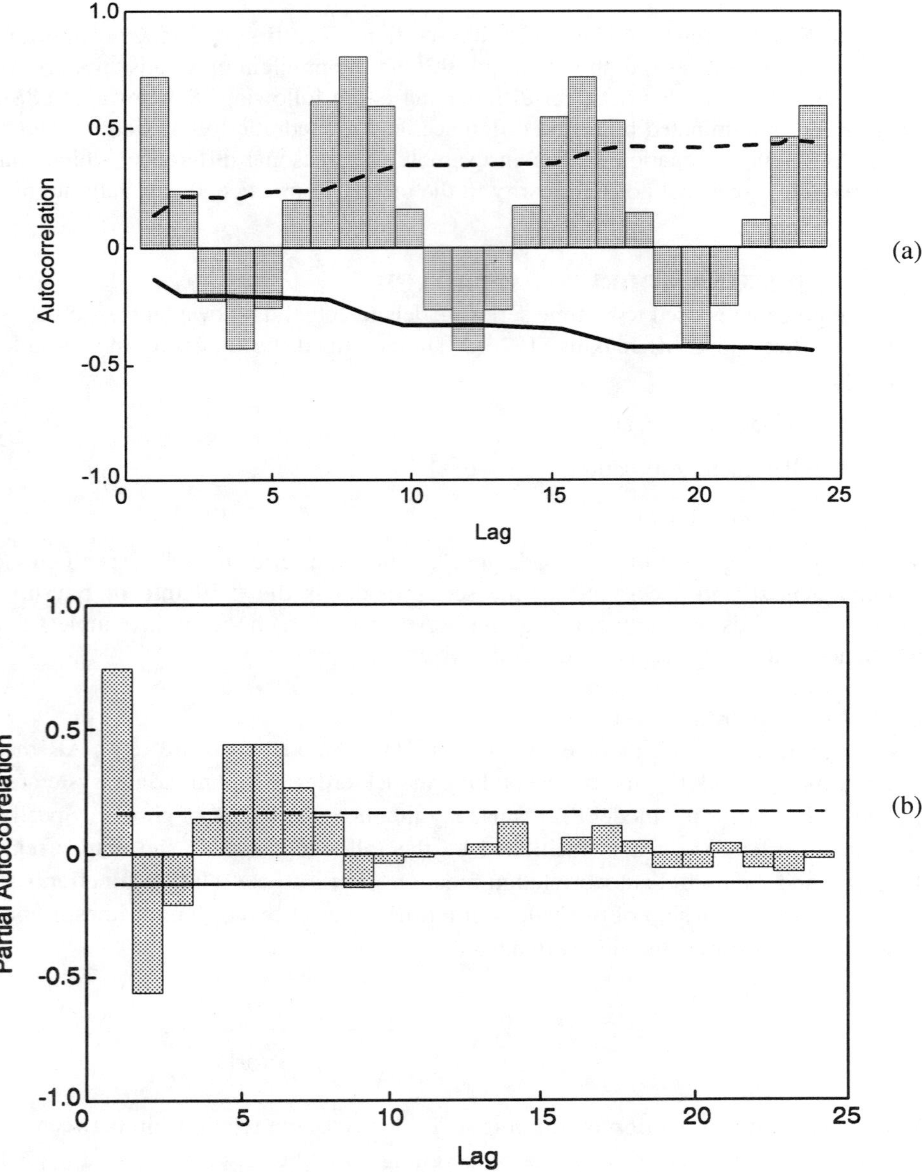

Figure 8. (a) Autocorrelation function (ACF) and (b) partial autocorrelation function (PACF) graphs for the blue crab data series (Table 1). The dashed (upper) and solid (lower) horizontal lines delineate the statistical significance threshold ($p < 0.05$) between zero and nonzero function values (i.e., function values extending beyond the statistical significance threshold in either the positive or negative direction indicate nonzero lags).

The typical representation of the ACF and PACF is illustrated in Figure 8, which presents bar graphs of these functions computed for lags k = 1 to 24 for the blue crab data series. The properties of the ACF and PACF as depicted in bar graphs provide guidance in recognizing the following time series "behaviors": (1) mean nonstationarity, (2) strong seasonality, (3) autoregressive behavior, and (4) moving average behavior. A synopsis of the ACF and PACF properties for each "behavior" is listed below (after Schumway 1988):

(1) **Mean Nonstationarity**: The ACF decays slowly; the PACF possesses a large positive or negative value at lag 1.

(2) **Strong Seasonality**: The ACF is zero except at the seasonal lags k = S, 2S, 3S, ... , and decays very slowly.

(3) **Autoregressive Behavior**: For a low order AR(n_a) model, the PACF is nonzero for lags k = 1, 2 ,..., n_a , and is zero thereafter. For a seasonal AR(n_a) model, the PACF is nonzero at lags k = S, 2S ,..., n_aS , and is zero elsewhere.

(4) **Moving Average Behavior**: For a low order MA(n_c) model, the ACF is nonzero for lags k = 1, 2 ,..., n_c , and is zero thereafter. For a seasonal MA(n_c) model, the PACF is nonzero at lags k = S, 2S ,..., n_cS , and is zero elsewhere.

When performing model identification, mean nonstationarity and strong seasonality should be corrected before specifying the AR and MA model orders.

2.2.2 Parameter Estimation

Once a model has been tentatively specified, the model parameters are estimated. For example, if an AR(2) model (equation 6) is specified, the parameters a_1, a_2, and σ^2 are then estimated. Various recursive statistical algorithms are employed to carry out the computations (Cryer, 1986; Ljung, 1987).

2.2.3 Model Validation

The tentative, parameterized model, at this point, is still a first-cut, "informed" guess. A critical step in fitting time series models is the validation of the model. As in linear regression analysis, it is a good practice to inspect the estimated model's residuals as a measure of goodness-of-fit. The most important validation procedure in the Box-Jenkins modeling approach is to inspect the ACF of the residuals for any significant lags that still remain. An "adequate" model in the Box-Jenkins view is one in which the ACF contains no significant lags and thus resembles a white noise process.

2.2.4 Fitting an Actual Time Series

The iterative modeling procedure of identification, estimation, and validation is best understood by constructing a time series model for an actual data series. In this section, we will use the Box-Jenkins modeling procedure to analyze the Maryland blue crab catch data.

The first and foremost tool of the time series analyst is the time series plot. Figure 2 is thus inspected for nonstationary variance, nonstationary mean, and strong seasonality. Although the latter portion of the series seems to have increased in terms of both mean and variance, the increase is not a linear trend but rather a "jump". Thus, employing a Box-Cox transformation or low order differencing will not alleviate this problem. Section 4 treats this type of situation; in this example, we shall consider the series to be adequately stationary. There does, however, seem to be a marked periodicity in the catch series, indicating possible strong seasonality. To further explore this possibility, we inspect the ACF and PACF (Figure 8). According to the guidelines in section 2.2.1, strong seasonality of lag 8 is detected in the ACF plot. A seasonal difference, ∇_8^1, is thus applied to the data series.

The ACF and PACF of the seasonally differenced series are now inspected (Figure 9). Significant lags are present in both plots. A rule of thumb in univariate modeling is to account for the autoregressive behavior first and the moving average behavior second. An AR(4) model is tentatively identified, and the estimation procedure results in the following model:

$$\nabla_8^1\left(1 - .5549q^{-1} + .0604q^{-4}\right)y_t = e_t \tag{38}$$

Inspecting the ACF of the model residuals, a significant autocorrelation is present at lag 8 (Figure 10a). An ARMA(4, 8) model is subsequently identified and estimated, yielding

$$\nabla_8^1\left(1 - .6335q^{-1}\right)y_t = \left(1 - .6849q^{-8}\right)e_t \tag{39}$$

Notice that the a_4 term was deemed insignificant, and was subsequently excluded from the model. The ACF of this model's residuals appear to be indicative of a white noise process (Figure 10b), and therefore this ARMA(1,8) model is accepted as being adequate.

2.2.5 Forecasting

An important use of a time series model is to forecast future values of an output series. The forecasts are usually statistically valid only for the next season of an output series, due to increasing prediction error as the forecasts continue into the future. Cryer (1986) presents a detailed discussion of forecast and associated prediction error estimation procedures. Figure 11 graphs the forecast values and confidence intervals for the next 8-month crabbing season for the blue crab series, based on the ARMA(1,8) model constructed above.

2.2.6 The Box-Jenkins Modeling Approach in Perspective

The Box-Jenkins procedure for constructing time series models outlined above is essentially a technique for systematically removing deterministic elements from an

(a)

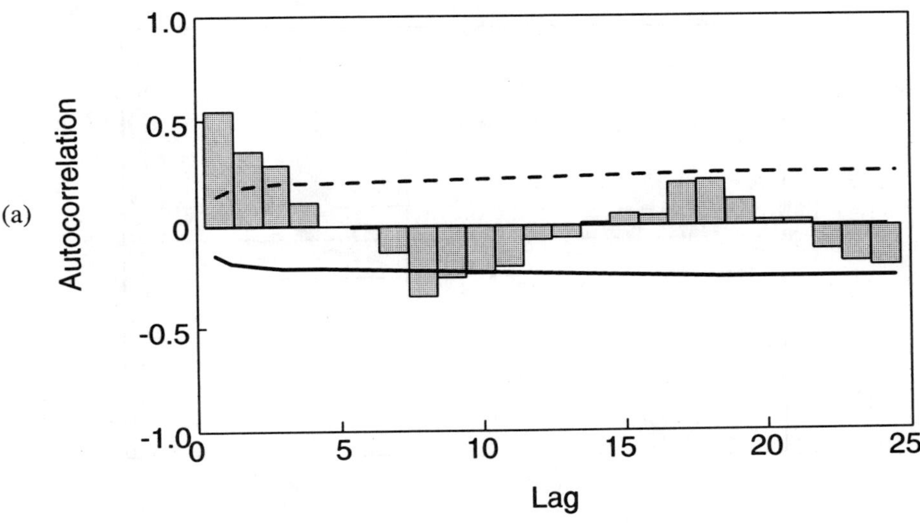

(b)

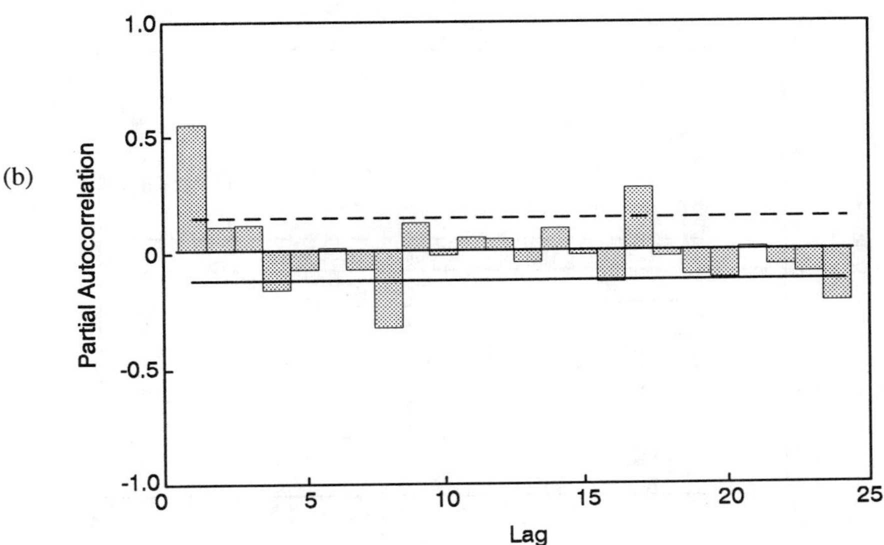

Figure 9. (a) ACF and (b) PACF graphs of the seasonally differenced blue crab catch data series.

(a)

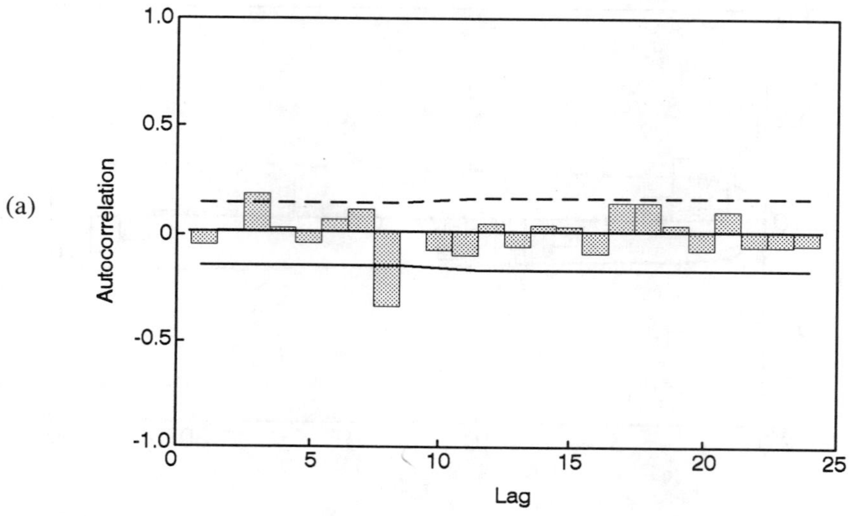

(b)

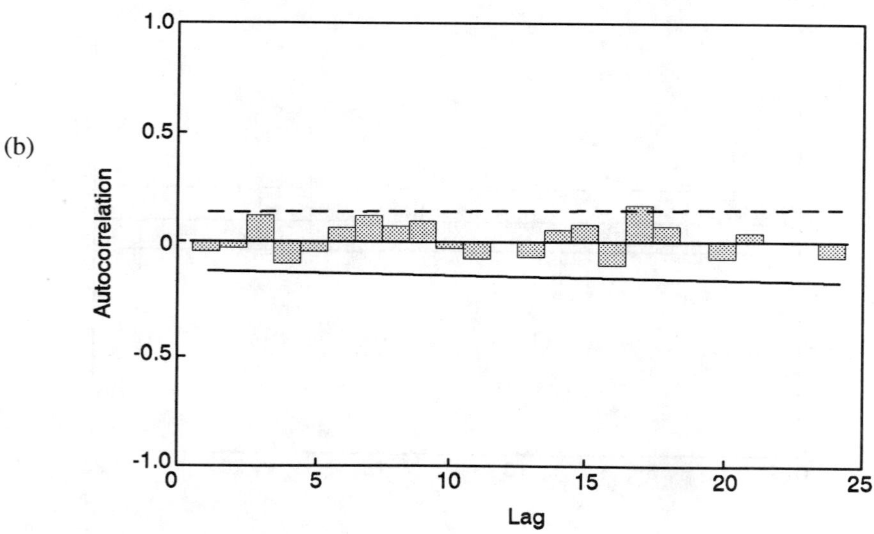

Figure 10. ACF graphs of error residuals of models (a) AR(4) and (b) ARMA (1, 8) fit to the seasonally differenced blue crab catch data series.

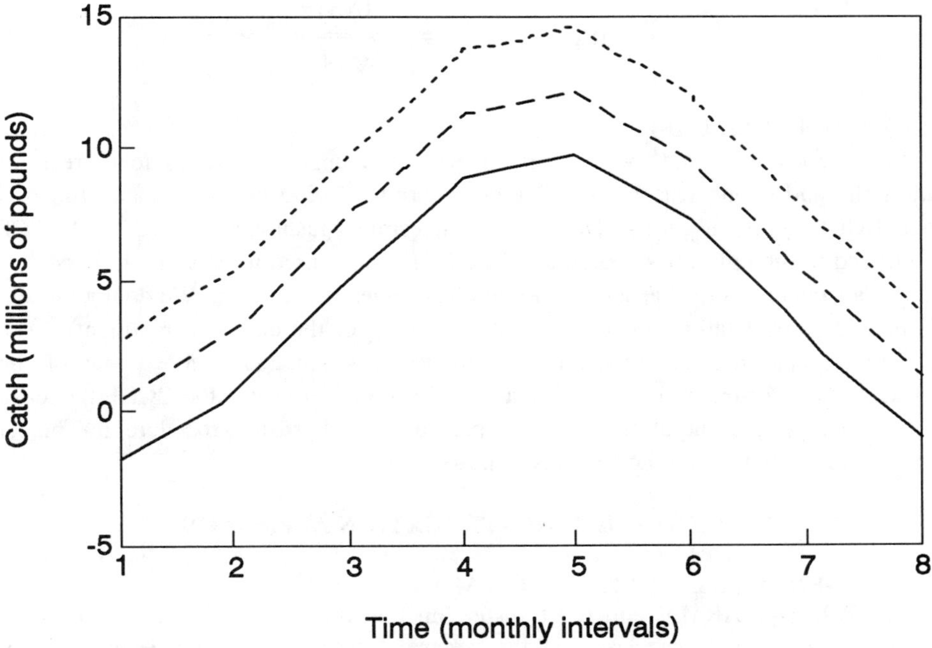

Figure 11. Forecast values for the next 8-month crabbing season for the blue crab catch time series (Figure 2). The middle line represents the mean forecast values, and the upper and lower lines are the corresponding 95% confidence intervals.

observed data series until only random error remains. One difficulty of applying this approach in practice, however, stems from the more or less "haphazard" model identification procedure. As illustrated above, our choice of a tentative model was by no means clear-cut. The guidelines for detecting autoregressive and moving average behavior outlined in 2.2.1 are based on theoretical properties of the ACF and PACF for pure AR and MA processes. The ACF and PACF patterns of mixed ARMA processes are not as clear, especially when dealing with environmental data that usually have a large noise component.

2.2.7 Box-Jenkins Transfer Function Modeling

The Box-Jenkins (B-J) procedure for modeling input-output data series is very similar to the univariate procedure described above. An additional tool, the **cross-correlation function** (CCF), is employed. The CCF is defined as

$$\rho_{xy}(k) = \mathrm{Corr}\left(x_{t-k}, y_t\right) = \frac{\mathrm{Cov}\left(x_{t-k}, y_t\right)}{\sqrt{\mathrm{Var}\left(x_t\right)\mathrm{Var}\left(y_t\right)}} \tag{40}$$

(see Box and Jenkins, 1976).

Each series, $\{x_t\}$ and $\{y_t\}$, is first treated as a univariate series to correct for stationarity and strong seasonality. The procedure for fitting the B-J transfer function model BCDF(n_b, n_c, n_d, n_f) utilizes the CCF in a similar fashion as the ACF and PACF are utilized in the univariate procedure. The CCF is also used to validate the model in that an "adequate" model has a CCF plot that resembles white noise. We do not present this procedure in detail (see Box and Jenkins, 1976; for the classical treatment). The difficulty in applying this procedure in practice is compounded above that of the univariate case, for many of the same reasons. Section 3 discusses the BCDF model in light of other input-output models, and presents an alternate procedure for model identification of both univariate and bivariate systems.

3. THE SYSTEM IDENTIFICATION APPROACH

3.1 THE GENERAL LINEAR SYSTEM MODEL

The AR, MA, ARMA, and B-J transfer function time series models discussed in section 2.0 are actually subclasses of the **general linear system** model described by Ljung (1987):

$$A(q)y_t = \frac{B(q)}{F(q)}x_t + \frac{C(q)}{D(q)}e_t \tag{41}$$

This model is also referred to as a ABCDF model of order (n_a, n_b, n_c, n_d, n_f).The black-box diagram of the general linear model is presented in Figure 12a. The model is composed of three filters: (1) an autoregressive filter, (2) a transfer function filter, and (3) a moving average filter. Both the transfer function and moving average components are described by a ratio of two polynomials. Inspection of equation (41) reveals that there are 8 possible univariate model structures and 24 possible bivariate model structures. The possible number of model orders that exist for each model structure ranges from tens to thousands. We describe this situation as the model structure/model order dilemma.

3.2 MODEL STRUCTURE SELECTION

Box and Jenkins (1976) chose the AC and BCDF model structures to describe univariate and bivariate systems, respectively. Upon reflection, the AC model seems an appropriate choice due to its flexibility in describing both autoregressive and moving average behavior in time series data. The BCDF model, on the other hand, models the autoregressive component separately within the transfer function and moving average filters (Figure 6b). Perhaps a more flexible transfer function model structure is the ABC or ARMAX model (Ljung, 1987):

$$A(q)y_t = B(q)x_t + C(q)e_t \tag{42}$$

The ABC black-box diagram illustrates its simplicity in modeling a dynamic input-output system as a function of the three basic filters (Figure 12b).

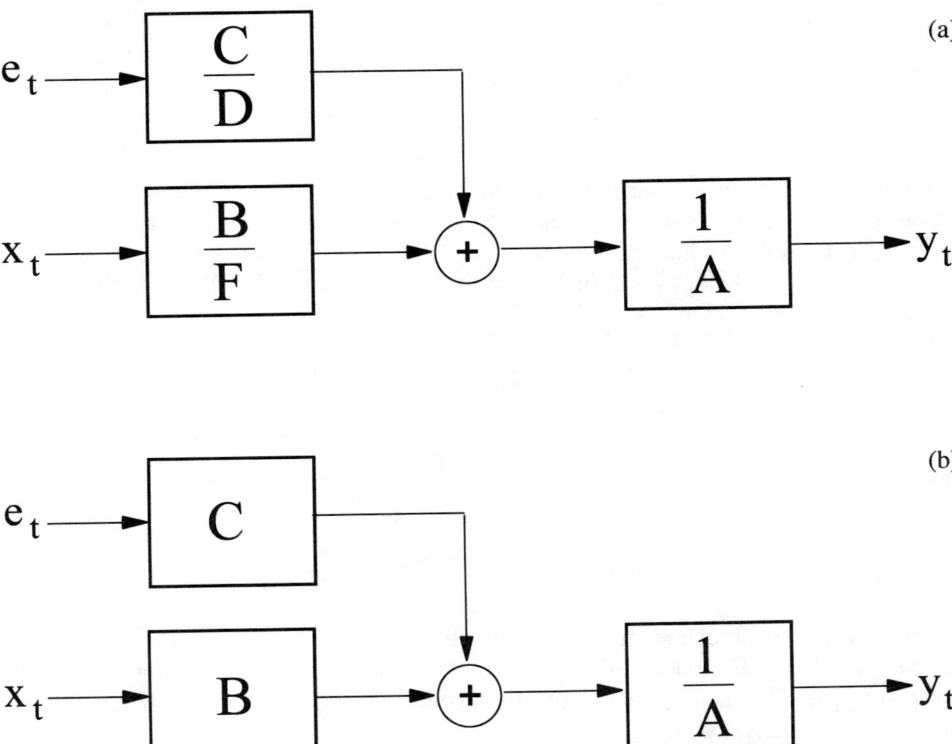

Figure 12. Black box diagrams of (a) the general linear system ABCDF model, and (b) the ABC input-output model.

To explore the consequences of choosing a model structure different from the "true" structure, a simulation experiment was performed. Three different input-output model structures were simulated at three different levels of random error. The input series was

13 years arranged in nonchronological order of the yellowfin tuna monthly effort series (Table 2a). Output series were generated for each model at each level of white noise and given to a time series analyst to fit. The analyst had no knowledge of the model structure of the 24 possible structures from which the data were generated, nor the model order. The analyst used the standard techniques of viewing the time series plot, computing the ACF and PACF, and much trial and error in attempting to fit various model structures and orders to the simulated output series. Akaike's FPE statistic was chosen to be the measure of goodness-of-fit.

Simulation series #1 was generated from the model structure BF of order (0, 1, 0, 0, 1), where

$$\begin{aligned} B(q) &= .3421 + .7530q^{-1} \\ F(q) &= 1 + .2112q^{-1} \end{aligned} \tag{43}$$

Abbreviated results of the analysis are presented in Table 3a. Note that the FPE increases dramatically with the increased noise level. Also note that the ABC and B model structures fit the data as well as the "true" BF model for each level of noise.

Simulation series #2 was generated from model structure BCDF, the Box-Jenkins transfer function model, of order (0, 24, 2, 1, 1) where

$$\begin{aligned} B(q) &= .8452 + .7521q^{-24} \\ C(q) &= 1 - .3235q^{-1} - .4172q^{-2} \\ D(q) &= 1 + .6204q^{-1} \\ F(q) &= 1 + .5217q^{-1} \end{aligned} \tag{44}$$

Table 3b lists the abbreviated results. Again, the ABC model fit the data equally as well as the "true" BCDF model. A slightly poorer fit was obtained using the AB model structure, perhaps an indication of its inability to describe the moving average component of the system.

Simulation series #3 was generated from the ABDF model of order (2, 0, 0, 1, 2) where

$$\begin{aligned} A(q) &= 1 - .4041q^{-1} + .7271q^{-2} \\ B(q) &= 1 \\ D(q) &= 1 + .8450q^{-1} \\ F(q) &= 1 + .2112q^{-1} + .6789q^{-2} \end{aligned} \tag{45}$$

Table 3. Estimation results for model structure/model order simulation experiments.

a.

Simulation Experiment #1
True Model Structure: BF
True Model Order: (0, 1, 0, 0, 1)

Model structure	Model order	Simulation error σ_e^2	FPE
BF	(0, 1, 0, 0, 1)	1	0.826
ABC	(1, 1, 1, 0, 0)	1	0.838
B	(0, 2, 0, 0, 0)	1	0.828
BF	(0, 1, 0, 0, 1)	9	7.44
ABC	(1, 1, 1, 0, 0)	9	7.54
B	(0, 2, 0, 0, 0)	9	7.38
BF	(0, 1, 0, 0, 1)	25	20.68
ABC	(1, 1, 1, 0, 0)	25	20.95
B	(0, 2, 0, 0, 0)	25	20.55

b.

Simulation Experiment #2
True Model Structure: BCDF
True Model Order: (0, 24, 2, 1, 1)

Model structure	Model order	Simulation error σ_e^2	FPE
BCDF	(0, 24, 2, 1, 1)	1	0.822
ABC	(1, 24, 1, 0, 0)	1	0.924
AB	(1, 24, 0, 0, 0)	1	1.133
BCDF	(0, 24, 2, 1, 1)	9	7.42
ABC	(1, 24, 1, 0, 0)	9	7.93
AB	(1, 24, 0, 0, 0)	9	10.29
BCDF	(0, 24, 2, 1, 1)	25	20.67
ABC	(1, 24, 1, 0, 0)	25	21.25
AB	(1, 24, 0, 0, 0)	25	28.80

Table 3. (continued)

c.

Simulation Experiment #3
True Model Structure: ABDF
True Model Order: (2, 0, 0, 1, 2)

Model structure	Model order	Simulation error σ_e^2	FPE
ABDF	(2, 0, 0, 1, 2)	1	0.851
ABC	(6, 2, 3, 0, 0)	1	0.874
ABC	(5, 1, 2, 0, 0)	1	0.873
ABDF	(2, 0, 0, 1, 2)	9	7.67
ABC	(6, 2, 3, 0, 0)	9	7.90
ABC	(5, 1, 2, 0, 0)	9	7.89
ABDF	(2, 0, 0, 1, 2)	25	21.32
ABC	(6, 2, 3, 0, 0)	25	21.97
ABC	(5, 1, 2, 0, 0)	25	21.98

The results indicate that the ABC model structure, once again, is able to fit data generated from a different structure (Table 3c).

It appears that many different model structures can be used to fit data generated from any one particular structure. The ARMAX ABC model seems to have the flexibility to model most types of input-output systems. We thus advocate the ABC model structure over the B-J BCDF model for three reasons: (1) the ABC model contains the essential three elements of the general linear system, namely, an autoregressive filter, a moving average filter, and a transfer function filter, all of which provide great flexibility in describing input-output systems; (2) since the ABC model contains one less polynomial, it is estimated more efficiently than the BCDF model; and (3) given that it contains no ratios of polynomials, the ABC model is easier to interpret than the BCDF model.

3.3 MODEL ORDER SELECTION

As discussed in Section 2, model identification entails both model structure and model order selection. We prefer the AC and ABC model structures for describing, respectively, univariate and bivariate dynamic systems. Even given the AC or ABC model structure, it is evident that a variety of model orders will represent a dynamic system with the same degree of accuracy (e.g., Table 3c). The Box-Jenkins modeling procedure outlined in Section 2.2 is one method of determining the model order.

Although an "adequate" model may be obtained by this procedure, it is only one of a number of "adequate" models that exist for a particular univariate or bivariate system.

An alternate procedure for model order selection is to estimate a variety of model orders for the same time series data set, and then choose the model which performs the best according to some criterion such as the FPE statistic (Gooijer et al., 1985; Fildes, 1988). A method of this type would alleviate the subjectivity and difficulty of the Box-Jenkins approach by making model order selection more "automatic"; however, as demonstrated in the simulation experiment results discussed above, the FPE values of a number of different model orders for the same data may be so similar as to be indistinguishable from a realistic point of view. Just because a model is statistically the "best", it may not make any sense with respect to the system being modeled. What is required is a balance between a statistical criterion and a practical criterion. The first can be provided to the analyst by computational methods; the second must be obtained from the analyst's knowledge of the system being modeled.

A model order selection method incorporating both statistical accuracy and system practicality is now presented. The method takes advantage of the following heuristic derived from time series analysis in practice: the significant lags of the various polynomials of a time series model applied to actual data are usually between 0 and 3 plus those occurring at seasonal intervals. Thus, a set of feasible model orders can be constructed according to these guidelines. To assist in this task, the time series plot, ACF, and PACF are employed as described in Section 2.2. All of the feasible models are estimated by the standard computational algorithms, and then sorted by the FPE statistic. From the set of top 10 to 15 models, the time series analyst chooses one which is both realistic and parsimonious. This model is subsequently validated, and then forecast if so desired.

This method, termed the **feasible model set search**, is summarized in the following steps:

(1) Plot the time series, ACF, and PACF. Inspect for nonstationarity and strong seasonality.

(2) Correct for nonstationarity and strong seasonality, if necessary.

(3) Plot the time series, ACF, and PACF again.

(4) Choose a model structure.

(5) Construct a set of feasible model orders. Orders 0–3 are automatically included. Inspect the plots from step 3 to determine the most likely seasonal lag $k = S$. Model orders (1/2)S, S, (3/2)S, and 2S are then included in the set.

(6) Estimate the feasible models, and sort them by the FPE criterion.

(7) Select an appropriate model from the top 10 or 15 in the sorted set.

(8) Validate the model. If valid, the model is ready for subsequent use. If nonvalid, return to step 7.

The blue crab catch series is now analyzed by this method. Steps 1 through 3 were performed in Section 2.2; thus, we begin with step 4. The AC model structure is chosen, and the feasible model orders are determined to be (0, 1, 2, 3, 4, 8, 12, 16) for both the A(q) and C(q) polynomials. The set of 64 models is estimated and sorted. Table 4 presents the top ten models. Model #1, AC(1, 8), seems reasonable and is the most parsimonious of the set.

Table 4. Top 10 AC models, sorted by Akaike's FPE goodness-of-fit criterion, estimated for the blue crab data series using the feasible model set search procedure. Listed values are model orders for the A and C model components.

$A(n_a)$	$C(n_c)$	FPE
1	8	0.0187
3	16	0.0189
2	8	0.0189
1	16	0.0190
2	16	0.0191
4	16	0.0191
8	8	0.0191
1	12	0.0192
3	12	0.0192
2	12	0.0192

4. TECHNIQUES FOR HANDLING REAL DATA

4.1 DATA PRETREATMENT

Since actual, raw data observations are usually not in a form which is required by time series computational algorithms, **data pretreatment** procedures are required. Two essential techniques are presented below: (1) missing value estimation, and (2) offset level adjustment.

4.1.1 Missing Values

Missing values are typical in temporal fishery data. Estimation of missing values can be accomplished through a variety of procedures. The most common approach is to average the adjacent values of actual observations. A similar technique that is applicable in data series exhibiting marked seasonal behavior is to average the values of corresponding time periods. For example, if a datum from November, 1952 is missing, all other Novembers could be averaged to estimate the missing value. These averaging approaches usually are acceptable only when scattered point values are missing. A more sophisticated missing value estimation technique, cubic-spline interpolation, is recommended if an entire string of data points are missing.

4.1.2 Offset Levels

"Raw" data observations, especially those of input-output systems, need to be adjusted in terms of numerical value before being utilized by computational algorithms. The first step is to scale the data by a suitable value. This is normally accomplished by dividing the entire time series by the corresponding power of 10 of the observation with the highest value. For example:

$$(\text{z - offset}) = \frac{(\text{z - raw})}{1 \times 10^{7}} \tag{46}$$

The second step is to subtract the mean from the series:

$$z_t = \left(\text{z} - \text{offset}_t\right) - \overline{(\text{z} - \text{offset})} \tag{47}$$

Now the data is ready for time series computational algorithms.

4.2 INTERVENTION ANALYSIS

A common occurrence in environmental and ecological time series is a natural or human-induced event which causes either a temporary or permanent change in the mean level of a data series. In fisheries time series, an extreme weather episode or other similar event may cause a change in the ecological conditions of the fishing grounds, which in turn may affect the time series of effort and catch data. Implementing gear, season, and/or size restrictions in a fishery also may cause a change in the effort and catch series. These induced changes are referred to as interventions, and they can be incorporated into time series models by the technique of **intervention analysis.** Some simple intervention analysis procedures are described below for univariate time series. For a more elaborate discussion, refer to Box and Tiao (1973) and Hipel et al. (1975).

Two basic types of interventions can occur: (1) **pulse interventions** which effect short-term changes in the mean level of a time series, and (2) **step interventions** which effect more or less permanent changes in the mean level of a time series. Both types of interventions can be detected by a careful inspection of the time series plot. Additionally, interventions may be detected in the model validation phase by inspecting the plot of the residuals for outliers after fitting a model to the time series.

If a pulse or step intervention is detected in a univariate time series, it can be incorporated into an AC model by creating an input intervention series and fitting an input-output model. For either type of intervention, the first step should be to fit an AC model to the preintervention data series. The next task is to create an intervention input series, which is a series consisting of 0's and 1's. A pulse intervention occurring at time t = T can be described as the following input series:

$$P_t(T) = \begin{cases} 0 & t \neq T \\ 1 & t = T \end{cases} \tag{48}$$

A step intervention occurring at time t = T can be described as

$$S_t(T) = \begin{cases} 0 & t < T \\ 1 & t \geq T \end{cases} \tag{49}$$

An ABC model of order $(n_a, 0, n_c)$ or $(n_a, 1, n_c)$ for a pulse intervention

$$A(q)y_t = B(q)P_t(T) + C(q)e_t \tag{50}$$

or a step intervention

$$A(q)y_t = B(q)S_t(T) + C(q)e_t \tag{51}$$

is then fit to the input-output series. A model of this type will adjust the time series to a new mean level caused by the intervention, and may result in a better-fitting model than if the intervention is ignored.

An alternate procedure for modeling a step intervention is to separate the time series into two series, one that occurs before the intervention and one that occurs after. Keep in mind that each series must consist of ≥50 data points. An example of this approach is the blue crab data series (Figure 2). Upon inspection of the time series plot, a step intervention seems to have occurred after the first eighteen crabbing seasons in which the mean level has increased for the last seven seasons of the series. The blue crab series can thus be split into two series, which we will refer to as blue crab series 1 and blue crab series 2, respectively. Each series can then be modeled separately.

These intervention analysis techniques are for single interventions only. If more than one intervention occurs in a time series, each intervention must be modeled as a separate input series. The resulting multiple input-single input series is beyond the scope of this chapter.

5. SOME USEFUL SPECTRAL ANALYSIS METHODS

To this point, the focus has been on time series models and model-building procedures based in the time-domain; however, time series analysis can also be performed in the frequency-domain. Frequency-space time series analysis is sometimes termed **spectral analysis**. The basic notion of spectral analysis is that a time series can be modeled by a series of sine and cosine waves of different frequencies. Although spectral analysis is an entire field of study in its own right (see Kay, 1988), our use of frequency-domain methods will be to apply them in building time-domain models.

The periodic nature of fisheries time series indicate that the fishery dynamic system is nonlinear. Spectral analysis methods can assist in both identifying and removing the strong seasonality component in nonlinear time series (MacDonald, 1989). The method presented below provides an alternative technique for modeling time series exhibiting strong seasonality.

5.1 TOOLS

In spectral analysis, the time series must first be converted from the time-domain to the frequency-domain. This is accomplished by applying one of the family of Fourier transforms to a data series of time-domain observations. Two spectral analytic tools that perform this transformation are (1) the periodogram and (2) the spectral density function.

The **periodogram** is computed with a discrete Fourier transform, and is defined as

$$I(f_h) = \frac{2}{N}\left[\sum_{t=1}^{N} z_t e^{-\frac{2\pi h t}{N}}\right]^2 \tag{52}$$

where N is the sample size and the frequency f_h is determined by

$$f_h = \frac{h}{N} \tag{53}$$

The value of h will differ whether N is even or odd:

$$h = \begin{cases} 1, \ldots, \dfrac{N}{2} - 1, & \text{N even} \\ 1, \ldots, \dfrac{N-1}{2}, & \text{N odd} \end{cases} \tag{54}$$

An example periodogram of the blue crab data series 2 is presented in Figure 13a. Note the dramatic peak corresponding to h = 7. The formula

$$s = \frac{N}{h} \tag{55}$$

converts the value of h to the seasonal time lag s where the peak occurs. In the above example, s = 8, which is the strong seasonality component identified previously for this data set.

The **spectral density function** or power spectral density is computed from a Fourier transformation of the autocovariance function (Schumway, 1988). Figure 13b presents the spectral density function for the blue crab series 2. Notice that a strong peak is also apparent at h = 7. Notice also that the spectral density function is smoother than the

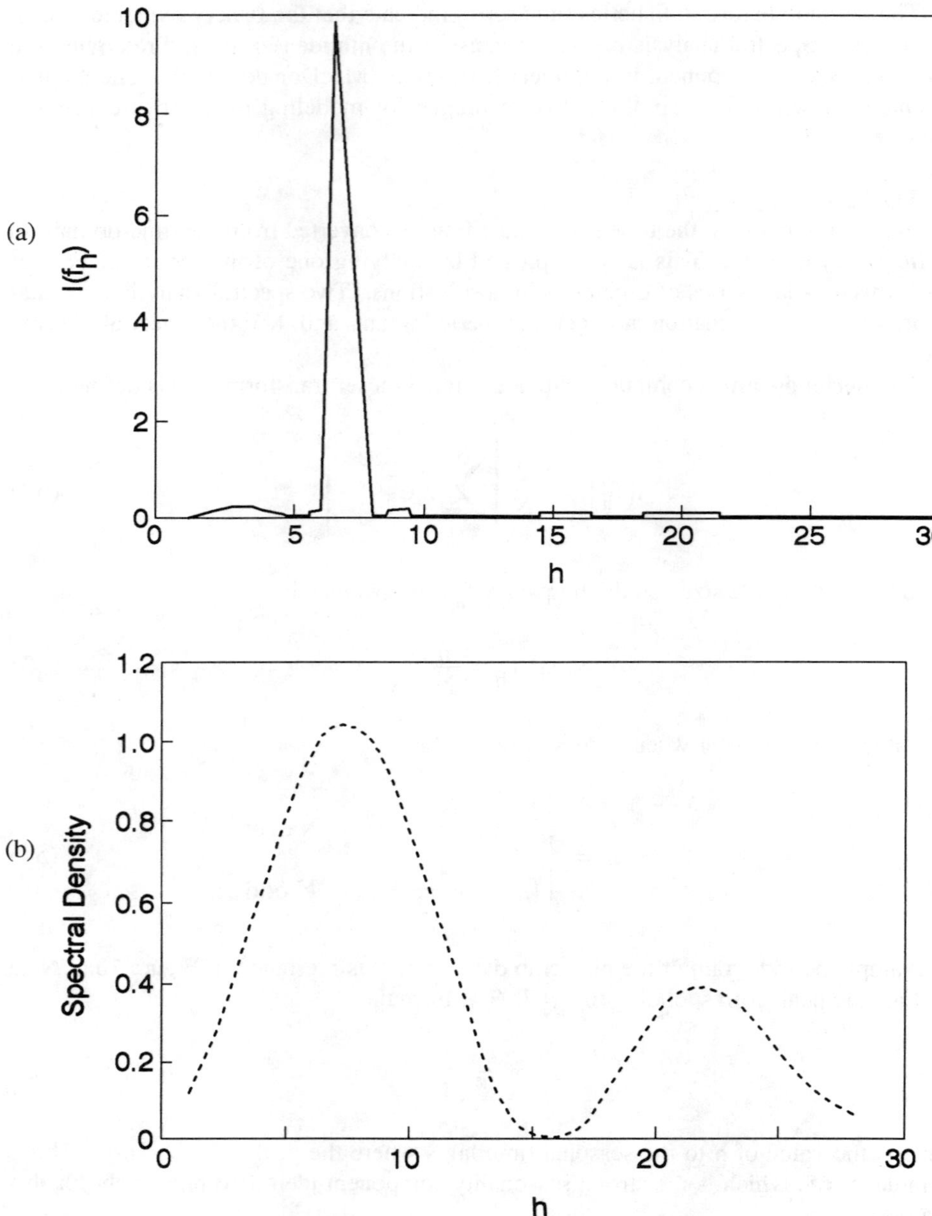

Figure 13. Graphs of the (a) periodogram and (b) spectral density functions computed for blue crab data series 2 (monthly catch values for crabbing seasons 1981–1987 from Table 1).

periodogram. Our use of the spectral density function is to corroborate the strong peaks identified in the periodogram, since in "noisy" systems the periodogram is subject to "statistical instability" which makes it difficult to identify strong periodicities (MacDonald, 1989).

5.2 REMOVING PERIODIC PEAKS

In Section 2, strong seasonality was corrected by performing a seasonal difference. An alternate method using spectral analysis techniques is described below using the blue crab data series 2 as an example.

The procedure begins with pretreated data from the methods of Section 4. The first step is to compute and plot both the periodogram and spectral density functions (Figures 13a and 13b). To identify the strong periodic peaks, a table is constructed consisting of the point values of h, $I(f_h)$, and the percent contribution of the peak determined by

$$\frac{I(f_h)}{\sum_h I(f_h)} \times 100 \tag{56}$$

From a table of this type computed for the blue crab series 2 (Table 5), the single strong peak is identified at h = 7, accounting for 89.4 percent of the periodicity in the data set. In the time domain, h = 7 corresponds to a seasonal periodicity of s = 8 (by equation 55). A deterministic periodicity is denoted by the function $\psi(s)$, where s is the seasonal time period.

To remove the deterministic periodicity $\psi(s)$ from the data set, the following procedure is used (MacDonald, 1989):

(1) Calculate the identified peak's inverse discrete Fourier transform (IDFT).

This yields the value of $\psi(s)$, which is essentially a set of s time-domain observations that repeat for the length of the original data series. Thus, for the blue crab data, the following eight values repeat for the entire length of the series (56 observations):

$$\Psi(8) = \begin{cases} -.5786, & t = 1, 9, \ldots \\ -.3326, & t = 2, 10, \ldots \\ .1082, & t = 3, 11, \ldots \\ .4856, & t = 4, 12, \ldots \\ .5786, & t = 5, 13, \ldots \\ .3326, & t = 6, 14, \ldots \\ -.1082, & t = 7, 15, \ldots \\ -.4856, & t = 8, 16, \ldots \end{cases} \tag{57}$$

Table 5. Periodogram function point values for blue crab data series 2.

h	$I(f_h)$	% contribution
1	0.0215	0.1979
2	0.1274	1.1742
3	0.1969	1.8152
4	0.0474	0.4371
5	0.0010	0.0091
6	0.0601	0.5538
7	9.7003	89.4266
8	0.0052	0.0481
9	0.0848	0.7818
10	0.0112	0.1032
11	0.0013	0.0120
12	0.0058	0.0532
13	0.0269	0.2478
14	0.0337	0.3106
15	0.0594	0.5474
16	0.0891	0.8215
17	0.0053	0.0489
18	0.0196	0.1804
19	0.0781	0.7204
20	0.0538	0.4961
21	0.0550	0.5070
22	0.0193	0.1781
23	0.0316	0.2914
24	0.0357	0.3293
25	0.0095	0.0879
26	0.0143	0.1316
27	0.0259	0.2390
28	0.0272	0.2505

A plot of these repeating values is presented in Figure 14a.

(2) Subtract the repeating data set from the original, pretreated data set, point by point.

A plot of the blue crab series with the periodic peak removed is shown in Figure 14b.

To fit a time-domain model to the residual blue crab series 2 data, the procedures outlined in section 3.0 are followed. Figure 15 plots the ACF and PACF for the residual series. It appears that removal of the periodic peak resulted in a series which is

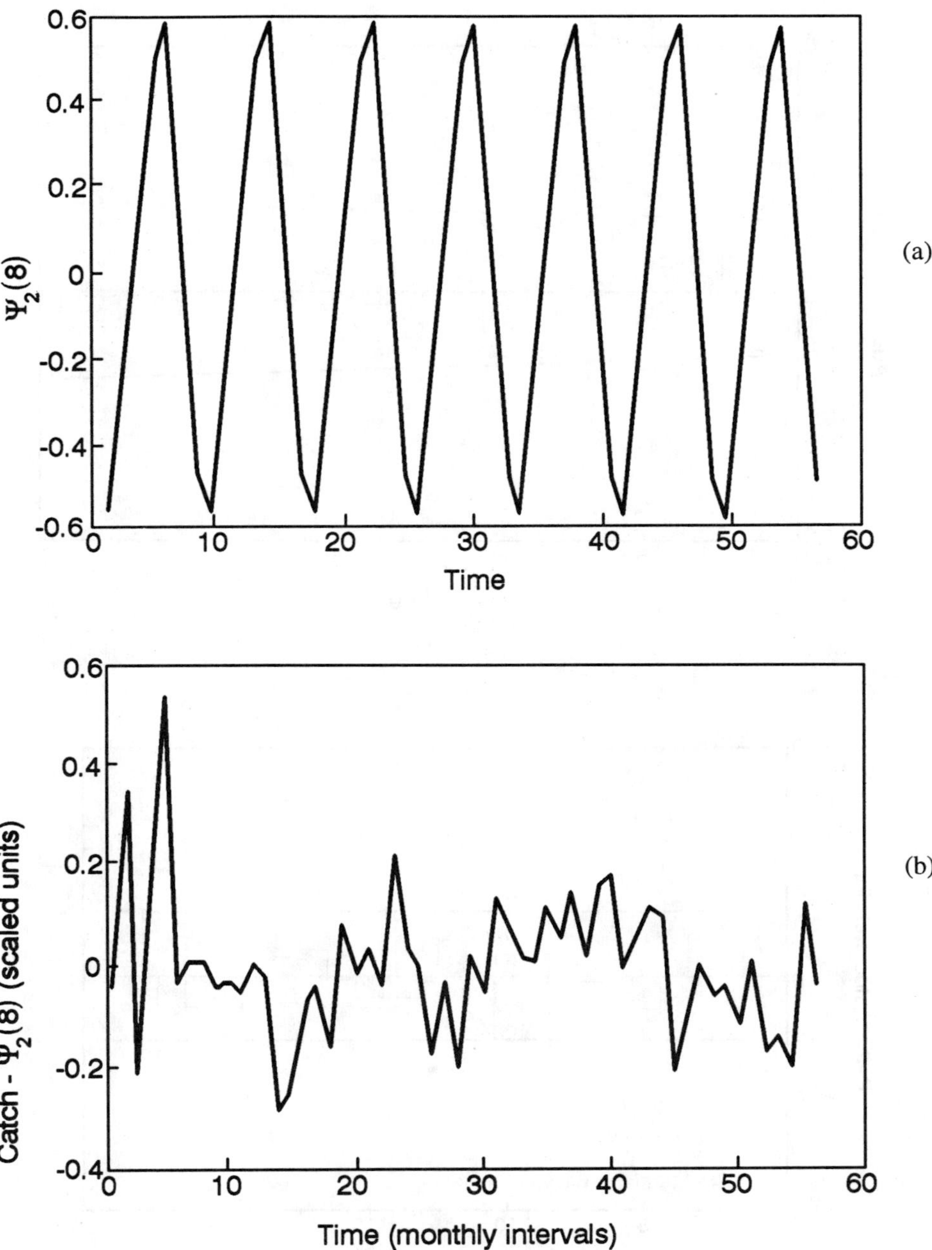

Figure 14. (a) Plot of the periodicity function $\psi(8)$, equation (57) in text, for blue crab data series 2 (in scaled units). (b) Plot of blue crab data series 2 (in scaled units) after subtracting the periodicity function.

(a)
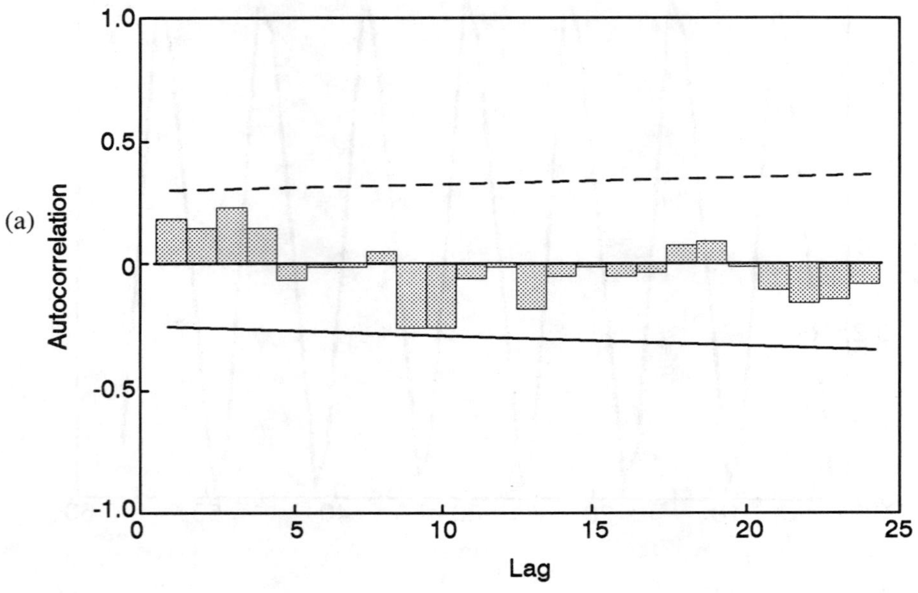

(b)
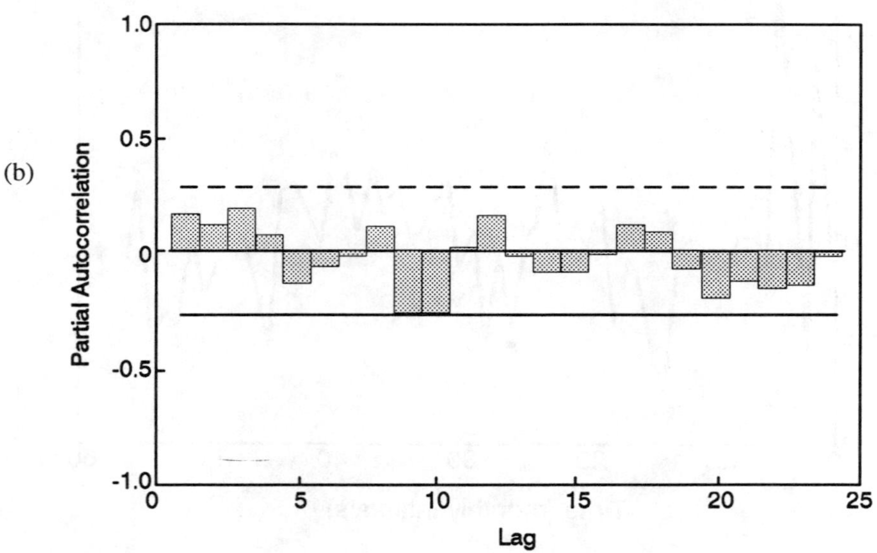

Figure 15. (a) ACF and (b) PACF graphs for blue crab series 2 with periodic peak removed (Figure 14b).

essentially white noise. The search for a feasible model set is performed, and an AR(1) model is chosen:

$$y_t - .1253y_{t-1} = e_t \tag{58}$$

To forecast the model, the forecasting procedures outlined in Section 2 are performed on the above model. The periodic peak values are now added back in, yielding

$$y_t = .1253y_{t-1} + e_t + \Psi(8) \tag{59}$$

This model is subsequently back-transformed into the original values.

To compare this method of correcting strong seasonality with the seasonal difference method, the model-fitting procedure outlined in Section 3 was performed on the blue crab series 2. The resulting model is also an AR(1) model:

$$y_t - .4151y_{t-1} = e_t \tag{60}$$

The final forecasting model is

$$\nabla_8^1 \left(1 - .4151q^{-1}\right) y_t = e_t \tag{61}$$

The r^2 of model (59) is .90, while that of model (61) is .86. The peak removal method results in a slightly better model. Two disadvantages of the seasonal difference method that are absent in the spectral peak removal method are: (1) seasonal differencing results in the loss of data observations (eight in the blue crab case); and (2) the chance of overdifferencing is quite high when seasonally differencing a time series which exhibits a moderate seasonality but not a strong seasonality. The main disadvantage of the spectral peak removal method is that it only is valid in the univariate case. Transfer function modeling must rely on the seasonal difference method to correct for strong seasonality.

6. METHODOLOGY FOR FISHERIES TIME SERIES ANALYSIS

The techniques for building time series models for fisheries data sets presented in Sections 2 through 5 are now synthesized into a comprehensive methodology. Figures 16 and 17 display respective methodological flow charts for univariate and bivariate time series analysis. To illustrate the methodology, both the univariate blue crab catch data series and the bivariate yellowfin tuna effort and catch series are analyzed below.

6.1 UNIVARIATE METHODOLOGY

The time series of blue crab catch has been used as the primary example in sections 2 through 5. Thus, the following analysis of this data set will refer in part to results presented previously. In Section 4, intervention analysis separated the blue crab series

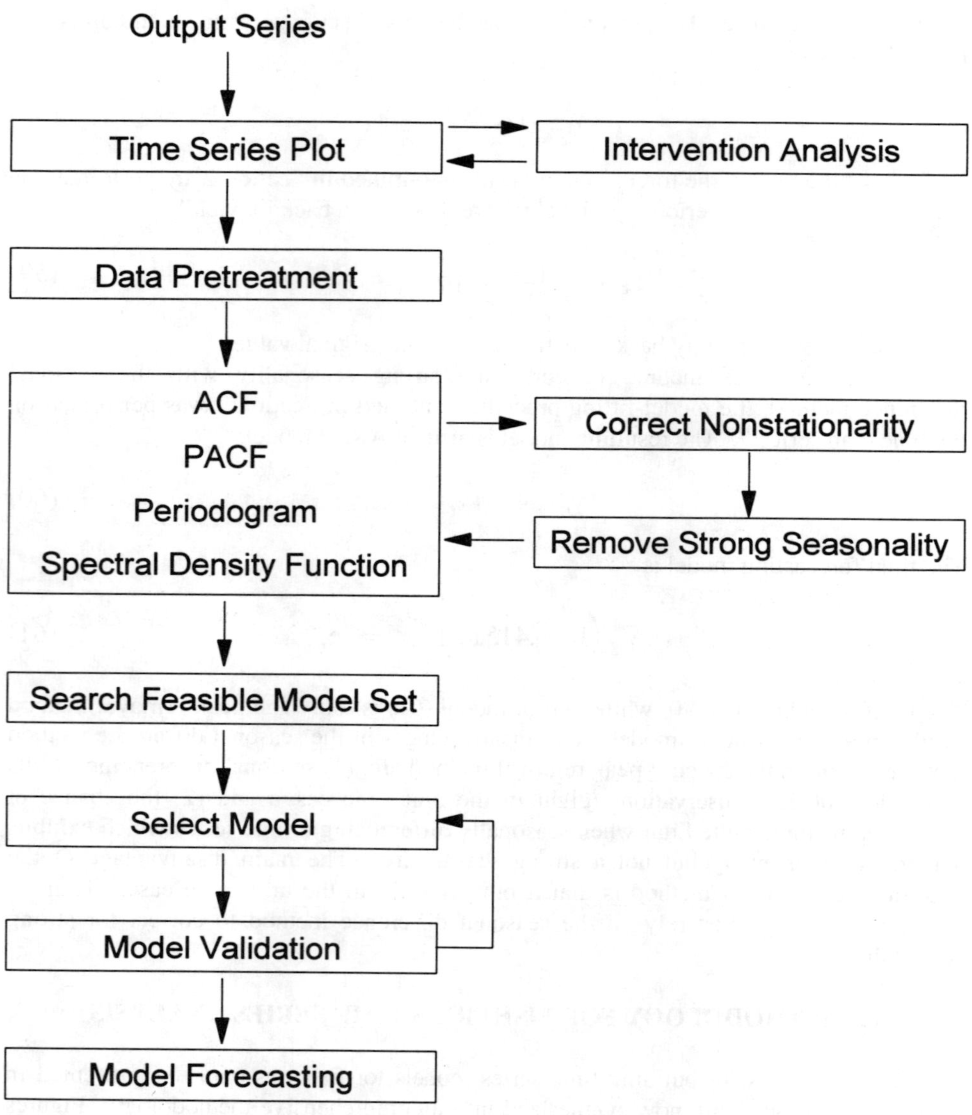

Figure 16. Methodological flow chart for univariate time series analysis.

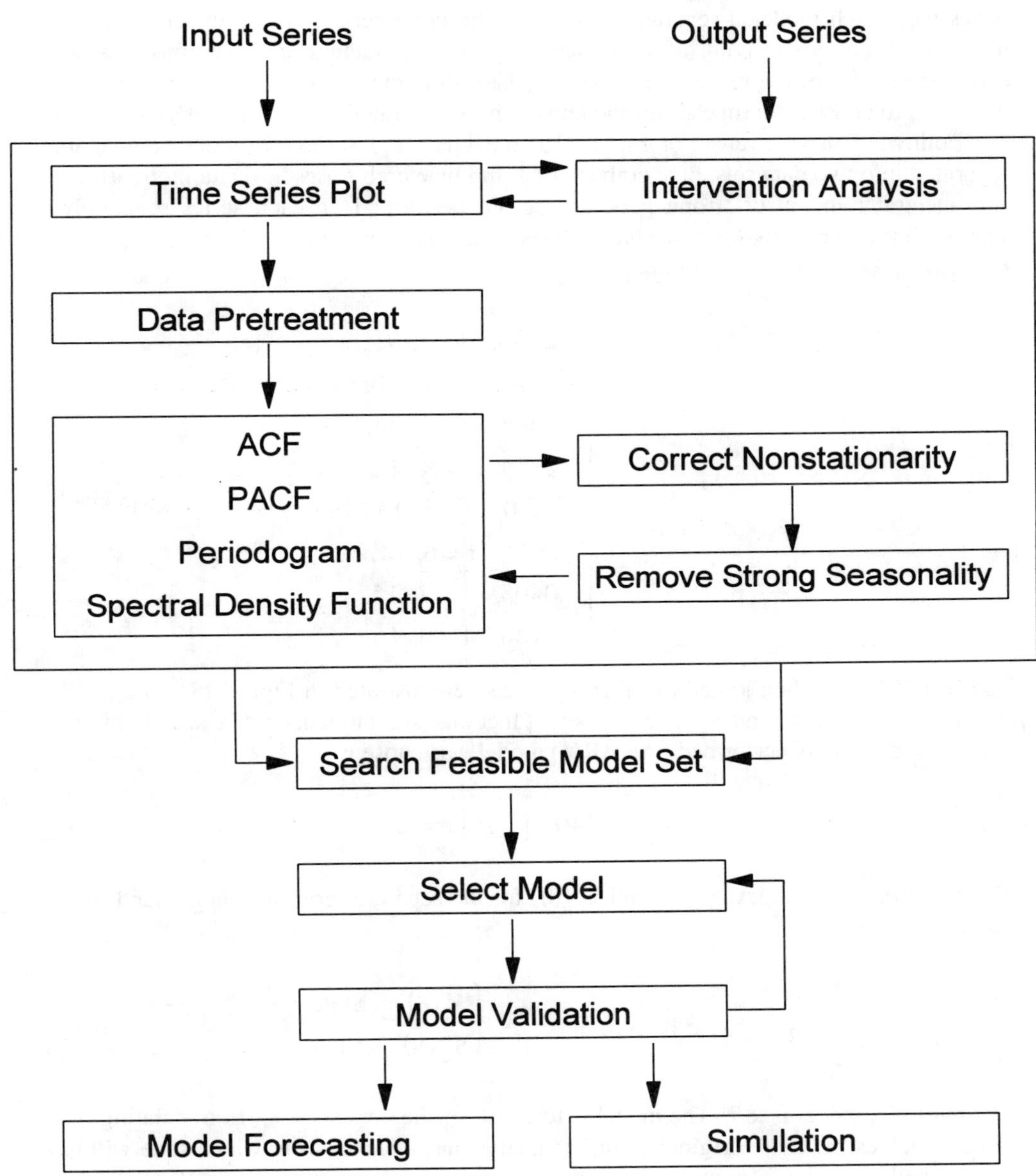

Figure 17. Methodological flow chart for bivariate (input-output) time series analysis.

into two parts, corresponding to an assumed "permanent" change in conditions. The series were each modeled separately, and only the latter series, blue crab series 2, was forecast. An alternate point of view may attest that the change in conditions was not "permanent" at all, but rather the series may be oscillating between the two states. We thus need to construct a time series model which incorporates this new perspective.

Following the flowchart of Figure 16, the blue crab series is plotted and again separated into two data sets, blue crab series 1 and blue crab series 2. Data pretreatment and spectral removal of strong periodicities are performed on each series separately. The periodic function $\psi_2(8)$ for blue crab series 2 was identified in Section 5 (equation 57). The periodic function for blue crab series 1 is

$$\Psi_1(8) = \begin{cases} -.2630, & t=1,\ 9,\ \ldots \\ -.1573, & t=2,\ 10,\ \ldots \\ .0405, & t=3,\ 11,\ \ldots \\ .2146, & t=4,\ 12,\ \ldots \\ .2630, & t=5,\ 13,\ \ldots \\ .1573, & t=6,\ 14,\ \ldots \\ -.0405, & t=7,\ 15,\ \ldots \\ -.2146, & t=8,\ 16,\ \ldots \end{cases} \tag{62}$$

The two series are then joined together again, as demonstrated in Figure 18. The ACF, PACF, periodogram, and spectral density functions are plotted, and a search of the feasible model set is performed. An AR(1) model was chosen:

$$(1 - .3402q^{-1})y_t = e_t \tag{63}$$

The removed peak values, $\psi_1(8)$ and $\psi_2(8)$, for the separate series are then added back into the model:

$$y_t = .3402y_{t-1} + e_t + \begin{cases} \Psi_1(8), & \text{State 1} \\ \Psi_2(8), & \text{State 2} \end{cases} \tag{64}$$

The r^2 of this model is .86. The model is forecast for the next season. In translating the forecast values back into original form, the analyst has a choice as to which state will be represented during the next fishing season, state 1 or state 2. Figure 19 plots the forecast values for the two states.

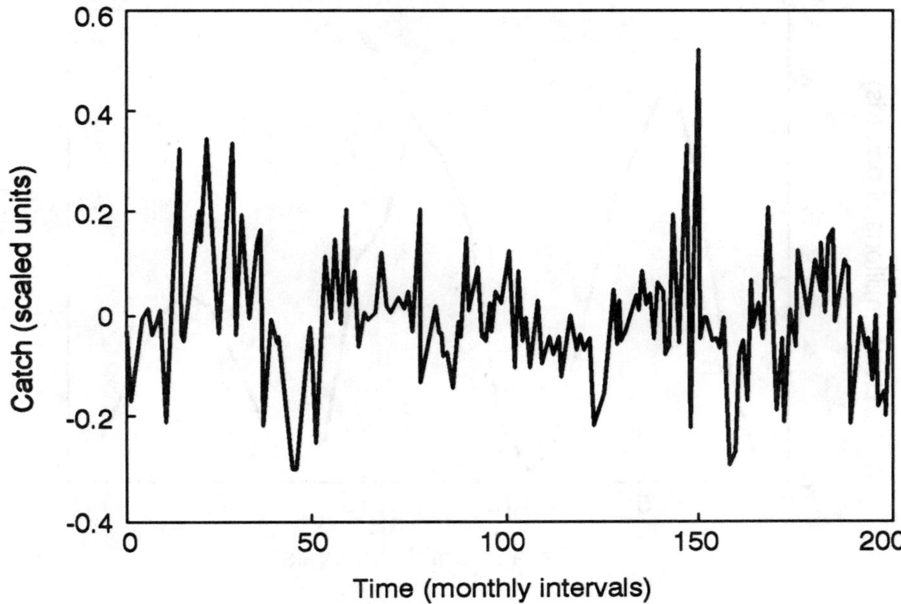

Figure 18. Time series plot of complete blue crab data series (in scaled units) after spectral removal of strong periodicities was performed separately on blue crab series 1 and 2.

6.2 TRANSFER FUNCTION METHODOLOGY

Time series plots of the yellowfin tuna effort and catch data series are presented in Figure 20. The effort and catch series are first treated separately to correct for nonstationarity and strong seasonality (Figure 17). The ACF and PACF are plotted for both the input and output series in Figures 21 and 22. From these plots, it appears that although a seasonal periodicity of k = 12 is apparent in each series, strong seasonality does not exist. The following feasible ABC model set was searched: A(0, 1, 2, 3, 6, 12, 18, 24); B(0, 1, 2,3, 6, 12, 18, 24); and C(0, 1, 2, 3). Table 6 presents the top ten models sorted by the FPE criterion. After various estimation and validation iterations of these models, an ABC(1, 18, 0) model was chosen:

$$\begin{aligned} y_t - .2035y_{t-1} &= \\ &.6935x_t - .1515x_{t-1} - .0896x_{t-6} - .0497x_{t-12} - .0567x_{t-18} + e_t \end{aligned} \tag{65}$$

The model's r^2 is .70.

In forecasting the above model, an input series must be provided. A common approach to providing a future input series is to forecast its univariate time series model.

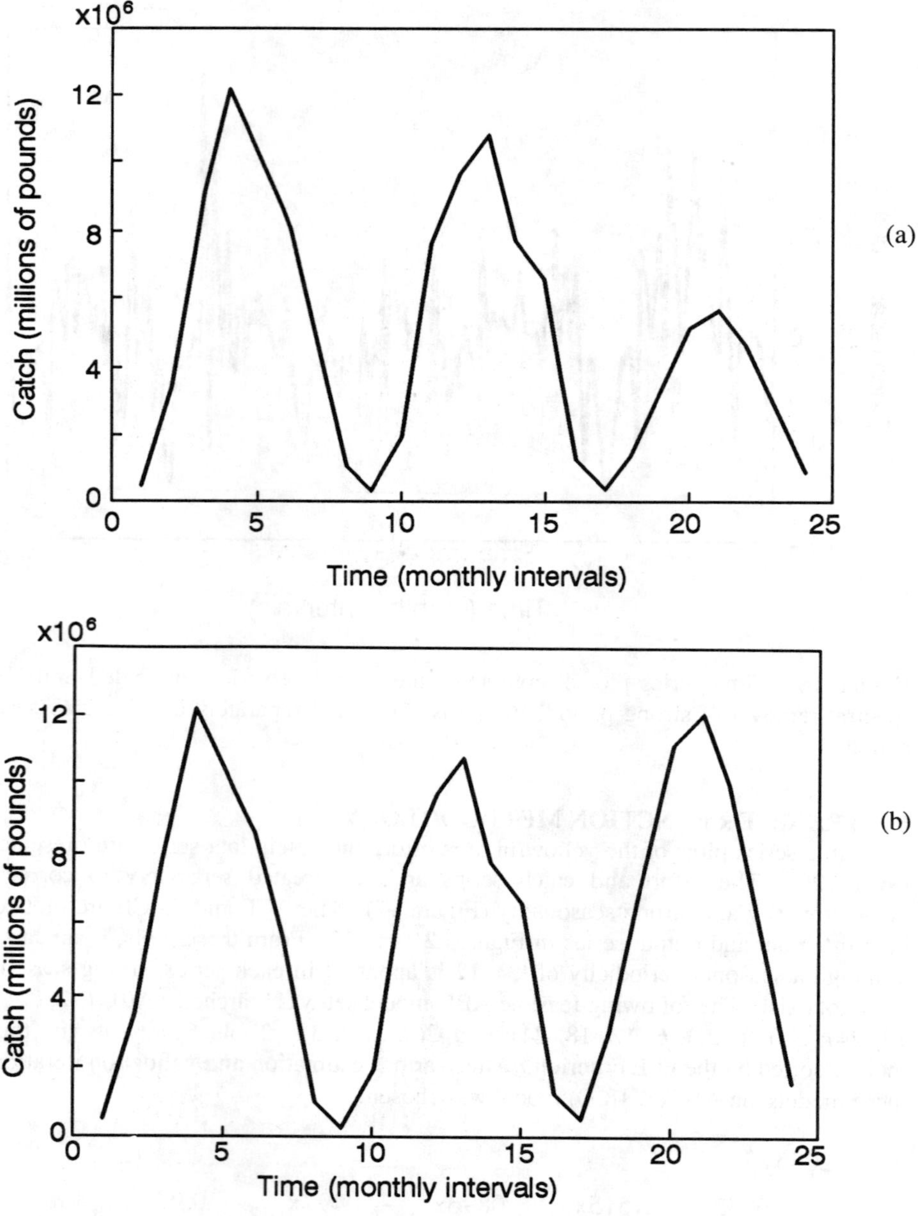

Figure 19. Plots of the last two crabbing seasons (months 1 to 16) of the blue crab data series and forecast values of the next crabbing season (months 17 to 24) for two different system states: (a) state 1 = conditions for blue crab series 1 are in effect; and (b) state 2 = conditions for blue crab series 2 are in effect.

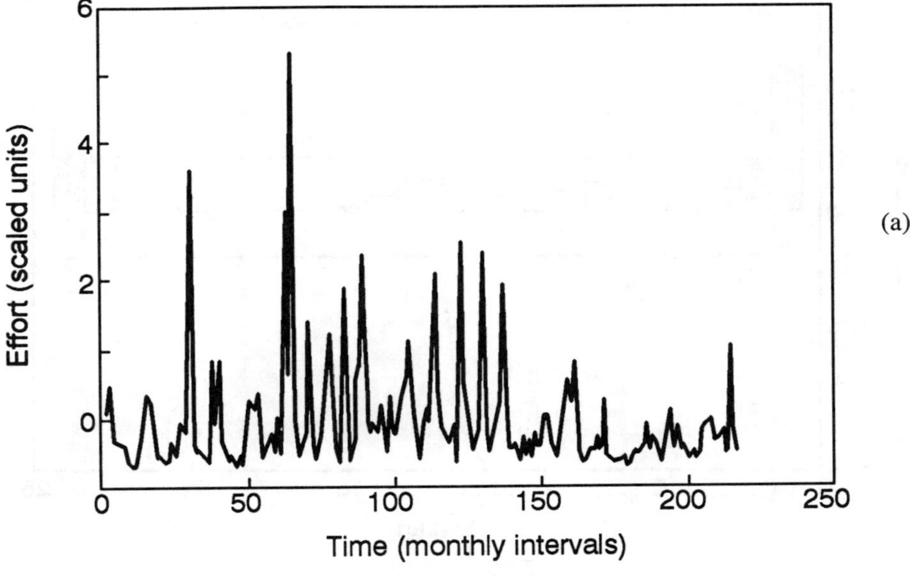

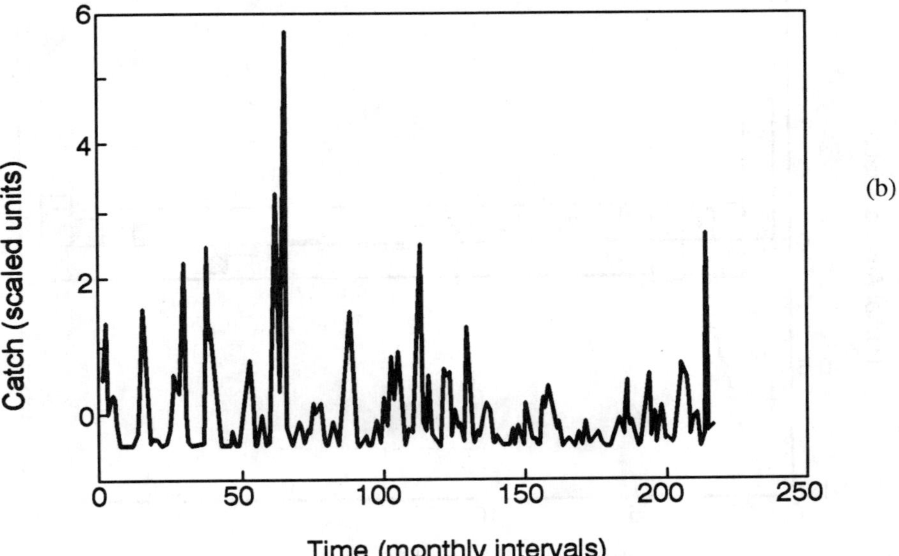

Figure 20. Time series plots (in scaled units) of yellowfin tuna (a) effort and (b) catch data series (Table 2).

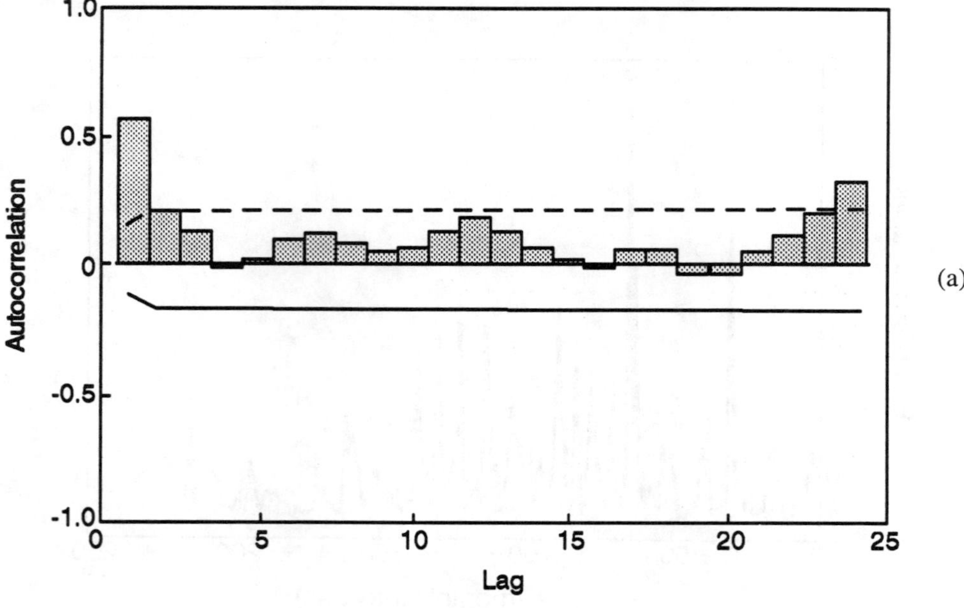

(a)

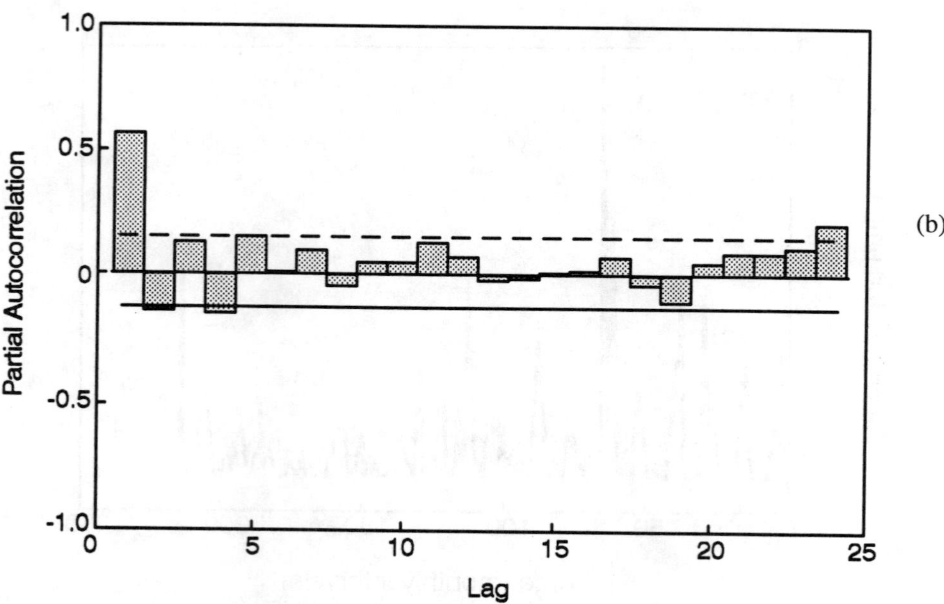

(b)

Figure 21. (a) ACF and (b) PACF graphs for the yellowfin tuna effort (input) series.

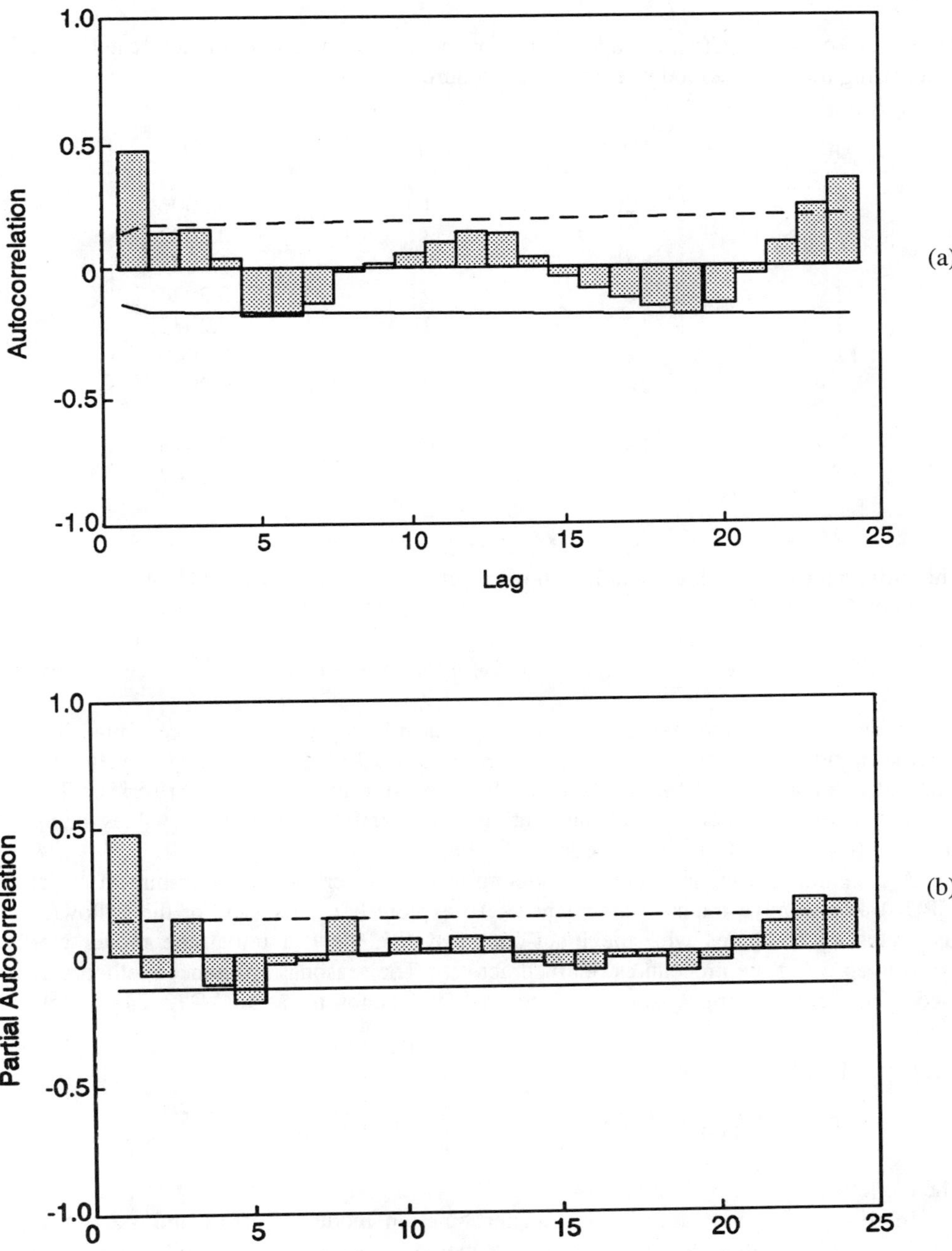

Figure 22. (a) ACF and (b) PACF graphs for the yellowfin tuna catch (output) series.

Table 6. Top 10 ABC(n_a, n_b, n_c) models for the yellowfin tuna effort and catch data series using the feasible model set search procedure.

$A(n_a)$	$B(n_b)$	$C(n_c)$	FPE
2	2	1	0.2006
2	2	3	0.2022
1	3	2	0.2023
2	2	2	0.2024
12	12	1	0.2032
12	12	3	0.2040
24	24	3	0.2041
3	3	1	0.2044
1	18	0	0.2045
12	12	2	0.2047

The univariate methodology detailed above resulted in the following AC(24, 1) model for yellowfin tuna effort:

$$y_t - .1983y_{t-1} - .2514y_{t-24} = e_t + .4621e_{t-1} \tag{66}$$

The model's r^2 was quite low, .37. Another approach is to provide a variety of plausible input scenarios for the coming fishing season. Figure 23 plots the forecast of yellowfin tuna catch for the next 12 months, using the forecast values of the effort series. The forecast seems less than spectacular, but this is a result of the input values being provided from a poor time series model.

A common practice in fisheries modeling is to consider the catch-per-unit-of-effort (CPUE) as a relative index of abundance. To explore how this view of the yellowfin tuna data set compares with the input-output system view, a univariate model was constructed using the procedures outlined above. The seasonal difference method was used to correct for strong seasonality. The final model chosen was an AC (2, 24) model:

$$\begin{aligned} &\nabla^1_{12}(y_t - 1.667y_{t-1} + .4826y_{t-2}) = \\ &\quad e_t - .8044e_{t-1} + .1728e_{t-2} + .1282e_{t-12} + .1247e_{t-18} + .1858e_{t-24} \end{aligned} \tag{67}$$

The model's r^2 is .32.

In this particular example, the transfer function model of effort and catch was significantly better than the univariate CPUE model. This is perhaps a reflection of the enhanced flexibility of the ABC model in adjusting to dramatic changes in the input effort series.

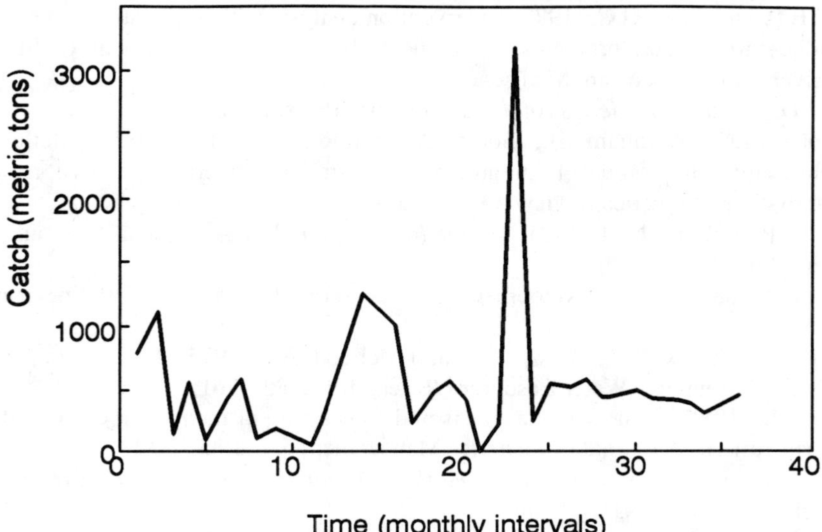

Figure 23. Plot of last 2 fishing seasons (months 1 to 24) of the yellowfin tuna catch series and forecast values for the next fishing season (months 25 to 36).

7. CONCLUSIONS

As demonstrated in this chapter, time series models have the potential to be a powerful tool for the fishery analyst and manager. The yellowfin tuna transfer function model of effort and catch not only significantly outperformed the univariate model of CPUE, it provided a greatly increased ability to adjust to fluctuating levels of effort. The transfer function of the dynamic effort-catch fishery system can be considered a generalized model of fishing mortality, since it accounts not only for the immediate effects of fishing effort, but also time-lagged effects of both effort and catch.

REFERENCES

Box, G.E.P. and Cox, D.R. 1964. An analysis of transformations. Proceedings of the Royal Statistical Society B 26: 211-252.

Box, G.E.P. and Jenkins, G.M. 1976. *Time Series Analysis: Forecasting and Control*. Holden-Day, San Francisco.

Box, G.E.P. and Tiao, G.C. 1973. Intervention analysis with applications to economic and environmental problems. Technical Report 335, Department of Statistics, University of Wisconsin, Madison.

Cryer, J.D. 1986. *Time Series Analysis*. Duxbury Press, Boston.

de Gooijer, J.G., Abraham, B., Gould, A. and Robinson, L. 1985. Methods for determining the order of an autoregressive-moving average process: a survey. International Statistics Review 53: 301-329.

Draper, N.R. and Smith, H. 1981. *Applied Regression Analysis*. John Wiley and Sons, New York.

Fildes, R. 1988. Recent developments in time series forecasting. OR Spektrum 10: 195-212.

Hipel, K.W., Lennox, W.C., Unny, T.E. and McLeod, A.I. 1975. Intervention analysis in water resources. Water Resources Research 11: 855-861.

Jensen, A.L. 1985. Time series analysis and forecasting of menhaden catch and cpue. North American Journal of Fisheries Management 5: 78-85.

Kay, S.M. 1988. *Modern Spectral Estimation: Theory and Application*. Prentice Hall, Englewood Cliffs, NJ.

Ljung, L. 1987. *System Identification: Theory for the User*. Prentice Hall, Englewood Cliffs, NJ.

MacDonald, G.J. 1989. Spectral analysis of time series generated by nonlinear processes. Review of Geophysics 27; 449-469.

Mendelssohn, R. 1981. Using Box-Jenkins models to forecast fishery dynamics: identification, estimation, and checking. Fishery Bulletin, U.S. 78: 887-896.

Ricker, W.E. 1975. Computation and interpretation of biological statistics of fish populations. Bulletin of the Fisheries Research Board of Canada 191.

Roff, D.A. 1983. Analysis of catch/effort data: a comparison of three methods. Canadian Journal of Fisheries and Aquatic Science 40: 1496-1506.

Saila, S.B., Wigbout, M. and Lermit, R.J. 1980. Comparison of some time series models for the analysis of fisheries data. Journal du Conseil, Conseil Permanent International pour l´Exploration da la Mer 39: 44-52.

Schaefer, M.B. 1957. A study of the dynamics of the fishery for yellowfin tuna in the eastern tropical Pacific Ocean. Bulletin of the Inter-American Tropical Tuna Commission 2: 245-285.

Schumway, R.H. 1988. *Applied Statistical Time Series Analysis*. Prentice Hall, Englewood Cliffs, NJ.

Stergiou, K.I. 1989. Modeling and forecasting the fishery for pilchard (*Sardina pilchardus*) in Greek waters using ARIMA time-series models. Journal du Conseil, Conseil Permanent International pour l'Exploration da la Mer 46: 16-23.

CHAPTER 8

EMPIRICAL METHODS AND MODELS FOR MULTISPECIES STOCK ASSESSMENT

Saul B. Saila[1], James E. McKenna[2], Sonia Formacion[3], Geronimo T. Silvestre[3], and John W. McManus[4]

[1] Graduate School of Oceanography, University of Rhode Island,
[2] Fisheries Statistics Section, Florida Marine Research Institute,
[3] College of Fisheries, University of the Philippines in the Visayas,
[4] Marine Science Institute, University of the Philippines

1. INTRODUCTION

1.1 BACKGROUND

There is relatively little theory available currently to predict the impacts of different levels of catch on exploited tropical fish assemblages. Effective management of a tropical fishery resource depends upon an ability to predict alterations in species composition and yields under various management regimes, so that it may be possible to maintain the exploited assemblage at adequate levels of both the density of organisms, as well as the desired diversity of species. This is perceived as an ultimate goal, one which has not yet been achieved and to which our efforts are being directed.

Pope (1979) developed some extensions of the stock-production model for multispecies fisheries. In general, an extension to the Schaefer stock production model can be stated as follows:

1-56670-151-1/96/$0.00+$.50

$$\frac{1}{P}\frac{dP}{dt} = a_1 - b_1 P - c_1 Q - q_1 E$$

$$\frac{1}{Q}\frac{dQ}{dt} = a_2 - b_2 Q - c_2 P - q_2 E$$

where:

P = population size of species P (biomass)
Q = population size of species Q (biomass)
b_1, b_2 = intraspecific competition coefficients
c_1, c_2 = interspecific competition coefficients
E = fishing effort
q_1, q_2 = catchability coefficients

If c_1 and c_2 are positive, the two species (P and Q) are competitors. If c_1 and c_2 have opposite signs, then the species with a negative sign for c is a predator and the species with a positive c value is the prey species. It is evident from the above that the problem of real world parameter estimation for this model is difficult even in the two species case. It becomes virtually intractable when the number of interacting species is even slightly increased. However, Pope (1979) used this model to show some interesting properties of the system under some simplifying assumptions. He showed that a strategy of reducing stock size of each species to about one half its original biomass will result in close to a global maximum yield, under the assumption that all interactions remain the same over time.

In this chapter, we attempt to consider some empirical methods and models which are designed to assess temporal and spatial changes in the composition of exploited multispecies fish assemblages. These will hopefully contribute to a more rational fisheries management program for tropical multispecies fisheries in the future.

1.2 RATIONALE

We recognize that although single-species deterministic models are still in use, they have serious limitations for certain applications, such as in tropical multispecies fisheries. We also believe that fisheries models, based on systems of differential equations, such as illustrated in the above example, have severe limitations for practical applications. This is primarily due to the virtual impossibility of obtaining reasonable estimates of model parameters, such as the interaction terms. We also think that system simulation models, which are difficult to construct and are often species and site specific, are not a completely appropriate or cost-effective approach. Thus, our emphasis is on empirical methods and models, in the hope that they will minimize some of the above-mentioned problems and that they will contribute to more effective management of fisheries.

More specifically, one example of our initial approach is a method for the comparison of faunal lists (such as fish assemblages) to evaluate their similarities and differences over space and/or time. This method involves a probabilistic index of faunal similarity (FAUNSIM) to be described in detail in a later section. This probabilistic index differs from many prior approaches to faunal similarity comparisons, which have relied on similarity or diversity indices, which are easily calculated but cannot be rigorously compared unless data are replicated. This and other related methods are considered initial steps to multispecies stock assessment in an adaptive mode in accordance with the general recommendation of Walters (1986).

We have chosen to apply so-called computer-intensive methods to assess the significance of a statistic in hypothesis testing whenever this seemed feasible. This approach coincides with that described by Noreen (1989), although we had attempted to apply some of the methods prior to the publication of the above-mentioned book. In general, the significance of almost any test statistic can be assessed by computer-intensive techniques, which do not involve the necessity of generating the probability distribution of the test statistic under the assumption that the null hypothesis is true. The specific techniques involve approximate randomization, which can be used to test the null hypothesis that one variable is unrelated to another and Monte Carlo sampling, as well as jackknife and bootstrap re-sampling. Software for these procedures (written in BASIC, FORTRAN, and PASCAL) is available in Noreen's book. This software can be utilized with little modification for developing computer-intensive techniques for diverse applications, including fisheries. In summary, we believe that computer-intensive techniques will sometimes avoid the pitfalls of the required assumptions in conventional statistical hypothesis testing and become increasingly used in multispecies studies.

Our approaches utilized to date for the development of empirical models for multispecies assemblages are primarily based on application of methods to develop transition probability matrices from aggregated data on fish assemblages obtained primarily from sequential research vessel surveys. Algebraic methods are then applied to project the transition probability matrices to make inferences about the long-term and short-term behavior of the multispecies fishery. These procedures are developed and illustrated in cases where they have not yet been published. In summary, we regard these models as potentially useful inasmuch as they do not require the extensive data needed by conventional time series methods, nor do they require the estimation of numerous parameters as required by simulation models. In spite of this, the methodology is capable of describing and predicting dynamics of components of an assemblage involving many species even when observational data are limited if assumptions of the method are accepted.

2. COMPARISONS BETWEEN MULTISPECIES FISH ASSEMBLAGES USING SIMILARITY AND DIVERSITY INDICES

2.1 INTRODUCTION AND PROBLEM STATEMENT

Although some may be more useful than others, it is our opinion that no diversity or similarity index can be considered to be totally effective in quantifying diverse responses of a multispecies fish assemblage or other assemblages to perturbations ranging from exploitation to environmental changes. Therefore, it seems that fisheries scientists must interpret their results in the context of the limitations of the index or indices employed.

The use of diversity and similarity indices has been in vogue for at least 20 years. There have been criticisms of these indices on both theoretical and practical grounds. See for example, Hurlbert (1971) and Murphy (1978).

It is believed that most, if not all, published diversity indices involve the following implicit assumptions when used in a biological context:

(a) all individuals of the same species are identical

(b) all pairs of differing species are equally different

It is evident that neither of these assumptions are true in biological applications. Therefore, it should be kept in mind that violation of these assumptions may have significant effects on computation and interpretation of diversity measures.

Our approach with respect to both similarity and diversity indices is to attempt to explicitly state the limitations of the methods provided, as well as their relative merits. We also try to evaluate indices derived from known or postulated probability distributions (such as Fisher's α, based on the logarithmic series distribution) and other distributions, such as the Poisson or gamma models. The indices derived from known distributions are parametric because they are embedded in the parameters of the distribution model. We contrast these with non-parametric indices, which have mostly been developed on an ad hoc basis for specific applications.

It should be noted that methods to be described (such as FAUNSIM) utilize a probabilistic approach to adjust for sample size differences between samples. Clearly, the effects of sample size must be considered in composition studies of species distributions in space or time. Koch (1987) has provided one such study which indicated that most species occur rather infrequently in samples of species assemblages, and therefore, sample size effects may be large relative to the faunal patterns often reported. This observation seems to be true in the case of our fisheries data examined to date. There are actually several aspects of sample size that must be considered. The most basic question concerns the relationship between sample size and variability in the sample or, how representative is the sample of the population. For example, examination of variances may suggest pre- or post stratification. A second question concerns comparisons among samples which differ greatly from each other and the associated statistical difficulties. The reader is referred to Chapter 4 for further information about sampling designs.

If it is assumed that two assemblages contain exactly the same fauna, we can then compute the predicted similarities and differences as a function of sample size. Then it is possible to judge whether the calculated values are due to sampling or whether they indicate other effects. For example, if a species occurs only once in 100 samples, we can estimate the probability of it being found in an additional sample from its proportion of occurrences n/m, when the species occurs n times in m samples. In this case the estimate of the probability (P_n) is equal to 1/100. Koch (1987) has presented a formula for the probability (P_n) of the species being found in at least one of r additional samples as $1 - (1 - n/m)^r$. If 100 additional samples were taken, then the value of $P_n = 0.766$. From this it can be concluded that if 25 species occurred only once in the 100 samples taken initially, then only 9 of these 25 species are expected to occur in 100 additional samples. Similar calculations are possible for species which occur 2, 3, ..., T times, to provide estimates of the number of species found in the original 100 samples that would be among the species found in an additional 100 samples. The "bottom line" is that if species occur very rarely in samples, the number of species occurring in the original sample and not found in an additional sample of the same size may be fairly large, due to sample size effects.

In summary, we believe it is important to understand the concepts and limitations underlying various measures of diversity and similarity, and we attempt to provide some of this information to the potential user. However, we also believe that much more research will be required before species diversity or similarity defined by a single index term can be related to the theory of community stability and ecosystem structure. Indeed, it may turn out that no real relationships exist.

2.2 FAUNSIM BACKGROUND

FAUNSIM is a computer program which calculates a probabilistic index of faunal similarity based on comparison of the number of taxa common to two stations (locations in space or time) with the number that would be expected to be in common if the taxa were distributed randomly. An important problem in adaptive tropical fishery science involves the critical examination of species assemblages to evaluate their similarities and differences over time and/or space. We restrict our attention in this case to two assemblages, with no replicates in either one. Clearly, if two lists of fish species have no taxa in common, it can be easily concluded that they are different. If the two lists (assemblages) are exactly the same, then no temporal or spatial changes are assumed at the appropriate scale. Problems of evaluation occur when some but not all taxa are shared. A common solution to this problem has been to calculate some index of similarity between the two assemblages. However, many indices of similarity have important shortcomings, primarily because most of them have not been derived rigorously, and the statistical properties of the indices are not well known. Therefore, no formal testing can be effectively done with them.

Saila (unpublished ms.) was aware of the limitations of common indices of similarity and attempted to develop a probabilistic model of fish assemblage similarity (coherence) by comparing two assemblages with the numbers expected if the taxa were distributed randomly. This initial effort was termed comparative coherence analysis. It was later found that this approach had already been developed by Henderson and Heron (1977) and was utilized and further described in the paleontological literature by Raup and Crick (1979). This brief explanation of the method follows Raup and Crick, and the interested reader is referred to them for further details.

2.3 FAUNSIM METHOD

The method involves a comparison between the observed number of species common to two assemblages and the probability distribution of the expected number of common species as a measure of the similarity of the two assemblages. The expected number of shared species and its probable variation constitute the null hypothesis for assessing faunal similarity. Monte Carlo simulations are utilized in testing the null hypothesis in the following manner. It is necessary to know the number of species in each of the two assemblages and the number of species shared by them. An estimate of the pool of species, reflecting the relative abundance of all species, must also be available or provided.

(a) For each real pair of species assemblages in the available data set, an imaginary pair of assemblages is constructed by randomly sampling species from the pooled data set. This is accomplished by means of a random number generator in the computer program. The analysis proceeds by randomly selecting individuals from the species pool and recording their species identification. Each individual is replaced after each draw. If an individual is drawn from the pool whose species identification is already recorded in the imaginary assemblage being generated, it is returned to the pool and another individual is drawn at random. This continues until the number of species in the generated assemblage is equal to the number of species in the first natural assemblage. This procedure is also applied similarly so that the second generated assemblage contains the same number of species as the second natural assemblage. At this point, the two generated assemblages are compared and the number of species shared is recorded. This value provides one point on an approximate probability distribution describing the number of species expected to be in common to the two assemblages.

(b) The above procedure is repeated (100 times in our case), and the number of taxa shared by each assemblage is again recorded for each repetition.

(c) A frequency distribution of the results is an estimate of the probability distribution of the number of taxa in common, under the conditions specified by the two samples.

(d) The number of taxa actually shared between the two real world assemblages is compared with the generated distribution.

(e) An index of similarity is computed as:

Index of similarity = 1 – the probability that the expected number of shared taxa will be greater than the observed number of shared taxa

= probability that k expected will be less than or equal to k observed.

2.4 SAMAR SEA EXAMPLE

FAUNSIM was used to compare the demersal fish communities at a number of stations within the Samar Sea, Philippines in both time and space. Comparisons in space were made for both March 1979 (cruise 96) and March 1980 (cruise 107). The species pool consisted of the sum of the abundances (measured by weight) of all the species sampled at all 28 stations of a given cruise. The species pool for temporal comparisons consisted of the mean abundance (by weight) of all the species sampled at the 28 stations for the two cruises of interest.

Community analyses using clustering methods (McManus, 1985) have shown a difference between the communities of deep and shallow stations in the Samar Sea. The division line between these communities fell at approximately the 30 m isobath. The same line was used here to define the deep and shallow communities. These communities were compared in March 1979 and March 1980. Each community was also examined for differences over time by comparing them from 1979 to 1980 (March). The same spatial and temporal comparisons were made of the communities at stations 2 and 18. These two stations were chosen for their proximity and completeness of data.

The null hypothesis being tested in all of these analyses was: the species are randomly "sprinkled" in space (or time) and thus all assemblages will be the same, given natural variability (Raup and Crick, 1979). Rejection of the null hypothesis was set at the 95% level or greater.

Table 1. Results of faunal comparisons in the Samar Sea using FAUNSIM.

Communities Compared	Year	Probability $K_{obs} < K_{exp}$	Similarity[1] index
Shallow and deep	1979	1.0	0.0
Shallow and deep	1980	1.0	0.0
Shallow	1979–1980	1.0	0.0
Deep	1979–1980	1.0	0.0
2 and 18	1979	0.2	0.69
2 and 18	1980	0.84	0.1
2	1979–1980	0.99	0.005
18	1979–1980	0.72	0.195

[1] Based on FAUNSIM.

Analysis of the shallow and deep communities by FAUNSIM showed there was a significant difference between these communities both in space and time (Table 1) (Figure 1). Comparison of stations 2 and 18 showed that the communities there were not significantly different in March of 1979 or 1980 (Figure 2). Temporally, station 18 did not change from 1979 to 1980, while station 2 did show a significant change (Table 1). The community structures are shown in Figure 2.

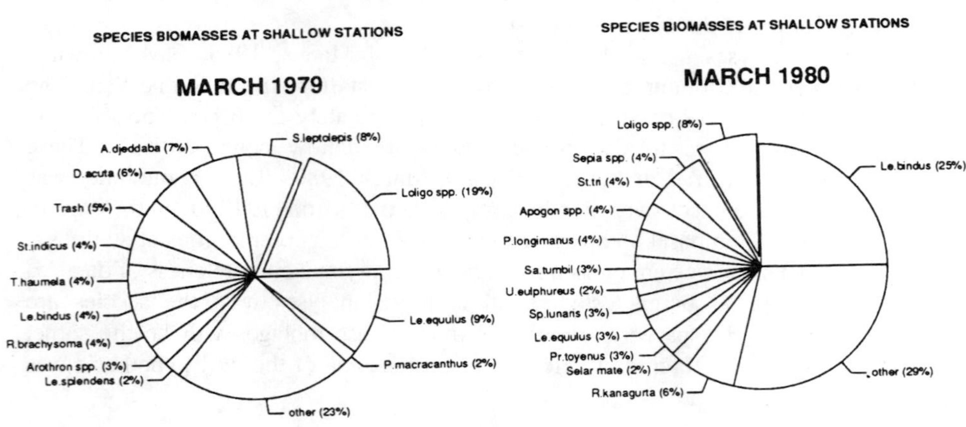

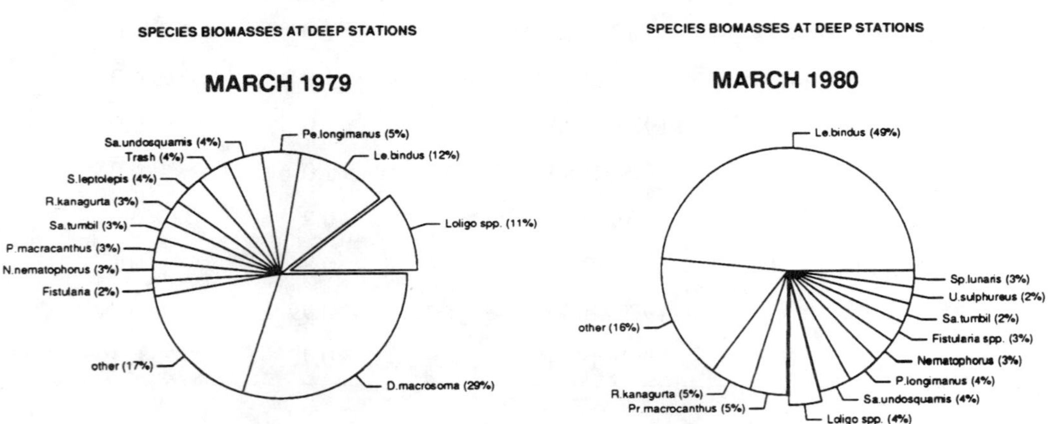

Figure 1. Pie charts showing the species biomasses at shallow stations and deep stations, March 1979 and March 1980.

The number of species observed to be common to the shallow and deep communities was far less than would be expected from a random distribution. Thus, the shallow and deep stations were significantly different, supporting McManus' (1985) conclusions. Also, over time each of these communities shared fewer species than would be expected. This demonstrates a significant change from March 1979 to March 1980. Some differences can be seen visually in the communities (Figure 1).

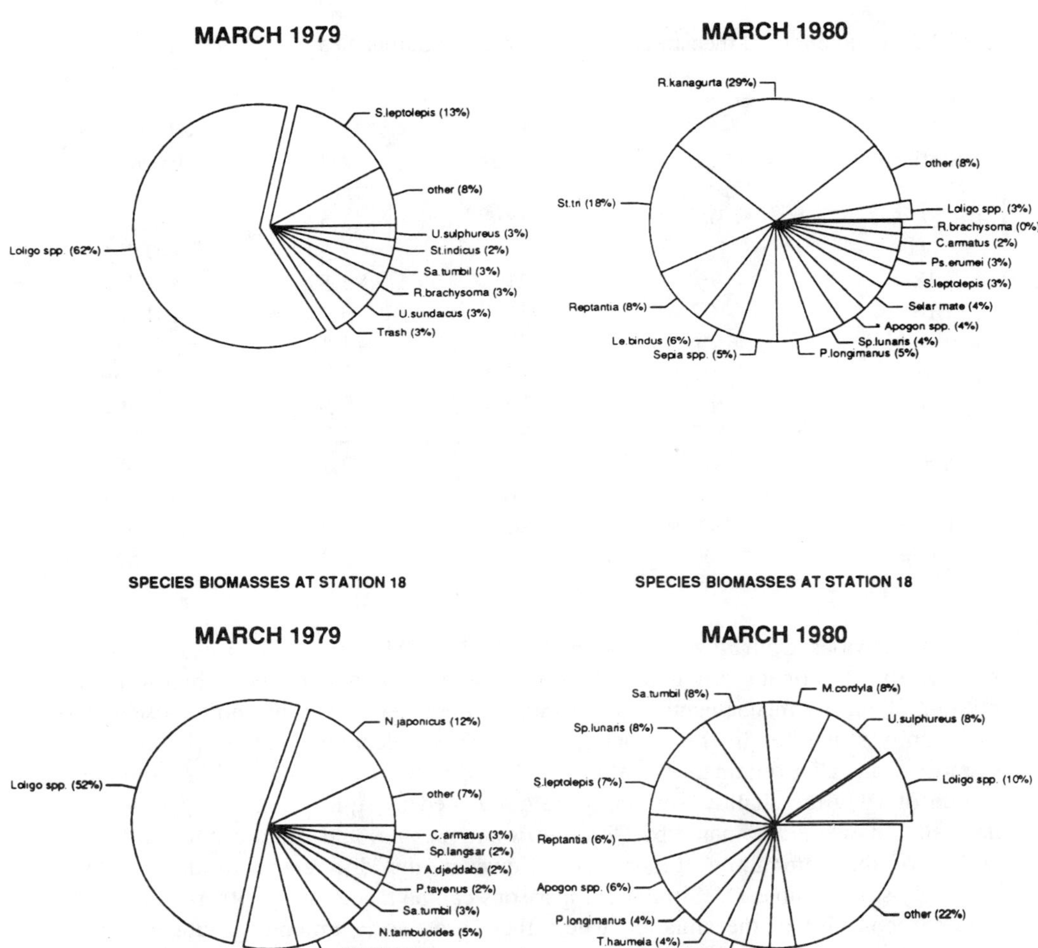

Figure 2. Pie charts showing the species biomasses at stations 2 and 18, March 1979 and March 1980.

Comparison of faunal similarities with some standard diversity indices shown in Table 2 provides some justification for the use of the FAUNSIM method. There were a number of cases where there were large changes in the diversity indices which were not significant (e.g., station 18, 1979–1980) and more moderate changes in these indices which were significant (e.g., station 2, 1979–1980). The standard diversity indices seem to integrate components of the assemblages which may counteract each other, resulting in little or no change in the diversity index, when in fact a significant change has occurred in the community.

Table 2. Conventional measures of diversity in the Samar Sea.

Station	Cruise	Richness	H′	[H′/ln (S*)] Evenness
2	96	18	1.53	0.335
2	107	38	2.54	0.556
18	96	21	1.87	0.410
18	107	46	3.13	0.686
4	96	7	1.12	0.245
4	107	43	2.94	0.644
5	96	19	1.68	0.368
5	107	38	2.77	0.607
Shallow	96	39	3.12	0.683
Shallow	107	107	3.23	0.708
Deep	96	80	2.77	0.607
Deep	107	94	2.39	0.524

It is obvious from this discussion that FAUNSIM seems effective at identifying statistically significant differences between samples of assemblages, but it tells us nothing about the mechanisms causing those differences. Looking more closely at the community structures in the Samar Sea, it becomes clear that not only richness and evenness, but also dominance ranking of the species is important. For example, in March of 1979 the shallow stations were dominated by squid (*Loligo* spp.), while the deep stations were dominated by *Decapterus macrosoma*, and squid accounted for only 10.8% of the community (Figure 2). However, the differences in diversity were relatively small (Table 2). Similar comparisons can be made for the other communities. One can conceive of the situation where there are two communities with exactly the same richness and evenness, but whose species are ranked differently by dominance. Standard diversity indices would indicate that they are identical, while FAUNSIM would be able to distinguish significant differences between them.

2.5 SKEWER ANALYSIS: GRADIENTS IN SPACE OR TIME

Pielou (1984) developed and described a test to indicate whether multivariate data collected at a sequence of sites (or over an environmental gradient) exhibit any trend in species composition. In examining this paper, we quickly reached the conclusion that the approach described by Pielou had considerable application in multispecies fisheries work.

A computer program (SKEWER) written in Microsoft Quick BASIC was developed from the paper by Pielou (1984), and much of the background for the program is adapted from her original description. The original paper provides further details, and the user is urged to consult it. The program implements a test to determine whether multivariate data collected from a sequence of sites along an abiotic gradient in space or time exhibit any trend in species composition. The program SKEWER is one of the computer programs provided on the diskette in this book. It, and a number of other programs, are described in the appendix.

The background for the test follows. Consider a swarm of n points in s-dimensional coordinate space. Imagine the points represent sampling units placed at a sequence of sites (or times) along some gradient. The points possess a predetermined order, that is, they are labeled, 1, 2, ..., n according to their position along the gradient. Let a randomly oriented line or "skewer" pass through the s-dimensional coordinate system. The orientation of the skewer is determined by selecting s random numbers, normalizing them to sum to unity, and taking their points, n square roots as the direction cosines of the skewer. Positive and negative signs are assigned randomly to the direction cosines. Perpendiculars are next dropped from each of the n points onto the skewer. The order of the feet of the perpendiculars along the skewer is noted. A coefficient of rank correlation (Kendall's τ) is computed between this ordering and the natural ordering of the numbers 1 to n. If the calculated value of τ is close to + 1, the data points must be non-randomly ordered, and the direction of ordering must have a component that is parallel (or anti-parallel) to the random skewer. A τ value near zero indicates random order of the points in all directions or, although the points are non-randomly ordered in one direction, the orientation of the random skewer does not have an appreciable component in this direction.

If the cluster of points is sampled with a large number of random skewers, the frequency distribution of τ is expected to have a mode of 0 if the points are unordered in all directions. The distribution would be bimodal with values of $\pm\,\tau$ if there were some preferred direction through the swarm along which the points were ordered.

To determine whether an observed bimodal distribution of τ is a chance event or a significant ordering, it is possible to construct and explore with skewers a number of comparison swarms of points, having random coordinates. For a 5% test, 20 such random swarms are necessary, each with the same number of coordinate axes, s, and the same number of points, n.

In our application the "folded" distribution of τ in the interval (0, 1) is examined. Under the null hypothesis, the probability density of (τ) decreases monotonically from some maximum at $\tau = 0$. If the points are non-randomly ordered, the distribution of τ

will have (τ) greater than zero. That is, a mode at some value other than zero will occur.

The test is distribution free. It can be applied to gradients observed in either time or space, although Pielou indicates only spatial gradients in her paper.

The user of this program should be aware that it is slow in execution for relatively large data sets. The time of execution is clearly related to the dimensions of s and n, as well as the number of random skewers utilized to determine a simulated distribution of τ.

This program can be extremely useful for the preliminary analysis of multispecies fish communities for changes which might be significant over time or space. We have utilized the program for this purpose in a preliminary way with very encouraging results.

The example we have chosen with which to illustrate the application of Pielou's skewer analysis to tropical multispecies fishery data involves a portion of the Thai trawl survey data set. The data set has been used primarily by Saila and Erzini and it has been updated by Pauly. It represents one of the most extensive tropical multispecies data sets available. The data were collected by the Division of Marine Fisheries, Government of Thailand, using a standardized stratified sampling design in space and one hour trawls in time. The gulf is a generally soft bottom environment which the trawls sampled from the benthos to several meters in height. The data span a period of about 17 years. Fish samples were sorted into species and species groups and are an excellent example of the high level of species diversity usually associated with tropical environments. Extensive fishing was also occurring in the gulf concurrent with the sampling periods. The catch composition (in kg/hour) of selected major species or species groups were converted to relative proportions by weight. All the available species were not used in the example primarily to ease the computational burden. The raw data used for this example are illustrated in Table 3. In each of the two 5-year periods, we are interested in whether or not the temporal gradient of 5-years represented any significant trend in the species composition expressed in weight per standardized effort.

The test was applied separately to the 1966–70 data and to the 1976–80 data. We used a sample of 500 random skewers and generated 20 random swarms to yield 20 simulated distributions of τ. We examined the "folded distribution" of τ, or the distribution of $|\tau|$ whose value lies in the interval (0, 1). Under the null hypothesis, $|\tau| = 0$. Under the alternative hypothesis that the points are non-randomly oriented in some direction, the distribution will have a mode at $|\tau| > 0$. The probability density of $|\tau|$ decreases monotonically from some maximum value at $|\tau| > 0$.

It seems evident from a visual inspection of the raw data of Table 3 for 1966–70 that there was some decrease in all species (species groups) except for Loligo which showed some increase. However, it is not immediately clear as to whether these data exhibit a significant trend. The lower data set (from 1976–80) seems to suggest some increase in Leiognathidae and possible declines in rays and Loligo spp. However, these changes are not very obvious. Visual inspection does not lead to a very firm conclusion regarding either data set.

Both data sets were subjected to skewer analysis, and 20 frequency distributions of tau based on 500 random skewers were generated. The results of these analyses are as follows. For the 1966–70 data set, 2 out of 20 frequency distributions showed modes of $\tau > 0$ for the raw data. For these data expressed as proportions, 1 out of 20 showed a mode of $\tau > 0$. In the case of the 1976–80 data set, 6 of 20 frequency distributions showed τ values > 0 for the raw data. We interpret these data as suggesting that some indication of trend was evident, but it was barely significant statistically in the early data set. The latter data (1976–80) had a stronger indication of deviation from a random distribution of field data than the earlier data set and was statistically significant. This was not at all evident from the visual inspection of the data.

We believe that an analysis of a larger number of species or species groups would be a very worthwhile endeavor with skewer analysis. These data were utilized herein primarily as examples of the kinds of data which are considered to be appropriate for analysis.

Table 3. Partial listing of the catch composition of major species (in kilograms per trawling hour) from surveys in the coastal areas of the Gulf of Thailand. Percent by weight is also included for the species groups.

Group/Year	1966		1967		1968		1969		1970	
	wt	%	wt	%	wt	%	wt	%	wt	%
Leiognathidae	20.02	35	10.87	27	14.37	37	10.59	30	10.25	32
Nemipterus spp.	15.31	26	11.78	29	7.46	19	7.40	31	8.61	26
Rays	9.03	17	4.77	12	2.17	6	2.99	8	7.86	9
Loligo spp.	8.04	14	9.13	22	10.60	28	11.01	33	8.55	26
Lutjanidae	4.76	8	4.02	10	3.83	10	3.01	8	2.25	7
Group/Year	**1976**		**1977**		**1978**		**1979**		**1980**	
	wt	%	wt	%	wt	%	wt	%	wt	%
Leiognathidae	(2.64)*	13	(2.76)	16	(2.87)	16	(2.98)	18	(3.09)	19
Nemipterus spp.	5.75	29	5.34	30	3.99	23	4.98	11	4.73	29
Rays	1.54	8	1.07	6	1.22	7	0.65	4	0.78	5
Loligo spp.	9.27	46	8.31	47	8.99	51	6.79	42	6.98	44
Lutjanidae	0.75	4	0.29	1	0.50	3	0.77	5	0.47	3

*Values in parenthesis were estimated by extrapolation

2.6 DIVERSITY INDICES

Understanding how to measure changes in species number and abundance of aquatic assemblages is a challenging problem which has received considerable attention but which is not yet fully resolved.

Changes in assemblages of organisms are sometimes quantified by means of diversity indices. A dictionary defines diversity in a somewhat circular fashion, as the condition of being diverse. Ecologists tend to define diversity of biological communities as the degree of uncertainty attached to the specific identity of any randomly selected individual. The larger the number of species and the more equal their proportions, the greater the uncertainty and hence diversity. Clearly, species diversity refers to both the number of species present (species richness or abundance) and the evenness with which the individuals are distributed among species (species equitability or evenness). See, however, the limitations described in the Introduction to the section on similarity and diversity.

There have been many reports of presumed relations between diversity and stability. It appears that a diverse system can be stable, but no real relationship exists where diversity causes stability or vice versa.

The following material lists a few of the large number of indices of diversity, and/or biotic conditions. The indices most frequently used are those based on information theory, but other indices have been shown to have desirable statistical properties.

2.7 SOME DIVERSITY INDICES BASED ON PROBABILITY DISTRIBUTIONS

Fisher's α, where

$$S_1 = \alpha \ln (1 + \alpha)$$

2.7.1 Based on Information Theory (Distribution Free)

Brillouin's H,

$$H = 1/N \cdot \ln N! \prod_{i=1}^{S} N_i!$$

Shannon's H′

$$H' = \sum_{i=1}^{S} n_i/n \cdot \ln(n_i/n)$$

Evenness E

$$E = H'/H'max$$

Redundancy R

$$R = H'max - H'min/H'max - H'min$$

2.7.2 Based on Theory of Runs

Keefe's TU

$$\mathrm{TU} = 1-(n/n-1)\left\{\sum_{i=1}^{k} P_i^2 - 1/n\right\}$$

List of symbols:

S = number of species on a sample or population
N = number of individuals in a population or community
N_i = number of individuals in species i of a population or community
n_i = number of individuals in a species i of a sample from a population
n = number of individuals in a sample from a population
P_i = n_i/n = fraction of a sample of individuals belonging to species i

For a review of diversity and other indices, see Washington (1984). It is regrettable that Washington's review did little to resolve important sampling properties of the indices which were considered. The debate about the relative merits of various diversity measures continues. For example, a recent paper (Gadagkar 1989) suggests that the diversity indices derived by Hill (1973) have an anomalous behavior under certain conditions.

Magurran (1988) summarizes some of the extensive literature on measurement of ecological diversity. We tend to agree with her advice that at the least the investigator should make serious efforts to assure that the sample sizes obtained are equal and/or they are large enough to be representative of the population. We also believe it is useful to draw a rank abundance graph which can provide some indication of which distributional form the data appear to follow.

Effective statistical comparisons of diversity indices have been limited, because the distributions of the indices are not usually known. George and Hanumara (1989) have recently illustrated a procedure developed by Rao to relieve this problem partially. This procedure develops diversity analysis as a generalization of hierarchical analysis of variance, for populations classified by combinations of factors. This procedure is called analysis of diversity (ANODIV) and is a generalization of analysis of variance (ANOVA) to qualitative data. The procedure partitions total diversity of a population into a number of additive components.

The choice of the diversity measure to be used in the ANODIV procedure is somewhat arbitrary, if the diversity measure satisfies the following. A function H is an approximate diversity measure if for a population of s species of proportion p_i (i = 1 to s) such that $p_i > 0$ and $\Sigma p_i = 1$,

(a) $H(p) > 0$

(b) $H(p)$ is concave

In the above, H(p) = 0 only if $P_i = 1$ for some species i and $P_i = 0$ for all others. Condition (b) insures that the diversity in a mixture of populations will not be smaller than the average diversity of the individual population. This is the condition identified by Gadagkar as being violated by Hill's indices. It is known that the indices mentioned above satisfy these conditions. We recommend the application of Rao's ANODIV procedure for replicated data sets, because in its simplest application (a one-way design) ANODIV partitions diversity into components existing within and between communities. The applications of the procedure are nicely detailed in George and Hanumara (1989).

The measurement of diversity by indices (sometimes called heterogeneity indices) can be developed by use of statistical sampling theory to investigate how communities or assemblages are structured. Among the distributions which have been applied to community samples, the logarithmic series, log normal, and negative bionomial distributions seem to have received the most attention by biologists. The logarithmic series was first applied by Fisher et al. (1943) to a range of community samples. It is our belief that although no distribution is expected to fit a given data set or similar sets with complete fidelity, the logarithmic series distribution seems to be a useful distribution for working with multispecies fish assemblages. A detailed account of the log-series distribution and its application follows.

In general, the logarithmic series distribution implies that the greatest number of species has nominal abundance, and that the number of species represented by a single specimen is always maximal. This is not always the case and Preston (1948) suggested examining the x-axis (number of individuals in the sample) on a logarithmic scale rather than an arithmetic one. When a proper conversion of scale is accomplished, the data approximate a normal distribution. The log normal distribution has been known to fit a variety of data rather well. However, we have placed emphasis on the logarithmic series distribution because of relatively good fits to our multispecies fishery data, and also because parametric indices of diversity and similarity can be derived from this distribution.

3. USING SIMILARITY INDICES

3.1 GENERAL

Various measures of similarity have been used in making pairwise comparisons between ecological communities, and they have also been frequently used as elements of a matrix of input coefficients for various clustering algorithms. Therefore, we consider it useful to consider similarity indices and their potential utility in the context of applications to empirical studies related to tropical multispecies fisheries.

The basic definition of a similarity index as used herein is restricted to a single number, which is a function of the pairwise comparison of values for each attribute for two samples. This use of the word "similarity" is not construed to exclude indices of dissimilarity, since one is the logical complement of the other. That is, similarity indices relate how "close" two samples are to each other, and dissimilarity indices refer to how "far apart" they are. For the purposes of this report, only those indices which are intermediate similarity scores defined on the pairwise comparisons of the values for each assemblage attribute in the two samples will be considered. This excludes any type of index which only summarizes the attribute values for each sample into a single index and then makes comparisons between these indices. Thus, this definition excludes so-called diversity or community structure indices as defined by Pinkham and Pearson (1976) and excludes the diversity indices reviewed by Washington (1984). The usage of the α diversity index in multispecies fisheries applications is considered in another section of this chapter.

As shown by Johnston (1976), a general similarity index can be defined on two N-dimensional vectors, say X_l and $X_{l'}$ where l and l' are two selections of location by time (ij) assemblage samples. The subscripts for the vectors are changed from ij to l in order to avoid a confusing array of subscripts. We then have:

$$S_{ll'} = F(s_{ll'k}) \,, \tag{1}$$

where F is some function of the N pairwise comparisons, $s_{ll'k}$, and:

$$s_{ll'k} = g(x_{lk}, x_{l'k}) \tag{2}$$

where g is some function of the attribute values observed in the two samples.

From equation (1) and equation (2), any particular similarity index ($S_{ll'}$) can be defined by specifying the functions g and F.

In summary, a similarity index is characterized as the result of a two-step process defined on a pair of vectors. In the first step, an attribute similarity score is obtained for each attribute by comparing the attribute values observed in the pair of vectors. The result is a vector of attribute similarity scores. These are combined in the second step to arrive at a similarity index. The operation in the first step was characterized as a function (g) defined on pairs of attribute values. The second operation was characterized as a function (F) defined on the vector of attribute similarity scores from the first step. In most instances, F is a simple score or weighted score of attribute similarity scores.

A number of similarity indices ($S_{ll'}$) have been based on a transformation of the original data to presence-absence form by:

$$\delta_{lk} = \begin{cases} 1 & if\ x_{lk} > 0 \\ 0 & if\ x_{lk} = 0 \end{cases}$$

The data matrix in this case, or in the case where binary data are collected, will consist of vectors of 0's and 1's. Another form of transformation is to reduce individual counts and continuous data to more than two categories by defining cells. These can be of equal, geometric, or logarithmic intervals. The data are then classified into these cells. This transformation is of the form:

$$C_{l,k} = \begin{cases} 1 & if \;\; 0 \leq x_{lk} < C_1 \\ 2 & if \;\; C_2 \leq x_{lk} < C_2 \\ & \vdots \\ m & if \;\; C_{m-1} \leq x_{lk} < C_m \end{cases}$$

The choice of a particular measure or index of similarity may influence the interpretation of the original data (Czarnecki 1979) in a significant manner. Different indices appear to be useful in indicating different changes in community composition. An example includes the fact that one kind of index may be useful for detecting changes in numerical abundance while another may be better at indicating changes in proportional abundance. In addition to the choice of similarity index, there is another and probably more important problem. This concerns the determination of what numerical value of a given similarity index is important for biological interpretation.

Many users of similarity indices have not taken sampling variability into account. Some suggestions for estimating the variance of a similarity index have been made, but it remained for Smith, Genter, and Cairns (1986) to present two alternate methods for estimating confidence intervals for comparing two multispecies groups (assemblages) that provide an accurate estimate of variability. The methods described above are appropriate for most measures of similarity computed from species counts, species biomass, or species density. This work has been extended by Nemec and Brinkhurst (1988) to the comparison of dendograms.

The purpose of this section is to briefly explain and illustrate the methods for estimating confidence intervals for generalized similarity indices as described by Smith et al. (1986), to illustrate these methods with an example applied to a tropical multispecies fishery, and to provide documentation and a program for computing a measure of similarity with confidence intervals for similarity comparisons.

3.2 SPECIFIC SIMILARITY INDICES

Johnston (1976) indicated that more than 50 similarity indices have been proposed or used. He investigated the characteristics of 25 similarity indices used in studies of ecological communities. Other similarity measures, such as PS_1, which is a measure of proportional similarity, have been recommended by Pielou (1978) and Kohn and Riggs (1982). Their reasons for this recommendation appear to be related to the fact that PS_1 is relatively independent of sample size. Washington (1984) reviewed diversity, biotic, and some similarity indices. However, his review included only five similarity indices used in aquatic systems. Heltshe (1988) has derived an exact expression for the two-sample jackknife estimate of the simple matching coefficient and the Jaccard index of similarity when quadrat sampling procedures are employed. He demonstrated that these estimators were not affected by quadrat size as long as the sampling area was fixed. In this report we confine our attention to the use of only one index which we believe has utility for the purposes at hand. However, we emphasize that other published indices are appropriate for the methodology to be demonstrated herein, but we strongly suggest using indices which are not affected by sample size (group size).

In terms of an example, our attention will be focused on SIMI, a similarity measure proposed by Stander (1970) and illustrated by Smith et al. (1986). Another index of similarity (I) described initially by Mountford (1962) is based on the logarithmic series distribution and the assumption that the samples are drawn from the same assemblage (community). Mountford's index has some desirable properties for certain applications. The same holds true for HR, the heterogeneity ratio introduced by Kobayashi (1987). Both are relatively insensitive to sample size.

Brock (1977) made a critical comparison of two community similarity indices and concluded that investigators should not rely on one index. Instead, he stated that they should consider several which have different areas of sensitivity. Kobayashi (1987) studied fifteen previously used similarity indices for the effects of variable sample size. He found only three (NESS, — Grassle and Smith 1976; C_λ — Morisita 1957; and C_λ' — Morisita 1971) to be relatively independent of sample size. He also described a new index, the heterogeneity ratio (HR), which was sample size independent but assumes the samples were taken from a community (assemblage) whose species abundances are reasonably described by the log series distribution.

In attempting to summarize the results of the above-mentioned studies, the following statements and suggestions are considered useful to users of similarity indices.

(1) Most similarity indices tend to ignore both the time series and response pattern information in the raw data. They are usually defined on an arbitrary pair of sample vectors. These vectors are identifiable as representing particular time and location samples. The samples may be selected for comparison by a similarity index to reflect a potential change between two time periods at the same location or a difference between two locations at the same time period. However, many indices are algebraically symmetric, so the fact that one was taken before the other has no impact on the calculation. The functions assigning pairwise attribute similarity scores are symmetric, so that:

$$g(x_{lk}, x_{l'k}) = g(x_{l'k}, x_{lk})$$

(2) Some similarity indices are relatively unaffected by differences in sample size. Such indices include Morisita's C_λ and C_λ', Grassle and Smith's NESS, Kobayashi's HR and Mountford's I, among others. The above indices are suggested for future application and further consideration because they have satisfied one important criterion since sample size is not necessarily fixed in the case of some fishery surveys, and comparative studies often involve variable sample sizes.

(3) Greater general use of jackknife and bootstrap methods for estimating confidence intervals is also recommended for the following reasons. Precise variance estimates can be developed which need not be symmetric but which yield accurate confidence intervals. Equal numbers of replicates are not required. However, some replication is required for both jackknife and bootstrap methods to be applied successfully.

(4) Jackknife and bootstrap methods may be used with virtually any data and similarity index as long as it is replicated. However, see item (2) above. When sample sizes are small and variability is high, the jackknife seems to perform better than the bootstrap.

(5) The variance of a similarity measure (index) depends on the number of replicates and the number of organisms counted per replicate. When variability between replicates is the largest component of variability, a large number of replicates is required to provide reliable estimates of similarity measures.

(6) When similarities are close to 0.0 or to 1.0, the distributions of the various estimates may be skewed. Skewness does not seem to be a problem if values of similarity range from 0.1 – 0.9, unless the sample sizes are small (< 6).

3.3 APPROACH AND EXAMPLE

For the purposes at hand, assume that there are two multispecies fish assemblages with m_i (i = 1, 2) replicate samples (trawl hauls, trap lifts, gillnet sets, etc.) from each and T species or species groups are considered. The analysis of the trawl hauls results in data of the form n_{ijk} = the abundance of species k in replicate j of community i.

Let $n_{ij\bullet} = \sum_k n_{ijk}$ be the total count for the jth replicate of the community i. Let $N_i = \sum_j n_{ij}$ be the total count of individuals in the sample from community i and $n_{i\bullet k}$ be the total number of species k in the sample from community i. The general data that arise in a specific experimental comparison are shown in Table 4.

The estimate of SIMI for each pair of assemblages is based on the total count vectors $n_1 = \left(n_{i\bullet 1}, n_{1\bullet 2}, \ldots, n_{i\bullet \mathrm{T}}\right)$, the abundance of each species summed over all replicates. Proportions, $P_{i\cdot k} = \frac{n_{i\cdot k}}{N_i}$ and SIMI are estimated as:

$$\text{SIMI} = \sum P_{1k} P_{2k} \Big/ \sqrt{\sum P_{1k}^2 \sum P_{2k}^2}$$

by

$$\widehat{\text{SIMI}} = \sum p_{1k} p_{2k} \Big/ \sqrt{\sum p_{1k}^2 \sum p_{2k}^2}$$

Capital letters indicate true proportions and lower case *p*'s indicate the estimates from the sample data.

Table 4. An example of the form of data to be collected for two multispecies fish assemblages, which could have been separated by space or time.

	Taxa for assemblage 1				
Replicate	1	2	...	T	Totals
1	m_{111}	m_{112}	...	n_{11T}	$n_{11\cdot}$
2	m_{121}	n_{122}	...	n_{12T}	$n_{12\cdot}$
3	.	.	.	.	.
.	.	.	.	.	.
.	.	.	.	.	.
.	.	.	.	.	.
m_1	n_1m_{11}	n_1m_{12}	...	n_1m_{1T}	$n_1m_{1\cdot}$
Totals	$n_{1\cdot1}$	$n_{1\cdot2}$	...	$n_{1\cdot T}$	N_1
	Taxa for assemblage 2				
Replicate	1	2	...	T	Totals
1	m_{211}	m_{212}	...	n_{21T}	$n_{21\cdot}$
2	m_{221}	n_{222}	...	n_{22T}	$n_{22\cdot}$
3	.	.	.	.	.
.	.	.	.	.	.
.	.	.	.	.	.
.	.	.	.	.	.
m_2	n_2m_{21}	n_1m_{22}	...	n_2m_{2T}	$n_2m_{2\cdot}$
Totals	$n_{2\cdot1}$	$n_{2\cdot2}$	...	$n_{2\cdot T}$	n_2

It is clear that an estimate of the mean score of SIMI is possible from Table 4 data, as well as a measure of variance. However, the jackknife and bootstrap are recommended alternate methods, because they do not depend upon use of the multinomial distribution, and the bootstrap can produce asymmetric confidence intervals.

Using the notation of Smith et al. (1986), we let $G = G(\bar{P}_1, \bar{P}_2)$ denote a general measure of similarity where $\bar{P}_1 = (P_{i1}, \ldots, P_{iT})$ is the vector of proportional abundances for community i and $P_{i.k}$ refers to the proportion of species k in community i. The estimate of similarity (G) is based on n_{ijk}, the abundance of species k in replicate j for community (assemblage) i. The estimate of P_{ik} is:

$$p_{ik} = \sum_{j=1}^{m_i} n_{ijk} \Big/ \sum_{k=1}^{T} \sum_{j=1}^{m_i} n_{ijk} = n_{i \cdot k} / N_i \tag{3}$$

where N_i is the total count for community (assemblage) i. To estimate the similarity between communities (G), use

$$\hat{G} = G(p_1, p_2)$$

3.4.1 Jackknife

The two-sample jackknife estimate is determined by calculating G with different replicates removed each time. Let G'_{-k} denote the estimate of G when the kth replicate is removed from community 1 and G''_{-k} denote the estimate where the kth replicate is removed from community 2. Pseudo-values are then generated by sequentially removing replicates from community 1 and computing

$$\hat{G}'_k = \left(m_1 - \tfrac{1}{2}\right)\hat{G} - \left(m_1 - 1\right)\hat{G}'_{-k} \tag{4}$$

and for community 2:

$$\hat{G}''_k = \left(m_2 - \tfrac{1}{2}\right)\hat{G} - \left(m_2 - 1\right)\hat{G}''_{-k} \tag{5}$$

The jackknife estimate of G is calculated based on these pseudo-values as:

$$\hat{G}_J = \tfrac{1}{m_1} \sum_{k=1}^{m_1} \hat{G}'_k + \tfrac{1}{m_2} \sum_{k=1}^{m_2} \hat{G}''_k \tag{6}$$

The estimate of the variance is:

$$s^2\left(\hat{G}_J\right) = \tfrac{1}{m_1} \hat{\sigma}_1^2 + \tfrac{1}{m_2} \hat{\sigma}_2^2 \tag{7}$$

where:

$$\hat{\sigma}_1^2 = 1/\left(m_1 - 1\right) \sum_{k=1}^{m_i} \left(\hat{G}'_k - 1/m_1 \sum \hat{G}'_k\right)^2 \tag{8}$$

and $\hat{\sigma}_2^2$ is defined similarly. Construction of the jackknife estimate in this way automatically adjusts for bias.

3.4.2 Bootstrap

This method estimates the sampling distribution of G by repeatedly sampling with replacement from the actual data. The bootstrap proceeds as follows:

(a) Using the data, sample with replacement m_1 replicates from the m_1 replications for community 1 and sample with replacement m_2 replicates from the m_2 replications for community 2 (i.e., sample from the original data, with replacement). This sample is called the bootstrap sample.

(b) Compute the bootstrap estimate of G based on the bootstrap sample which is denoted $\hat{G}_{(i)}$.

(c) Repeat a) and b) b times to get estimates of G.

(d) Calculate the bootstrap estimate of G and its sampling variance, i.e.,

$$\hat{G}_B = \frac{1}{b}\sum_{i=1}^{b}\hat{G}_{(i)} \tag{9}$$

and

$$s_B^{\,2} = (b-1)^{-1}\sum_{i=1}^{b}\left(\hat{G}_{(i)} - G_{(B)}\right)^2 \tag{10}$$

It has been suggested that b values lie in the range of 50–200 for estimating the variance. For realistic confidence intervals, b should be ~1000. The bootstrap method also gives an estimate of the bias in estimating $\hat{G} \bullet \text{Bias}\left(\hat{G}\right) = \hat{G}_B - \hat{G}$. To adjust for bias, form:

$$\hat{G}^* = \hat{G} \cdot \text{Bias} - \left(\hat{G}\right) = 2 \cdot G - G_B \tag{11}$$

3.5 CONFIDENCE INTERVALS

For the jackknife, confidence intervals are symmetric and are given by:

$$\text{SIMI}_J \pm t_c \cdot s\,(\text{SIMI}_J) \tag{12}$$

where t_c is the critical value from the t-table with $m_1 + m_2 - 2$ degrees of freedom and confidence level $1 - \alpha$.

Symmetric or asymmetric intervals may be computed for the bootstrap methods described by Smith (1985). Due to the simplicity of the derivatives of the proportional similarity measures:

$$\mathrm{PS} = \sum_{j=1}^{r} \min\left(P_{1j}, P_{2j}\right) = 1 - \frac{1}{2}\sum_{j=1}^{r}\left|P_{1j} - P_{2j}\right| \tag{13}$$

This formula can be reduced if the weights are all equal. Because $g_{1j} = 0$ when $P_{1j} > P_{2j}$, contribution to the variance comes only from those proportions for species 1 that are less than or equal to those for species 2. Hence, these proportions can be combined. If $P_{1j} \neq P_{2j}$ then:

$$\mathrm{V(PS)} = \mathrm{V}_j \sum_{J} + \mathrm{V}_j \sum_{K} P_{2k} \tag{14}$$

Here J refers to indices when $p_{1j} < p_{2j}$ and K to those indices where $p_{2k} < p_{1k}$. Hence, to compute the variance simply combine the data into two groups.

3.6 THE SIMCONF.BAS PROGRAM DOCUMENTATION

The SIMCONF.BAS program was developed to apply both jackknife and bootstrap techniques to generate critical comparisons of similarity indices. The program is based on the work of Smith et al. (1986). It was written in Microsoft's QuickBASIC version 4.5. It is designed to calculate similarity values for any two communities. The SIMI similarity coefficient is used at present, but any other similarity coefficient could be implemented, if desired.

$$\mathrm{SIMI} = \sum_{k=1}^{\mathrm{T}} p_{1k} \cdot p_{2k} \Bigg/ \sqrt{\left(\sum_{k=1} p_{1k}^{\ 2} \cdot \sum_{k=1} p_{1k}^{\ 2}\right)} \tag{15}$$

where T is the total number of species included, p_{1k} and p_{2k} are the proportions of species k (based on the sum of that species abundances over all replicates) in communities 1 and 2, respectively. This SIMI value is symbolized by G in most of the following discussion.

The SIMI values and their variances are determined by jackknifing and bootstrapping. Ninety-five percent confidence intervals are also generated by the jackknife and bootstrap methods. For comparison, the simple SIMI values, comparing each possible combination of replicates and the mean of those values, are also generated.

3.6.1 Program Operation

The proportional representation of each species from each community is calculated by summing the abundance of each species over all replicates within a community and dividing by the total number of individuals observed in that community. These proportions are estimates of the true proportions of each species in each community and are used in equation (11) to calculate the SIMI value.

The jackknife estimate proceeds by removing one replicate from community 1, while leaving community 2 intact, and recalculating the SIMI value (G'_{-k}). It then replaces that replicate in community 1, removes a replicate from community 2, and calculates a second SIMI value (G''_{-k}). Each of these values is then used to calculate a "pseudo" value according to the following:

$$G'_k = (m_1 - 1/2)G - (m_1 - 1)G'_{-k} \tag{16}$$

$$G''_k = (m_2 - 1/2)G - (m_2 - 1)G''_{-k'} \tag{17}$$

where m_1 is the number of replicates from community 1, m_2 is the number of replicates from community 2, and G is the original SIMI value calculated by equation (11).

This process is repeated, each time replacing the previous one and removing the next consecutive replicate, until all replicates have been removed once. A pair of "pseudo" values is generated at each iteration. These are then used to calculate the final jackknife estimate of SIMI:

$$G_J = 1/m_1 \sum_{k=1}^{m_1} G'_k + 1/m_2 \sum_{k=1}^{m_2} G''_k \tag{18}$$

They are also used in the determination of the variance of this estimate:

$$\sigma_1^2 = 1/(m_1 - 1) \sum_{k=1}^{m_1} \left(G'_k - 1/m_1 \sum_{k=1}^{m_1} G'_k \right)^2 \tag{19}$$

$$\sigma_2^2 = 1/(m_2 - 1) \sum_{k=1}^{m_2} \left(G''_k - 1/m_2 \sum_{k=1}^{m_2} G''_k \right)^2 \tag{20}$$

estimating the variance by:

$$s^2(G_J) = 1/m_1 \, \sigma_1^2 + 1/m_2 \, \sigma_2^2 \tag{21}$$

The 95% confidence interval is symmetric and is determined by the following:

$$G_J \pm t_c \cdot s(G_J), \tag{22}$$

where t_c is the critical value from the t-table with $m_1 + m_2 - 2$ degrees of freedom and a confidence level of $1 - \alpha$ (i.e., $\alpha = 0.95$). Smith et al. (1986) describe a transformation using the inverse cosine which can be applied when the SIMI estimate is near zero or one, to account for the skewed distribution.

The bootstrap estimate proceeds by generating two sets of random numbers. These random numbers are uniformly distributed between 1 and the number of replicates for each community. Each number represents the number of a replicate (from the original data) that will be used as a given replicate in the bootstrap sample. Each bootstrap sample contains the same number of replicates as the original data set. However, some of those original replicates may be represented more than once in each bootstrap sample.

After these randomly chosen representatives of each community are generated, they are used to calculate a new SIMI value ($G_{(i)}$). This process is repeated many times (1000 or more is best). The bootstrap estimate of the true SIMI value (G_B) is then calculated as the mean of these values. The variance of this estimate is determined by the following:

$$s_B^2 = (b-1)^{-1} \cdot \sum_{i=1}^{b} \left(G_{(i)} - G_B \right)^2 \qquad (23)$$

where b is the number of bootstrap samples used to calculate G_B.

The 95% confidence interval is asymmetric and is bounded by the 2.5th and 97.5th percentiles of the sorted values. This program uses a bubblesort to order the bootstrapped values of SIMI ($G_{(i)}$). It is a dependable algorithm, but is probably the slowest step in the program. A shellsort or quicksort routine may prove to be faster, but these have not been explored.

A symmetric confidence interval can also be generated based on the t-statistic. However, that is not included in this program at present.

Smith et al. (1986) also discuss bias encountered in the bootstrap estimate. The program provides adjusted and biased values for both the bootstrap and jackknife estimates.

3.6.2 Data

As input, the program uses estimates of the abundance of each species in each replicate from each of the two communities in question. These data may be entered interactively at run time with the keyboard or via a previously constructed file. If a file is used, it must conform to the following format:

LINE 1: TITLE (enclosed in double quotation marks ["])

LINE 2: The number of species included in the data set (= number of columns of data)

LINE 3: The number of replicates from community 1 and community 2 (in that order and separated by a comma or a blank).

LINE 4: The abundance of each species observed in the FIRST replicate from the FIRST community (each separated by a comma or a blank).

LINE 5: The abundance of each species observed in the SECOND replicate from the FIRST community (each separated by a comma or a blank).

.
.
.

LINE X: The abundance of each species observed in the FIRST replicate from the SECOND community (each separated by a comma or a blank).

.
.
.

A file of this form may be constructed by the program after keyboard entry of the data.

The data may be in the form of either absolute abundances or proportions. The program makes frequent reference to the original raw abundance data throughout its operation. Due to this, abundance data is generated from any data entered as proportions. The assumption is made that exactly 1000 individuals were sampled by each replicate. The abundance data generated in this way can also be saved to a file if the original proportion data were entered through the keyboard.

The number of replicates taken is not usually the same from different communities. Therefore, this program allows the use of data sets which represent communities containing different numbers of replicates.

Many of the operations are iterative and can consume a considerable amount of time, even on a fast microcomputer. For this reason, there are many symbols displayed on the screen to reassure the user that the program is operating and not caught in an infinite loop.

3.6.3 Output

Output may be sent to the screen, the printer, or a file. Three sets of output are generated by this program (Table 5). The first is the simple SIMI values without variance measurements. A table filled with a SIMI value for each possible combination (pairwise) of a replicate in community 1 with a replicate in community 2 is printed. The mean for all the values in this table is printed. Next, the SIMI value, based on the sum of species abundances over all replicates (G), is printed. This is followed by the output of the jackknife and bootstrap procedures. Each contains its SIMI estimate, that estimate adjusted for bias, the associated variance, and the 95 % confidence interval. For the jackknife estimate, this interval is around the adjusted value. The jackknife estimate also includes the t-value used and the degrees of freedom associated with it. The bootstrap output displays the number of bootstrap samples used to provide the estimate.

Table 5. Hypothetical test data from Table 2 of Smith et al. (1986) and comparative results with the program described herein.

		Species type					Total
Community	Replicate	A	B	C	D	E	
1	1	75	75	250	100	0	500
	2	175	75	175	75	0	500
	3	100	100	150	125	25	500
	Totals	350	250	575	300	25	1500
2	1	125	100	125	150	0	500
	2	125	75	75	75	150	300
	3	100	100	75	75	150	500
	Totals	350	275	275	300	300	1500

Pairwise SIMI measures for above table

		Community 1		
Community	Replicate	1	2	3
2	1	0.863	0.918	0.980
	2	0.008	0.732	0.780
	3	0.615	0.700	0.790

3.7 COMPARISON OF PROGRAM PERFORMANCE WITH TEST DATA

3.7.1 TEST.DTA

Bootstrap estimates used 1000 bootstrap samples. Table 2 of Smith et al. (1986) (which is shown as our Table 6) offers some hypothetical data (TEST.DTA) and their results. These results include the simple SIMI values from all possible pairs of replicates from each community, as well as the mean of these values and the SIMI based on the abundances summed over replicates (G).

The SIMCONF program exactly reproduced those results (see Table 6). Smith et al. also give bias adjusted similarity measures for the jackknife and bootstrap estimates and values for the associated 95% confidence intervals. In Table 6 Smith et al.'s values of the unadjusted estimates were back-calculated from the bias-adjusted values that they reported.

The program's jackknife and bootstrap estimates of SIMI closely matched those of Smith et al. However, the confidence intervals were narrower than those reported. In the case of the bootstrap, the values were close to those reported, and the random nature of the process or differences in machines used may explain most of this difference.

Table 6. Test data from Smith et al. 1986 and results compared.

SIMI values for each combination of replicates

	Community 1		
	1	2	3
1	0.863911	0.919329	0.980409
2	0.607785	0.732422	0.780644
3	0.618412	0.700841	0.789669

Mean of these similarity coefficients: .776647
Similarity coefficient based on the averages of all replicates: .852024

Jackknife estimate of similarity with 95% confidence interval:

t-value = 2.132, with 4 df.

Similarity coefficient: 0.814571
Similarity coefficient adjusted for bias: 0.889477
Variance: 0.015759

0.6218373 < JACKSIMI < 1.157116

Bootstrap estimate of similarity with 95% confidence interval:

1000 bootstrap samples used

Similarity coefficient: 0.826716
Similarity coefficient adjusted for bias: 0.877332
Variance: 0.007648

0.662016 < BOOTSIMI < 0.96700

Comparative results

Data set	Analysis	Result	SIMCONF	Smith et al.
TEST.DTA	Simple	MEANSIMI	0.78	0.78
	SSIMI	XSIMI	0.85	0.85
	jackknife	similarity	0.814	0.811*
		adj.-sim.	0.889	0.889
		variance	0.0157	N/A
	t = 2.132	lower bound	0.622	0.54
	(4, 0.05)	upper bound	1.157	1.23
	bootstrap	similarity	0.827	0.82*
		adj.-sim.	0.877	0.876
		variance	0.0157	N/A
	t = 2.132	lower bound	0.622	0.73
	(4, 0.05)	upper bound	0.967	1.04

*Estimated from bias-adjusted value.

Machine differences might also contribute to the deviations of the jackknife estimate, but it was the t-value that seemed to be most responsible for these.

3.7.2 Samar Sea Data

The Samar Sea data set (SAMAR.DTA, Table 7) consisted of three replicates (cruises) of demersal fish species abundances from one station (#22) at two points in time. The estimates of similarity between these two communities displayed a wide range, from 0.282 (smallest simple SIMI value) to 0.873 (largest simple SIMI value) (Table 8). The jackknife (unbiased 0.545, biased 0.701) and bootstrap (0.822) estimates indicated that the true value was probably in the upper end of this range, but their confidence intervals (0.627, 1.06 [jackknife]; 0.466, 0.846 [bootstrap]) were wide. Smith et al. (1986) state that in situations where the variability is large and the sample size is small, the jackknife performs better than the bootstrap. This situation clearly exists for the Samar Sea data set used here. However, the jackknife did not perform better than the bootstrap. Both clearly showed the high variability in the system. These results suggest that the community at station 22 did not change much over the time period between surveys. However, the variability was high and the true similarity may be lower than is indicated from this analysis. A larger data set with more replicates would help clarify the situation.

If one looks at the species abundances graphically (see Figure 3), it is clear that the dominance of the community shifted from species 67 to species 106, while the proportions represented by the other species remained relatively constant.

Table 7. Data from six cruises at two different sites in the Samar Sea, Philippines (five major species).

Assemblage	Cruise	Spp67	Spp77	Spp106	Spp117	Spp179	Totals
1	97	375	920	453	277	89	2114
	98	718	110	372	8	99	1307
	99	1736	203	448	0	54	2441
Totals		2829	1233	1273	285	242	5862
2	107	36	983	999	30	84	2132
	108	2514	337	3878	355	369	7453
	109	796	458	2196	453	498	4401
Totals		3342	1778	7073	838	951	13986

Species Codes:

67 — *Lelognathus bindus*
77 — *Loligo* spp.
106 — *Pentaprion longimanus*
117 — *Priacanthus tagenus*
179 — *Upeneus moluceensis*

Table 8. Data from six cruises at two different times at station 22 in the Samar Sea, Philippines.

SIMI values for each combination of replicates

	Community 1		
	1	2	3
1	0.873	0.445	0.282
2	0.596	0.867	0.736
3	0.646	0.737	0.572

Mean of these similarity coefficients: 0.639382

Similarity coefficient based on the averages of all replicates: 0.773120

Jackknife estimate of similarity with 95% confidence interval:

t-value = 2.132, with 4 df.

Similarity coefficient: 0.701589

Similarity coefficient adjusted for bias: 0.844650

Variance: 0.010389

$0.627338 < \text{JACKSIMI} < 1.061961$

Bootstrap estimate of similarity with 95% confidence interval: (1000 bootstrap samples used)

Similarity coefficient: 0.728795

Similarity coefficient adjusted for bias: 0.817444

Variance: 0.007888

$0.471423 < \text{BOOTSIMI} < 0.857882$

3.8 CONCLUSIONS

Similarity values are important tools for gauging change in complex natural (as well as artificial) communities. When replicate samples from the communities being compared are available, we may take advantage of some powerful, but computationally intensive methods to more accurately measure their similarity and the variability around that measure.

This variability can be used to judge the reliability of the similarity measure and to test for significant differences between the similarity of other community pairs.

As in all research, the extent to which the available data accurately describe the system of interest must be considered. Despite the powerful nature of these techniques

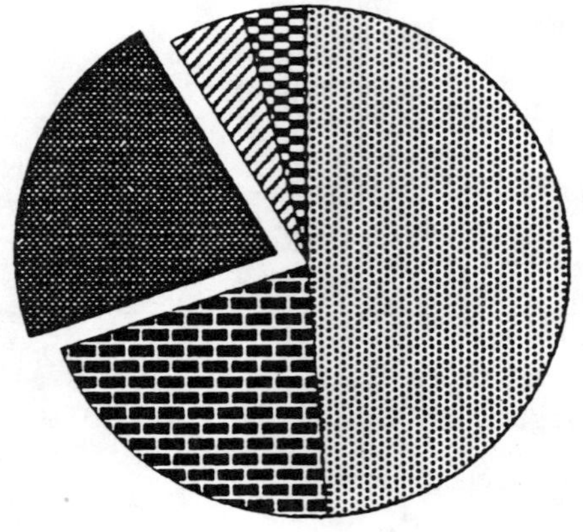

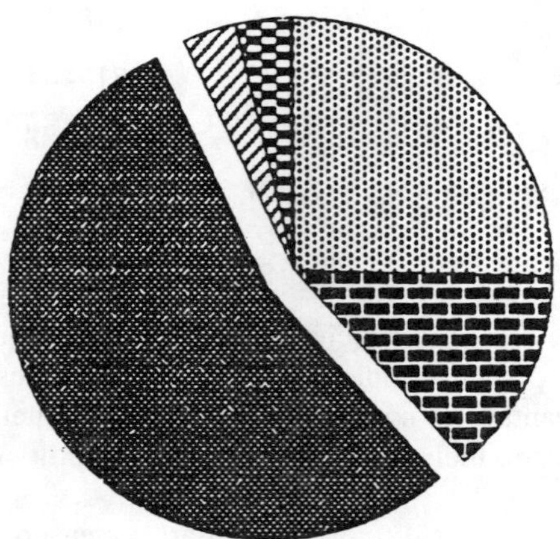

Figure 3. Graphical portrayal of species abundances from a portion of the Samar Sea fish community taken by a trawl sampling.

to get the most out of the data, the results will only be as good as those data. If the data set is inadequate, then variability will probably be too large to draw any definitive conclusions about the communities being investigated.

4. THE LOGARITHMIC SERIES DISTRIBUTION AND SOME APPLICATIONS TO THE ANALYSIS OF TROPICAL MULTISPECIES FISHERIES DATA

4.1 INTRODUCTION

The logarithmic series distribution was first described by Fisher, Corbet and Williams (1943). They considered the relative numbers of individuals of different species obtained when sampling at random from an animal population. This problem was examined in relation to the distribution of butterflies on the Malay peninsula and to data on the number of moths of different species caught in light traps. Williams (1964) provided many later applications for which this distribution was found to be useful. Recently, Dial and Marzluff (1989) demonstrated that hollow curve distributions, as exemplified by the logarithmic series, are common in nature and are a result of assemblages being dominated by one or a few taxa.

The logarithmic series distribution is a special case of the more general power series distribution. Noack (1950) and Khatri (1959) studied the analytic properties of this distribution and Patil (1961) investigated estimation procedures for it.

One of the several available derivations for the logarithmic series distribution is provided below. It is hoped that this material will provide a general overview of the logarithmic series distribution and that it will provide some rationale for an interest in its applications as they relate to tropical multispecies stock assessment methodologies. Of special relevance are the suitability and fit of the log-series distribution to tropical multispecies fish abundance data, as well as the potential utility of Fisher's index of diversity (α), which is derived from this distribution, and Mountford's index of similarity (I), also derived from this distribution.

The concern here is with changes in multispecies assemblages exemplified by the distribution of species (group) abundances. This is thought to be useful because these changes permit some investigation of species balance independent of which species (group) may be present. In choosing a statistic that characterizes this distribution as a measure of diversity, it is important to recognize that any such statistic should be independent of sample size and should show relatively small variability in replicated temporal samples. The latter property would then allow good spatial discrimination among sampling locations.

4.1.1 Derivation

Fisher et al. (1943) are responsible for the first mathematical derivation of the logarithmic series distribution. Alternative derivations of this distribution are due to many authors. In this material the derivation by Kendall and Stuart (1958), in which the logarithmic series distribution is considered as a limiting form of the binomial distribution, will be briefly described. This derivation is considered intuitively appealing, especially for the application used herein.

Let q represent the probability for the presence of an attribute and p represent the probability for its absence. Then consider the Pascal form of the negative binomial:

$$\binom{x+r-1}{x} p^r q^x$$

where x is the number of attributes present and r is the number of attributes absent.

Now consider the above form as a model for the number x of different species found in standardized trawl hauls, for example. If the trawl haul gear is at all effective, the tow is of reasonable duration and the gear is not torn or broken, there will be at least one species present. Under this assumption, the complete absence of the attribute ($x = 0$) can be excluded.

We are then concerned with a negative binomial truncated at zero, with frequencies proportional to:

$$p^r \left[rq, \frac{r(r+1)}{2!} q^2, \frac{r(r+1)(r+2)}{3!} q^3, \ldots \right]$$

Since the total frequency is $1 - p^r$, the distribution may be written as:

$$\frac{rp^r}{1-p^r} \left[q, \frac{r+1}{2!} q^2, \frac{(r+1)(r+2)}{3!} q^3, \ldots \right]$$

The case where $r = 0$ will occur if the size of the samples is large enough so that it is virtually certain that every possible species in the area is present in at least one of the trawl hauls. Then, it is possible to consider the limiting form, with $r = 0$ of the present equation:

$$\lim_{r \to 0} \frac{r}{p^{-r} - 1} = \lim_{r \to 0} \frac{r}{e^{-r \ln P} - 1}$$

Applying L'Hospital's Rule we have that:

$$\lim_{r \to 0} = \lim_{r \to 0} \frac{1}{(-\ln P) e^{-r \ln P}} = \frac{1}{-\ln P}$$

Therefore, the limiting form of the first equation is as follows if we set r = 0 in the expression in the brackets:

$$\frac{1}{\ln(1-q)}\left[q, \frac{q^2}{2}, \frac{q^3}{3}, \ldots\right]$$

For $x = 1, 2, 3, \ldots$, the frequency is the coefficient of t^x in the frequency generating function:

$$P(t) = \frac{1}{\ln(1-q)}\ln(1-qt)$$

The above equation is an expression for the logarithmic series distribution.

4.1.2 General Properties

The logarithmic series is derived mathematically from expansion of $\ln(1+x)$. To avoid negative numbers in the series, it may be written as:

$$-\ln(1-x) - x + x^{2/2} + x^{3/3} + x^{4/4}$$

Since there are no logarithms of negative numbers, it follows that x must be less than unity.

In using the log-series to express the frequency distribution of species containing different numbers of individuals, it is written as:

$$n_1, n_1x/2, n_1x/3, n_1x/4 \ldots$$

with the successive terms being numbers of species with 1, 2, 3 ... individuals. No zero term in included.

The series in this form has two parameters, (n_1) which is the number of species with one individual and (x) which is a number less than 1. The series has an infinite number of terms. It is discontinuous with only integer values of individuals per species and is convergent. The sum of all species to infinity is:

$$S = \frac{n_1}{x}[-\ln(1-x)]$$

The corresponding series of individuals (individuals of all species in the same abundance class) is:

$$n_1, n_1x, \mathrm{n}_1x^2, n_1x^3, n_1x^4 \ldots$$

This is a geometric series with a constant multiple x. As $x < 1$ the series is convergent, and the sum to infinity (N) or the total number of individuals is:

$$N = n_1(1 - x), \text{ or } n_1 = N/(1 - x)$$

The ratio of the number of species to the number of individuals is:

$$\frac{S}{N} = \frac{1-x}{x}[-\ln(1-x)]$$

Thus, for any average number of individuals per species (the reciprocal of the above), there is only one value of x, and from this n_1 can be obtained. If N and S are known, the series is fixed.

If sampling by individuals is done from a population graduated by a log-series distribution, then the smaller the sample the smaller the average number of individuals per species and the smaller the value of x. The larger the sample the larger the average number of individuals per species and the value of x will also become larger. However, in all random samples of any size from one population, the ratio of n_1 to x is a constant and is called the index of diversity or α or Fisher's index of diversity:

$$n_1/x = \alpha \quad \text{or} \quad n_1 = \alpha x\ .$$

With increasing sample size, the value of x increases and approaches unity. With increasing sample size, the number of species with one individual increases and gradually approaches the value of α, but never (theoretically) exceeds it.

The log-series can also be written as:

$$\alpha x\,,\ \frac{\alpha x^2}{2},\ \frac{\alpha x^3}{3},\ \ldots\text{etc.}$$

where αx is the number of species (groups) represented by one individual, $\frac{\alpha x^2}{2}$ is the number of species represented by two individuals, etc. The sum of this, the total number of species is:

$$S = -\ \alpha \ln(1 - \mathrm{x})$$

where α is a property of the sampled population, and x is a property of the sample and depends on its size. Some definitions of symbols are:

x = sampling parameter of log-series distribution: $0 < x < 1$

α = parameter of log-series distribution: $\alpha > 0$ (diversity index)

S = the number of species encountered

N = the number of individuals encountered

n_1= the number of species encountered with one individual.

Some relations among the above include:

$$(1)\ \alpha = n_1/x$$

$$(2)\ S = \alpha[-\ln(1-x)]$$

$$(3)\ x = N(N+\alpha)$$

$$(4)\ \alpha = N(1-x)/x$$

$$(5)\ n_1 = N\alpha/(N+\alpha)$$

$$(6)\ n = N(1-x)$$

$$(7) N/S = x/(1-x)[-\ln(1-x)]$$

$$(8)\ S = \alpha\ln(1+N/\alpha)$$

For times when N/S is large in relation to 1, then:

$$(9)\ S = \alpha\ln(N/\alpha)$$

The large sample variance of α is given by Anscombe (1950) as:

$$(10)\ \mathrm{var}(\alpha) = \frac{\alpha}{-\ln(1-x)}$$

4.2 FIT TO SAMAR SEA FISHERIES DATA

Various models have been applied to data sets involving diverse taxa in an effort to find appropriate descriptions of species abundance relationships. The comprehensive works of Pielou (1975, 1977) and Williams (1964) describe many of these models. This brief review only relates to applications of the log-series model to fisheries and other aquatic applications of this model. Based on a study of benthic macro-invertebrates, Shepard (1984) suggested that the logarithmic series distribution is the most appropriate for studying species abundance patterns in aquatic ecosystems. Kobayashi (1987) developed a new index of similarity which is independent of sample size, and it is based on the assumption that the communities from which samples are derived are described by the log-series distribution. Koch (1987) examined large marine data sets and concluded that the species occurrence frequency distribution for such data sets fit the log-series distribution well. Wolda (1981) stated that Fisher's α is the best available measure of species diversity. In the case of multispecies trawl data or any other assemblage of organisms, it is not expected that any one simple mathematical model will fit all data with complete fidelity, especially in view of sampling variability and behavioral differences. However, if Fisher's logarithmic series distribution is found to provide reasonable fits to available empirical data on the distribution of species (groups) in standardized trawl samples from tropical multispecies fisheries, this could provide a probability distribution-based model for quantifying species abundance patterns in these complex ecosystems. Although use of the logarithmic series must be judged on its utility for given applications, there seems to be increasing evidence for use of this distribution to describe species abundance relations.

Taylor, Kempton, and Woiwood (1976) made an exhaustive study of macrolepidoptera collected at several sites, and showed that although the log-series model was not an ideal description of population structure, diversity as measured by the parameter α generally behaved more predictably and consistently than other diversity statistics tested. In the above study, α was found to be a normally distributed property of the populations studied.

The data used in our example consisted of the Samar Sea, Philippines trawl survey data base. These data include a total of 267 individual trawl hauls which collected more than two million marine organisms (mostly fish) identified to 179 taxonomic groups, primarily at a species level. Temporal replications were obtained by sampling at ten quarterly intervals using 28 sampling stations. However, not all stations were completely sampled. Description of the gear, standardized hauls, processing procedures and other details are given in Saeger (1983). The Samar Sea data set is believed to be one of the largest tropical fisheries data sets examined in this detail to date.

Clearly, sampling errors ascribed to a single sample depend on the population it is supposed to represent. Following Taylor, Kempton, et al. (1976), in order to allow for a certain amount of natural population variability in addition to intrinsic sampling variation, we assume that the number of species S, rather than the number of individuals N, in replicate samples follows a Poisson distribution. This is in contrast to Fisher et al. (1943) who assumed that the number of individuals summed over all species should conform to a Poisson model. Empirical evidence for the validity of the assumption is presented in Table 9. This table shows the mean and the variance-to-mean ratios of quarterly cruise totals for both the number of species and the number of individuals caught at 28 trawl stations for which at least five successive hauls were available. It is evident from Table 9 that the variance of N (the number of individuals captured) increases roughly as the square of the mean and is substantially greater than the mean in all instances. This provides some evidence that Fisher's assumption is not valid for individuals sampled over time. In contrast to the variance of N, the variance of S (species) is quite close to the mean (see last column), suggesting that the total number of species behaves as a Poisson variate. This justifies to some extent the assumption made earlier that the number of species, not individuals, conforms to the Poisson model.

It is desirable to extend the assumption that the total number of species S is a Poisson variate to the case where the number of species (groups) with a particular abundance $r(n_r)$ is also a Poisson variate. As Anscombe (1950) has pointed out, it then follows that the total number of species $\left(S = \sum n_r\right)$ is a Poisson variate. Table 10 illustrates the distribution of species (group) frequencies for ten cruises covering stations 1–5 in the Samar Sea. Note (last column) that each cruise involved between 17 – 75x10^3 individuals. The data have been grouped into abundance classes of approximately equal range on a logarithmic scale. The pattern of variation shown by cruises should show independent Poisson variation. However, the variation in the first three abundance clauses (namely, 1, 2^+, 4^+) is considerably higher than expected. It is

Table 9. Mean and variance-to-mean ratio of individuals and species caught at 28 stations in the Samar Sea, Philippines.

		No. of individuals (N)		No. of species (S)	
Station no.	Sample size	$\bar{x}$	$s^2/\bar{x}$	$\bar{x}$	$s^2/\bar{x}$
1	10	15,213	21,821	38	2.01
2	9	6,434	5,152	36	1.12
3	10	6,118	2,876	33	2.04
4	10	8,130	8,152	37	1.29
5	10	5,165	1,979	33	1.23
6	10	6,970	2,360	31	0.96
7	10	6,276	2,646	31	0.59
8	10	7,270	5,734	29	0.43
9	10	7,017	4,008	30	1.15
10	5	37,222	26,601	24	1.31
11	10	34,892	33,560	29	1.54
11	10	17,508	1,133	30	0.96
13	9	7,817	4,955	33	1.28
14	10	5,111	4,509	36	1.17
15	10	5,170	1,342	38	0.41
16	10	8,654	14,415	37	1.04
17	10	6,829	4,255	38	0.34
18	10	5,486	6,330	40	0.48
19	10	7,372	8,477	33	1.11
20	10	5,886	19,258	34	0.70
21	9	8,272	5,300	34	1.18
22	10	9,125	8,843	39	1.16
23	10	3,872	1,254	38	2.20
24	9	12,852	14,327	34	0.74
25	9	10,527	5,573	33	1.58
26	10	13,573	13,137	23	0.46
27	10	13,460	8,668	33	0.97
28	10	10,020	2,298	31	2.50

believed that this situation is the result of sampling variability for rare species. The analysis of Table 10 suggests that those species which occur as individuals or as a very few individuals in this type of sample are subject to a great deal of sampling variability, as they probably have a highly contagious spatial distribution and/or are not vulnerable to the gear used. Koch (1987) has shown that for those species which occur infrequently, sample size effect may be very large relative to the faunal patterns reported. At this point it is suggested that the rare species be eliminated before formal testing of specific hypotheses. Whether sampling strategies could be refined to permit

Table 10. Distribution of species frequencies for successive cruises in the Samar Sea, Philippines at stations 1–5

	Abundance class											Total	
Cruise#	1	2+	4+	8+	16+	32+	64+	128+	256+	512+	1024+	spp.	No.'s
97	8	15	15	13	10	22	15	11	4	8	0	67	17,913
98	28	28	23	22	15	17	12	13	8	3	5	77	22,053
99	7	18	18	19	21	18	17	12	7	5	6	62	24,424
100	12	22	19	20	1	9	14	14	7	6	5	68	70,296
102	8	20	24	23	27	17	17	12	13	5	5	80	34,038
104	6	6	11	19	21	19	12	8	5	10	7	63	49,354
106	2	15	30	15	25	20	22	25	11	11	6	83	31,721
107	3	0	50	22	17	25	16	22	7	9	13	80	47,677
108	6	21	16	24	31	21	13	10	7	8	7	71	30,873
110	6	0	9	38	37	27	24	22	18	12	13	78	75,608*
$\bar{x}$	8.5	14.5	21.5	21.5	22.5	19.5	16.2	14.9	8.7	7.7	6.7	72.9	40,396
s^2	53.4	90.7	138.9	45.6	62.1	24.5	16.4	34.5	17.6	8.4	14.9	58.3	
s^2/x	6.2	6.2	6.5	2.1	2.8	1.3	1.0	2.3	2.0	1.1	2.2	0.8	

* Minimum value because station #2 had no total available.

more consistent representation of rare species is not known at present. Koch (1987) developed methods to test for sample size effects on distributional patterns of species. However, the paleontological studies he addressed dealt exclusively with nonmotile species. In our case the number of species (groups) with r individuals in a sample (n_r) tends to show independent Poisson variation with expectations (E_r) described by a logarithmic series distribution, except in the case of the first three abundance classes, which show considerable variability.

A computer program for calculating the parameters of the log-series distribution is found in the Appendix. The method used for calculating the parameters of the log-series distribution and their variances are based on Shepard (1984). That is, for a sample with N individuals and s species, the value of α is calculated from:

$$\alpha\,[\exp - \mathrm{s}/\alpha - 1] - N = 0 \tag{24}$$

by means of an iterative procedure. To obtain the expected number of species with n individuals each (s_n), the following relation is used:

$$s_n = (\alpha/n)x^n$$

However, it is first necessary to calculate

$$x = N(N + \alpha)$$

The variance associated with s and α are:

$$\mathrm{var}(s) = \alpha \ln[(2N + \alpha)/(N + \alpha)] - [\alpha^2 N/(N + \alpha)^2]$$

$$\mathrm{var}(\alpha) = [\alpha^3\{(N + \alpha)^2 \ln[(2N + \alpha)/(N + \alpha)] - \alpha N\}]$$

For testing the fit of observed to expected number of species, a test described by Fisher is used. A value of the difference:

$$\left[\alpha n\left(1 + 1/2 + 1/3 + \ldots + \frac{1}{n-1}\right)\right] - \frac{s^2}{2a}$$

is evaluated. In the expression αn is the number of observed species with n individuals each and s is the number of species. Solution of the above in the first term represents the score of observed values and the second term represents the expected total score. Significance is tested by estimating the standard deviation by

$$\text{s.d.} = \left[0.908\left(\log\frac{N}{s}\right)^{1.673} \times s\right]^{-2}$$

Non-significance is accepted if the difference is less than its standard deviation. If this test indicates a significant difference between observed and expected scores, the difference is divided by its variance to obtain a value of k which provides an exact fit.

Table 11 illustrates estimates of the log-series distribution parameters to data from three cruises chosen at random from the Samar Sea file. These three cruises are 97, 98, and 108. The values of α range from less than 2 to more than 8. The goodness-of-fit of the distribution to the data is demonstrated by high values of x and by the values of k. Values of k less than unity are considered to provide reasonably good fits to the data. It is clear from examination of the last column that most of the calculated k values are greater than unity. However, if the data are truncated by removing the most variable abundance classes (namely, those with seven or less individuals) then the recalculated k values are significantly reduced and most of them are less than unity. The parameter k varies inversely with the unevenness in species abundance represented by each term in the series for which the distribution is a continuous function. When the abundances of each species are roughly similar, k is expected to have a high value. When the abundances vary greatly, k is expected to have a small value. This approach to describing the fit of the log-series model is identical to that done by Shepard (1984) for stream macro-invertebrates. It is thought that this method is more appropriate than the usual chi-square goodness-of-fit test. The reason is that that power of the chi-square test is low, and samples may be accepted in fitting the logarithmic series using this test when in fact the fit is not good. Krebs (1989) has recommended that so-called Whittaker plots be made to demonstrate conformity to a log-series model. These plots involve plotting relative abundance (percent) on a logarithmic scale on the y-axis and species in rank order on the x-axis. The result is a relatively straight line for the logarithmic series, if the data conform to the distribution.

The log-series fitting program provides an option to calculate the theoretical Whittaker plot for a logarithmic series. This methodology is described by Krebs (1989) and involves solving the following equation:

$$R = \alpha E_1\left(n \ln\left(1+\frac{\alpha}{N}\right)\right)$$

where R = species in rank order (x-axis of plot)
α = Fisher's index of diversity
n = number of individuals expected for a specified value of R (y-axis of plot)
N = total number of individuals in sample
E_1 = standard exponential integral (Abramowitz and Stegun, 1964, Ch. 5)

Solving the above equation for n using integer values of R provides values for the expected Whittaker plot. These can be compared with the original data.

Table 11. Log-series distribution parameters for three cruises selected at random and for 28 stations sampled during each cruise, Samar Sea, Philippines. Note that some station data were incomplete and were listed as "missing data."

Cruise/Station	N	S	α	$SE\alpha$	x	k
97-01	3,180	26	3.874	.286	.999	0.210
97-02	5,384	35	5.014	.311	.999	0.181
97-03	1,346	20	3.331	.303	.998	2.996
97-04	2,251	28	4.504	.328	.998	3.025
97-05	5,772	29	3.984	.264	.000	2.934
97-06	3,904	26	3.740	.198	.999	4.835
97-07	3,388	28	4.180	.198	.998	0.166
97-08	2,002	28	4.608	.351	.997	N.S.
97-09	2,920	19	2.723	.219	.999	1.818
97-10	10,936	31	3.905	.237	.999	1.371
97-11	105,220	19	1.724	.109	.999	1.842
97-12	51,855	23	2.294	.140	.999	2.417
97-13	1,628	29	5.012	.388	.999	7.269
97-14	878	36	7.557	.601	.991	9.962
97-15	2,396	35	5.810	.398	.997	0.233
97-16	1,160	30	5.624	.453	.995	1.078
97-17	(missing data)					
97-18	1,767	33	6.760	.421	.997	1.971
97-19	3,087	23	3.372	.262	.999	2.987
97-20	4,102	29	4.114	.290	.999	3.336
97-21	1,156	22	3.855	.346	.997	N.S.
97-22	3,587	37	5.747	.366	.998	0.893
97-23	8,114	31	4.081	.255	.999	1.774
97-24	8,283	33	4.373	.266	.999	2.834
97-25	18,018	29	3.379	.202	.999	3.208
97-26	7,663	26	3.263	.222	.999	5.478
97-27	36,397	26	2.738	.162	.999	4.276
97-28	18,292	23	2.596	.171	.999	N.S.
98-01	11,196	40	5.213	.285	.999	N.S.
98-02	2,029	39	6.848	.462	.997	1.882
98-03	1,748	42	7.744	.521	.996	0.212
98-04	3,754	36	5.518	.353	.998	N.S.
98-05	3,326	38	5.015	.383	.998	1.008
98-06	4,234	32	4.703	.311	.998	2.245
98-07	2,511	38	6.352	.420	.997	3.371
98-08	5,153	26	3.574	.251	.999	4.891
98-09	3,436	30	4.522	.314	.998	N.S.
98-10	36,101	25	2.613	.158	.999	0.462
98-11	6,545	33	4.536	.282	.999	1.543
98-12	760	23	4.474	.422	.994	7.153
98-13	4,231	30	4.361	.295	.999	3.282
98-14	1,799	36	6.396	.450	.992	5.472
98-15	2,591	40	6.713	.434	.997	5.619
98-16	2,250	36	6.086	.416	.997	7.902
98-17	1,156	42	8.456	.616	.993	10.881
98-18	3,244	37	5.855	.378	.978	8.407

Table 11. (continued)

Cruise/Station	N	S	α	$SE\alpha$	x	k
98-19	5,331	32	4.524	.291	.999	8.926
98-20	11,851	34	4.251	.249	.999	N.S.
98-21	3,029	28	4.264	.301	.998	1.409
98-22	3,345	32	4.903	.333	.998	2.301
98-23	3,444	33	5.057	.338	.998	N.S.
98-24	6,016	31	4.275	.275	.999	1.092
98-25	4,123	33	5.963	.366	.998	2.183
98-26	6,428	36	5.032	.303	.999	3.084
98-27	9,988	39	4.421	.224	.999	3.192
98-28	6,781	33	4.511	.280	.999	5.646
108-01	2,429	42	7.214	.462	.997	0.138
108-02	2,588	40	6.715	.434	.997	1.525
108-03	3,277	35	5.471	.360	.998	0.250
108-04	12,678	40	5.118	.276	.999	.983
108-05	9,901	39	5.159	.288	.999	2.200
108-06	11,211	36	4.618	.263	.999	3.453
108-07	9,347	35	4.594.	.269	.999	4.857
108-08	14,075	36	4.467	.249	.999	5.149
108-09	4,751	33	4.781	.308	.999	9.196
108-10	(missing data)					
108-11	23,172	36	4.176	.223	.999	6.169
108-12	14,958	38	4.713	.256	.999	0.130
108-13	7,935	36	4.867	.287	.999	1.480
108-14	6,076	40	5.743	.334	.999	2.694
108-15	373,614	37	5.739	.365	.998	1.527
108-16	8,205	35	4.686	.278	.999	4.715
108-17	10,200	38	4.660	.251	.999	0.110
108-18	3,199	42	6.827	.421	.998	1.613
108-19	5,561	40	5.830	.342	.999	2.474
108-20	7,244	38	5.201	.301	.999	3.537
108-21	9,678	37	4.822	.278	.999	4.424
108-22	9,840	43	5.779	.310	.999	4.856
108-23	2,595	47	8.151	.497	.997	0.163
108-24	4,463	42	6.415	.379	.998	1.166
108-25	27,249	27	2.958	.176	.999	1.920
108-26	19,768	36	4.164	.231	.999	2.434
108-27	8,599	34	4.500	.269	.999	4.524
108-28	8,780	38	5.100	.291	.999	5.181

Figure 4 illustrates the distribution of alpha for 267 stations from the Samar Sea. The fit of the normal distribution to the observed frequency distribution looks reasonably good. The null hypothesis is not rejected at the 0.05 probability level. The evidence at hand suggests that use of the α diversity statistic with symmetric intervals for comparative studies of tropical multispecies fish assemblages seems justified on the basis of these data.

Taylor, Kempton, and Woiwood (1976) also suggested that the logarithmic series parameters α and x may be given biological interpretations. They indicated from their work that the parameter α remains reasonably constant over time for the same station, but that x changes somewhat more. From the Samar Sea data, a two-way randomized

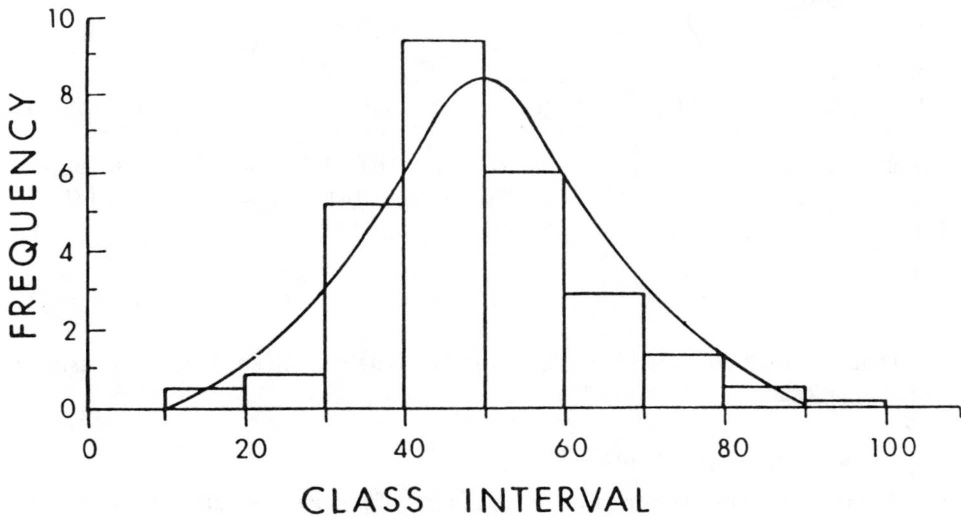

Figure 4. Distribution of α for 267 stations in the Samar Sea, Philippines with fitted normal curve. Chi-square (d.F. = 5) 11.0197.

block analysis of variance design was used to compare temporal vs. spatial variability of α.

The analysis consisted of 15 stations x 10 samples at each station taken at quarterly intervals. These stations and samples were chosen to permit orthogonal comparisons. Table 12 illustrates the results of this analysis. From this table it is evident that both temporal and spatial variability were statistically significant. This would be expected, since the area was in the early stages of recovery from a trawling restriction on larger vessels. However, the variation within stations over the ten quarterly samples seemed to be somewhat smaller than that observed over the 15 stations. This lends modest support to the premise that α seems somewhat less variable within assemblages at a given location (environment) than among locations.

4.3 LOGSRFIT

4.3.1 Data Requirements and Program Flow

This program is designed to satisfy two end user needs for examining the fit of species abundance data to the log-series model. The first is the quick generation of the parameters of the log-series model fitted to the data from a single sampling site. The second is the generation of these parameters for each of a large number of sampling sites.

Table 12. Analysis of variance table for a 15 x 10 randomized block experiment testing for the temporal and spatial changes in the parameter α; Samar Sea data for stations 1–5 over 10 cruises.

Source of Variation	d.f.	Sum of squares	Mean square	F_o	Prob ($F > F_o$)
Among stations	14	84.260	6.019	4.935	0.05E–04
Within stations	9	46.270	5.141	4.215	0.15E–03
Error	116	153.671	1.220		
Total	149	284.201			

Data for multiple sampling sites must be contained in an ASCII file using the following format:

LINE 1 a title (in quotation marks)

LINE 2 the number of sites and the total number of species from all sites (separated by a comma)

LINE 3 label for each column of data (separated by commas; quotation marks are not necessary) (e.g., STATION, SPP1, SPP2, SPP3, SPP4, etc.)

LINE 4 (and all following lines) site id, abundance of species 1, abundance of species 2, etc. (separated by a comma)

If a species was absent from a particular location, a zero should be used to represent its abundance.

Data for a single sampling site may be contained in a data file formatted as described above or entered through the keyboard at run time. There are two options for processing the data from a single site, if those data are entered through the keyboard. Species abundance data can be entered as the user is prompted. This will produce the same output as that when data is read from a file. The other option is to enter only the number of species and total number of individuals. This will provide estimates of the α and x parameters, but will not estimate *k* (goodness of fit descriptor).

The initial screen gives the program name and a brief description of its use. A menu will appear next, asking the user if his/her data are from multiple sites or a single site only. The user responds with 1 or 2, accordingly.

Choosing single site data leads to a second menu, which provides the user with the choice of entering species abundance data or simply the total number of species and individuals. If species abundance data are used (option 1), the program will ask for the number of species at that site. It will then proceed to prompt the user for the abundance of each species, one at a time. If only number of species and individuals are used (option 2), then the program will request only those two values. Choosing multiple sites data will be followed by a description of the necessary file format and a prompt for the name of the data file.

The first output is to the screen. A hard copy can be acquired by using SHIFT-PrtSc to dump the screen to a printer. This initial output includes estimates of the α and x parameters, as well as the variance and standard error of α.

The initial output is followed by a statement giving the user the option of generating the data for a Whittaker plot (linearized form of the theoretical distribution that best fits the data) for that site. This is a time-consuming routine and should only be chosen when it is truly necessary (the code for this routine was modified from Krebs, 1989). If this option is chosen, the user is prompted for the name of the file in which to record this data. A different name should be used each time this option is chosen, or else the file will be overwritten. The output consists of a list of the species rank, the abundance and proportion at that rank (predicted by the log-series model), and the observed abundance at that site.

If the input data were simply the total number of individuals and species, then the program will end here. For other data, the program proceeds with a second output screen. This output includes the estimate of *k*. A hard copy can be acquired by using SHIFT-PrtSc.

If the data set included more than one site, the calculations will be repeated for the next site (after a pause). For each site the two output screens and the prompt for Whittaker plots will be generated.

After all of the sites have been processed, the program provides three options for the display or storage of the output from each site. The results for each site may be listed to the screen (choosing *S*), sent to the printer (choosing *P*), or stored in a file (choosing *F*). If storage in a file is chosen, the user will be prompted for a file name. These data are stored in simple ASCII form and may be edited or printed at a later time with a word processor. This output includes a summary of *N*, *S*, α, *k*, and a breakdown for each site of the number of species that fall into each abundance class.

4.4 THE DIVERSITY INDEX (α) BASED ON THE LOG-SERIES DISTRIBUTION

Fisher et al. (1943) have demonstrated that for all samples (or summation of samples) taken from the same population by the same methods, $\alpha = n_1/x$ is constant. These symbols have been previously defined as:

x = sampling parameter of the log-series distribution $0 < x < 1$

α = parameter of the log-series distribution: also Fisher's diversity index

n_1 = the number of species encountered with one individual.

Fisher's demonstration of the above is of significance to studies of multispecies fisheries and to ecology in general. Because α is high in populations which have many groups (species) relative to the number of individuals and low in populations which have a small number of groups (species) relative to the number of individuals, Williams

(1947) termed α an index of diversity. It measures the extent to which individuals are apportioned into groups (species).

Some characteristics of this index of diversity listed by Williams (1944) are briefly mentioned below.

(a) If many samples are taken from the same population, not only will each have the same index of diversity but also the combined sample from two or more original samples will have the same index of diversity.

(b) For samples of differing sizes from the same population (x), the sampling parameter of the log-series distribution approaches unity as the sample size increases. Also, since $n_1 = \alpha x$, it is evident that α , the index of diversity, is the theoretical limiting value of n_1. That is, α is the maximum number of species (groups) with one individual obtained from one sample regardless of its size.

(c) If two samples of considerable size are taken from the same populations and the larger population is p times the smaller in size, then the number of species in the larger sample will be α ln$p$ more than in the smaller sample. This follows from relationship (8) of the General Properties section where:

$$S = \alpha \ln(1 + N/\alpha)$$

If N is very large relative to α, we have that:

$$S = \alpha \ln(N/\alpha).$$

Therefore, if two samples from the same population contain N and PN species (groups):

$$S_{PN} - S_N = \alpha(\ln PN/\alpha - \ln N/\alpha) = \alpha \ln P.$$

By way of illustration, if one sample is twice as big as the other, it will contain αln2 = .69 more species. Similarly, if the size of the sample is multiplied by e = 2.718, the number of species added = α. This relation may be useful in multispecies trawl surveys using two trawl hauls whose time durations (hence areas swept) are in the ratio of 1 to 1.65 (which is $\sqrt{e}$). From this, the average increase in species between samples of the two sizes should be a direct measure of the index of diversity (α). To the best of our knowledge, this approach has not be tested in the field with fisheries data and may be worthy of future consideration.

The index of diversity (α) may also be used in a comparison of different populations under certain restrictive assumptions and conditions. Williams (1947) presents results which follow, but these have been modified by substitution of an example from a tropical multispecies fishery. Specifically, the method relates to the question of the number of species (groups) that would be common to two areas if they were from the same population. A necessary assumption is that the density of the population be the same for the two areas. That is, the number of individuals (but not species) must be proportional to the area under consideration. Also, the two areas must be of different size. Let the two areas be of size a and size b, and the number of species in each be S_a and S_b. If the areas are from the same population, then they will have the same index of diversity. If the samples are large, the 1 in the following equation can be neglected:

$$S = \text{a} \ln(1 + N/\alpha).$$

Let S_{ab} be the total number of species in the two samples. The increase in species by adding b to a is:

$$S_{ab} - S_a = \alpha \ln\left(\frac{a+b}{a}\right) - \alpha \ln\left(\frac{a}{\alpha}\right) = \alpha \ln\left(\frac{a+b}{a}\right) \tag{25}$$

Similarly, the increase in species by adding a to b is:

$$S_{ab} - S_b = \alpha ln\left(\frac{a+b}{b}\right) \tag{26}$$

The unknowns S_{ab} and α may then be calculated by solving the two equations (equation 25 and equation 26) for the two unknowns. It is evident from the two equations above (equation 25 and equation 26) that it is required that the area sizes a and b must be different in order for the model to work. The number of species (groups) expected to be common to both samples is:

$$\text{Species (groups) in common} = S_a + S_b - S_{ab}$$

The following example is taken from trawl survey data acquired from the Samar Sea, Philippines, to demonstrate the application of the above-mentioned method. Stations 1–4 are located in the shallow north area of the Samar Sea. These stations had 56 species or species groups in an assumed area of 50 km^2. Stations 24–25, located in the shallow south portion of the Samar Sea had 39 species in an assumed area of 25 km^2. In this case, the proportions of the two areas are approximately correct, but the absolute sizes have been assumed. However, this does not affect the results. On the assumption of identity of origin, we have that:

$$S_{ab} - 56 = \alpha \ln\left(\frac{75}{50}\right)$$

$$S_{ab} - 39 = \alpha \ln\left(\frac{75}{25}\right)$$

Hence, $S_{ab} = 66$, $\alpha = 24.52$, and the expected number common to the two areas is 29. The actual number of species (groups) observed to be in common was 25. This result does not suggest any significant deviation from the expected number. However, no formal statistical test of significance is available. In addition, it should be recalled that this exercise is based on the assumption that the number of individuals be proportional to the two areas under consideration. This assumption is clearly difficult to test in practice and may not be realistic.

Sironmoney (1961) introduced the concept of entropy for the logarithmic series distribution. He considered two types of logarithmic series distributions (Type I and Type II). Type I involves randomization by groups and Type II involves randomization by units.

Type I or randomization by groups may be illustrated by a sample of parasitized fish, classified according to the number of parasites on each fish. An increase in the sample of fish will not add any parasites to the fish already counted. That is, the new units (individual parasites) will fall into the new groups exclusively. In the case of Type II or randomization by units, consider the number of fish caught in a trawl haul which are classified according to species (groups). In this case, an increase in the sample may add new individuals to species already represented. That is, new units may fall into old groups in this case.

Our further interest is restricted to the Type II situation. Consider the following problem. Let f_r represent the frequency of species (groups) of exactly r fish each in a population of fish classified according to species (groups). The Type II distribution equates f_r to $\frac{aB^r}{r}$; $r = 1, 2, 3, \ldots$ where B is a constant. This is a distribution on the set of all species. Thus, if a fish caught belongs to a species with exactly r fish, its probability is:

$$\frac{r}{\text{total number of fish}} = \frac{r}{\sum rf_r} = \frac{r(1-B)}{\alpha B}$$

and there are $\frac{aB^r}{r}$ species with exactly r fish. Let H_α denote the entropy of this distribution. H_α measures the average uncertainty per fish with reference to the species. Sironmoney showed that H_α increases with B, when α is fixed and also that H_α is greater for larger samples from the same biological population.

Assuming a logarithmic series distribution for species abundance, Kempton and Taylor (1979) have shown that the expected number of species in a sample of size m is:

$$S_m(\alpha) = \sum_{i=1}^{m} \frac{\alpha}{\alpha + i - 1}$$

The above authors conclude that using $S_m(\alpha)$ from the log-series model enables all information in the sample to be used, with the consequent reduction in temporal variability and an increase in site (location) discrimination.

4.5 SUMMARY OF THE LOG-SERIES MODEL AND DIVERSITY STATISTICS

Assuming that the number of species (not the number of individuals) in replicate samples follows a Poisson distribution, and using Fisher's log-series distribution as a description for the expected number of species (E_r) with r individuals:

$$E_r = \alpha \frac{X^r}{r}, \; r = 1, 2, 3, \ldots$$

where $\alpha > 0$ and $0 < x < 1$. These are parameters of the log-series distribution. This is a simple model that gives a description of species frequency distributions.

The expected number of species in the sample is given by:

$$E(S) = -\alpha \ln(1 - X).$$

The expected number of individuals in the sample is:

$$\mathrm{E}(N) = \alpha X/(1 - X)$$

The maximum likelihood estimate $\hat{\alpha}$ of α is given by the solution of:

$$S = \hat{\alpha} \ln(1 + N/\alpha).$$

The variance of $\hat{\alpha}$ for large samples is:

$$\mathrm{var}(\hat{\alpha}) = \frac{\alpha}{-\ln(1-x)}$$

Note: This variance can be used for comparing diversities at different sites or times.

The basic model is: the number of species with r individuals in a sample n_r, show independent Poisson variation with expectations E_r, which are described by a logarithmic-series distribution.

The information statistic H, proposed by Shannon and Weaver, is a measure of the information content per symbol in a coded message.

Expressed in terms of species abundance, it is defined as:

$$H = \ln N - \frac{1}{N}\sum_{i=1}^{\infty}\left(\frac{r}{n_r}\right)n_r$$

Asymptotically, for large sample size:

$$H \approx V + \ln\left(\alpha + \frac{1}{2}\right)$$

where V = 0.577721, which is Euler's constant, with a relative error less than 10^{-4} when alpha ≥ 5. Thus, for data fitting the log-series model, the statistic

$$I = 0.561\ e^{H} - 0.5$$

may be equated with α with small bias when the sample size is large. The expression e^H is equivalent to the minimum number of species the sample must contain for the information statistic to have value *H*.

Definition of Symbols

- S = total number of species encountered in replicate samples: $S = \Sigma\ n_r$.
- N = total number of individuals in replicate samples.
- n_r = number of species with a particular abundance *r*.
- E_r = expected number of species with r individuals.
- α = parameter of log-series distribution: $\alpha > 0$ (diversity index).
- X = parameter of log-series distribution: $0 < X < 1$ (sampling parameter).
- $E(S)$ = expected number of species in the sample.
- $\hat{\alpha}$ = estimate of α
- H = The Shannon/Weaver information statistic.

5. LOGARITHMIC SERIES INDEX OF SIMILARITY AND ITS APPLICATION TO CLASSIFICATION OF STATIONS

Mountford (1962) derived an index of similarity termed I, which is based on the logarithmic series distribution. The material which follows is largely taken from the above reference.

Let: a = the number of species in a sample from one site
b = the number of species in a sample from a second site
j = the number of species common to both sites.

An index of similarity based on the above three quantities must be a combination of these quantities and:

(a) should be independent of sample size, and

(b) should increase with increasing j and decrease with increasing a and b.

In order to meet these requirements, it is necessary to postulate a theoretical distribution of the species frequencies. The logarithmic series distribution (Fisher et al., 1943) is postulated as:

$$\alpha x, \frac{\alpha x^2}{2}, \frac{\alpha x^3}{3}, \ldots$$

where αx^n is the number of species with n individuals, and α is a constant for all samples from the same population. The constant α is construed as an index of diversity. The relation between the expected number of individuals (N) and the expected number of species (S) in a sample from a log-series population with the constant α is:

$$S = \alpha \ln\left(1 + \frac{N}{\alpha}\right)$$

If two samples having A and B individuals and a and b species, respectively, are taken from the same population, then:

$$a = \alpha \ln\left(1 + \frac{\mathrm{A}}{\alpha}\right) \tag{27}$$

$$b = \alpha \ln\left(1 + \frac{\mathrm{B}}{\alpha}\right) \tag{28}$$

If j is the number of species common to both samples, then if the two samples are considered as one large joint sample, the total number of individuals is now A + B and the total number of species is $a + b - j$. Then:

$$a+b-j = \alpha \ln\left\{1+\frac{(\mathrm{A}+\mathrm{B})}{\alpha}\right\} \tag{29}$$

Taking antilogarithms and eliminating A and B from the above three equations gives:

$$\frac{e^a}{\alpha} + \frac{e^b}{\alpha} = 1 + \frac{\mathrm{A}}{\alpha} + 1 + \frac{\mathrm{B}}{\alpha}$$

$$= 1 + \left\{1 + \frac{(A+B)}{\alpha}\right\} \quad (30)$$

$$= 1 + \frac{e^{(a+b-j)}}{\alpha}$$

The similarity index $I = 1/\alpha$ and it is the positive root of the equation:

$$e^{aI} + e^{bI} = 1 + e^{(a+b-j)I}, \quad (31)$$

which is the defining relationship for I. Equation (30) can be solved iteratively for I. A good approximation to I is obtained from:

$$\frac{2j}{2ab-(a+b)j}$$

Since α is a constant for the population, then I, which is the inverse of α, is independent of sample size for samples drawn from the same population.

Mountford's method for classifying sampling stations into groups of similar stations uses both an index of similarity between a pair of stations and an index of similarity between two groups of stations. The index of similarity between station B and a group composed of stations A_1 and A_2 is defined as:

$$I(A_1A_2, B) = \frac{I(A_1B) + I(A_2B)}{2}$$

where $I(A_1B)$ is the index of similarity between the pair of stations A_1 and B. In general terms, the index of similarity between a station B and a group composed of m stations: A_1, A_2, ... A_m; B is:

$$I(A_1, A_2, ...A_m; B) = \frac{I(A_1B) + I(A_2B) + ... + I(A_mB)}{m}$$

The index between a group composed of stations A_1 and A_2 and a second group composed of stations B_1 and B_2 is:

$$I(A_1A_2; B_1B_2) + \frac{I(A_1B_1) + I(A_1B_2) + I(A_2B_1) + I(A_2B_2)}{4}$$

In general terms, the index between groups A_1, A_2, ... A_m and B_1, B_2, ... B_m is:

$$\frac{1}{mn}\sum_{i=1}^{m}\sum_{j=1}^{n} I\left(A_i B_j\right)$$

The following rule is used for classification. From a derived table of indices of similarity select the biggest value. The pair corresponding to this value is combined to form a single group. The indices of similarity between this new group and each of the other stations are evaluated according to the definition of the index between groups of sites.

Shepard (1984) illustrated the application of Mountford's index I to testing the statistical significance of clusters. This approach to testing significance of clusters seems worthy of further consideration in cases where the log-series distribution is considered to provide a reasonable fit in species abundance data.

6. EMPIRICAL MULTISPECIES MODELS

6.1 INTRODUCTION

The necessity for considering alternatives to conventional models for multispecies fisheries has been indicated by Saila and Erzini (1987) and by Yang (1989), as well as in the Introduction to this chapter. These papers also summarize some of the more recent developments in the modeling of multispecies fisheries.

In view of the fact that our empirical multispecies model has been published (Saila and Erzini, 1987), only a limited description of the method is provided herein. Briefly, we developed a method for analyzing a data set consisting of a sequence of trawl survey data providing the catch per standard effort of fish and other organisms at stations which were sampled with similar gear and effort over the entire time sequence. These data were termed an aggregated time series. These aggregated (macro) data were converted to unaggregated (micro) data by use of quadratic programming and a least squares estimator (LSE). These micro data provided a transition probability matrix which could then be projected over time to provide some insights about the future stability of the system and probable changes in the various components of the fish assemblage. The preliminary results of this work were encouraging, and they were generally similar to prior inferences made from the data using other methods of analysis.

Much of the material which follows is based on Yang(1989), which has contributed further to the analysis of aggregated (macro) data in the context of multispecies fisheries.

6.2 MARKOV TRANSITION PROBABILITY MATRIX MODEL

There are two major approaches to determining the qth order transition probabilities (p^q) from an aggregated time series of data, such as research vessel trawl survey data. The two approaches involve the use of the maximum likelihood estimator (MLE) and the least square estimator (LSE).

In their initial study Saila and Erzini (1987) used the LSE for deriving a transition probability matrix from aggregated data. In this approach each survey cruise in the time series represents one state, so that the maximum number of states N must be equal to or greater than the number of species or species groups in the assemblage of interest.

Let us define an appropriate model as:

$$y_k = A_p q + \varepsilon$$

where:

y_k = a column vector of observations of species k from $t = q, q + 1, \ldots T$

A = a matrix of size ($T = q$+1) x N with N column vectors, each representing the time series of observations from $t = 0, 1, \ldots, T - q$ for one of the N species

p^q = a qth order transition probability matrix

ε = error between observed and actual values, which must be minimized by solving $(A'A)^{-1}$ constrained by the conditions that p^q are non-negative and the row sum of p^q equals unity.

The necessary conditions for obtaining $(A'A)^{-1}$ are that $T \geq N + q$ given that $(A'A)^{-1}$ exists.

In many data sets (especially in the tropics), the number of species N is high, so that $(N + q)$ is larger than the length of the time series T. In our applications the time series were sufficiently short so that some species had to be grouped together in order to permit a more suitable data set for analysis. It should also be pointed out that in our LSE analysis, the assumption is made that the process is a first-order Markov process, i.e., $q = 1$.

Yang (1989) has improved our approach in several ways. Perhaps most important, transition states are defined in quantitative space rather than in "species" space as was done by Saila and Erzini (1987). Furthermore, the MLE allows the definition of transition states in a more flexible manner, and also permits one to statistically validate the order of the transition probability matrix. However, in order to make use of the MLE, it is necessary to transform the aggregated data in a suitable manner.

The method is briefly as follows. Consider an ordered quantitative set (b_i) with O < $b_1 < b_2 < \ldots < b_M$, which divides the range of species abundances into M categories. These are referred to as states, S_i, for $T = 1, 2, \ldots$, M. If an observation of population k at time t, $\bar{y}_k(t)$, falls into the interval between $b_{i=1}$ and b_i, then this is rewritten as $\bar{x}_k(t) = i$. This indicates that population k at time t resided in the ith state, S_i. The set $\{b_i$: $i = 1, 2, \ldots$, M$\}$ is chosen so that all the observations $\overline{Y}$ over time and species, are distributed as evenly as possible in the M states. In this manner the original aggregated time series is transformed into a new series $\bar{x}_k(0), \ldots, \bar{x}_k(1), \ldots, \bar{x}_k(T)$. This is now in micro (disaggregated) form because the series now describes the time path for population k, for $k = 1, 2, \ldots, N$. Then p^q is a set of transition probability parameters for the community (assemblage) rather than for any individual population.

6.3 DEVELOPMENTS REGARDING MARKOV MODELS

Clearly, additional work remains to be completed in critically comparing and evaluating Markovian models which are defined in quantitative state space with those which have been defined in qualitative state space. It is expected that these comparisons will become a part of our projected research for forthcoming years. In addition to the above, we are also interested in exploring simpler models based on Markov processes. One such approach to a simpler model is briefly described below. It is emphasized that this material represents research which is still in progress rather than completed work. Therefore, substantial changes may be made prior to formal publication.

It is a common observation that the composition of fish communities, in terms of species composition and their relative abundance by weight or numbers, does not remain constant over space and time. Indeed, if one were to sample at various points in a given area at the same time of the day or of the year, one would find that not only are the relative abundances of the various species changing, but also that the species caught would vary from sample to sample. Our use of the term "species" is broader than that used in taxonomy. Here, species can refer to species or to larger taxonomic groups or even to ecologically similar species lumped into a single category.

A fish community is in reality a collection of various species which is in constant dynamic change as various fish species move in and out of the community. However, it is possible, by focusing attention on a particular fishing area and observing what is the most abundant species caught in the area over periodic samplings, to think of the fish community in the area as entering various states, where a state is characterized by the most abundant species caught in the area at a particular time. That is, suppose that the species A is the most abundant observed in the fishing area during the first period of observation or sampling. During the second period, the most abundant is species B. We could say that the fish community present in the fishing area has moved from being in state A and has entered state B. If m species are observed to predominate in the fishing area over successive periods of observation or sampling, then we can imagine the fish community to be in any one of the m states at any particular time of observation or sampling. They enter and leave any state in a probabilistic manner, where the changes in state that are taking place are governed by probabilistic laws.

By thinking of the dynamic changes taking place with respect to the most abundant species present in the fish community inhabiting a particular fishing area in a probabilistic manner, a Markov Chain model may be formulated to assess the predominant species present in the area, as well as to make predictions as to what species will predominate in the area at some future time, knowing that a particular species is most abundant at present. Such an assessment will, of necessity, assume that the dynamics of change occurring within the fish community probabilistically remain the same during the period of observation. In cases when an event took place that is suspected to have drastically altered the probabilistic structure of the dynamic changes that are occurring, a comparison of the transition probabilities generated by the Markov processes before and after the given event may allow for a determination of the significance of the event with respect to altering the composition of the group of species that are predominant in the fishing area. For example, the effects of the introduction of a destructive fishing practice such as blast fishing, the opening of a chemical plant discharging effluent in the fishing area, the banning of certain-sized fishing boats from exploiting the fish community, etc. may be initially assessed through such a discrete-state, discrete-time Markov chain formulation.

6.3.1 Theoretical Considerations

Consider a fish community inhabiting a particular fishing area, such as the Visayan Sea in the Philippines. Let the fishery be sampled once per unit time interval, say once a month; and let there be a record of the predominant species or higher taxonomic group observed at each sampling.

Define a state *i* at time *t* of the fishery so as to mean that species *i* is the predominant species present in the fishery at time *t*; and let there be *m* species that predominate in the fishery at one time or another during the various time periods under study.

Let the one-step transition probability, p_{ij}, be defined as

P_{ij} = Pr (fishery is at state *j* now | fishery was at state *i* in the immediately preceding period).

That is, P_{ij} is the probability that the predominant species in the fishery is species *j*, given that the predominant species in the fishery during the immediately preceding period was species *i*.

The matrix of one-step transition probabilities for all the *m* states of the fishery is given by:

$$P = \begin{pmatrix} P_{11} & P_{12} & \#\#\# & P_{1m} \\ P_{21} & P_{22} & \#\#\# & P_{2m} \\ & \vdots & & \\ P_{m1} & P_{m2} & \#\#\# & P_{mm} \end{pmatrix}$$

where

P_{ii} = Pr [the predominant species in the fishery from one period to the next remains species *i*]

P_{ij} = Pr [the predominant species in the fishery changes from one period to the next from species *i* to species *j*] $(i \neq j)$

The *m*-step transition probability is given by:

$$\Pr\left[x_{m+n} = K \middle| Xn = j\right] = P_{jk}{}^{(m)},$$

This states that the probability that from state j at the nth trial, state k is reached at the $(m{+}n)^{th}$ trial in m steps. That is, if the predominant species in the fishery at one sampling is species j, the probability that after m samplings the predominant species in the fishery will be species k, is given by $P_{jk}^{(m)}$.

Higher step transition probabilities may be obtained from the one-step transition probabilities through a special case of the Chapman-Kolmogorov equation satisfied by the transition probabilities of a Markov chain given by the following:

$$P_{j+k}^{(m+n)} = \sum_{r} P_{rk}^{(n)} P_{jr}^{(m)} = P_{jr} P_{rk}.$$

In matrix notation, this is given by $P^{m+n} = P^m P^n = P^n P^m$ where $P^S = P_{ij}^{(s)}$, $s > 1$, is the matrix of the S order transition probabilities and where the two-step transition probabilities are directly obtained from the one-step transition probabilities by setting $m = n = i$ in the special Chapman-Kolmogorov equations.

A maximum likelihood estimator of the transition probability, p_{ij}, is given by:

$$\hat{P}_{ij} = \frac{n_{ij}}{n_{i.}}$$

where

n_{ij} is the number of observed direct transitions from state i to state j in one step (i.e., the number of observed changes of the fishery from having species i as the predominant species to having species j as the predominant species i the next period).

$n_{1.} = \sum_{j=1}^{m} n_{ij}$ = total number of observed transitions from state i to the other states in one step.

A test of whether an observed realization comes from a Markov chain with a given transition matrix P^0 is provided by the statistic:

$$\sum_{j=1}^{m} \sum_{k=1}^{m} \frac{n_{j.}\left(\hat{P}_{jk} - P_{jk}^{0}\right)^2}{P_{jk}^{0}},$$

which has a chi-square distribution with $[m(m-1) - d]$ degrees of freedom, where d is the number of zeros in p^0, and the summation is taken only over (j, k) for which $p^0{}_{jk} > 0$.

6.3.2 Application

Monthly data on purse seine catches from January 1983 to December 1987 from the Visayan Sea, the Philippines, were summarized in terms of the top six predominant families of fishes by weight present in each of the catches. The data showed the following families as being predominant at one time or another:

Thunnidae	=	A	Clupeidae	=	D
Scombridae	=	B	Trichiuridae	=	E
Carangidae	=	C	Meneidae	=	F

The data was defective in that there were months with no catches. The Markov Chain formulation requires that the time periods considered to be equidistant from each other. With these constraints, three realizations were obtained in the following manner:

Realization I: monthly samplings from June 1983 to December 1984

J J A S O N D J F M A M J J O S O N D(months)
C D D B D B D C C C D A A A A D E F C

Realization II: samplings every two months from February 1983 to April 1985

F A J A O D F A J A O D F A(months)
C B C D D D C D A A E C E B

Realization III: sampling every three months from August 1983 to November 1985

A N F M A N F M A N(months)
D B C A A F E C D B

or: sampling every three months from September 1985 to December 1987

S D M J S D M J S D(months)
B C A B D D B D D D

From each of these realizations, a matrix of transition counts was obtained. The matrix of transition counts will allow for a maximum likelihood estimation of the transition probability matrix for each of the three modes of sampling. The tally matrices of transition counts are:

$$F_I = \begin{array}{c|cccccc|c} & A & B & C & D & E & F & n_{i\cdot} \\ A & 3 & 0 & 0 & 1 & 0 & 0 & 4 \\ B & 0 & 0 & 0 & 2 & 0 & 0 & 2 \\ C & 0 & 0 & 2 & 2 & 0 & 0 & 4 \\ D & 1 & 2 & 1 & 1 & 1 & 0 & 6 \\ E & 0 & 0 & 0 & 0 & 0 & 1 & 1 \\ F & 0 & 0 & 1 & 0 & 0 & 0 & 1 \\ \hline n_j & 4 & 2 & 3 & 6 & 1 & 1 & \end{array}$$

$$F_{II} = \begin{array}{c|cccccc|c} & A & B & C & D & E & F & n_{i\cdot} \\ A & 1 & 0 & 0 & 0 & 1 & 0 & 2 \\ B & 0 & 0 & 1 & 0 & 0 & 0 & 1 \\ C & 0 & 1 & 0 & 2 & 1 & 0 & 4 \\ D & 1 & 0 & 1 & 2 & 0 & 0 & 4 \\ E & 0 & 1 & 1 & 0 & 0 & 0 & 2 \\ F & 0 & 0 & 0 & 0 & 0 & 0 & 0 \\ \hline n_j & 2 & 2 & 3 & 4 & 2 & 0 & \end{array}$$

$$F_{III(a)} = \begin{array}{c|cccccc|c} & A & B & C & D & E & F & n_{i\cdot} \\ A & 1 & 0 & 0 & 0 & 0 & 1 & 1 \\ B & 0 & 0 & 1 & 0 & 0 & 0 & 3 \\ C & 1 & 0 & 0 & 1 & 0 & 0 & 1 \\ D & 0 & 2 & 0 & 0 & 0 & 0 & 4 \\ E & 0 & 0 & 1 & 0 & 0 & 0 & 0 \\ F & 0 & 0 & 0 & 0 & 1 & 0 & 0 \\ \hline n_j & 2 & 2 & 2 & 1 & 1 & 1 & \end{array}$$

$$F_{III(b)} = \begin{array}{c|cccccc|c} & A & B & C & D & E & F & n_{i\cdot} \\ A & 0 & 1 & 0 & 0 & 0 & 0 & 1 \\ B & 0 & 0 & 1 & 2 & 0 & 0 & 3 \\ C & 1 & 0 & 0 & 0 & 0 & 0 & 1 \\ D & 0 & 1 & 0 & 3 & 0 & 0 & 4 \\ E & 0 & 0 & 0 & 0 & 0 & 0 & 0 \\ F & 0 & 0 & 0 & 0 & 0 & 0 & 0 \\ \hline n_j & 1 & 2 & 1 & 5 & 0 & 0 & \end{array}$$

The estimates of the transition probabilities from row sums under each of the three modes of sampling are:

$$P_I = \begin{array}{c|cccccc|} & A & B & C & D & E & F \\ A & 3/4 & 0 & 0 & 1/4 & 0 & 0 \\ B & 0 & 0 & 0 & 1 & 0 & 0 \\ C & 0 & 0 & 1/2 & 1/2 & 0 & 0 \\ D & 1/6 & 2/6 & 1/6 & 1/6 & 1/6 & 0 \\ E & 0 & 0 & 0 & 0 & 0 & 1 \\ F & 0 & 0 & 1 & 0 & 0 & 0 \end{array}$$

$$P_{II} = \begin{array}{c|cccccc|} & A & B & C & D & E & F \\ A & 1/2 & 0 & 0 & 0 & 1/2 & 0 \\ B & 0 & 0 & 1 & 0 & 0 & 0 \\ C & & & & & & \\ D & 1/4 & 0 & 1/4 & 2/4 & 0 & 0 \\ E & 0 & 1/2 & 1/2 & 0 & 0 & 0 \\ F & 0 & 0 & 1 & 0 & 0 & 0 \end{array}$$

$$P_{III(a)} = \begin{array}{c|cccccc|} & A & B & C & D & E & F \\ A & 1/2 & 0 & 0 & 0 & 0 & 1/2 \\ B & 0 & 0 & 1 & 0 & 0 & 0 \\ C & 1/2 & 0 & 0 & 1/2 & 0 & 0 \\ D & 0 & 1 & 0 & 0 & 0 & 0 \\ E & 0 & 0 & 1 & 0 & 0 & 0 \\ F & 0 & 0 & 0 & 0 & 1 & 0 \end{array}$$

$$P_{III(b)} = \begin{array}{c|cccccc|} & A & B & C & D & E & F \\ A & 0 & 1 & 0 & 0 & 0 & 0 \\ B & 0 & 0 & 1/3 & 2/3 & 0 & 0 \\ C & 1 & 0 & 0 & 0 & 0 & 0 \\ D & 0 & 1/4 & 0 & 3/4 & 0 & 0 \\ E & 0 & 0 & 0 & 0 & 0 & 0 \\ F & 0 & 0 & 0 & 0 & 0 & 0 \end{array}$$

The transition probabilities under sampling every 2 months and sampling every three months were tested against the transition probabilities under sampling every month in order to see whether the more spaced out manner of sampling would drastically change the values of the transition probabilities.

The values of the test statistic are as follows:

(i) when testing II vs. I

$$\sum_{j=i}^{m}\sum_{k=i}^{m} n_{j\cdot}\frac{\left(\hat{P}_{jk}-p_{jk}^{0}\right)^{2}}{p_{jk}^{0}} = 10.667$$

(ii) when testing III vs. I

$$\sum_{j=i}^{m}\sum_{k=i}^{m} n_{j\cdot}\frac{\left(\hat{P}_{jk}-p_{jk}^{0}\right)^{2}}{p_{jk}^{0}} = \begin{array}{l} 8.667 \text{ for (1)} \\ 12.583 \text{ for (2)} \end{array}$$

When $\alpha = 0.01$, the χ^2 value for $m(m-1) - d = 7(6) - 37 = 5$ d.f. is 15.10. Thus, at $\alpha = 0.01$, there is no reason to say that the three estimates of the transition probability matrix are different from each other.

6.3.3 Markov Matrix and Entropy

Any sequence of dominant species can be regarded as a series of discrete states (dominant species). It is possible to express the transition count from any state (column) to its successor (row) with a tally matrix illustrated in the previous section as F_I, F_{II}, $F_{III(a)}$, $F_{III(b)}$. Each element of each of the above matrices can also be expressed as a probability simply by dividing that entry by the row total. These are illustrated in the previous section as P_I, P_{II}, $P_{III(a)}$, $P_{III(b)}$. Entropy across the row with respect to i is defined as:

$$E_i^{(\text{row})} = -\sum_{j=1}^{6} p_{ij\cdot}\cdot\log_2 p_{ij\cdot}$$

We sum to six in the above because there are six dominant species in the tally and transition probability matrices indicated above. If a calculated value of $E_i^{(\text{row})} = 0.0$, then one of the $p_{ij}(j = 1, 2, ..., 6)$ is unity and all the others are 0.0. In this case it can be said that i exerts a decisive influence on the selection of states of its successors. On the other hand large $E_i^{(\text{row})}$ values indicate that the memory effect of i is obscure, and, therefore, many states (dominant species) can follow i.

By dividing the entry by the column totals (m_j) in the tally matrices ($F_I - F_{III(b)}$), another type of Markov matrix is obtained. The previous definition of entropy can be extended to this case as:

$$E_i^{(col)} = -\sum_{j=1}^{6} q_{iji\cdot} \cdot \log_2 q_{ji\cdot}$$

In this case the q's are the entries of the transition matrix Q. $E_i^{(col.)}$ is entropy along the columns of Q with respect to i. If $E_i^{(col.)}$ is null, i is preceded by a particular state. Large $E_i^{(col.)}$ indicates that i is occurring independently of the preceding state.

Estimates of the transition probabilities from column sums under each of the three modes of sampling are:

$$Q_I = \begin{array}{c|cccccc|} & A & B & C & D & E & F \\ \hline A & .75 & 0 & 0 & .17 & 0 & 0 \\ B & 0 & 0 & 0 & .33 & 0 & 0 \\ C & 0 & 0 & .50 & .33 & 0 & 0 \\ D & .25 & 1.00 & .25 & .17 & 1.00 & 0 \\ E & 0 & 0 & 0 & 0 & 0 & 1.00 \\ F & 0 & 0 & .25 & 0 & 0 & 0 \end{array}$$

$$Q_{II} = \begin{array}{c|cccccc|} & A & B & C & D & E & F \\ \hline A & .50 & 0 & 0 & 0 & .50 & 0 \\ B & 0 & 0 & .13 & 0 & 0 & 0 \\ C & 0 & .50 & 0 & .50 & .50 & 0 \\ D & .50 & 0 & .33 & .50 & 0 & 0 \\ E & 0 & .50 & .34 & 0 & 0 & 0 \\ F & 0 & 0 & 0 & 0 & 0 & 0 \end{array}$$

$$Q_{III(a)} = \begin{array}{c|cccccc|} & A & B & C & D & E & F \\ \hline A & .50 & 0 & 0 & 0 & 0 & 1.00 \\ B & 0 & 0 & .50 & 0 & 0 & 0 \\ C & .50 & 0 & 0 & 1.00 & 0 & 0 \\ D & 0 & 1.00 & 0 & 0 & 0 & 0 \\ E & 0 & 0 & .50 & 0 & 0 & 0 \\ F & 0 & 0 & 0 & 0 & 1.00 & 0 \end{array}$$

$$Q_{III(b)} = \begin{array}{c|cccccc|} & A & B & C & D & E & F \\ \hline A & 0 & 0.5 & 0 & 0 & 0 & 0 \\ B & 0 & 0 & 1.00 & .40 & 0 & 0 \\ C & 1.00 & 0 & 0 & 0 & 0 & 0 \\ D & 0 & 0.5 & 0 & .60 & 0 & 0 \\ E & 0 & 0 & 0 & 0 & 0 & 0 \\ F & 0 & 0 & 0 & 0 & 0 & 0 \end{array}$$

$E_i^{(row)}$ and $E_i^{(col.)}$ serve as indications of the variety of species transitions immediately after and before the occurrence of i, respectively. If $E_i^{(row)}$ and $E_i^{(col.)} = 0$, states preceding and succeeding i are predicted to be in common. If both $E_i^{(row)}$ and

$E_i^{(col.)}$ are > 0, i precedes and is succeeded by more than one state. If $E_i^{(col.)} > E_i^{(row)}$, i can possibly occur after different states and is also followed by them. However, in the latter case, their degree of dependence on i are not even. In summary, the dominant species following i may be conjectured with greater certainly than those preceding i. On the other hand, if $E_i^{(row)} > E_i^{(col.)}$, the dependency of i on its predecessor is greater than the influence of i on its successor.

Consider, for example, the case of tally matrix P_I, which relates to monthly samplings from June 1983 to December 1984. Entropies, $E^{(row)}$ and $E^{(col.)}$ are calculated by applying the appropriate entropy formulas to the Markov matrices P_I and Q_I. These are displayed in Table 13.

Table 13. Entropies with respect to rows and columns for six dominant fish groups sampled monthly in the Visayan Sea, June 1983, December 1984.

$E^{(row)}$	State	$E^{(col.)}$
0.3114	A	0.8113
0.0000	B	0.0000
1.0000	C	1.5000
2.2520	D	1.9250
0.0000	E	0.0000
0.0000	F	0.0000

where, A = *Thunnidae*
B = *Scombridae*
C = *Carangidae*
D = *Clupeidae*
E = *Trichiuridae*
F = *Meneidae*

$E^{(row)}$ and $E^{(col.)}$ values for state D (*Clupeidae*) are the largest for this data set. This suggests that the influx of Clupeid fishes into the fishery was the most random event. For state A (*Thunnidae*) $E^{(col.)} > E^{(row)}$. This indicates that the dominant species following i may be conjectured with greater certainty than those preceding i. That is, the dominant species following Thunnidae is more predictable than the species preceding Thunnidae.

Clearly, the same exercise as indicated for the first data set could be done for the remaining three. However, the purpose of this has been primarily to demonstrate the potential utility of entropy tables for interpreting data. At this point, we do not speculate about the meaning or the causes of the observed similarities or differences.

6.4 OTHER APPROACHES TO FISHERIES MODELS AND ASSESSMENT PROJECTIONS

The abstract of CRSP Working Paper No. 61, entitled "On Measuring Ecological Stress: Variations on a Theme by R.M. Warwick", follows. This work represents a new and potentially useful approach to assessing changes in ecological communities under stress.

"Two new indices are presented which reflect quantitatively the changes to be expected in an ecological community under stress, as previously described by R.M. Warwick (Mar. Biol. 92: 557–562, 1986). The indices summarize information which Warwick presented graphically and permit analyses of trends and inferential tests. We suggest that these indices should be tested with a wide variety of ecological time series data in order to evaluate the feasibility of inferring ecological stress from static data."

The abstract of another working paper, CRSP Working Paper No. 5, "Effect of Incorporating Sigmoid Selection on Optimum Mesh Size Estimation for the Samar Sea Multispecies Trawl Fishery" by G.T. Silvestre and M.L. Soriano, follows.

"The evaluation of optimum mesh size for multispecies trawl fisheries relies primarily on the aggregation of individual yield-per-recruit response surfaces. The analytic model expression incorporated in these procedures assumes knife-edge selection, an assumption recently demonstrated to generate considerable bias in single-species assessment of short-lived tropical fish species. The present study examines the effect of replacing the usual knife-edge selection assumption with empirically based sigmoid selection in the evaluation of the optimum mesh size for the Samar Sea multispecies demersal trawl fishery. Relaxation of the knife-edge assumption in favor of sigmoid selection results in the increase of the optimum mesh size for the mix of 12 trawl-caught species considered in the study from 3.5 to 5.5 cm. In addition, sigmoid selection leads to other more conservative results or measures (e.g., lower optimum exploitation levels and catch rate expectations) than would otherwise have been obtained with knife-edge selection."

There are many possible extensions of the concepts in this chapter. These include further studies of Markov models based on empirical fisheries data, continuation of work on ecosystem stress measurements, and further studies on mesh regulations for multispecies fisheries. Of special interest and emphasis are new approaches to studying and measuring ecological succession and the probable effects of using explosives for fish capture in coralline and shallow water environments. New conceptual approaches to succession in multispecies assemblages using fuzzy graph theory have also been published.

REFERENCES

Abramowitz, M. and Stegun, I.A. (Eds.) 1964. *Handbook of Mathematical Functions with Formulas, Graphs, and Mathematical Tables.* Applied Math. Series 55. National Bureau of Standards, Washington, DC.

Anscombe, F.J. 1950. Sampling theory of the negative binomial and logarithmic series distributions. Biometrika 37: 358-382.

Brock, A.R. 1977. Comparison of community similarity indexes. Journ. W.P.C.F. 48: 2488-2493.

Czarnecki, D.B. 1979. Epipetic and epilithic diatom assemblages in Montezuma Wall National Monument. Arizona. J. Phycol. 15: 246-352.

Dial, K.P. and Marzluff, J.M. 1989. Non-random diversification within taxonomic assemblages. Systematic Zoology 38: 26-37.

Fisher, R.A., Corbet, A.B. and Williams, C.B. 1943. The relation between the number of species and the number of individuals in a random sample of an animal population. Journal of Animal Ecology 12: 42-58.

Gadagkar, R. 1989. An undesirable property of Hill's diversity index N_2 Oecologia 90: 140-141.

George, Y.S. and Hanumara, C. 1989. Assessing change through analysis of diversity. Journal of Environmental Management 28: 25-41.

Grassle, J.F. and Smith, W. 1976. A similarity measure sensitive to the contribution of rare species and its use in investigation of variation in marine benthic communities. Oecologia 25: 13-22.

Heltshe, J.F. 1988. Jackknife estimate of the matching coefficient of similarity. Biometrics 44: 447-460.

Henderson, R.A. and Heron, M.L. 1977. A probabilistic model of paleogeographic analysis. Letheia 10: 1-15.

Hill, M.O. 1978. Diversity and evenness: A unifying notation and its consequences. Ecology 57: 427-432.

Hurlbert, S.A. 1971. The nonconcept of species diversity: A critique and alternative parameters. Ecology 52: 577-586.

Johnston, J.W. 1976. Similarity indices I: What do they measure. Battelle Northwest Labs. BNWL-LIS2 NRC-1. 100 pp + appendix.

Kempton, R.A., and Taylor, N.R. 1979. Some observations on the yearly variability of species abundance at a site and the consistency of measures of diversity. In *Contemporary Quantitative Ecology and Related Econometrics.* (G.P. Patil and Rosenzweig, M., Eds.) International Cooperative Publishing House, Fairland, MD.

Khatri, C.G. 1959. On certain properties of power series distributions. Biometrika 46: 480-490.

Kobayashi, S. 1987. Heterogeneity ratio: A measure of beta-diversity and its use in community classification. Ecol. Res. 2: 101-111.

Koch, C.F. 1987. Prediction of sample size effects on the measured temporal and geographic distribution pattern of species. Paleobiology 13: 100-107.

Kohn, A.J. and Riggs, A.C. 1982. Sample size dependence in increases of proportional similarity. Marine Ecology Progress Series 9: 149-151.

Krebs, C.A. 1989. *Ecological Methodology.* Harper and Row, New York.

Magurran, A.E. 1988. *Ecological Diversity and Its Measurement..* Princeton University Press, Princeton, N.J.

McManus, J.W. 1985. Descriptive community dynamics: Background and an application to tropical fisheries management. Ph.D. dissertation. University of Rhode Island, Narraganset, RI.

Morisita, M. 1959. Measuring of interspecific association and similarity between communities. Mam. Fac. Science Kyushu University Series E (Biology) 3: 65-80.

Morisita, M. 1971. Composition of the Il-index. Res. Popul. Ecol. 13: 1-27.

Mountford, M.D. 1962. An index of similarity and its application to classification problems. In *Progress in Soil Zoology.* (Murphy, P.W., Ed.) Butterworths, London.

Murphy, P.M. 1978. The temporal variability in biotic indices. Environmental Pollution 17: 227-236.

Nemec, A.F. L. and Brinkhurst, R.O. 1988. Using the bootstrap to assess statistical significance in the cluster analysis of species abundance data. Canadian Journal of Fish. Aquatic Science 45: 965-970.

Noack, A. 1950. A class of random variables with discrete distributions. American Mathematical Statistics 21: 127-132.

Noreen, E.W. 1989. *Computer Intensive Methods for Testing Hypotheses.* John Wiley and Sons, New York.

Patil, G.P. 1961. Asymptotic bias and variance of ratio estimates in generalized power series distributions and certain applications. Sankhya A. 23: 269-280.

Pielou, E.C. 1975. *Ecological Diversity.* John Wiley and Sons, New York.

Pielou, E.C. 1977. *Mathematical Ecology.* John Wiley and Sons, New York.

Pielou, E.C. 1979. Interpretation of paleoecological similarity indices. Paleology 5: 435-443.

Pielou, E.C. 1984. Probing multivariate data with random skewers: A preliminary to direct gradient analysis. Oikos 42: 161-165.

Pikitch, E.K. 1988. Objectives for biologically and technically interrelated fisheries. In *Lecture Notes on Coastal and Estuarine Studies. Fisheries Science Management: Objectives and Limitations.* (Wooster, W.S., ed.) Springer-Verlag, New York.

Pinkham, C.F. and Pearson, J.G. 1976. Application of a new coefficient of similarity to pollution surveys. J. Water Pollution Control Federation. 48: 717-723.

Pope, J. 1979. Stock assessment on multispecies fisheries with special reference to the Gulf of Thailand. South China Seas Fisheries Development and Coordinating Program. SCS/DEY/29/13.

Raup, D.M., and Crick, R.E. 1979. Measurement of formal similarity in paleontology. Journal of Paleontology 53: 1213-1227.

Saeger, J. 1983. Results of the Samar Sea trawl survey. Department of Marine Fisheries Technology Report 3: 1-191.

Saila, S.B. and Erzini, E. 1987. Empirical approach to multispecies stock assessment. Transactions of the American Fisheries Society 110: 601-611.

Shepard, R.B. 1984. The log-series distribution and Mountford's similarity index as a basis for the study of stream benthic community structure. Freshwater Biology 14: 53-71.

Shepherd, J.G. 1988. An exploratory method for the assessment of multispecies fisheries. Journal du Conseil, Conseil Permanent International pour l' Exploration de la Mer 44: 189-199.

Sironmoney, G. 1961. Entropy of logarithmic series distributions. Sankhya A. 24: 419-420.

Smith, E.P. 1985. Statistical comparison of weighted overlap measures. Transactions of the American Fisheries Society 114: 250-257.

Smith, E.P., Genter, R.B., and Cairns, J., Jr. 1986. Confidence intervals for the similarity between algal communities. Hydrobiologia 139: 237-245.

Stander, J.M. 1970. Diversity and similarity of benthic fauna off Oregon. MS. Thesis Oregon State University, Corvallis, OR.

Taylor, L.R., Kempton, R.A. and Woiwood, I.P. 1976. Diversity statistics and the log-series model. Journal of Animal Ecology 45: 255-272.

Walters, C. 1986. *Adoptive Management of Renewable Resources.* Macmillan, New York.

Washington, H.G. 1984. Diversity, biotic and similarity indices. Water Res. 18: 651-694.

Williams, C.B. 1944. Some applications of the logarithmic series and the index of diversity to ecological problems. Journal of Animal Ecology 36: 1-44.

Williams, C.B. 1947. The generic relations of species in sample ecological communities. Journal of Animal Ecology 16: 11-18.

Williams, C.B. 1964. *Patterns in the Balance of Nature.* Academic Press, London.

Wolda, H. 1981. Similarity indices, sample size and diversity. Oecologia 50: 296-302.

Yang. Chi-Wen. 1989. A methodology for multispecies stock assessment with application to analyzing population dynamics of the groundfish community on the Scotian Shelf. Ph.D. Thesis. Dalhousie University, Halifax, Nova Scotia.

APPENDIX. OVERVIEW OF PROGRAMS

The programs described in this chapter were written in Microsoft QuickBasic version 4.5 and both source code and executable versions for IBM-compatible computer systems are included on the diskette. The executable files are stand-alone programs which can be run by simply typing the name of the program at the DOS prompt. Versions 5 and 6 of Microsoft's DOS operating system include Qbasic, which can be used to examine the source code and run the programs in an interpreted mode. Test data sets and example output files for each program are also included on the diskette. In addition, the text of this overview is contained in the file, OVERVIEW.DOC, which is a Microsoft Word for Windows v6.0 file. A simple ASCII version is contained in OVERVIEW.TXT.

Anyone with versions of DOS older than v5.0 and familiar with both QuickBasic and BASICA should be able to convert one to the other with a little effort. However, these are computationally intensive methods and the compiled programs run much faster than the interpreted versions. We recommend that the QuickBasic versions be used whenever possible, particularly in the executable form. The executable files contained on the diskette should run under DOS version 3 or more recent without modification.

The following files are on the diskette:

FAUNSIM.BAS — Source code for faunal similarity program

FAUNSIM.EXE — Executable file for faunal similarity program

FAUNSIM.OUT — Output from faunal similarity program

LOGSRFIT.BAS — Source code for log-series fitting program

LOGSRFIT.EXE — Executable file for log-series fitting program

LOGSRFIT.OUT — Output from log-series fitting program

WHITAKR.OUT — Data for creating a Whittaker Plot, generated by LOGSRFIT.EXE

SIMCONF.BAS — Source code for similarity confidence intervals program

SIMCONF.EXE — Executable file for similarity confidence intervals program

SIMCONF.OUT — Output from similarity confidence intervals program

SKEWER.BAS — Source code for skewer analysis program

SKEWER.EXE — Executable file for skewer analysis program

SKEWER.OUT — Output from skewer analysis program

CONFTEST.DTA — Test data set for similarity confidence intervals program

FAUNREEF.DTA — Test data set for faunal similarity program

TEST.DTA — Test data set for skewer analysis and log-series fitting programs

OVERVIEW.DOC — Word v6.0 document with text of program overview.

OVERVIEW.TXT — ASCII file with text of program overview.

All of the input data files consist of species abundance data in a simple ASCII form. Each program contains a routine which allows the user to enter data at run time (and optionally, save it to a file). However, if LOGSRFIT is being used to examine more than one site at a time, input must come from an external file. LOGSRFIT.BAS and SKEWER read data from the same file format. One can use SKEWER to create a data set that can be used as input to LOGSRFIT, though any of these data files may be created with a text editor. The input data set used by SKEWER and LOGSRFIT (e.g., TEST.DTA) is formatted in the following manner:

LINE 1: a title in quotes (")

LINE 2: number of stations, number of species

LINE 3: observation label header (e.g. "STATION" or "OBS", etc.) and a label for each column (=species); each must be in quotes and separated by a space.

LINE 4: observation 1 label (in quotes), abund SPP 1, abund SPP 2, ..., abund SPP S

LINE 5: observation 2 label (in quotes), abund SPP 1, abund SPP 2, ..., abund SPP S

.

.

LAST LINE: observation N label (in quotes), abund SPP 1, ..., abund SPP S

It is best to limit column labels to four or five characters. Species abundance values must be separated by a space or a comma (or both). Any number of comments may be recorded on the lines following the last line of data. This is a very simple format, but punctuation and spacing are critical and failure of the program to run properly can often be attributed to tiny errors in the data structure.

The input data set read by SIMCONF (e.g., CONFTEST.DTA) is similar to that described above, but differs as follows:

LINE 1: TITLE (in quotes)

LINE 2: # OF SPECIES (total in the samples being compared)

LINE 3: # OF REPLICATES FROM COMMUNITY 1, # OF REPLICATES FROM COMMUNITY 2

LINE 4: THE ABUNDANCE OF EACH SPECIES FROM REPLICATE 1, COMMUNITY 1

LINE 5: THE ABUNDANCE OF EACH SPECIES FROM REPLICATE 2, COMMUNITY 1

LINE 6: THE ABUNDANCE OF EACH SPECIES FROM REPLICATE 3, COMMUNITY 1

. **...**

. **THE ABUNDANCE OF EACH SPECIES FROM REPLICATE 1, COMMUNITY 2**

. **THE ABUNDANCE OF EACH SPECIES FROM REPLICATE 2, COMMUNITY 2**

. **THE ABUNDANCE OF EACH SPECIES FROM REPLICATE 3, COMMUNITY 2**

. **...**

No observation labels are provided. It is vital that all of the observations which represent replicates for the first community are together and occur first in the data set.

In both of the above file structures (i.e., TEST.DTA and CONFTEST.DTA), all species in the data set must be represented in the same order on each line of data (i.e., observation). A species' abundance should be represented by a zero if it was absent from an observation.

Finally, the input data set for FAUNSIM (i.e., FAUNREEF.DTA) is a simple list of the relative abundances of each species in the region of interest. Nothing else should be contained in this data file.

CHAPTER 9

A SYSTEMS SCIENCE APPROACH TO FISHERIES STOCK ASSESSMENT AND MANAGEMENT

Brian J. Rothschild[1], Jerald S. Ault[2] and Steven G. Smith[1]

[1]Center for Environmental and Estuarine Studies, University of Maryland, [2] Rosenstiel School of Marine and Atmospheric Science, University of Miami

1. INTRODUCTION

Presently, capture fisheries of the world are directed towards a remarkable array of species types and assemblages that occur over a wide range of geographical locales and aquatic habitats. Examples of this diversity are the eastern tropical Pacific pelagic tuna fishery, the Antarctic krill fishery, the northwestern Atlantic multispecies bottom fishery, and tropical coral reef fisheries. The associated industrial and artisanal fisheries employ a multitude of fishing gear-vessel combinations, ranging from dugout canoes equipped with handlines to high technology ocean-going fishing fleets with enormous purse seines or trawls. In addition, the number of landing sites varies with the type of fishing operation, ranging from a few ports in industrial single-species fisheries to literally hundreds of port locations in multispecies artisanal fisheries. However, regardless of the fishery type, there exists one overall goal of fisheries management: the creation and implementation of an optimal harvest policy which jointly maximizes the constituency requirements from the resource. The exact definition of "optimal" depends, of course, on the various objectives of management. Common multiple-objective management includes the determination of harvest levels which either optimize the biological production of the fish stock or maximize the net present value of

1-56670-151-1/96/$0.00+$.50

the catch, or some joint optimization of the two. Other associated or alternative objectives may be to provide high quality recreational angling, or even to preserve the traditional way of life of a longtime fishing community. All of these objectives are obfuscated because the fishery becomes embedded within the inherent complexity of the aquatic ecosystem. The associated fish stocks comprise but one part of the natural system, whereas the fishery constitutes an interaction between the human socio-economic and ecological systems.

Given the complexity of the decision making environment, realization of the objectives of management is best accomplished by integrating numerous areas of expertise. These disciplines include biology, physics, chemistry, geology, meteorology, oceanography, mathematics, statistics, engineering, food science, economics, management science, operations research, management information systems, computer science, sociology, political science, and even law. The successful practice of fisheries management science can thus be envisioned as a system in which the many fields of study combine to perform an orderly progression of activities that will ultimately result in the creation of optimal fish stock harvest strategies.

We consider "fishery management systems" to be comprised of eleven components: (1) the natural-fishery system complex (i.e., some definition of the actual temporal and spatial dynamics of any arbitrary number of stocks), (2) sampling design and data collection, (3) data processing, (4) data management, (5) a descriptive overview of the biological, physical, and fishery characteristics, (6) demography, (7) systems identification and principal components analysis, (8) perception of the natural-fishery complex, (9) a metasystem description (i.e., synthesis of better understood stocks), (10) optimization and decision analysis, and (11) implementation of optimal management policy. A schematic representation of the general fisheries management system is shown in Figure 1. As practiced today under the guise of fishery management, the activities which comprise components 2 through 11 are predominately carried out independently. This is perhaps due in part to the inherent interdisciplinary nature of examining exploited ecosystems; the analyses carried out in one element are sometimes considered abstract and undecipherable to workers in other elements. Whatever the reason, the current practice of fisheries management is neither integrated nor coordinated for achieving the desired management objectives.

Thus, the current challenge to fisheries management is to devise methods for realizing specific objectives for any given fishery setting (i.e., single- or multispecies, tropical or boreal, commercial or recreational, artisanal or industrial) and at any information level (i.e., sparse to complete data). In meeting this challenge, we advocate a systems science approach to fisheries stock assessment and management. The approach diverges from traditional practices by linking together three fundamental phases of activity: (i) "systems description" which includes the collection and processing of data; (ii) "systems analysis" which provides parameter estimates from data through a well structured format; and, (iii) "systems optimization-implementation" which, based on the estimates, provides the means for enactment of fishing policy designed to achieve optimal harvest levels.

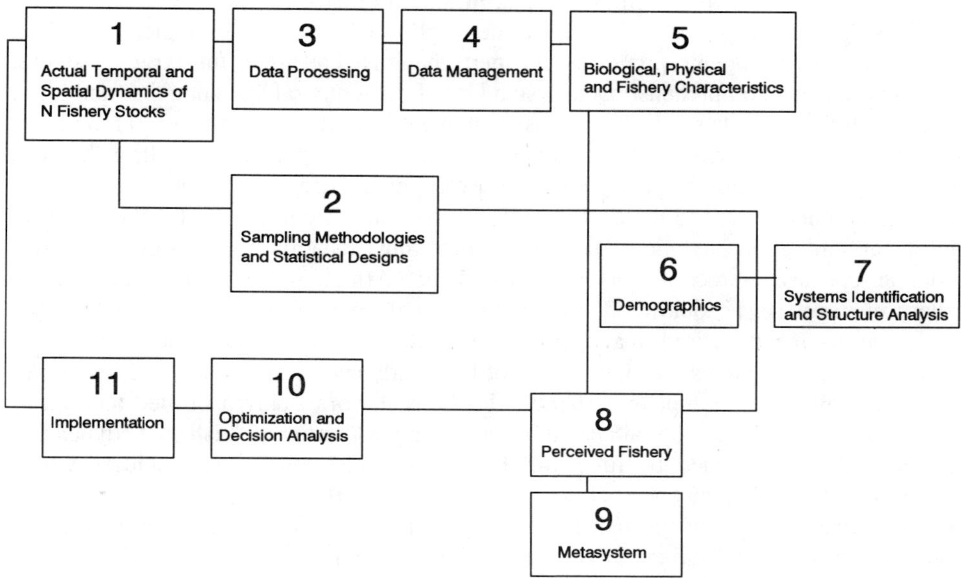

Figure 1. Schematic diagram showing the eleven components of the conceptual fishery management system.

This chapter's intent is to demonstrate the utility of considering fisheries stock assessment and management from a structured systems viewpoint. We present an in-depth description of the components and linkages of the general fisheries management system. We then introduce the design of a prototype computer system based on the systems science approach.

2. THE STRUCTURED FISHERIES MANAGEMENT SYSTEM

The structured fisheries management system can be divided into three complexes which follow along the lines of the three fundamental phases of activity: (i) system description; (ii) system analysis; and, (iii) system optimization-implementation.

2.1 SYSTEM DESCRIPTION COMPLEX

Components 1 through 5 form a coordinated subunit of the fisheries management system termed the "system description complex". This complex performs the basic data collection, processing, and descriptive analyses that will eventually provide information to the rest of the system.

2.1.1 Actual Temporal and Spatial Dynamics of N Stocks

This component provides a complete description of all of the biological, physical, fishing and anthropogenic processes and perturbations that affect the targeted stocks. The complete and comprehensive understanding of the temporal and spatial dynamics of the commercially viable stocks in an ecosystem would entail knowledge of: 91) the exact location of every individual of each population at every instant of time from birth to death; (2) the physiological outcome of every encounter with conspecifics, prey, predators, competitors, and mates during the individual's lifetime; and, (3) the physical, biological, and anthropogenic processes that determine location, physiology, and intra- and inter-species interactions of each individual. Armed with this information, the fishery manager (FM) could then choose the optimal time and location to harvest each individual to meet desired management objectives. Obviously, even when given unlimited fiscal resources, our knowledge of fishery dynamics at this level of detail will never be known with complete certainty. In the real world, only a limited amount of time and resources can be allocated to assessing a particular fishery's dynamics. Consequently, constrained by time and budget, the FM must first elucidate which variables of the ecological and socio-economic systems yield the information paramount to proper management of the fishery. Secondly, the FM must develop sampling and analysis procedures to estimate the primary variables at an acceptable level of precision. The FM's third task is then to optimize the system response to meet management objectives. Finally, the FM must implement specific policies to reach these objectives. The FM's responsibilities and function occur within components 2 through 11 in Figure 1.

2.1.2 Sampling Methodologies and Statistical Designs

The first step in assessing the dynamics of any particular fish stock begins with the collection of data. An expanded view of component 2 is shown in Figure 2. Fisheries data sampling is separated into two basic categories: fishery dependent and fishery independent. Fishery-dependent sampling involves the collection of data from the fishery. This data can be collected in either a "passive" fashion in which anglers or landing site operators report catch, effort, and perhaps fishing cost and revenue to the management agency; or "actively" by sending out specialized personnel on fishing vessels or to landing sites. Fishery-independent sampling methods generally involve research group studies which use their own personnel, sampling equipment, and vessels to collect specific information on the targeted stocks. Three classes of sampling programs are used. The most common programs of this type are surveys which are directed towards any life stage of a particular species and employ many diverse sampling gears such as plankton nets, fish traps, gill nets, bottom trawls, electric currents or chemical anesthetics and poisons. Some stocks may also be sampled visually via transects performed by underwater divers or observers in airplanes, hydroacoustically with sonar technology, or with remote sensing devices deployed on airplanes or satellites. In each case the organisms are not subject to capture. A third class of "mixed" sampling techniques incorporates both fishery dependent and fishery

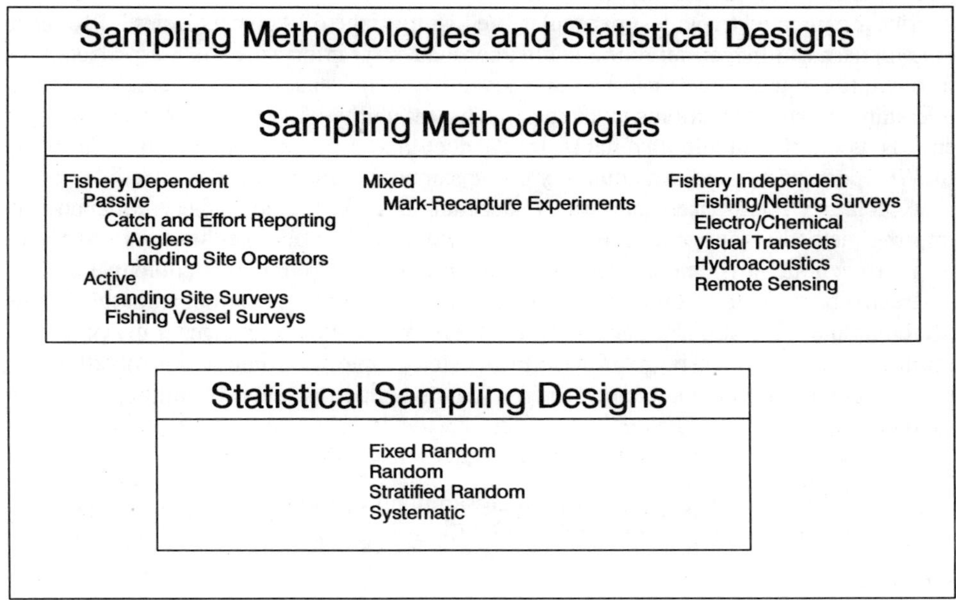

Sampling Methodologies and Statistical Designs
Sampling Methodologies
Fishery Dependent
Passive
Catch and Effort Reporting
Anglers
Landing Site Operators
Active
Landing Site Surveys
Fishing Vessel Surveys
Mixed
Mark-Recapture Experiments
Fishery Independent
Fishing/Netting Surveys
Electro/Chemical
Visual Transects
Hydroacoustics
Remote Sensing
Statistical Sampling Designs
Fixed Random
Random
Stratified Random
Systematic

Figure 2. Sampling methodologies and statistical designs employed in the fishery management system.

independent methodologies. Typical examples of this hybrid class are release-recapture studies (Burnham et al., 1987).

Two criteria that determine which sampling methodologies are relevant for a given fishery situation are: (1) the quality of the data to be obtained, and (2) the feasibility of the particular methodology. The quality of data which results from the application of a certain methodology class varies according to data completeness, accuracy, coverage, and bias. Data completeness refers to the number of variables that are measured relative to the full set of variables. This can range from a few, such as catch and effort, to a complete suite of biological and physical variables, such as those which may result from a fishery-independent trawl program. Data accuracy may also vary widely. In most cases, data collected by trained personnel tend to be much more accurate than those reported by anglers or landing site operators. This is especially true if conditions exist which may favor biased reporting. Sampling program coverage refers to how many observations of a given variable are sampled. That is, if information is being collected from a commercial fishery, the number of sampling units (i.e., fishing trips) will be very high; whereas, in contrast, the number of trips taken by a research team during a fishery independent sampling program will likely be very low due to budgetary constraints. Sampling bias is present when a variable is measured only within a certain range. This is very common in fishery-dependent and mixed sampling methodologies, since most fisheries are directed towards a particular life history stage of a given species. The most common case is a fishery targeting adults; consequently, information is not collected on the other life stages such as larvae and juveniles. The lapses may be crucial for the understanding of population's dynamics. The feasibility of a sampling methodology depends upon the monies available to conduct the program, the training level of sampling personnel, and the availability of the sampling units to the survey.

A relative comparison of fishery dependent, mixed, and fishery independent sampling methodologies according to the criteria mentioned above is shown in Table 1. From inspection, it is readily apparent that no one sampling methodology class is universally best. For example, even though fishery independent methodologies may yield complete, accurate, and unbiased data, their sample coverage is generally poor and sampling costs are very high. Thus, FMs often employ a combination of all three sampling classes to implement a program which is both comprehensive and of balanced data quality.

Table 1. Relative comparisons of the three sampling methodology classes in terms of their expected data quality and manpower and fiscal requirements.

	Data quality				Requirements	
Sampling class	Completeness	Accuracy	Coverage	Life-stage bias	Personnel training	Sampling costs
Fishery-Dependent						
Passive	Low	Low	High	Yes	Low	Low
Active	Moderate	High	Moderate	Yes	Moderate to High	Moderate
Mixed	High	Moderate	Moderate	Yes	High	High
Fishery-Independent	High	High	Low	No	High	High

Data must be collected in a fashion which ensures statistical validity and accuracy when applying a sampling methodology of any class to an actual fishery. Thus, a proper statistical sampling design must be employed. The most common statistical designs used in collecting fisheries data are: random, stratified random, fixed site, and systematic (Figure 2). Two main issues involved in choosing the appropriate statistical sampling design are: (1) the inherent statistical bias in estimates of the mean and variance of sampled variables, and (2) the acceptable level of sampling variance. Addressing the first issue requires that the variables of interest meet basic statistical design assumptions. For example, suppose we employ a random sampling design to measure the abundance of a certain species. We thus assume that this particular stock is evenly distributed throughout the sampling domain; that is, no "patchiness" exists in the spatial abundance pattern of the population. We also assume that each sample is independent and identically distributed. In most cases when these assumptions are met, implementation of a statistical sampling design will yield unbiased estimates of the mean and variance. The issue then becomes the acceptable level of sampling variance. If the variance is high for a particular variable, such as stock abundance, the FM may not be able to detect statistically a true population increase or decline. Thus, the design which minimizes sampling variance is desirable.

2.1.3 Data Processing

The information flow of the system begins in component 3 where raw sampling data is compiled and processed. A representative list of some of the variables that are typically sampled and some common associated data processing procedures are presented in Figure 3. Sampling variables are divided into four categories: (1) spatio-temporal, (2) physical, (3) biological, and (4) fishery. Spatio-temporal variables pin-point the location in time and space where physical, biological, and fishery variables are collected. Physical sampling variables are those which provide oceanographic and meteorologic information. Biological variables provide information about individual organisms: species or stock identification, size, age, sex, reproductive status, and food habits. Fishery variables are related to the process of capturing organisms. These variables not only provide information on the abundance of species and/or stocks, but also on the size, complexity, economics, and politics of the fishing industry, regardless of sector. Some of this information, such as the age or species identification of an individual fish, is not readily obtained by rapid visual observation and may require laboratory or other data processing techniques for its determination.

2.1.4 Data Management

This component collects and processes raw data and is essentially a database of all of available information on the fishery. This database is separated into two smaller entities; one containing historical information, and the other containing current information (Figure 4). The information within these entities is further subdivided according to the three sampling classes discussed in Section 2.1.2. Two hierarchies of data quality are recognized: the first relates to time, and the second relates to sampling class. In terms of the first hierarchy, no exact definition of what constitutes "historical" and "current" exists. The time boundary could vary from weeks to seasons to years to decades. It is important to recognize that the underlying conditions of both the aquatic ecological and human socioeconomic systems will change over time. The rate of these changes will, of course, vary by system and sampling variables. Current information may thus be "weighted" for reliability more heavily than historical information. The second hierarchy of data quality is dependent to some degree upon the sampling class chosen (cf. Section 2.1.2). Given that the data provide the basis for the decision-making power of the FM and ultimately support recommendations for implementation of a management plan, it is essential that its origins be kept in clear perspective.

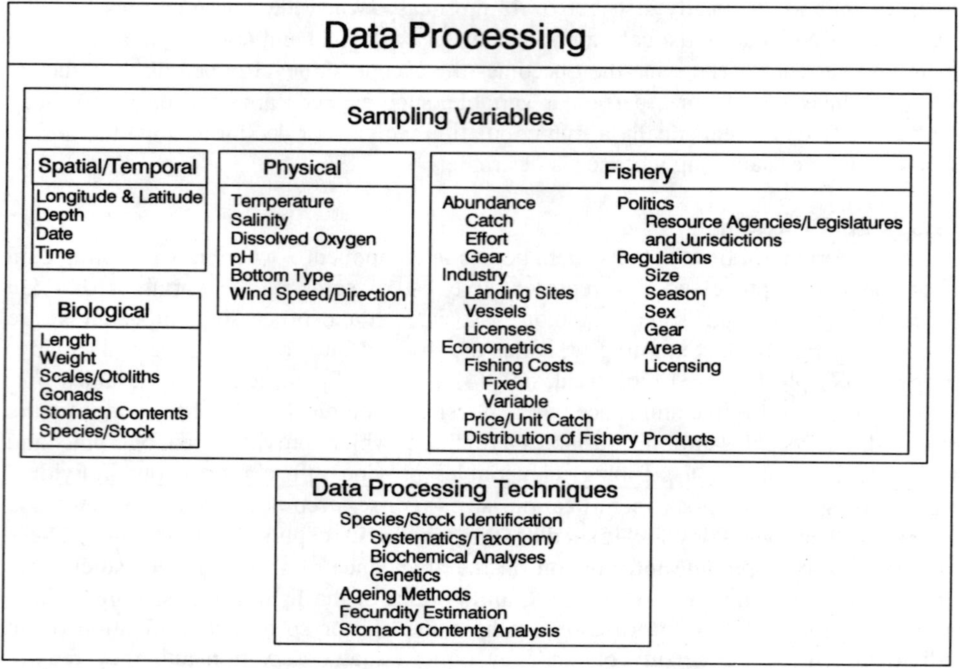

Data Processing
Sampling Variables
Spatial/Temporal
Longitude & Latitude
Depth
Date
Time
Physical
Temperature
Salinity
Dissolved Oxygen
pH
Bottom Type
Wind Speed/Direction
Fishery
Abundance
Catch
Effort
Gear
Industry
Landing Sites
Vessels
Licenses
Econometrics
Fishing Costs
Fixed
Variable
Price/Unit Catch
Distribution of Fishery Products
Politics
Resource Agencies/Legislatures
and Jurisdictions
Regulations
Size
Season
Sex
Gear
Area
Licensing
Biological
Length
Weight
Scales/Otoliths
Gonads
Stomach Contents
Species/Stock
Data Processing Techniques
Species/Stock Identification
Systematics/Taxonomy
Biochemical Analyses
Genetics
Ageing Methods
Fecundity Estimation
Stomach Contents Analysis

Figure 3. Sampling variables and data processing techniques employed in the fishery management system.

2.1.5 Biological, Physical, and Fishery Characteristics

This component's fundamental task is to furnish the FM with a basic idea of the processes of the ecological and socio-economic systems within which a particular fishery operates through descriptive analyses of the data. The three main descriptive categories and their interactions which comprise component 5 are illustrated in Figure 5. The understanding of aquatic ecosystems requires knowledge of characteristics of both the physical and biological processes which constitute the "natural" environment. An ecosystem's physical environment can be generally characterized by several important parameters (e.g., temperature, salinity, dissolved oxygen, pH, density) with respect to time (time of day, time of year) and location (latitude, longitude, depth). Further comprehension is facilitated by familiarity with major seasonal oceanographic processes such as depth stratified currents, tidal flows, and upwelling locations, concurrent with meteorological patterns of wind, rainfall, barometric pressure, and air surface temperatures.

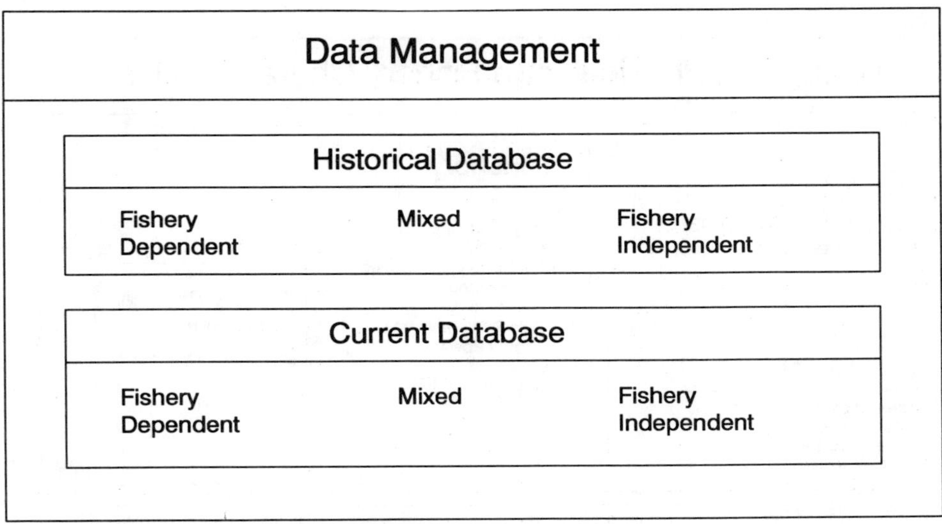

Figure 4. Historical and data processing components involved in the data management function of the fishery management system.

The biological system is embedded within the physical environment. The FM is primarily concerned with the harvested species population biology to include: (i) age-size-sex composition; (ii) the timing and location of spawning; (iii) the age (size) at sexual maturity; (iv) larval ecology; (v) feeding habits and food compositions; (vi) habitat preference; and, (vii) migrational patterns. However, since the harvested population must interact with both predator and prey organisms, the biology of the interacting species are also important. Biophysical interactions must be explored to complete our general knowledge of the natural system. Thus, the physical factors which regulate biological processes (e.g., growth and reproduction) and determine seasonal abundance patterns are also of interest.

Once a general knowledge of the aquatic ecosystem has been obtained, the FM must then assess the impact of the human system (i.e., fishery) on this natural system. This includes a description of the fishery technology employed to harvest fish stocks, the historical catch and fishing effort time series, the species composition of the catch, and the economics of the fishing industry. By exploring the seasonal patterns of fishing effort and catch, the fishery manager can determine how both the physical and biological environments regulate the behavior of the fishing industry. These types of relationships are classified as fishery-bio-physical interactions.

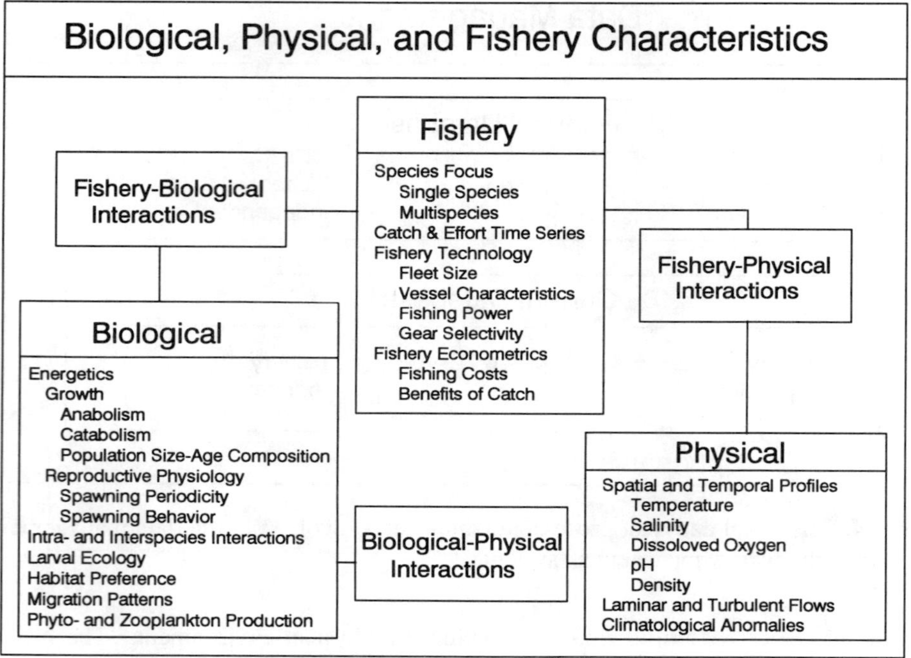

Figure 5. Biological, physical and fishery characteristics monitored by the fishery management system.

2.2 SYSTEM ANALYSIS COMPLEX

Components 6 through 9, coupled with components 2 and 5 previously discussed, form a second coordinated subunit of the fisheries management system here termed the "system analysis complex". This complex utilizes data produced from the primary natural-fishery (component 1) or a "mirror" system to provide analyses of the demographics and stochastic structure of the stock(s), the aim of which is to ultimately produce a complete representation of the natural fishery dynamics.

2.2.1 Demographics

Mathematical models of reproduction, growth, and survivorship have traditionally been the central focus of fisheries science (Beverton and Holt 1957; Ricker 1975). Various statistical procedures are used to estimate the parameters of these models, with age-based and more recently length-based methods being the two most common approaches. Population vital rates including per capita births, growth, and deaths are determined for the harvested species within component 6 (Figure 6). Per capita birth

Demographics
Parameter Estimation
Reproduction/Fecundity
Growth
Age-based methods
Length-based methods
Mortality
Age/size specific natural and fishing mortality
Age/size specific survivorship
Population Age/Size/Sex Distribution
Interaction Models
Cohort interaction
Species interaction
Biophysical interaction

Figure 6. Elements of the demographics component of the fishery management system.

rates are estimated by considering the expectation of births at a given age (size) in the context of several traditional stock-and-recruitment models and decision theoretic approaches. Upon estimation of these vital rates, the FM can then calculate the age-size-sex specific population abundance. Additional demographic analysis entails the construction and estimation of interaction models, many of which are simulation-based, in which cohort, species, and/or bio-physical relationships are mathematically evaluated and projected.

2.2.2 Systems Identification and Structure Analysis

Component 7 contains two main modeling approaches that explore the empirical structure of population data (Figure 7). The first approach, time-series analysis, evaluates the stochastic structure of a time stream of data. Time series analysis produces a model defining the stochastic processes which govern periodic trends over time (Box and Jenkins 1976). The extracted information provides bases for exploring biological and physical processes which give rise to a particular temporal pattern. In addition, time-series models can be used to provide forecasts of variables over short time frames. The second approach, principal components analysis (PCA), examines the eigenstructure and orthogonality of a system of variables. PCA is a tool for the analysis of spatial or temporal variability of physical and biological fields (Preisendorfer, 1988).

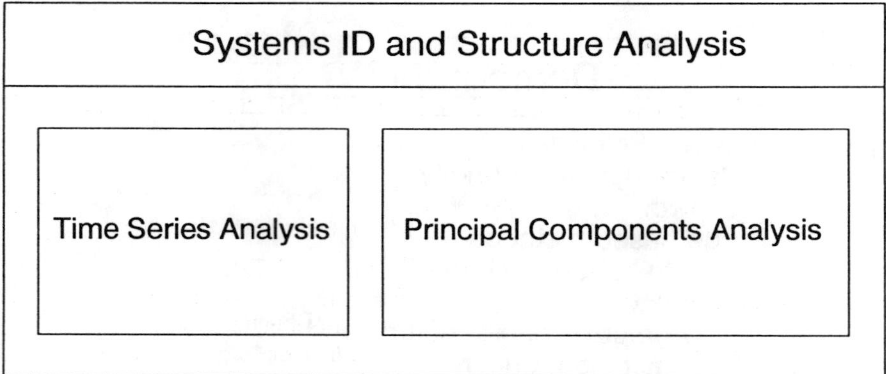

Figure 7. Systems identification and structure analysis component of the fishery management system.

The system eigenstructure provides insight into which variables account for the greatest proportion of system variance.

2.2.3 Perceived Fishery

The most substantial analytical results from the descriptive, demographic, and systems structure analyses are compiled and linked in component 8. The Perceived Fishery functionally represents the best perception of the natural fishery system complex depicted as component 1.

2.2.4 Metasystem

A metasystem is literally a system in the midst of systems. The metasystem component contains a catalog of various perceived fisheries from other ecological systems distributed circum-globally. When the FM is confronted with a particularly sparse data situation, the metasystem can be searched to provide information on other "similar" well-studied stocks. The synthesized descriptive and analytical results from these "mirror" fisheries may provide the FM with guidance for managing the relevant fishery.

2.3 SYSTEM OPTIMIZATION-IMPLEMENTATION COMPLEX

Components 10 and 11, together with components 1, 2 and 8, form the third and final coordinated subunit of the fisheries management system, the "system optimization-implementation complex" (Figure 8). This complex visualizes the perceived fishery relative to various sets of management scenarios, optimizes the system subject to various

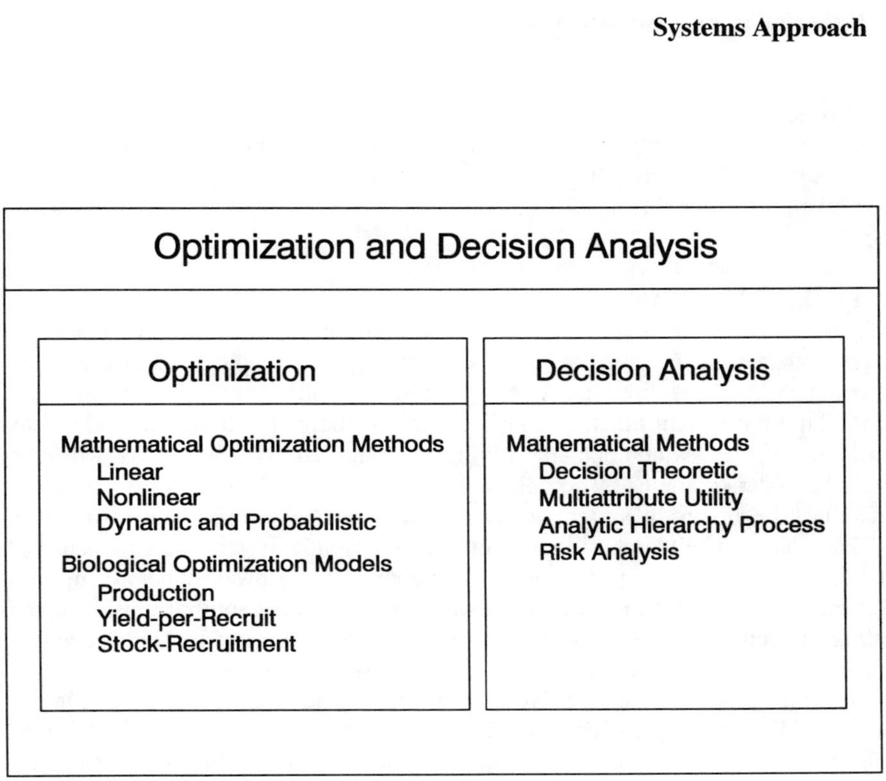

Optimization and Decision Analysis

Optimization

Mathematical Optimization Methods
- Linear
- Nonlinear
- Dynamic and Probabilistic

Biological Optimization Models
- Production
- Yield-per-Recruit
- Stock-Recruitment

Decision Analysis

Mathematical Methods
- Decision Theoretic
- Multiattribute Utility
- Analytic Hierarchy Process
- Risk Analysis

Figure 8. Optimization and decision analysis component of the fishery management system.

constraints, and provides a set of policy implementation procedures for realization of management objectives.

2.3.1 Optimization and Decision Analysis

Component 10 provides the pathway to local and global system optimization through a large array of mathematical programming optimization and decision analytic techniques. Optimization entails construction of linear, nonlinear, dynamic and probabilistic mathematical programming models to determine various optimality criterion such as harvest, economic revenue, and/or constituency utility. Decision analysis refines the optimization relative to various time-dependent objectives, sources of risk, and satisfaction with system outcomes. Jointly, the elements of this component address specific issues such as size-specific mortality rates, gear and season restrictions, and proportional allocation of harvest benefits.

2.3.2 Implementation

From the specific optimal outcomes provided by component 10, component 11 formulates policy designed to achieve all objectives simultaneously while overlaying the proper levels of enforcement and compliance necessary to reach those management goals.

2.4 SYSTEM FUNCTION

There remains the requirement to provide efficient, cohesive and unambiguous linkages both within the components of each individual complex and also among the three complexes: (i) description, (ii) analysis, and (iii) optimization-implementation. To illustrate how the system functions, suppose we are charged with the task of managing an unfamiliar fishery which operates within an unfamiliar ecosystem. We will proceed through the system complex by complex to highlight the system function.

The task begins in the system description complex (Figure 9). Since we have no a priori knowledge of either the natural or socio-economic systems (component 1), we must begin in the sampling component (component 2). However, before initiating a field sampling program, we must first assess any historical data that exists. Our first sampling "expedition" is then to locate relevant published and technical documents, examine data sets, conduct interviews, and so forth, that may provide information on many of the sampling variables for inclusion to the data processing component (component 3). This information is then grouped according to the original classes of sampling and then organized within the historical section of the data management component (component 4). We explore the physical, biological, and fishery characteristics (component 5) and their interactions to gain a basic understanding of the structure of the ecological and socio-economic functions with respect to the targeted stock(s). The background knowledge gleaned from the data hopefully provides a general picture of the ranges of the variables in the physical environment, the biology of the targeted species, effective sampling gears, fishing patterns, industry technology, and seasonal currents and weather patterns. This information should prove to be a very useful guide as to how we proceed in the present situation; thus, the historical data flows back to component 2 and assists us in designing a sampling program to collect current data. We then proceed through the current information flow path of the system description complex, eventually concluding in component 5 with an understanding of both the historical and present states of the fishery and its ecosystem.

Data from the systems description complex, now capsulized in component 5, are transferred to components 6 and 7 for population dynamics and systems identification analytical work (Figure 10). A portion of the descriptive analyses data is passed directly to the perceived fishery module (component 8), and some of the data is passed back to the sampling component (2).

Within the system analysis complex "simulation modeling" becomes a prevalent part of the systems structure. While a system simulation component could be located within the general system, components 2, 6, and 7 each contain at least one or more specialized simulation modules to perform local optimization study. The data used for parameterizing simulation models are derived from component 5 in the typical case, and

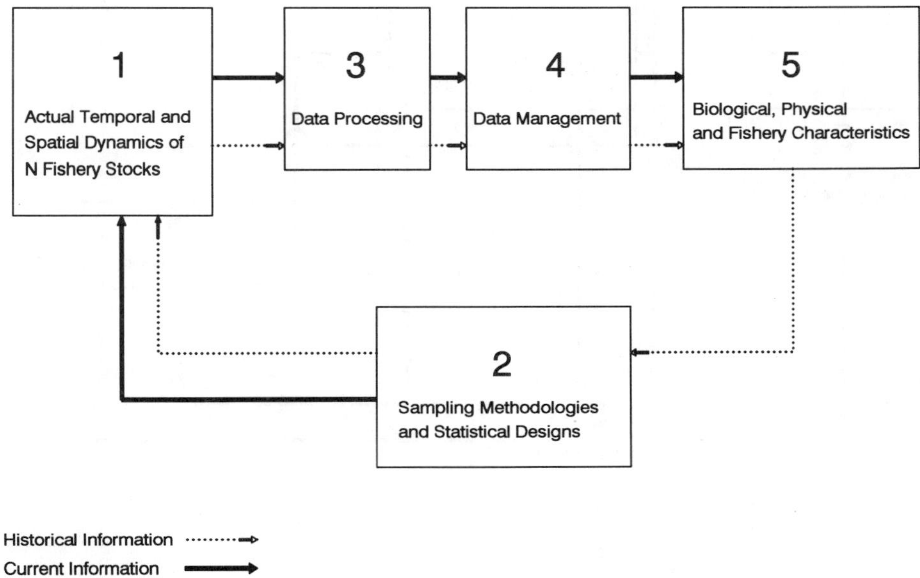

Figure 9. Systems description complex component of the fishery management system.

component 9 for the sparse data situation. In component 2 simulation is used to evaluate the efficacy of various sampling regimes; in component 6 the performance of demographic models is tested under various population dynamics scenarios; and in management system.

Additionally, operations research methods are integral to the successful performance of components 2, 6, and 7. For instance, a linear programming model could be developed to describe optimal route dynamics for vessels conducting the sampling program; and in both components 6 and 7 decision theory could be employed to evaluate the best analytical model. In using the combined forces of simulation and operations research local optimization occurs independently within components 2, 6 and 7. The local optimizations serve to provide a general optimization of the analytical results contained within the perceived fishery (component 8). If results obtained from component 8 are deemed incomplete or suboptimal, the system then returns to component 2 to obviate the deficiency. The return flow is obligated to pass through the descriptive and analytic complexes once again for reevaluation. Please bear in mind that the information contained within component 8 may be derived from both real and simulated data (Figure 11). In fisheries with sparse data the percentage of information occupying component 8 which is derived from simulation modeling will be high. As

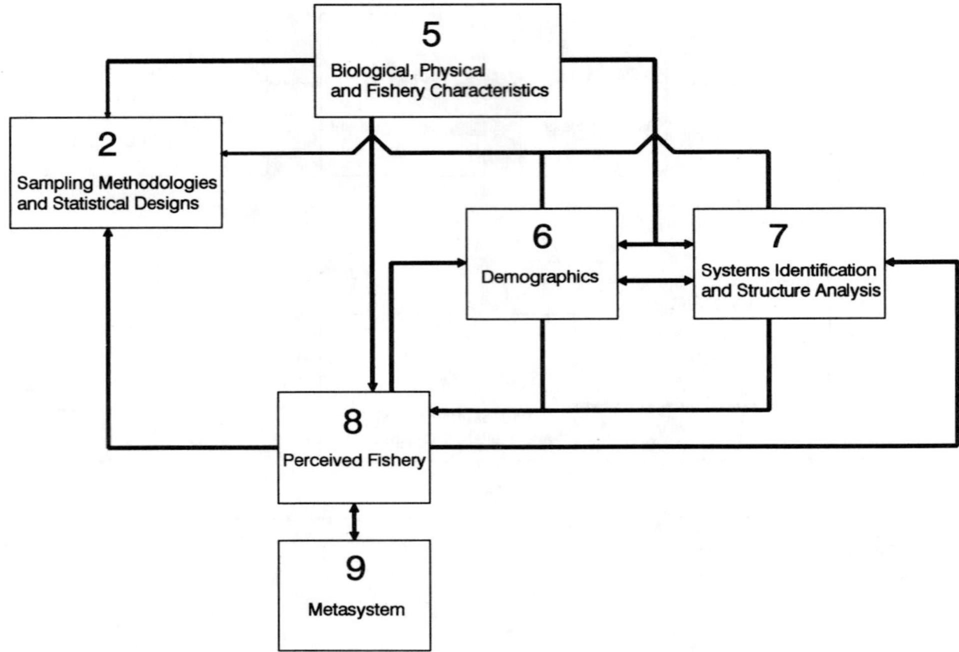

Figure 10. Population dynamics and systems identification function of the fishery component 7 time series and random data matrices may be simulated by Monte Carlo methods.

time goes on, and more effective sampling programs are put in place, analytic results derived from real data will occupy the predominant portion of the database.

While local and semiglobal optimizations were performed in the previous complex, component 10 of the system optimization-implementation Complex (Figure 12) is exclusively dedicated to integrating system results and providing a pure **global** system optimization. This global optimum is then forwarded to the implementation component (11). Again, if the analyses are deemed inconclusive or deficient, then the flow is passed back to component 2 and reiterated through the descriptive and analytic complexes. If optimal policy is derived, and a harvest policy is invoked, then a direct action occurs upon component 1, the natural fishery system. Subsequent evaluation of the merits of the implemented policy launches another pass through the entire fisheries management system.

As presented, we began our network flow analysis in component 2 and worked forward through the system. In reality, we need to start in components 10 and 11,

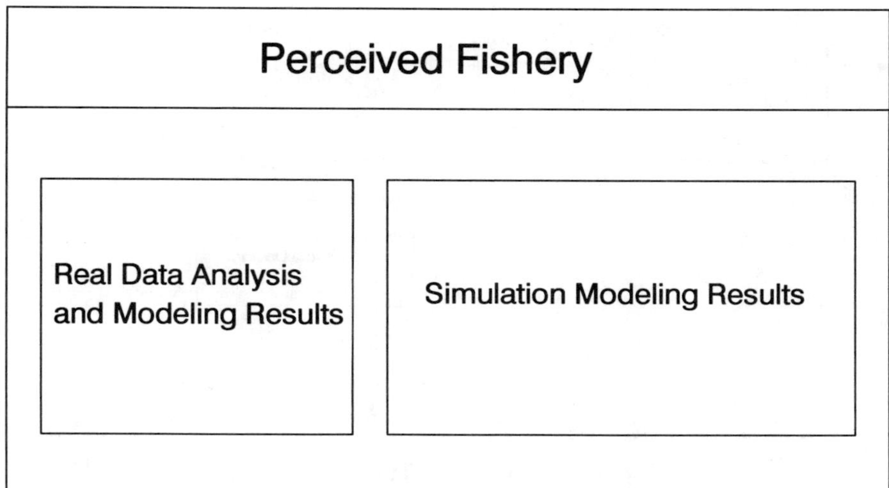

Figure 11. Components of the perceived fishery in the fishery management system.

defining clear objectives and the feasibility of their implementation. This requires a recursive pass through the network. Thus, we determine the information needed to perform the global system optimization within component 10. Given that insight, we then return to component 8 in which we evaluate the important descriptive and analytical results required from components 5, 6 and 7. Subsequently, we must ultimately regress to component 3 and elucidate the sampling variables required to perform the analysis of components 5, 6, and 7. At this point, we then return to the forward progression through the descriptive, analytic and optimization-implementation complexes (cf. Figures 9, 11, 12) as described above.

3. CANONICAL FISHERY MANAGEMENT COMPUTER SYSTEMS

Given that we have defined definite linkages between complexes in the system, and components within complexes, it is clear that the information load is significant for any one individual to process and sort out in a real-time basis. To facilitate the practical implementation of the systems science approach, a prototype of an expert decision support system was designed patterned after the structured fisheries management system delineated above. Of the eleven system components, eight (2, 4, 5, 6, 7, 8, 9, 10) are specifically suited for incorporation as discrete modules within the expert computer

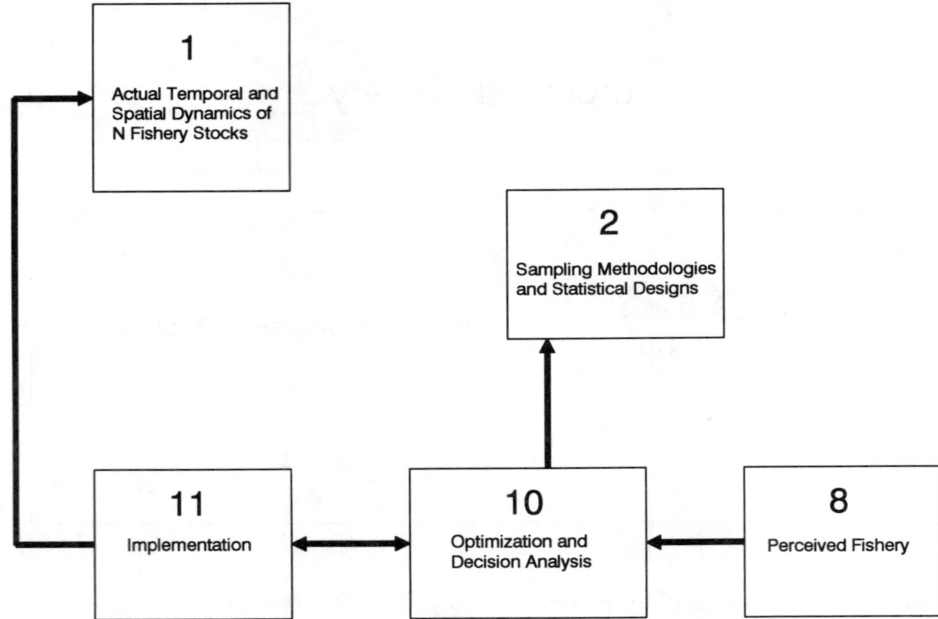

Figure 12. Systems optimization and implementation complex of the fishery management system.

system, while the three other components (1, 3, 11) lie outside the computer's domain (Figure 13).

The expert decision support system, CANOFISH (CANOnical FISHery) is composed of three main sectors: (1) the decision support system (DSS), (2) the metasystem, and (3) the front-end processor or user interface. The DSS is a functionally integrated and automated software package which provides data linkage between seven of the eight essential component modules. Their individual tasks parallel the discussions presented in Sections 2.1.2, 2.1.4–2.1.5, 2.2.1–2.2.3, and 2.3.1. The meta-system is a discrete module which provides a rich database condensed to computer code covering species, ecosystems and fishing industries which are distributed circum-globally (cf. Section 2.2.4).

The user interface (UI) is an advanced artificial intelligence entity that serves to modulate the workings and the interfacings of the DSS and metasystem. At each level of linkage, feedback, and interaction the UI makes the internal decision concerning quality of the data, quality of the decision, and the logical path on which the system shall proceed. Additionally, at each level the UI provides the user with informative

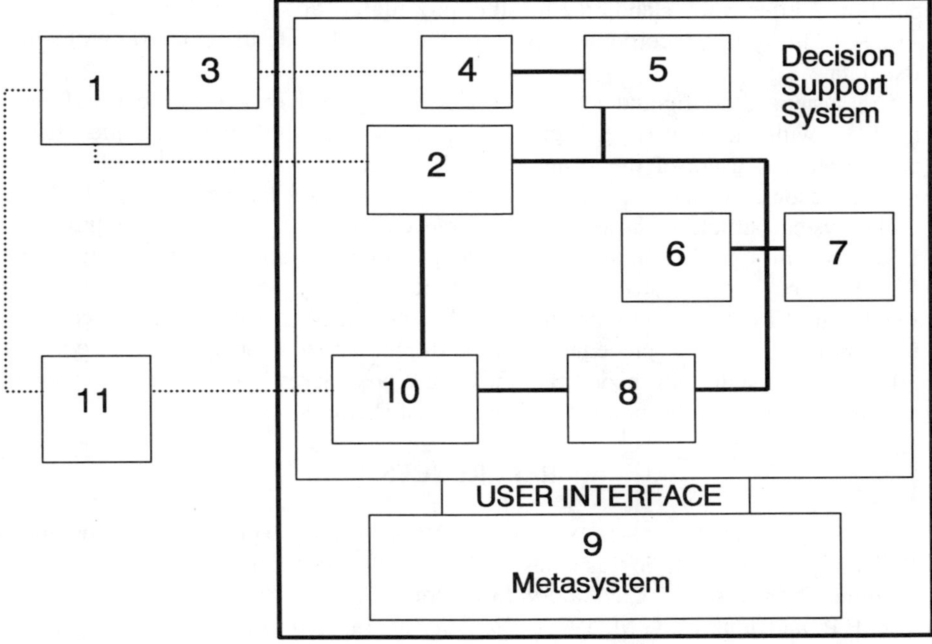

Figure 13. Interactions of the eleven components of the canonical computer-based fishery management decision support system.

updates on status of the process, and various alternatives to be considered, and asks relevant questions as to whether the user feels he personally is competent to enter at that point into the decision-making process. In cases of inadequate user experience or background, or in default, the UI will provide a menu of most reasonable choices based on the consensus of the thousands of experts embedded and embodied within the UI's knowledge base.

The expert DSS CANOFISH has to this point proceeded through several increasing levels of sophistication, the first iteration serving principally as a prototype demonstration system with simple linkages among the components (Stagg, 1990).

4. CONCLUSIONS

The main features of the systems science approach demonstrated the strength, necessity and utility of an expert DSS for fishery management. All the activities involved in the practice of fishery management are integrated into three cooperative working units: the system description, analysis and optimization-implementation complexes. CANOFISH II provides a systematic approach to managing exploited populations, effectively standardizing the procedural flow of fishery management activities. The system incorporates many features to alleviate the common problem of sparse data.

The system is configured to compensate for any user experience level, allowing individuals with minimal experience to deal effectively with highly complex fishery issues. Potential users of the system include academicians, scientific and technical personnel, resource managers, industry representatives, analysts, and policy makers.

The system structure shown here as three complex units serves to optimize the activities within a given complex, as well as providing an optimal path for the entire system function. The system is fully dynamic. As any new information becomes available, the UI automatically updates the already extensive database, and recomputes every optimal analytical procedure. In addition, as new analysis techniques are developed, the system will incorporate those new procedures. A unique feature of the system is that it is flexible enough to adapt to any fisheries situation.

REFERENCES

Beverton, R.J.H. and Holt, S.J. 1957. On the dynamics of exploited fish populations. Ministry of Agriculture Fisheries and Food. Fisheries Investigations, Series II, Volume XIX. Fisheries Laboratory, Lowestoff, Suffolk, UK.

Box, G.E.P. and Jenkins, G.M. 1976. *Time Series Analysis: Forecasting and Control.* Holden-Day, San Francisco.

Burnham, K.P., Anderson, D.R., White, G.C., Brownie, C. and Pollack, K.H. 1987. Design and analysis for fish survival experiments based on release recapture. American Fisheries Society Monograph 5: 1-437.

Preisendorfer, R.W. 1988. *Principal Component Analysis in Meteorology and Oceanography*. Developments in Atmospheric Science 17. Elsevier, Amsterdam.

Ricker, W.E. 1975. Computation and interpretation of biological statistics of fish populations. Bulletin of the Fisheries Research Board of Canada 191.

Stagg, C.M. 1990. The expert support system as a tool in fishery stock assessment and management. Proceedings of the NATO Advanced Study Institute on Operations Research and Management in Fishing, Povoa de Varzim, Portugal: 299-314.

ANNEX A

FISHERIES STOCK ASSESSMENT COLLABORATIVE RESEARCH SUPPORT PROGRAM WORKING PAPERS

*** University of Rhode Island/University of the Philippines**
**** University of Maryland/University of Costa Rica**
***** University of Washington/University of Costa Rica**

No.	Date	Title / Author(s)
1*	5/87	**An Empirical Approach to Multispecies Stock Assessment** Saul B. Saila and Karim Erzini
2*	5/87	**Geometric Programming Applied to Some Optimal Harvesting Problems** Saul B. Saila and Karim Erzini
3*	5/87	**A Comparison of the Relationship Between Optimal Harvesting Strategies and Reproductive Values in Four Marine Species Different Life History Characteristics** Saul B. Saila and Karim Erzini
4*	6/87	**Status of Philippine Stocks: An Overview** G.T. Silvestre and S. Ganaden
5*	6/87	**Effect of Incorporating Sigmoid Selection on Optimal Mesh Size Estimation for the Samar Sea Multispecies Trawl Fishery** G.T. Silvestre and M.L. Soriano
6**	5/87	**Feasibility of Relating Recruitment to Environmental Variables** B. Rothschild
7**	5/87	**Expert Systems, Microcomputers, and Operations Research** Arjang A. Assad and Bruce L. Golden

1-56670-151-1/96/$0.00+$.50

8***	5/87	**Some Impacts of Remote System Technology on Future Fisheries Management** R.E. Thorne
9***	5/87	**Hydroacoustics and Ground Truth** R.E. Thorne
10***	10/87	**The Use of Stationary Hydroacoustic Transducers to Study Diel and Tidal Influences of Fish Behavior** R.E. Thorne, John Hedgepeth and Jorge Campos
11***	10/87	**Optimal Sampling for the Estimation of Age-Length Key: Analysis and the Program, 'AGECOMP'** Han-Lin Lai and V.F. Gallucci
12***	10/87	**Users' Guide for the Program 'LCAN' (Length Cohort Analysis) and an Overview of the Underlying Analysis** Han-Lin Lai and V.F. Gallucci
13*	12/87	**Comparison of Two Length Frequency-Based Packages Used to Obtain Growth and Mortality Parameters Using Simulated Samples with Varying Recruitment Patterns** M. Castro and Karim Erzini
14***	12/87	**Hydroacoustic Observations of Fish Abundance and Behavior Around an Artificial Reef in Costa Rica** R.E. Thorne, Gary L. Thomas and Jorge Campos
15***	12/87	**Hydroacoustic Observations of Fish Abundance and Behavior Around Reefs and Structures** R.E. Thorne and Gary L. Thomas
16*	3/88	**Extreme Value Theory Applied to the Statistical Distribution of the Largest Length of Fish** S.P. Formacion, J.M. Rongo and C. Sambilay
17**	3/88	**Report of Workshop: San Jose, Costa Rica, January 21-24, 1986** C. Stagg and Jorge Campos
18**	3/88	**A Simulation Study of Marine Population Dynamics** B.L. Golden, B.J. Rothschild and H. Trivedi
19**	3/88	**Continuum-Distribution Population Dynamics** J.M. Gracia-Bondia
20**	3/88	**A Strategy for Organizing and Analyzing Golfo de Nicoya Fishery Data** J. Campos and C. Stagg
21**	3/88	**A Bibliography of Optimization Models in Fisheries Management** A.A. Assad and G.T. DiNardo
22**	3/88	**A Difference Equation Simulation Model of Marine Population Dynamics** R.A. Estrada
23**	3/88	**A Simple Renewal Model of Larval Fish Feeding** J. Lobo

24**	3/88	**Biodynamics of the Sea: Preliminary Observations on High Dimensionality** B.J. Rothschild and T.R. Osborn
25**	3/88	**Fish Stock Assessment Programs for the IBM-PC and Compatibles** V.R. Restrepo, D. Die, W.W. Fox, Jr. and J. Chavarria
26**	3/88	**A Discrete Dynamic Fisheries Utilization Model: Towards the Practical Application of Modern Fisheries Economics** L.G. Anderson
27**	3/88	**Using Decision-Analysis to Manage Maryland's River Herring Fishery: An Application of AHP** G.T. DiNardo, D. Levy and B. L. Golden
28**	3/88	**Spawning-Stock Biomass: A Source of Error in Recruitment-Stock Relationships and Management Advice** B.J. Rothschild and M.J. Fogarty
29**	3/88	**The Physical Basis for Recruitment Variability in Fish Populations** B.J. Rothschild, T.R. Osborn, T.D. Dickey and D.M. Farmer
30**	3/88	**Parameter Uncertainty and Simple Yield Per Recruit Analysis** V.R. Restrepo and W.W. Fox
31**	3/88	**Program REPSIM: A REcruitment Pattern SIMulator** D.Die, V.R. Restrepo and W.W. Fox, Jr.
32**	3/88	**The Spawning and Recruitment Patterns of Tropical Marine Fish Stocks** J.S. Ault and W.W. Fox, Jr.
33**	3/88	**The Tropical Fishery System Model: CORECS: Continuous Recruitment Simulation** J.S. Ault and W.W. Fox, Jr.
34**	3/88	**User's Guide to CORECS** J.S. Ault and W.W. Fox, Jr.
35**	3/88	**Abundance/Density Equations for Age-Structured Multicohort Populations: MCON (Multiple Cohort, N-dimensional model)** J.S. Ault and W.W. Fox, Jr.
36**	3/88	**User's Guide to MCON (Multiple Cohort, N-dimensional model)** J.S. Ault and W.W. Fox, Jr.
37**	3/88	**On the Efficacy of Some Techniques to Assess the Tropical Fishery System** J.S. Ault and W.W. Fox, Jr.
38**	3/88	**Small-Scale Turbulence and Plankton Contact Rates** B.J. Rothschild and T.R. Osborn
39*	4/88	**Philippine Coral Reef Fisheries Management** J.W. McManus and C.C. Arida
40**	4/88	**The Significance of Physiologically Structured Models for Fish Stock Dynamics** J.M. Gracia-Bondia and J.C. Varilly

41*	4/88	**Coral Reef Fisheries at Cape Bolinao, Philippines: An Assessment of Catch, Effort, and Yield** A.R. Acosta and C.W. Recksiek
42*	5/88	**User's Manual for SIMULPOP, a Program to Simulate Series of Length Frequency Distributions of an Exploited Fish Stock** M. Castro
43*	6/88	**An Improved Method for Fitting Gillnet Selectivity Curves to Predetermined Distributions** Saul B. Saila and Karim Erzini
44*	8/88	**Sigmoid Selection and the Beverton and Holt Yield Equation** G.T. Silvestre, M. Seriano and D. Pauly
45***	12/88	**Effects of Parameter Variability on Length-Cohort Analysis** Han-Lin Lai and V.F. Gallucci
46***	12/88	**Inferring the Distribution of the Parameters of the von Bertalanffy Growth Model from Length Moments** R.L. Burr
47***	12/88	**Comparative Study of Postlarval Life-History Schedules in Four Sympatric Species of Cancer (*Decapoda*: *Brachyura*: *Cancridae*)** J.M. Orensanz and V.F. Gallucci
48*	3/89	**Philippine Marine Capture Fisheries Exploitation, Potential, and Options for Sustainable Development** G.T. Silvestre
49*	3/89	**Stock Assessment of the Bolinao Reef Flat Fishery: Yield Estimates and the Use of Dominant Species in Assessing Coastal Multispecies Resources** W.L. Campos, J.B.P. Cabansag, A. del Norte, C.A. Nañola and R.B. Reyes, Jr.
50*	3/89	**Virtual Population Estimates of Monthly Recruitment in Biomass of Rabbitfish, Siganus canaliculatus, off Bolinao, Northern Philippines** A. del Norte and D. Pauly
51*	3/89	**Zonation Among Demersal Fishes of Southeast Asia: The Southwest Shelf of Indonesia** J.W. McManus
52*	3/89	**Overfishing on a Philippine Coral Reef: A Glimpse into the Future** A. del Norte, C.L. Nañola, J.W. McManus, R.B. Reyes, W.L. Campos and J.B.P. Cabansag
53***	4/89	**Age Determination in Biology: A Retrospective and Presentation of Contemporary Methods** Han-Lin Lai, V.F. Gallucci and D.R. Gunderson
54*	4/89	**A Study of Band Noise Associated with Landsat Bathemetric Images** P.A. Roe

55**	5/89	**Development of a Sampling Expert System: FISHMAP** B.J. Rothschild, G.T. DiNardo and M. Bhandary
56**	5/89	**Strategy for Management Modeling** B.J. Rothschild and C.M. Stagg
57*	8/89	**A Village-Level Approach to Coastal Adaptive Management and Resource Assessment (CAMRA)** J.W. McManus, E.M. Ferrer and W.L. Campos
58*	8/89	**Comparisons Between Multi-Species Fish Assemblages Using Similarity Indices** Saul B. Saila and J.E. McKenna
59*	8/89	**Spatio-Temporal Variations in Community Structure on a Heavily Fished Slope in Bolinao, Philippines** C.L. Nola, Jr., J.W. McManus, W.L. Campos, A. del Norte, R.B. Reyes, Jr., J.B.P. Cabansag and J.N. Pasamonte
60*	8/89	**Earth Observing System and Coral Reef Fisheries** J.W. McManus
61*	8/89	**On Measuring Ecological Stress: Variations on a Theme by R.M. Warwick** J.W. McManus
62*	8/89	**The Logarithmic Series Distribution and Some Applications to the Analysis of Tropical Multispecies Fisheries Data** Saul B. Saila and J.E. McKenna
63*	8/89	**Sample Size and Grouping of Data for Length Frequency Analysis** Karim Erzini
64***	9/89	**Merging Aggregate Catch Data with Uncertain Prior Knowledge to Approximate Age and Size Distributions and Selectivity Functions** R.L. Burr
65***	9/89	**Hydroacoustic Assessment of Fish Stocks in the Gulf of Nicoya, Costa Rica** J.B. Hedgepeth, R.E. Thorne and V.F. Gallucci
66***	9/89	**A Catch-at-Length Analysis That Incorporates a Stochastic Model of Growth** P.J. Sullivan, Han-Lin Lai and V.F. Gallucci
67***	9/89	**Age Determination and Growth for Two Corvinas (*Cynoscion stoltzmanni*) and (*Cynoscion squamipinnis*), in the Gulf of Nicoya** Han-Lin Lai and J. Campos
68*	6/90	**Ecological Computer Programs: The Importance of Being Friendly** J.W. McManus
69*	6/90	**Preliminary Model Development for Coral Reef Fish Assemblage Recovery from Blast Fishing** J. McLeavy and Saul B. Saila

70* 12/90 **Table Analysis by Classification of Ordinations (TACO) A Method for the Simultaneous Division**
J. McManus

71*** 12/90 **User's Guide to Program 'Cohort' Pope's Cohort Analysis**
B. Amjoun, Han-Lin Lai and V.F. Gallucci

72*** 12/90 **Minimum Cross-Entropy Spectral Analysis of Time-Varying Biological Signals**
R.L. Burr and D.W. Lytle

73*** 12/90 **Effects of Ageing Errors on Estimates of Growth, Mortality and Yield Per Recruit for Walleye Pollock (*Theragra chalcogramma*)**
Han-Lin Lai and D.R. Gunderson

74*** 12/90 **Age Determination of Pacific Cod, Gadus Macrocephalus, Using Five Ageing Methods**
Han-Lin Lai and D.R. Gunderson

75*** 12/90 **User's Guide to Program CASA A Catch-At-Size Analysis**
B. Amjoun, V.F. Gallucci and Han-Lin Lai

76*** 12/90 **Improving Population Estimates Obtained from Highly Variable Trawl Data**
R.A. McConnaughey and L.L. Conquest

77* 3/91 **A Compilation of Data on Variability in Length-Age in Marine Fishes**
Karim Erzini

78* 4/91 **Application of Fuzzy Graph Theory to Successional Analysis of a Multispecies Fishery**
Saul B. Saila

79* 3/91 **Destructive Coral Reef Fishing: Seeking Perspectives**
J.W. McManus, C.L. Nañola and R.P. Reyes

80* 2/91 **Evaluation of Tuna Resources Potential in Indonesian Waters with Emphasis on East Indonesia**
Saul B. Saila and J. Uktolsega

81** 6/91 **FINMAN: A Fishery Institution Management Computer Model for Simulating the Decision-Making Environment in Tropical and Subtropical Regions**
J. Ault and W.W. Fox, Jr.

82** 6/91 **FINMAN: Simulated Decision Analysis with Multiple Objectives**
J. Ault and W.W. Fox, Jr.

83** 6/91 **Simulation of the Effects of Spawning and Recruitment Patterns in Tropical and Subtropical Fish Stocks on Traditional Management Assessments**
J. Ault and W.W. Fox, Jr.

84** 6/91 **The Decline of the Chesapeake Bay Oyster Population: A Century of Habitat Destruction and Overfishing**
B.J. Rothschild, J. Ault, P. Goulletquer, W.P. Jensen and M. Heral

102***	3/93	**User's Guide to CASA, Catch-at-Size Analysis, Version 1.1, for the Microcomputer** Han-Lin Lai and B. Amjoun
103***	3/93	**Age Determination of Corvina Reina (Cynoscion albus) in the Gulf of Nicoya Based on Otolith Surface Readings and Microincrement Analysis** M. Mug Villenueva, V.F. Gallucci and Han-Lin Lai
104***	3/93	**Harvest Strategies Based on Biological Reference Points with an Application to the Native Littleneck Clam** Han-Lin Lai and V.F. Gallucci
105***	3/93	**Programs for the Simulation and Estimation of Fisheries Acoustics Parameters for Size Based Analysis** J. Hedgepeth

ANNEX B

STOCK ASSESSMENT COMPUTER ALGORITHMS

Jerald S. Ault[1], Richard McGarvey[2], Brian J. Rothschild[3], and Juan B. Chavarría[4]

[1]Rosenstiel School of Marine and Atmospheric Science, University of Miami, [2]South Australian Research and Development Institute, [3]Center for Environmental and Estuarine Studies, University of Maryland, [4]CIMAR, Department of Statistics, University of Costa Rica

1. INTRODUCTION

Considered at its simplest, the dynamics of a fishery balance the natural rate of increase by fish stocks, which grow through reproduction and individual growth, with harvesting by fishermen. Managing this system requires knowledge of the population, obtained from commercial statistics reported by fishermen and by research survey sampling.

Clearly, in all real fisheries, the amount of information available will be limited and incomplete and, by the statistical nature of data collection, of finite precision. The methods of sampling, particularly the research survey under control of fishery managers, can be chosen to maximize the quality of information derived. Survey sampling in multispecies tropical fisheries is discussed in Chapter 4 of this volume. In this annex, we present a package of stock assessment calculation routines including some of the most widely applied and well-known techniques for analyzing data for stock assessment. The programs that perform these data-analysis calculations have been compiled into a PC-based software package with a simple user interface. Also included with this package is a User's Guide of instructions for running each program module, which

1-56670-151-1/96/$0.00+$.50

includes an introduction to the software and a detailed description of each program. The introduction to the User's Guide comprises a list of one-sentence summaries of each algorithm, information about how to install the software, and instructions for using the menu-driven user interface for choosing programs and entering data. The individual program descriptions cover the purpose of each stock assessment method, the nature and format of input data employed, and the output that each calculation generates. This documentation is included on the diskettes in Wordperfect 5.1 format and as regular DOS (i.e., ASCII) text files. The classes of stock assessment methodologies in this system include length-based stock assessment, stock-recruitment analysis, cohort analysis, yield-per-recruit, and surplus production. Simulation modeling tools, including fishery management training simulation, are also provided.

This compendium of stock assessment techniques contains some of the standard repertoire of managers of industrialized temperate fisheries. These are combined with results from the CRSP program and from other tropical programs with the objective of extending existing, and developing new techniques for tropical habitats where species composition is more diverse and reproductive dependence upon season is less pronounced. Some of the computer programs were produced under the aegis of the Food and Agriculture Organization (FAO) of the U.N., e.g., Gulland (1969), Abramson (1971), Sims (1985), and Sparre et al. (1989), others were produced under CRSP, and some were produced under other scientific projects.

2. GENERAL PROGRAM STRUCTURE

Source code for all programs except FINMAN (Ault and Fox 1989) is written and compiled in Microsoft FORTRAN 77. All programs are executed with calls for input and output files. Data input files are ASCII. The number and types of input data are determined by program needs. The user only needs to edit the sample data input files provided for successful application. Output files are automatically created at execution time by simply naming the filename and extension type such as "output.dat".

General machine requirements are a DOS-based microcomputer with a mathematics coprocessor chip installed.

The stock assessment package, both the documentation and the menu-driven hierarchy of program selection, is organized into eight categories of fishery statistics and modeling: growth, mortality, general nonlinear models, selection and fishing power, stock-recruitment, logistic assessment or surplus production, analytical yield, simulation and training.

2.1 MODEL DESCRIPTIONS

The description of each model is broken into three categories: model development motivation, mathematical theory, and overview of output interpretation.

Each program is structured into five major components: source code, structured ASCII data file, user's guide, output and graphics, and interpretation of the outputs.

2.1.1 Growth

Growth routines are of two types: (1) those that compute the parameters of a growth function in terms of length-at-age or weight-at-age for an indeterminate growth pattern when data on age and size is available, and (2) those that compute model parameters for mark-recapture data.

BGC2

BGC2 computes the three-parameter length-at-age von Bertalanffy growth model for length-dependent-on-age data using a least squares procedure with weights proportional to the sample size described by Tomlinson and Abramson (1961). The program always yields estimates when a least squares solution exists and immediately terminates the run when there is no solution. A constant time interval between ages is required; but the numbers of lengths in the age groups may be unequal. Output provides estimates of L_∞, K, t_0, and their standard errors. Fitted lengths, sample mean lengths and standard errors, variance-covariance matrix and standard error of the estimates are considered accurate to at least four significant digits. A maximum of 30 age groups with at least 2 and at most 1500 lengths in each group can be handled.

GROFIT

A generalization of the von Bertalanffy growth model was proposed for plant growth by Richards (1962) and later for fish growth by Chapman (1961). GROFIT following Causton (1969), uses length-dependent-on-age data input to the Chapman-Richards growth function to estimate the four model parameters and their variances for the extended Von Bertalanffy growth function:

$$W_t = W_\infty\left[1 - e^{-K(t-t_0)}\right]^{\frac{1}{1-m}} \qquad (1)$$

WVONB

WVONB, due to Pienaar and Thomson (1973), handles the analysis of the four-parameter generalized von Bertalanffy equation with an option to determine the maximum likelihood ratios of the desired parameters for length-dependent-on-age data.

FABENS

FABENS uses mark-recapture data following the method of Fabens (1965) to calculate least squares estimates of the parameters A and K for a modified von Bertalanffy growth curve of the form

$$Y = X + (A - X)\times\left(1 - e^{(1-K\times D)}\right) \qquad (2)$$

where, X is the length at release; Y is the length at recapture; and D is the elapsed time between release and recapture. Another program option allows for the least-squares estimation of $B = e^{(K \times t_0)}$ and t_0 using length-at-age data.

LCANAL

Usually used in conjunction with program CHTL, this program enables calculation of the effects of long-term changes in fishing effort and mesh size from catch-length composition data. The program uses the method described by Jones (1961), modified in Jones (1974).

ELEFAN

ELEFAN, using the method of Pauly (1984), implements a computing procedure that traces through a series of length-frequency samples sequentially arranged in time, a multitude of growth curves and selects the single curve which, by passing through a maximum of peaks, best describes the peaks. The method uses only length-frequency data; it requires no additional inputs.

2.1.2 Mortality

In this section we introduce algorithms for estimating mortality rates in a population. The usual methods describe the life-span transition of a cohort through time. If equilibrium conditions persist then this is equivalent to following a single cohort through all ages at a fixed point in time. Estimation methods usually strive to separate the total force of mortality into two components or competing risks of death: fishing and natural mortality.

CATCURV

CATCURV addresses the general problem of analyzing catch curves discussed by Chapman and Robson (1960) and Robson and Chapman (1961). CATCURV computes the total instantaneous mortality rate and a number of statistical measures associated with an age vector of catch in numbers of individuals at age.

COHORT

COHORT estimates age-specific fishing mortality rates, exploitation rates and population sizes from catch data for specific year classes or cohorts. The program uses the generalized Murphy (1965) catch equation from the computations given by Murphy (1965) and later Tomlinson (1970), and allows for zero catch intervals.

JOLSEB

JOLSEB derives total instantaneous mortality rate estimates from multiple mark-recapture histories (and catch data) for multiple cohort releases using the equations given by Jolly (1965) which gives an excellent description, history and application of the model.

LBAR

LBAR provides a method for estimating the total instantaneous mortality rate (Z) from a truncated equation for average length in the catch. Because the assumption of an infinite exploitable life span does not hold in many fisheries, Ehrhardt and Ault (1992) developed and fully exploited a modified mortality formulation reexpressing the Beverton and Holt (1957) model for the average length of fish in the catch as a censured distribution truncated with a bounded exploitable life span. In their new model, the hypothetical infinite life span assumed by Beverton and Holt (1957) is restricted such that the upper bound of the life span is defined by an age t_1 corresponding to a length L_1, while the lower bound is defined by an age t′ corresponding to the length L′, i.e; the lowest size at which the probability of the length of capture is equal (or at least very near) to one.

MCARLO

MCARLO was used by Ehrhardt and Ault (1992) to perform an assessment of the length-based total mortality model of Beverton and Holt (1957), and their own model (Ault and Ehrhardt 1991), assuming that catch length-frequency distributions in tropical artisanal fisheries are constrained by (1) the selective properties of multiple gears used, and (2) the spatiotemporal availabilities of certain size classes of fish to operationally restricted fleets. Their new estimator has zero bias at equilibrium and they determined variance of the new *Z*-estimator by bootstrap techniques.

CHTL

CHTL performs a cohort analysis on catch data grouped into length classes. Lengths are converted to ages using an assumed von Bertalanffy growth curve and the cohort mortality model is that described by Pope (1972). Given a known natural mortality rate *M*, and an estimate of the value of the instantaneous fishing mortality rate divided by total instantaneous mortality rate, *F/Z*, for the largest group of fish, the numbers in the population and *F/Z* values for each length group are calculated. The program allows calculations for a range of the parameter values L_∞ and *M/K*, where L_∞ and *K* are parameters of the von Bertalanffy growth function, and *M* is the instantaneous natural mortality rate.

2.1.3 Nonlinear Models

NLSQ

NLSQ is a generalized nonlinear least squares algorithm for fitting functions of any number of parameters using Marquardt's (1963) algorithm when adjusting *M* parameters, in array *P*, to minimize the residual squared between *N* observed data values, in array *Y*, and the *N* corresponding values, in array *FX*, obtained from an arbitrary function of the *M* unknown parameters and any number of additional fixed parameters, in array *X*.

2.1.4 Selection and Fishing Power

SECHIN

SECHIN computes the probabilities of capture for the Sechin selectivity-at-size model. The model uses data on opercular and maximum girth by size, retention girth, net elasticity, and mesh size dimensions. Outputs are size-specific selection probabilities.

FPOW

FPOW estimates the fishing powers of individual vessels relative to a standard vessel, and also the logarithm of fish density for time-area strata relative to a standard stratum using the methods described by Robson (1966). Input data are the compositions of commercial catches, in particular, the catches of sampled tows, broken down by vessel and fishing area. Conversion from log-relative fishing power and log density to the original scales is accomplished by employing a bias-correcting factor. The bias which is corrected derives from the fact that log catches are assumed to be normally distributed, and it follows that P_i, the estimate of log-relative fishing power for the ith vessel, is also normal with expectation P_i. However, the expected value of $\exp(P_i)$ does not equal $\exp(P_i)$ and consequently the uncorrected estimate is biased.

2.1.5 Logistic Assessment

PRODFIT

PRODFIT uses the equilibrium approximation model of Fox (1975) to estimate the parameters of the generalized surplus production model.

GENPROD

GENPROD fits the generalized stock production model described by Pella and Tomlinson (1969) to a time series of catch and effort using the transition prediction approach. The model estimates equilibrium yield as a function of effort or population size. The production curve is allowed to be skewed.

2.1.6 Analytical Yield

BHYIELD

Routine estimates the yield-per-recruit surface dependent on two control parameters, age of first capture and fishing mortality rate, following the model of Beverton and Holt (1957).

YPER

In the original form of the yield-per-recruit equation developed by Beverton and Holt (1957) the parameters of fishing and natural mortality, growth, and age at recruitment and at entry to the exploited phase, all appear separately. The number of permutations needed to cover all combinations of values of these parameters likely to be encountered in practice is too large for convenient tabulation. YPER combines the seven input parameters of the original Beverton-Holt yield equation into three derived parameters, making visual presentation of the entire range of parameters possible.

INCBETA

INCBETA uses an incomplete beta function approximation to describe the yield-per-recruit surface dependent on fishing mortality and age of first capture when the weight dependent on length function is allometric (i.e., the power of the curve varies from 3.0).

MGEAR

MGEAR computes estimates of yield-per-recruit and several parameters for fisheries that are exploited by several gears which may have differing vectors of age specific fishing mortality utilized by Lenarz et al. (1974). The Ricker (1954) yield equation is used for the computations.

2.1.7 Stock-Recruitment

RICURVE

RICURVE computes parameters and outputs associated with the family of curves proposed by Ricker (1954) and given in detailed tabular form by Ricker (1975). Ricker's (1954) model can be written in at least two ways. One is:

$$R = \alpha S e^{-\beta S} \tag{3}$$

where, R is the number of recruits, S is the size of parental stock (measured in numbers, weight, egg production, etc.), α is a dimensionless parameter, β is a parameter with dimensions of $1/S$.

The second form, with R and S in the same units, assumes there is a replacement abundance at which $R = S$; the replacement level is S_r. The expression (3) can then be modified by introducing a new parameter $a = S_r\beta$, and substituting

$$\beta = a/S_r \quad \text{and} \quad \alpha = e^a \tag{4}$$

in (3) leads to

$$R = Se^{\alpha(1-S/S_r)} \tag{5}$$

In this form, replacement abundance S_r appears as an explicit parameter, which is a convenience whenever S_r can be estimated.

2.1.8 Simulation and Training

POPSIM

POPSIM (Walters 1969) is an optimization routine for planning harvesting strategies over an extended planning horizon. The model is adaptable to the life history characteristics of fish populations which have been traditionally studied in fisheries science.

GXPOPS

GXPOPS is a generalized exploited population simulator designed for use on a wide variety of aquatic life history patterns developed by Fox (1973). Population processes programmed into the present version are (1) month-specific and density-independent mortality rates on the recruited population, (2) density-independent growth, (3) sex-and age specific, but density-independent, maturation, (4) reproductive success as related to random mating, and (5) density-dependent or density-independent recruitment.

DSPOPS

DSPOPS is a discrete-time stochastic population simulation model developed by substantial modification of the basic GXPOPS model. DSPOPS can be used for evaluating various growth, mortality and recruitment scenarios for any age-structured life history. The model is configured to compute stochastic Monte Carlo simulations of various catch, effort and management policies.

FINMAN

FINMAN (Ault and Fox 1989) is a fishery institution management model which simulates decision-making responses at three levels within the institution: fishery management rules, the fishery agency's general budget allocations, and research budget allocations. The program also allows for a variety of fishery types, rule development structures, and levels of authority over the fishery. Management effectiveness is defined as a complex function of the attributes cost, rate of profit, degree of enforcement, population capacity, and time to achieve goals.

2.2 DISCUSSION

Fishery data come from two basic sources: the commercial fishery statistics and population sampling surveys carried out by fishery scientists. The commercial data consist of reports by the skipper, usually submitted at the time and port of landing, sometimes verified by fishery inspectors. The two basic commercial data available are the weight of fish for each species captured, and the amount of fishing effort expended on that fishing trip to capture the catch reported. It is widely assumed that some fishermen on some occasions land fish which are not reported, though this is difficult to quantify. Underreporting is far more likely in fisheries where catch quotas are imposed, making the landing of captured fish of that species illegal or subject to fine. Commercial data are therefore often incomplete, with a strong bias toward underestimate of total catch. However the catch-per-unit-effort, as a relative (non-absolute) measure of stock density is not biased by underreporting since the ratio of catch to effort is calculable on a trip by trip basis and does not require knowing the total amount of fish captured. Catch-per-unit-effort is therefore a generally more reliable statistic than total catch.

In part because of underreporting, random sampling surveys are also often carried out independently of the fishery by scientists, providing a more rich and relatively unbiased source of information about the fish population. Generally, a net or dredge is used to sample the water or bottom from a research vessel. Commercial fishing gear is often employed, but modified with a smaller net mesh to capture smaller size fish than the commercial gear. This yields a sample of stock sizes which hopefully includes at least one age class not yet reduced by fishing. The fish from each tow are commonly identified by species, and their sex is determined. Each fish is measured for length and classified into one of several size classes (e.g., 0–5 mm, 5–10 mm, 10–15 mm, etc.). This data is recorded for each tow, at each sampling location. Thus, the raw data from a fish stock survey are often "size-frequency" histograms of the population, meaning they yield the numbers of individuals captured per tow in each size class for each species. It is also valuable to measure reproductive organs of sampled fish, if possible, to obtain a measure of reproductive output of the population.

Whether the samples be commercial or survey, it is crucial that the method of sampling be designed so that the samples obtained be random. With scientific surveys this is widely understood. In commercial samples, this can be more difficult, but attempts should be made to obtain commercial samples that are as random, with as complete a coverage of the population/fishery, as possible.

It is valuable to determine the age of captured fish, whether the samples are obtained from the commercial fishery or by survey, since estimations of growth and mortality, and to a lesser extent, recruitment depend on knowledge of age of the sampled individuals. Ageing methods exist for temperate fish stocks, and research funded by the CRSP program has extended these techniques for use with tropical fish (see Chapter 3 of this volume). These methods are labor intensive and often make use of imaging technology to determine age, by counting growth rings on otoliths, or by more expensive radio isotope techniques. Often however, especially in multispecies tropical fish stocks, ageing is not technically or practically feasible. When ageing is possible, many of the SACA programs can be applied directly, employing the input data of numbers-at-age. These include growth estimation for the parameters of the von Bertalanffy growth curve (WVONB, GROFIT, and BGC2), estimation of total mortality (CATCURV), and cohort analysis (COHORT).

Yield-per-recruit analysis uses these as inputs to determine the optimum levels of fishing effort and the optimum age of capture to obtain maximum weight of harvest for each recruit into a fish stock, given the competing influences of mortality and growth (Beverton and Holt 1957). If natural mortality is high, it is plainly better to harvest sooner, while if growth is rapid, it is advisable to wait since the harvested yield in weight from each fish increases. The Beverton-Holt method (BHYIELD), as well as a number of variations on that basic approach (YPER, INCBETA, and MGEAR) are included with the SACA package.

Besides commercial and survey data, mark-recapture experiments can be undertaken. Fish are collected and measured for all important variables, in particular, size when growth is being estimated. By tagging a range of sizes of fish, or marking them by staining or some other method, and releasing them at a specific time, their size at the time of release is known. When they are recaptured, their size is remeasured and because they were marked, the time elapsed, and thus the calculated rate of growth is very reliably ascertained. SACA provides a program for analyzing von Bertalanffy growth parameters (FABENS) and total mortality (JOLSEB) from mark-recapture data.

Each algorithm in the SACA package is based on a specific mathematical model of fish population dynamics. The algorithms input one of the forms of data described above to yield estimates for important parameters of fishery population dynamics implicit in that model. The models are written and solved incorporating underlying assumptions about the fish population and the data. In all real fisheries, these assumptions, which often take a simple mathematical form, are at best approximations. We therefore review some of the classes of stock-assessment calculation embodied in the SACA algorithms, their data requirements, assumptions, and underlying motivation. In the remaining discussion we shall focus on non-ageing-dependent stock assessment programs which are likely to be more useful in most tropical fisheries.

Some methods exist to infer growth and/or mortality directly from size-frequencies alone. The most widely employed (ELEFAN) is due to Pauly, which assumes that at least two age class peaks are visually evident in the size-frequency histogram. A comprehensive review and new techniques for length-based stock assessment are presented in detail in Chapter 2 of this book.

In addition, stock-recruitment analyses can also be undertaken with minimal knowledge of age. Recruits can be estimated by the numbers in the first (i.e., smallest-size) clearly defined age class peak in the size-frequency histogram. For example, the fish under the peak observed, say, from sizes 30 to 60 mm may be determined to be 2-year-olds. Egg production in each survey year can be calculated if there is a measure of fecundity (and sex-ratio, when necessary) as a function of fish size. If a fecundity-at-size vector is not available, it can be obtained by measuring gonad biomasses during the same survey undertaken to sample for size-frequencies. Annual egg production is calculated as the sum over all age classes of eggs-at-size times numbers-at-size. Thus, with no measure of age, except the age class of the fish in the first size-frequency peak, accurate measures of stock and recruitment can be obtained. These estimates of recruitment and egg production are likely to be more accurate than those derived from age-based analysis such as cohort or VPA, since there are no implicit assumptions (like the value of natural mortality, often taken as constant for many species and all ages, or that all fish caught were reported) and no errors are introduced in ageing.

Besides length-based methods such as ELEFAN and stock-recruitment analysis, a third category of fishery analysis can be undertaken without ageing capabilities. Employing commercial time series of catch and effort, these are called "surplus production" or simply "production" models. The original production model due to Schaefer (1954, 1957) is based on two fundamental assumptions: (1) that the fishable biomass grows logistically, and (2) that the system (the fishery, meaning both stock biomass and fishing effort) is at equilibrium. Assumption (1) does not limit the usefulness of this technique and subsequent production models have been developed to accommodate different density-dependent formulas for the growth dynamics of fish stocks. However more questionable is the assumption of equilibrium.

The Schaefer model and essentially all production models that followed seek to determine the optimum level of fishing effort. "Optimum" can be chosen by fishery managers to be either maximum long-term average harvest (called maximum sustainable yield, or MSY), maximum long-term average profit (maximum economic yield, or MEY), or "net present value" (Clark, 1985). Surplus production analysis is thus directly applicable to wise management of a fish stock. It provides recommendations about how to adjust levels of fishing effort, a component of the fishery system that managers can sometimes control, unlike the stock itself. And the outcome it can predict when the assumptions are correct is one of direct utility, namely how to achieve these three socially desired optimums.

The Schaefer model (Schaefer, 1954, 1957; Clark, 1976, 1985) is based on the idea that maximum fish harvest will be extracted when the fish population is maintained at the level where population growth is fastest. The goal is to deduce the corresponding optimum level of effort which balances the natural growth of the fish population at that desired optimum. Only the "surplus" biomass production, the increase in population above the chosen steady population level can be harvested year after year, since if more is taken, the population will, by definition, decline. A sustainable level of harvest is sought, and that is MSY. When the stock is overfished, reducing effort increases the reproductive output of the population by increasing (long-term) adult population size, and thus total annual egg production. Thus, implicitly, all production models assume the general population growth paradigm due to Pearl (1925). At low fish stock levels (below MSY) increasing stock size is thought to increase the rate of population growth, because more and bigger adult fish will produce more eggs and thus greater recruitment. At high stock densities, intraspecific competition for food in the marine habitat will reduce the reproductive success of each female, resulting in a lower rate of population growth. Thus the rate of growth will go to zero at the two extremes of stock abundance, namely, when there are no adult fish left, and when the population reaches its carrying capacity. This is why the fish population does not have to follow strict logistic growth in order to satisfy the required assumptions of surplus production. An optimum level of stock and effort will exist, if at low stock densities, below the optimum, there is a positive stock-recruitment relationship, and if, at high stock densities, above the optimum, population growth slows with increasing abundance. By adjusting effort and thus stock size to lie roughly at the level between these, average fish population growth and thus sustainable rate of harvest will be maximized.

The case of interest to managers, which includes the majority of fisheries that require regulation, are those where the stock is overexploited, lying below the optimum level of abundance. In these cases, reducing fishing effort, on average, increases harvest. This is an economic opportunity that is virtually impossible outside the realm of natural resource exploitation where, by investing less capital, expending less in operating costs and doing less work, the national economy, and the local coastal communities obtain greater production in food and revenue.

The drawback of surplus production analysis is that the assumption of equilibrium employed in the mathematical solution is rarely met in real fisheries. However, though equilibrium is assumed in the mathematical derivations of the classic Schaefer curve and most of its model descendants, it is not implicit in the underlying reasoning behind why lower effort can yield higher long-term yields. The fishery may be undergoing a wide range of dynamical behaviors. It may be far from equilibrium and be affected by any number of exogenous influences, such as variations in the climate, currents, or productivity, anthropogenic erosion of the marine habitat, or changes in the level of effort itself, and yet for the overfished case, still give rise to higher yields, on average, when average levels of fishing effort fall. The difficulty lies in estimating this optimum level of effort when data is not obtained under conditions of fishery equilibrium.

In one recent comprehensive review of tropical multispecies fisheries, a variation of the Schaefer model performed well. Bayley (1988) examined 59 tropical fisheries that were artisanal and multispecies in character and found that fishing effort was the principal determinant of long-term yield. In the largest subset of the 59 fisheries examined, namely, from 31 African lakes, a production model employing an aggregated measure of multispecies and multigear effort, explained 75% of the variance in log(yield). The "morphoedaphic index", based on environmental indicators of biological productivity, only marginally increased the predictive power compared with overall levels of fishing effort per unit area of lake surface. The second subset of fisheries examined by Bayley were river floodplains and coastal lagoons. Again, Bayley's effort-dependent model explained 60% of the variation in yields. Moreover, the ranges of maximum yields (per unit area) from tropical lakes and lagoon-floodplains overlapped. That two distinct classes of fisheries conformed to the same production model of catch as a function of effort, both yielding highly significant effort coefficients, is notable for its potential applicability in tropical fisheries.

The limitation of assuming equilibrium in the estimation of optimum effort was, in part, overcome in the PRODFIT model (Fox, 1975), included in the SACA package. The Schaefer production model assumes that the harvest in any particular year corresponds to the level of effort in that year. Generally, however, the harvest in a given year is more directly due to the recruitment success of the dominant year class or year classes in the fishable (adult) stock. There will be a time lag between lower effort and the anticipated increased harvest in the overfished case. This time lag, equal to or greater than the average age at which fish are harvested, is not incorporated in the standard production models which assume equilibrium. By choosing the PRODFIT option 2, the analysis will take the weighted sum of effort from specific years preceding the year of harvest and compare that to the observed harvest to seek the relationship between effort and catch hypothesized in the (generalized) production model. This explicit incorporation of the time lag is a significant improvement on the original Schaefer production model formulation.

For more advanced applications, fishery simulation programs are also included in SACA, specifically, POPSIM, GXPOPS, DSPOPS, and FINMAN. These tools may be used to quantify the extent of inaccuracy in fishery population estimates derived under assumptions of equilibrium. Simulations are dynamical and thus no equilibrium is assumed. Sets of synthetic non-equilibrium data can be obtained from the simulations by choosing reasonable values for the input parameters. Running the simulation yields time series of catch, catch-at-age, catch-at-length, or any other measure of abundance desired, that should be chosen to correspond as well as possible to what can be measured in the non-equilibrium fishery under study. These synthetic time series, i.e., the simulation output, may then be used as input in the same calculations, for instance surplus production, being applied to analyze real fishery data. This can serve as a test of how severely the assumption of equilibrium biases the results of the surplus production calculation which are taken from a non-equilibrium fishery. In this test the non-equilibrium fishery is the simulation model. Since the original parameters that generated the synthetic time series are known (they are supplied by the modeler to the simulation at the outset), direct comparison is possible of the results from surplus production calculations performed on the synthetic data with their original (simulation-input) values.

ACKNOWLEDGMENTS

The authors appreciate the programming and technical assistance provided by Huaixiang Li and Ambrose Levi.

REFERENCES

Abramson, N.J. 1971. Computer programs for fish assessment. FAO Fisheries Technical Paper no. 101.

Ault, J.S. and Ehrhardt, N.M. 1991. Correction to the Beverton and Holt Z-estimator for truncated catch length-frequency distributions. Fishbyte 9: 37-39.

Ault, J.S. and Fox, W.W. Jr. 1989. FINMAN: simulated decision analysis with multiple objectives. In *Mathematical Analysis of Fish Stock Dynamics* (Edwards, E.F. and Megrey, B., Eds.) American Fisheries Society Symposium 6: 166-179.

Bayley, P.B. 1988. Accounting for effort when comparing tropical fisheries in lakes, river-floodplains, and lagoons. Limnol. Oceanogr. 33: 963-972.

Beverton, R.J.H. and Holt, S.J. 1957. On the dynamics of exploited fish populations. Mininstry of Agriculture Fisheries and Food. Fisheries Investigations, Series II, Volume XIX. Fisheries Laboratory, Lowestoff, Suffolk, UK.

Causton, D.R. 1969. A computer program for fitting the Richards function. Biometrics 25: 401-409.

Chapman, D.G. 1961. Statistical problems in dynamics of exploited fisheries populations. Proceedings of the 4th Berkeley Symposium Math. Statist. Prob. 4: 153-168.

Chapman, D.C. and Robson, D.S. 1960. The analysis of a catch curve. Biometrics 16: 354-368.

Clark, C.W. 1976. *Mathematical Bioeconomics: the Optimal Management of Renewable Resources*. John Wiley and Sons, New York.

Clark, C.W. 1985. *Bioeconomic Modelling and Fisheries Management.* John Wiley and Sons, New York.

Conquest, L., Burr, R., Donnely, R., Chavarria, J.B. and Gallucci, V.F. 1994. Sampling methods for stock assessment of small-scale fisheries in developing countries: with a case history in the Gulf of Nicoya, Costa Rica. Chapter 4, this volume.

Ehrhardt, N.M. and Ault, J.S. 1992. Analysis of two length-based mortality models applied to bounded catch length frequencies. Transactions of the American Fisheries Society 121: 115-122.

Fabens, A.J. 1965. Properties and fitting of the von Bertalanffy growth curve. Growth 29: 265-289.

Fox, W.W. 1973. A general life history exploited population simulator with pandalid shrimp as an example. Fishery Bulletin, U.S. 71: 1019-1028.

Fox, W.W. Jr. 1975. PRODFIT: a least squares equilibrium approximation method for fitting the generalized production model. Fishery Bulletin, U.S.

Gallucci, V.F., Amjoun, B., Hedgepeth, J. and Lai, H.-L. 1994. Size-based methods of stock assessment of small-scale fisheries. Chapter 2, this volume.

Gulland, J.A. 1969. *Manual of Methods for Fish Stock Assessment Part I. Fish Population Analysis.* FAO Management of Fisheries Science 4.

Jolly, G.M. 1965. Explicit estimates from capture-recapture data with both death and immigration-stochastic model. Biometrika 52: 225-47.

Jones, R. 1961. The assessment of the long term effects of changes in gear selectivity and fishing effort. Mar. Res. Scot. 2, 19.

Jones, R. 1974. Assessing the long term effects of changes in fishing effort and mesh size from length composition data. ICES CM. 1974 Demersal Fish (Nthn) Ctte, Doc. F:33 (mimeo)

Lai, H-L., Gallucci, V.F. and Gunderson, D.R. 1994. Age determination in fisheries in developing countries. Chapter 3, this volume.

Lenarz, W.H., Fox, W.W Jr., Sakagawa, G.T. and Rothschild, B.J. 1974. An examination of the yield per recruit basis for a minimum size regulation for Atlantic yellowfin tuna, *Thunnus albacares*. Fishery Bulletin 72: 37-61.

Marquardt, D.W. 1963. An algorithm for least squares estimation of nonlinear parameters. Journal for the Society of Industrial and Applied Mathematics 11: 431-441.

Murphy, G.I. 1965. A solution of the catch equation. Journal of the Fisheries Research Bioard of Canada 22: 191-202.

Pauly, D. 1984. *Fish Population Dynamics in Tropical Waters: A Manual for Use with Programmable Calculators.* ICLARM Studies and Reviews 8, Manila.

Pearl, R. 1925. *The Biology of Population Growth.* Knopf, New York.

Pella, J.J. and Tomlinson, P.K. 1969. A generalized stock production model. Inter-American Tropical Tuna Commission Bulletin 13: 419-496.

Pienaar, L.V. and Thomson, J.A. 1973. Three programs used in population dynamics: WVONB-ALOMA-BHYLD. Fisheries Research Board of Canada Technical Report no. 367.

Pope, J. G. 1972. An investigation of the accuracy of virtual population analysis using cohort analysis. ICNAF Bulletin no. 9.

Richards, F.J. 1962. A flexible growth function for empirical use. Journal of Experimental Botany 10: 290-300.

Ricker, W.E. 1954. Stock and recruitment. Journal of the Fisheries Research Board of Canada 11: 559-623.

Ricker, W.E. 1975. Handbook of computations and interpretation of biological statistics of fish populations. Fisheries Research Board of Canada Bulletin. 191.

Robson, D.S. 1966. Estimation of the relative fishing power of individual ships. ICNAF Research Bulletin 3: 5-14.

Robson, D.S. and Chapman, D.C. 1961. Catch curves and mortality rates. Transactions of the American Fisheries Society 90: 181-189.

Schaefer, M.B. 1954. Some aspects of the dynamics of populations important to the management of commercial marine fisheries. Bulletin of the Inter-American Tropical Tuna Commission 1: 27-56.

Schaefer, M.B. 1957. Some considerations of population dynamics and economics in relation to the management of the commercial marine fisheries. Journal of the Fisheries Research Board of Canada 14: 669-681.

Sims S.E. 1985. Selected programs in FORTRAN for Fish Stock Assesment. FAO Fisheries Technical Paper 259.

Sparre, P., Ursin, E. and Venema, S.C. 1989. Introduction to tropical fish stock assessment Part 1-Manual. FAO Fisheries Technical Paper 306.

Tomlinson, P.K. 1970. A generalization of the Murphy catch equation. Journal of the Fisheries Research Board of Canada 27: 821-825.

Tomlinson, P.K. and Abramson, N.J. 1961. Fitting a Von Bertalanffy growth curve by least squares. California Fish Game Fisheries Bulletin no. 116.

Walters, C.J. 1969. A generalized computer simulation model for fish population studies. Transactions of the American Fisheries Society 98: 505-512.

INDEX

A

B

D

J

K

L

M

N

O

P

T

U

V

W

Y

Z